THE GANGES RIVER BASIN

The Ganges is one of the most complex yet fascinating river systems in the world. The basin is characterized by a high degree of heterogeneity from climatic, hydrological, geomorphological, cultural, environmental and socio-economic perspectives. More than 500 million people are directly or indirectly dependent upon the Ganges River Basin, which spans China, Nepal, India and Bangladesh. While there are many books covering one aspect of the Ganges, ranging from hydrology to cultural significance, this book is unique in presenting a comprehensive inter-disciplinary overview of the key issues and challenges facing the region.

Contributors from the three main riparian nations assess the status and trends of water resources, including the Himalayas, groundwater, pollution, floods, drought and climate change. They describe livelihood systems in the basin, and the social, economic, geopolitical and institutional constraints, including transboundary disputes, to achieving productive, sustainable and equitable water access. Management of the main water-use sectors and their inter-linkages are reviewed, as well as the sustainability and trade-offs in conservation of natural systems and resource development such as for hydropower or agriculture.

Luna Bharati is a Senior Researcher and the Head of the Nepal Program at the International Water Management Institute (IWMI).

Bharat R. Sharma is Scientist Emeritus (Water Resources) at IWMI, based in New Delhi, India.

Vladimir Smakhtin was Head of the Water Availability, Risk and Resilience Research Department at IWMI, Colombo, Sri Lanka. He is now the Director of the Institute for Water, Environment and Health (UNU-INWEH), based in Hamilton, Ontario, Canada.

EARTHSCAN SERIES ON MAJOR RIVER BASINS OF THE WORLD

Large river basins are dynamic and complex entities. Defined by hydrological boundaries, they are nearly always shared by more than one country. Encompassing a diverse range of landscapes with often huge temporal and spatial variability of resources, they are put to different and often conflicting uses, and managed by a range of institutions and organizations. While an intrinsic part of nature, many have been extensively engineered and used by people, often with adverse consequences. Each major river basin has its own development trajectory and, often, a fascinating history. Bringing together multidisciplinary teams of experts, this series explores these complex issues, identifies knowledge gaps and examines potential development pathways towards greater sustainability.

The Nile River Basin
Water, Agriculture, Governance and Livelihoods
Edited by Seleshi Bekele Awulachew, Vladimir Smakhtin, David Molden and Don Peden

The Volta River Basin
Water for Food, Economic Growth and Environment
Edited by Timothy O. Williams, Marloes L. Mul, Charles A. Biney and Vladimir Smakhtin

The Ganges River Basin
Status and Challenges in Water, Environment and Livelihoods
Edited by Luna Bharati, Bharat R. Sharma and Vladimir Smakhtin

www.routledge.com/series/ECMRBW

THE GANGES RIVER BASIN

Status and challenges in water, environment and livelihoods

Edited by
Luna Bharati, Bharat R. Sharma and Vladimir Smakhtin

Routledge
Taylor & Francis Group
LONDON AND NEW YORK

IWMI
International Water Management Institute

earthscan
from Routledge

First published 2016 by Routledge

2 Park Square, Milton Park, Abingdon, Oxon, OX14 4RN
605 Third Avenue, New York, NY 10017

Routledge is an imprint of the Taylor & Francis Group, an informa business

First issued in paperback 2020

British Library Cataloguing-in-Publication Data
A catalogue record for this book is available from the British Library

Library of Congress Cataloging in Publication Data
Names: Bharati, Luna, editor. | Sharma, B. R. (Bharat R.), editor. | Smakhtin, Vladimir, editor.
Title: The Ganges River Basin : status and challenges in water, environment and livelihoods / edited by Luna Bharati, Bharat R. Sharma and Vladimir Smakhtin.
Description: London ; New York : Routledge, 2016. | Series: Earthscan/ IWMI series on major river basins of the world | Includes bibliographical references and index.
Identifiers: LCCN 2016007977| ISBN 9781138900325 (hbk) | ISBN 9781315707310 (ebk)
Subjects: LCSH: Water resources development – Ganges River Watershed (India and Bangladesh) | Water-supply – Ganges River Watershed (India and Bangladesh) | Ganges River Watershed (India and Bangladesh) – Economic conditions. | Ganges River Watershed (India and Bangladesh) – Environmental conditions.
Classification: LCC HD1698.G36 G37 2016 | DDC 333.730954/1 – dc23
LC record available at https://lccn.loc.gov/2016007977

ISBN: 978-1-138-90032-5 (hbk)
ISBN: 978-0-367-73690-3 (pbk)

Typeset in Bembo
by Florence Production Ltd, Stoodleigh, Devon, UK

CONTENTS

NOTES ON AUTHORS

Upali A. Amarasinghe serves as Senior International Researcher at the International Water Management Institute (IWMI), Sri Lanka. He holds a PhD in Statistics from the University of Illinois Urbana-Champaign, USA. He has over 22 years of working experience as a researcher and a project leader in water resources management in an international research organization and 10 years of teaching experience as a lecturer in universities. His research focuses on various issues of water management, including water productivity and water footprints estimations of crops and river basins; food and water supply and demand assessments and projections at river basin, national and global scales; benchmarking and performance assessments of irrigation systems; poverty assessments and poverty mapping using small area estimation; water security assessment of industries; impact assessments of interventions in water resources management. He has extensive experience in managing research projects in India and Sri Lanka and short-term work experiences in other countries, including Pakistan, China, Malaysia, Thailand and Vietnam. He has published over 110 research papers and four edited books.

Giriraj Amarnath works as Senior Researcher–Remote Sensing and Water Resources at the International Water Management Institute (IWMI), Sri Lanka. He specializes in the application of remote sensing and geographic information systems to the study of risk assessment across a range of natural hazards and in monitoring land and water resources in Asia and Africa. He holds a doctoral degree in remote sensing from the National Remote Sensing Centre in Hyderabad, India. He is interested in studying the relationships between land cover/land use changes, hydrology and flood intensity – and their impact on food security and livelihood. He is applying his expertise to geospatial technology in a recent project that assesses South Asia's exposure and vulnerability towards climate hazards and development of drought monitoring systems in South Asia, and on agriculture flood insurance in India and Bangladesh to address flood risk reduction, poverty, enhancing agriculture resilience and sustainable development. Amarnath previously worked as a remote sensing specialist for the International Centre for Integrated Mountain Development (ICIMOD) in Nepal and as a research associate at the University of Bayreuth in Germany. He has published more than 50 research papers on various disciplines by integrating remote sensing data, tools and models.

Luna Bharati has over 10 years of post-PhD experience as a scientist and research manager. She is currently Senior Researcher and the Head of the Nepal programme at the International

Water Management Institute (IWMI). The key areas of her interests and expertise are in integrated and sustainable natural resources management. She has also worked extensively in assessing climate change risks and impacts as well as planning adaptation strategies from large river basins to small mountain watersheds. She has a multidisciplinary background with a Bachelors majoring in Biology and a minor in Economics from Luther College, USA and a Masters in Water Resources from Iowa State University. She conducted her doctoral research at the Department of Hydrological Modelling at the Helmholtz Center for Environmental Research – UFZ in Germany, focusing on catchment modelling of surface hydrology, erosion and pollutant transport. Prior to joining IWMI, she worked for the Centre for Development Research (ZEF) in Bonn, Germany on the Global Change and Hydrological Cycle (GLOWA) project in the Volta Basin, Africa. She has authored 90 publications including conference papers and more than 35 articles in peer-reviewed journals. She has worked on projects in North America, Europe, South Asia and West Africa.

Xueliang Cai is a senior lecturer at UNESCO-IHE. He has over 10 years' international working experience in Asia, Africa and Netherlands. Cai's key research interests are in land and water resources management for sustainable agriculture development. He started as a postdoctoral fellow at the International Water Management Institute (IWMI) in Colombo, Sri Lanka, where he was involved in global irrigated area mapping and water productivity assessment in South Asia, Central Asia and Southern Africa. In 2010, Cai became an international researcher and moved to IWMI regional office in Pretoria, South Africa, where he worked on a number of topics including integrated modelling of rural water resources for irrigation and fisheries; participatory irrigation management; hydrological function of ecosystems, and floods plain management. He was leader of several research projects including the ecological footprint of food security, which mapped irrigated, rain-fed, and water managed agricultural areas across Asia and Africa. Cai obtained his PhD degree in irrigation management from Wuhan University, China. He authored about 50 papers including 25+ in peer-reviewed journals.

Dipankar Chakraborti is Research Director of the School of Environmental Studies, Jadavpur University, Kolkata, India. He and his team members have been working on the problems of groundwater arsenic and fluoride in West Bengal, other states of India and Bangladesh since 1988. For nearly a decade, Chakraborti worked in various countries as Research Fellow and Visiting Professor of Environmental Analytical Chemistry. After discovering the gravity of the arsenic problem in West Bengal, he decided to return permanently to India in order to actively investigate this issue. He holds three patents on arsenic removal, including an international patent. He has joined Harvard University as a collaborator to mitigate groundwater arsenic contamination in Bangladesh. He has published 20 books and monograph chapters, and almost 200 papers (including 98 publications on arsenic and fluoride) in peer-reviewed journals. He has organized five international conferences (one in West Bengal and four in Bangladesh) on groundwater arsenic. He was a former advisor to the Asia Arsenic Network, Japan. At present, he is an advisor to the American Chemical Society's environmental journal, *Environmental Science & Technology*.

Ravi Chopra is Director of the People's Science Institute, Dehra Doon and a Managing Trustee of the Himalaya Foundation, New Delhi. He received his technical education in Metallurgical and Materials Engineering from the Indian Institute of Technology, Mumbai and Stevens Institute of Technology, New Jersey. As a researcher, Chopra has focused on the interactions between technology and society and environment and development. He first received national recognition when he helped produce the first citizens' report on *The State of India's Environment* in 1982.

This report was critically acclaimed as "a unique effort anywhere in the world". His recent analysis of India's water requirements in the twenty-first century has also attracted attention. It specifies the nature of India's water crises and sustainable methods for resolving them. He has worked for nearly four decades in the field of development and has helped establish several pioneering organizations. He has been a member of several committees of the Ministries of Rural Development and Water Resources, Government of India and the Planning Commission. He is a consultant to various national and international agencies. He has authored or co-authored 14 books and special reports. He has published more than 20 papers.

Floriane Clément works as a researcher in institutional and policy analysis at the International Water Management Institute (IWMI) Nepal Office. She has a multidisciplinary background, with an Engineering degree, an MSc in Environmental Sciences and a PhD in Geography/ Political Sciences. Her research has focused on the analysis of environment and development interventions, exploring the gap between institutional models and discourses on the one hand, and practices and outcomes in the field on the other hand. Her work cuts across disciplines and combines institutional and discourse analyses. She has work experience in France, Vietnam, India, Nepal and Sri Lanka. She has published in international peer-reviewed geography and political science journals, with recent publications on environmental justice, water security and the analysis of social-ecological systems.

Arica B. Crootof is a PhD student in the School of Geography and Development at the University of Arizona. With an MS in Hydrology and now pursuing a PhD in Geography, she is studying the complex and compounding challenges associated with climate change, water resources and energy systems relating to small hydroelectric power in mountain communities. She is a graduate research associate with Dr Christopher A. Scott at the Udall Center for Studies in Public Policy on a CGIAR/Water, Land and Ecosystems-funded project entitled 'The Irrigation-Hydropower Nexus in the Ganges Headwaters' in Uttarakhand, India. Her professional memberships include the Association of American Geographers, American Water Resources Association, Nevada Water Resources Association and Association of Pacific Coast Geographers.

Dwarika N. Dhungel, a senior researcher, is an executive member of the Tanka Prasad Memorial Foundation, and member of a number of organizations, including the Nepal Red Cross Society, the Nepal Council of World Affairs and the Nepal Association of Hubert H. Humphrey Fellows. He holds a PhD in Social Sciences from Nagpur University, India. He has served as Secretary to a number of ministries of the Government of Nepal, including the Ministries of Land Reforms and Management and Water Resources. Upon leaving the government service in 1998, Dhungel became more engaged in social research, including water resource relationships among the countries of South Asia. He has written numerous books and papers on different fields of social science. He is former Executive Director and Senior Researcher at the Institute for Integrated Development Studies (IIDS), a Kathmandu-based think tank. He has been a regular participant in the Ganga-Mekong Dialogue for the past three years, and jointly organized the programme of the MPower, ORF New Delhi and The Asia Foundation.

Mark Giordano is Director of the Program in Science, Technology and International Affairs, the Cinco Hermanos Chair in Environment and International Affairs, and Associate Professor of Environment and Energy in Georgetown's School of Foreign Service. He holds a PhD (2001) in Geosciences from Oregon State University and an MS in Applied Economics (1991) from the University of Minnesota. He is well known for his work on conflict and cooperation over resources, international water law and agricultural water management in developing countries, as well as his sometimes provocative challenges to common natural resources management paradigms. His passion is students, but he maintains an active research programme, advises a variety of organizations on water issues and is a frequent speaker at international events. Prior to joining Georgetown in 2013, he was a Theme Director at the Sri Lanka-based International Water Management Institute, winner of the 2012 'Nobel Prize for Water', the Stockholm Water Prize. He is from eastern Washington but has spent most of his career outside the US.

Dipak Gyawali is a Pragya (academician) of the Nepal Academy of Science and Technology and Chair of the non-profit Nepal Water Conservation Foundation as well as that of private research firm Inter Disciplinary Analysts. He is Vice-chair of the Technical Advisory Committee of UNESCO's World Water Assessment Program and a member of the Advisory Board of the STEPs Centre of the Institute for Development Studies (IDS), Sussex University. A Moscow-trained hydroelectric power engineer and a University of California at Berkeley-trained political economist, Gyawali initiated reforms in the electricity and irrigation sectors during his time as Nepal's Minister for Water Resources in 2002/2003. He initiated the first national review of Nepali laws with the guidelines of the World Commission on Dams and brought about policy change in irrigation through enacting an Irrigation Policy 2060 that provided more say to the informal farmer-managed irrigation systems. He has been conducting interdisciplinary research on the interface between technology and society; and has published numerous articles on the topics of water, energy, dams and climate change issues. He has been involved, inter alia, as guest scholar and researcher at various institutions such as the Queen Elizabeth House in Oxford and the Norwegian Center for Research in Organization and Management in Bergen.

A.K.M. Saiful Islam has been serving as a professor at the Institute of Water and Flood Management (IWFM) of the Bangladesh University of Engineering and Technology (BUET) since April 2012. He holds a PhD in Civil Engineering, Drexel University, Philadelphia, USA. He was a visiting researcher at various institutions, including IIS, Bangalore, UNESCO-IHE, the Netherlands, National Oceanographic Center (NOC), Liverpool, UK, Plymouth Marine Lab (PML), Plymouth, UK and Tufts University, Boston, USA. His recently completed research work includes *Climate Change Impacts, Vulnerability and Adaptation: Sustaining Rice Production in Bangladesh*; *Assessment of the Impact of Anticipated Key Drivers of Change on Water Resources of the Coastal Zone*; *Institutional Strengthening of Climate Change Study Cell at BUET for Knowledge Generation and Human Resource Development*; *Development of Flood Hazard and Risk Maps with Effect of Climate Change Scenarios*; and *Change Review of Warning System for Climate Disasters*. He is a member of the editorial board of the *International Journal of Oceanography and Marine Ecological System*; Technical Editor of the *Asian Journal of Applied Sciences*; and Coordinator of Climate Change Study Cell, BUET. He has authored a book, several book chapters, journal articles and discussion papers.

Sharad Kumar Jain is a senior scientist at the National Institute of Hydrology (NIH) and leads the Water Resources Systems Division and Climate Change Cell at NIH. He is also Coordinator of the Western Himalayan Regional Center (Jammu) of NIH. He has over 33 years of experience in research and teaching in the field of hydrology and water resources. He did his PhD at the University of Roorkee, India in 1990, and Master of Technology in Hydraulics and Water Resources at the Indian Institute of Technology (IIT), Kanpur. His research interests include *Surface Water Hydrology*, *Water Resources Planning and Management*, and *Application of Advanced Tools such as Artificial Neural Networks, Remote Sensing, GIS, and Decision Support Systems*. He has developed a software package for Reservoir Analysis and a web-based course 'Hydrologic measurement and analysis of data' (e-learning program of IITs). He has served as a member of various editorial boards of national and international journals, including the *Hydrological Processes* journal, Wiley-Blackwell, UK and the Indian Society for Hydraulics' *Journal of Hydraulic Engineering*. He has authored books on water resources system planning and management and written numerous technical papers in journals.

Marc A. Jeuland is an assistant professor at Sanford School of Public Policy & Duke Global Health Institute, Duke University, USA. He holds a PhD in Environmental Sciences and Engineering from the University of North Carolina-Chapel Hill. Since 2014, he has been associated as a Senior Research Fellow with the International Water Management Institute (IWMI) and as an Adjunct Senior Research Fellow with the National University of Singapore. He has provided consultancy services to World Bank-supported research projects that covered the Ganges Basin, eastern Nile, the Middle East and North Africa. He has written a number of peer-reviewed publications on irrigation efficiency, water quality, impact on health and wastewater reuse, amongst others.

Zahirul Haque Khan is Principal Specialist at the Institute of Water Modelling (IWM) in Bangladesh. A river and coastal expert, he did his Masters in Water Resources Engineering from Bangladesh University of Engineering & Technology, Dhaka in 1995. He works for the Formulation of the Bangladesh Delta Plan 2100 project, the Coastal Embankment Improvement project and the Char Development and Settlement project. Under the Delta Plan project, he has led four themes: disaster management, water resources, river management and water supply and sanitation on behalf of IWM. He is also involved in feasibility, detailed design, and implementation and monitoring studies for drainage and flood management, river management, integrated coastal zone management and land reclamation in the Bay of Bengal navigation of port and harbour, environmental impact assessment, disaster management and climate change impact assessment. He has provided his expertise to various government agencies, including the Bangladesh Development Board, the Planning Commission (General Economic Division) and the Department of Disaster Management in Bangladesh. He has written a number of conference and journal papers.

M. Dinesh Kumar is Executive Director of the Institute for Resource Analysis and Policy (IRAP) in Hyderabad, India. He holds a PhD in Water Management, and has 25 years of professional experience in research, action research, field implementation, consultancy and training in the field of water resources. He has around 160 publications to his credit, including five books, three edited volumes and several papers in international peer-reviewed journals. His latest book is *Thirsty Cities: How Indian Cities Can Manage their Water Needs*, published by Oxford University Press, and latest edited volume is *Water, Energy, Food Security Nexus: Lessons from India for Development*, published by Earthscan/Routledge, London. He is an associate editor of *Water Policy* and member of the editorial board of the *International Journal of Water Resources Development*.

Niki Maskey-Amatya works as a development programme coordinator at iDE Nepal. Her expertise is in natural resource management, sustainable agriculture and participatory development, climate change adaptation and integrated water resource management, and project management. She did her Masters in Environment and Sustainability Science in 2008 from Lund University, Sweden. From 2012 to 2015, she worked as a research officer (Social Science) at the International Water Management Institute (IWMI) Nepal Office. During her three-year stint at IWMI, she coordinated and implemented various donor-funded research activities on agricultural water management issues, including gender, climate change and vulnerability. She has co-authored research reports and journal articles on water management topics. Prior to joining IWMI, she had managed conservation projects in Lantang Buffer Zone, Sagarmatha Buffer Zone, Kachenjunga Conservation Area and Koshi Basin on behalf of WWF Nepal. She contributed to establishing a monitoring and evaluation system for agriculture production (farm-based and off-farm-based) through irrigation technologies and marketing for rural income generation for a project led by Winrock International.

Javier Mateo-Sagasta is a senior researcher at the International Water Management Institute (IWMI). He has extensive experience as Sri Lanka agricultural engineer and environmental scientist in programmes and projects aiming at the control of water pollution, the safe use of marginal quality water in agriculture, integrated water resources management and sustainable agricultural production in both developed and developing countries. Before joining IWMI, Javier gained experience in water management at research centres, in Jordan and the Netherlands, in the private sector, in Spain, and in the UN system with FAO, as a water quality officer. He has worked in multi-disciplinary teams in MENA, Latin America, Europe and South Asia. He has organized, facilitated and participated in international seminars and conferences, has published technical guidelines, global assessments and scientific papers and has provided technical backstopping to projects in the field. With his work in research, development and the private sector, he has forged a deep understanding of the knowledge generation to implementation pathways, vital for impact-oriented research.

M. Monirul Qader Mirza works with the Climate Research Division (CRD), Environment and Climate Change Canada (ECCC) as a physical scientist. He studied in Bangladesh and New Zealand. A water resources engineer, he received his PhD from the International Global Change Institute (IGCI), University of Waikato, Hamilton, New Zealand on climate change and flood modelling in Bangladesh. Since 1997, he has been associated with Working Group II of the Intergovernmental Panel on Climate Change (IPCC) – the winner of the 2007 Nobel Peace Prize – and contributed as a coordinating lead author (CLA) and lead author to its various assessments. Apart from the IPCC Assessments, he participated as a CLA in the Millennium Ecosystem Assessment and the Integrated Assessment of Agriculture, Science, Technology for Development (IAASTD) commissioned by the World Bank. He extensively researched the transboundary water issues of South Asia, particularly the Ganges, Brahmaputra and Meghna Basins. His major research interests are: climate change and risks, extreme weather events, transboundary water and energy issues, quantification of vulnerability and adaptation and statistical hydrology.

Sonali Mittra is an associate fellow at the Observer Research Foundation, a New Delhi-based think tank. She is also a consultant with Chatham House for a project on documenting case studies on successful management of transboundary rivers in South Asia. Her work in the past few years has centred on energy, water and environment security, and transboundary water management. Her current research interests include: water governance strategies, political sociology of water, water diplomacy, energy–water nexus and climate change. She has authored multiple policy recommendation reports and academic papers on the Ganga and Indus basins. She graduated in Botany (Hons) from Sri Venkateswara College, Delhi University. She completed her Masters in Environmental Impact Assessment and Management from the University of Manchester, UK. Previously, she has worked as a junior research associate at the Council on Energy, Environment and Water (CEEW), India. Before joining CEEW, she worked with Chatham House, London as a project consultant for a Sino-Indian Water Collaboration project.

Lal Mutuwatte is a hydrologist by training with specialization in hydrological modelling. His research and professional background includes over 20 years of experience in the fields of GIS, remote sensing and hydrological modelling. He has worked in multiple river basins across Asia and Africa, including Ganges, Salween, multiple river basins in Sri Lanka, and Karkheh in Asia; and Volta, Tana, Nile and Naiwasha in Africa. He has worked on issues related to hydrological assessments, climate change impacts, and water resource management with a focus on the agriculture sector. He has contributed to more than 40 research articles published in various scientific journals. In 2010, 2014 and 2015 he was awarded with the prestigious Sri Lanka presidential award for his research work. He has been a visiting lecturer of various universities in Sri Lanka. Lal holds a PhD in water resources management and hydrological modelling from the University of Twente, The Netherlands. His research interests are to work with remote sensing and hydrological modelling to study climate change impacts in a river basins/regions.

Rajmohan Natarajan, a hydrogeochemist, works as a Special Project Scientist (Vadose Zone Modelling) for the International Water Management Institute (IWMI). He has been working in the field of hydrogeochemistry and modelling since 1995. He holds a PhD in Hydrogeochemistry from Anna University, Chennai, India, and a Masters degree in Chemistry from Bharathidasan University, India. He has worked at a variety of universities in different countries (Dong-A University and Kunsan National University in South Korea; Sultan Qaboos University in the Sultanate of Oman; and French Atomic Research Center [CEA] in France). He has wide experience in groundwater quality monitoring, contamination assessment, contaminants transport in the unsaturated zone, and groundwater, geochemical and unsaturated zone modelling. He has published 29 research papers in various journals, book chapters and reports, and has 15 research papers in conferences to his credit. He has participated in and cooperated to conduct various professional training programmes during his stay in different universities, such as a capacity building programme for the officers of the state's water resources organization (PWD) under the World Bank-funded hydrology project and international workshops sponsored by UNESCO's International Hydrology Programme. He has also carried out a number of consultancy projects for international and Indian companies related to geochemical modelling, hydrogeological survey and bore wells installations.

Asghar Nawab is Senior Manager–Biodiversity with the Rivers, Wetlands and Water Policy Programme, WWF-India. He is involved in the conservation assessment of endangered aquatic fauna. He was awarded the WWF-India Meritorious Award 2008 for his outstanding contribution towards otter conservation research work in India. His research interests include landscape ecology with particular reference to habitat modelling, biology of small carnivores and integrated river basin management.

Ainun Nishat is working as Professor Emeritus at the Centre for Climate Change and Environmental Research, BRAC University, Dhaka. A graduate of BUET, in Civil Engineering, he was a professor of Water Resources Engineering for more than 26 years, until he moved to the International Union for Conservation of Nature (IUCN), in 1998, as its Country Representative. Then he worked as Vice Chancellor of BRAC University from 2010 to 2014. He obtained a PhD in Civil Engineering (Hydraulics) from the University of Strathclyde, Glasgow, UK in 1981. As an expert in management of international transboundary rivers, Prof. Nishat played an important role in the formulation of the Ganges Water Treaty that was signed between India and Bangladesh in 1996. Prof. Nishat worked as a member of the Indo-Bangladesh Joint River Commission, from 1981 to 1999, which is responsible for management of common water resources between the two countries. Professor Nishat also worked as a Member of the

Bangladesh National Water Council, the Bangladesh National Agricultural Commission and the National Council on Science and Technology and took an active part in the formulation of various national policy, strategy and action plans. His current research focuses on adaptation to climate change and related policy advocacy. For decades, he has worked as a catalyst, advocate, educator and facilitator, championing the wise use of natural resources and sustainable development in Bangladesh.

Paul Pavelic, a groundwater hydrologist, is based at the International Water Management Institute (IWMI), South East Asia Regional Office, Vientiane, Laos. He has a PhD from Flinders University of South Australia. He has a broad research interests that cover water resources, food security and rural livelihoods. His specialties include groundwater resources management, managed aquifer recharge, agricultural water management and water reuse. He has published extensively in these fields. In more recent years he has been particularly involved in the assessment and management of groundwater resources and conjunctive water use from multidisciplinary perspectives in Asia and Africa.

Paras Pokharel has been working at Water Aid Nepal in Kathmandu as a programme officer since January 2015. In this capacity, he focuses on monitoring of rural service delivery through development projects and advises emergency response for the organization. An MSc graduate in Sustainable Environmental Management from Middlesex University, London, he is interested in research and development of sustainable resources management. He has contributed to various research and development projects related to sustainable development, community mobilization and natural resources management, the findings and recommendations of which have contributed to policy formulation processes. He worked as a research consultant at the International Water Management Institute (IWMI) in 2014. He contributed to a Study of Sustainability of Multiple Water-Use Systems and Market Collection Centres under the Market Access and Water Technology for Women Project led by iDE Nepal. He has done climate change vulnerability assessments of some rivers in Nepal in collaboration with WWF Nepal and local NGOs.

Prachanda Pradhan, a senior social researcher, is a patron of the Farmer-Managed Irrigation System Promotion Trust (FMIST) based in Nepal. He holds a PhD in Public Administration and Development Administration from Claremont Graduate School, Claremont, California, USA. He was a Professor of Public Administration in Tribhuvan University and Visiting Fulbright Professor at the Center for International Studies, Cornell University in 1979. He has designed and organized international workshops and conferences in participatory irrigation management. His recent contributions to research include water users' associations in relation to out-migration and the role of women in water management in Nepal. He was appointed Participatory Development Advisor to an ADB-supported Community Managed Irrigated Agriculture Sector. He has worked in the International Irrigation Management Institute (now IWMI), at the Nepal Office, and also as Irrigation Specialist in Nepal, Bhutan, northern Pakistan, Sri Lanka, Nigeria and Niger. He has worked as an advisor to various donor-supported irrigation and livelihood projects within and outside Nepal. He helped the Hadejia Jammare River Basin Development Authority in Nigeria to introduce participatory irrigation management. He has written many articles, research papers and books on irrigation management.

Sanmugam Prathapar is Manager of Groundwater Modelling, DPI Water, NSW, Australia. He obtained his PhD in Agricultural Engineering from Texas A&M University in 1986. He specializes in modelling unsaturated and saturated transport in porous media and hydro-economic modelling of farming systems. His expertise extends into artificial recharge to aquifers, estimation of sustainable yields of aquifers, groundwater monitoring network rationalization, reclaimed water use in agriculture, irrigation, drainage and salinity management. He was Theme Leader at the International Water Management Institute (IWMI) from 2011 to 2014. He worked as Director of IWMI, Lahore, Pakistan for three years in the late 1990s and was also Research Coordinator of IWMI's Pakistan Programme. He worked as a senior research scientist in the CISRO Division of Water Resources, Australia from 1987 to 1996. He is also a Chartered Professional Engineer, Institution of Engineers, Australia.

Santa Bahadur Pun, an electric engineer, has a utility background. He started his career in 1967 with Nepal Electricity Corporation, which amalgamated with the Electricity Department in 1985 to become the Nepal Electricity Authority (NEA). During his eight years of work in the border towns of Birgunj and Biratnagar, one exporting power to Raxual and the other importing power under the Indo-Nepal Koshi Agreement, he had close relationships with power sector officials of India. After a short stint as the Managing Director of NEA, he was at the Ministry of Water Resources as an Officer on Special Duty. It was at the Water Resources Ministry that he came across the nuances of Nepalese water through the lens of various governments (India and Bangladesh in particular) and multilateral institutions like the World Bank and Asian Development Bank. After his retirement in 2000, he maintains his interest particularly on transboundary water issues and writes intermittently.

Golam Rasul, a development economist, is Head of the Economic Analysis Division after serving as a policy development specialist at the International Centre for Integrated Mountain Development (ICIMOD) for approximately five years. He holds a PhD in Regional and Rural Development Planning from the Asian Institute of Technology (AIT), Thailand. He worked for more than a decade in the Bangladesh Civil Service in different ministries and in field administration in different capacities where he was involved in the formulation and implementation of development planning and programming. Along with development work, he has been actively involved in research in areas that include agriculture, natural resource management, poverty alleviation, and sustainable development in Bangladesh and the South Asian region. His research findings have been published in many international journals and four of his papers have appeared as 'most read papers' in their respective journals in Science Direct. He has written on sustainable food security in mountains, economic perspective on water for growth and development in river basins, economic value of wetland ecosystem services, strengthening value chains, and gender and social equity in adaptation to climate change.

Ram Pratap Sah is a senior agriculture, food and nutritional security, livelihoods and climate change professional with over 25 years of experience. He is currently working as Rice Technical Expert for the Adaptation to Climate Change project, which is funded by the International Financial Corporation (IFC) and Practical Action. As a regional team member, he has also contributed to the 'Initiatives of seeds and fertilizers for transformation of agriculture' involving 12 countries in Asia. He is former Executive Director of the Nepal Agricultural Research Council. His areas of specialization include sustainable agriculture, climate smart technologies, livelihood, participatory research and development interventions, biodiversity and natural resources management. He worked as Technical Advisor to the Ministry of Agriculture and Cooperatives, the Government of Nepal, for three years. In his capacity as a consultant for policy research, he has analysed policy issues with regard to natural resources, climate change and biodiversity management for addressing poverty, social exclusion and livelihood challenges in South Asia on behalf of the International Centre for Integrated Mountain Development. He has undertaken research on seed business and private sector promotion with support from international donor agencies in Nepal.

Dipankar Saha, a hydrogeologist, is a member of the board of a department at the Central Ground Water Board, Ministry of Water Resources, River Development & Ganga Rejuvenation, India. He is spearheading the ambitious National Aquifer Mapping Programme of the Government of India. He has more than 35 years of experience in working on various aspects of ground water science and its sustainable management in different parts of India. He has obtained a PhD on Ground Water Management from the Indian School of Mines, Dhanbad. He is trained in the application of remote sensing on ground water investigation. He is professionally trained in applied quaternary geology by the Asian Institute of Technology, Bangkok, and on integrated water resources management by JICA, Tokyo. He has authored more than 20 hydrogeological survey reports, and State Ground Water Development Plans for two states of India. He has more than 35 papers in peer-reviewed journals. He has presented papers and key note addresses at various national and international seminars in India and abroad. He has represented the government department in workshops organized by the World Bank at Kathmandu, the International Atomic Energy Agency at Ho-Chi Minh City and World Water Week at Stockholm.

Panchali Saikia works as Scientific Officer–Social Science at the International Water Management Institute (IWMI), New Delhi. She completed her Masters in Political Science from the University of Delhi and two years' Postgraduate Diploma in Human Rights from the Indian Institute of Human Rights, New Delhi. Her primary areas of research include trans-boundary water governance and management, gender issues in water management, institutional arrangement in water governance, human migration and cross-border regional cooperation. She has extensive field-level experience in the regions of South Asia (India, Nepal and Bangladesh) and Southeast Asia (Thailand). She has also received the 'best early career scientist presentation runner-up award for oral presentation' at an international conference organized by CPWF Ganges Basin Development Challenge (GBDC) and CGIAR Research Program on Water, Land and Ecosystems (WLE).

Christopher A. Scott, PhD, is Research Professor of Water Resources Policy at the Udall Center for Studies in Public Policy, and Professor of Geography and Development, both at the University of Arizona. He is engaged in research and outreach on mountain water towers, river-basin water security, water–energy–food nexus, water reuse, climate adaptation, and social-ecological systems' resilience with particular emphasis on the Andes, Colorado Basin Sky Islands, and recurring interests in the Himalayas. He served as Asia Regional Director for the International Water Management Institute (Colombo, Sri Lanka, but based in Hyderabad, India, where he also coordinated IWMI's Nepal programme). He completed his schooling at Woodstock School in Mussoorie, India, and received his BS in Water Resources and BA in Asian Studies from Swarthmore College and his MS and PhD in Hydrology from Cornell University. He is a member of the International Water Resources Association, American Geophysical Union, Association of American Geographers and American Association for the Advancement of Science.

Tushaar Shah is Senior Fellow in the Colombo-based International Water Management Institute (IWMI). He is an economist and public policy specialist, serves on the boards of several academic institutions and NGOs and is a director on the board of the ICICI Bank. He works out of Anand in western India. Over the past 35 years, Shah's main research interests have been in water institutions and policies in South Asia, a subject on which he has published extensively. His notable contributions have been in comparative analyses of groundwater governance in South Asia, China and Mexico. More recently, his interests have lain in comparative analyses of water institutions and policies across Asia and between South Asia and sub-Saharan Africa.

Shah has also worked extensively on energy–irrigation nexus in India. Shah was honoured with the Outstanding Scientist award of the Consultative Group of International Agricultural Research (CGIAR) in 2002. The IWMI-Tata Water Policy Program led by Shah was chosen for the 2014 UN Water-for-Life Award for best practices in water management research. His most recent publication is *Taming the Anarchy: Groundwater Governance in South Asia*, published by the Resources for the Future Press, Washington. He is a former director of the Institute of Rural Management at Anand, India.

Bharat Sharma is Scientist Emeritus (Water Resources) at the International Water Management Institute (IWMI), New Delhi office, India. He holds a PhD in Soils and Water Management and an MSc (Hons) from Punjab Agricultural University. He worked at the Indian Council of Agricultural Research as Researcher (Water Management), Project Coordinator and eventually as Assistant Director General (IWM) where he was responsible for managing and supervising water management research at the national level. During his 14 years' tenure at IWMI as Principal Researcher and India-Coordinator, he implemented a number of research for development projects in South Asia, China and Africa. Some of the more recent projects include groundwater governance in Asia, strategic analysis of India's National River Linking project, basin focal project for the Indo-Gangetic Basin, integrated rainwater management systems for the Blue Nile Basin, Ethiopia and Smart ICT for Water and Weather Information in Africa. Management of natural resources and poverty alleviation are his main interest areas of research. He has published extensively on these subjects and is presently engaged with a number of grassroot and international organizations and the Government of India in improved management of the water and land resources.

Subodh Sharma, a professor at the Department of Environmental Science & Engineering, Kathmandu University, Nepal, specialises in bio-assessment, climate change, water quality, sanitation and health. Sharma holds a PhD degree in Biological Assessment of Water Quality in the Rivers of Nepal from the University of Agriculture, Forestry and Renewable Natural Resources (BOKU), Vienna, Austria. He was Post-Doctoral Senior Fulbright Research Scholar (2009–2010) at Towson University, Maryland, USA. He is best known in the region for his involvement in documentation of a macroinvertebrates taxa catalogue and applying this knowledge to formulation of a bio-assessment tool useful in water quality monitoring of high altitude lakes and mountain rivers in the Hindu Kush-Himalayan region (Bhutan, Bangladesh, Nepal, northern India and northern Pakistan). He has authored two books and a number of articles and research papers for national, regional and international publications.

Mandira S. Shrestha joined the International Centre for Integrated Mountain Development (ICIMOD) as Water Resources Specialist in 2001. At ICIMOD she coordinates regional programmes on transboundary flood risk reduction in the Hindu Kush-Himalayan (HKH) region and application of satellite-based rainfall estimates to flood prediction. She holds a Doctorate in Engineering from Kyoto University, Japan, an MSc in Civil Engineering from the University of Washington, Seattle, USA, and a Bachelors in Civil Engineering from the University of Roorkee (now IIT Roorkee). She is also an accredited professional engineer (PE) for Washington State, USA. She has 20 years of experience in water resources management including transboundary flood risk management, hydraulic and hydrologic modelling and hydropower projects. She worked for several organizations – including from 1988 to 1993 in the USA – as a water resources engineer for hydroelectric and water resources projects, and later for the Association for Research and Management, Norplan, Butwal Power Company and Care Nepal. Her publications focus on regional flood information systems in the Hindu Kush Himalayas, estimation of glacier and snow melt-driven streamflow, assessment and evaluation of satellite-based rainfall estimation products in river basins, water infrastructure, regional water cooperation and flood risk reduction.

Rashmi Kiran Shrestha is a researcher at the International Centre for Integrated Mountain Development (ICIMOD). Her research focuses on present energy scenarios in the Hindu Kush-Himalayan region, specifically hydropower and benefit-sharing in Nepal. She is a team member of the project 'The Irrigation-Hydropower Nexus in the Ganges Headwaters' led by the University of Arizona. She has six years of research and implementation experience on water supply, sanitation and health projects (NGO Forum, Lumanti) and water management options (NGO Forum for Urban Water and Sanitation). Her work on institutional analysis of the 2008 Koshi flood is published in *Economic & Political Weekly*. She reviewed the Asian Development Bank policy on the Tanahu Upper Seti Hydropower project under the auspices of the NGO Forum. She was a recipient of a Netherlands Fellowship for her MSc in Water Management from UNESCO-IHE, Delft, the Netherlands, prior to which she completed an MA in Anthropology from Tribhuvan University, Nepal.

Surendra Raj Shrestha leads the Branch Office of the Ground Water Resources Development Board (GWRDB), the Government of Nepal, in Dhangadhi, far-west Nepal. He has 19 years of experience in the field of hydrogeology. From 2012 to 2015, he worked as a senior divisional hydrogeologist at GWRDB. His research and other work experience includes groundwater exploration and management, feasibility study of groundwater construction of deep and shallow tube wells, training on groundwater irrigation management to water user groups, and monitoring and evaluation of groundwater irrigation projects. He completed his Masters degree in Geology in 1992 from Tribhuvan University, Kathmandu. He has written research reports and articles on various aspects of groundwater resource.

Alok K. Sikka is now the Head of IWMI-India Program. He was Deputy Director General (Natural Resource Management) at the Indian Council of Agricultural Research (ICAR). He is also Technical Expert (Watershed Development) at the National Rainfed Area Authority (NRAA), Planning Commission, Government of India, New Delhi. Before joining NRAA, he was Director of the ICAR Research Complex for Eastern Region, Patna and Basin Coordinator for Indo-Gangetic Basin under the CGIAR Challenge Program on Water and Food. He holds a PhD in Civil and Environmental Engineering with specialization in Hydrology and Water Resources Engineering from Utah State University, Logan, Utah, USA. He has a rich and diverse experience of over 34 years in research, institutional and policy issues, teaching, training, extension and consultancy in the areas of soil and water conservation, watershed management, water harvesting, hydrologic modelling, drought studies, climate change, water management, water productivity and farming systems. He has been associated with many internationally supported projects. He has written more than 250 articles and papers for international and national journals, book chapters, bulletins, reports and books. He is Fellow of the National Academy of Agricultural Sciences, the Indian Association of Soil and Water Conservationists and the Soil Conservation Society of India.

Ravindra Kumar Sinha is a professor and Head of the University Department of Zoology, Patna University, Patna. He has been working on various aspects of ecology and biodiversity of the River Ganga and its tributaries since 1980. He is best known for his pioneering research and conservation efforts to save the Ganges River dolphin from extinction. He has 105 research articles in national and international journals, 40 technical reports and four books to his credit. He was awarded the Order of the Golden Arc, the most excellent Order of the Netherlands, by His Royal Highness Prince Bernhard of the Netherlands in 1999 as international recognition of his pioneering work to save the Ganges dolphin in particular and Gangetic biodiversity in general.

Uttam Kumar Sinha is a fellow at the Institute of Defense Studies and Analyses (IDSA), New Delhi and holds an adjunct position at the Malaviya Centre for Peace Research, Benares Hindu University. A doctoral candidate from Jawaharlal Nehru University, New Delhi, he worked in the daily *Pioneer* and wrote the weekly space, 'Strategic Eye'. He is also Managing Editor of *Strategic Analysis* (Routledge). In 2006, he was a Visiting Fellow to the International Peace Research Institute Oslo and in 2008 a Chevening 'Gurukul' Scholar at the London School of Economics and Political Science. He has been actively involved with the Strategic Studies Network of the National Defense University, Washington, DC and was the Chair of the Working Group on Water Dispute Resolution Mechanism from 2012 to 2014. In 2015 he was at the Harvard Kennedy School, Harvard University on a US–South Asia Leaders Engagement Programme. His recent work includes edited volumes: *Non-Traditional Security Challenges in Asia: Approaches and Responses* (Routledge, 2015); *Arctic: Commerce, Governance and Policy* (Routledge, 2015); *Emerging Strategic Trends in Asia* (Pentagon Press, 2015). He co-authored the IDSA Report on *India and Water Security: External Dynamics* (2010) and also the IDSA Report on *Security Implications of Climate Change for India* (2009).

Vladimir Smakhtin, at the time of writing, led the Water Availability, Risk and Resilience Research Department at the International Water Management Institute (IWMI), based in Colombo, Sri Lanka. He also co-led the research flagship on Managing Variability and Competing Uses of the CGIAR Research mega-program on Water, Land and Ecosystems (WLE). He holds a PhD in hydrology and water resources from the Russian Academy of Science. His research and management experience spreads across agricultural and environmental water management, low-flow and drought analyses, assessment of basin development and climate change impacts on water availability and access, provision of hydrological information for data-poor regions, and global water availability and scarcity. He has initiated, managed and contributed to numerous research initiatives in over 20 countries worldwide, authored and co-authored over 200 publications, and consulted for a number of national and international organisations, including the Department of Water Affairs of South Africa, the Ontario Ministry of Natural Resources and the World Commission on Dams, IUCN and UNEP. He has been an active promoter of environmentally sustainable water resources management globally through science and policy gatherings in over 50 countries. He was involved in the state programme for mitigating the consequences of the Chernobyl accident, actively contributed to the development of ecological reserve methods in South Africa, and pioneered a ground-breaking analysis of ecosystem water requirements at the global scale. Vladimir is currently Director of the Institute for Water, Environment and Health of the United Nations University (UNU-INWEH), based in Hamilton, Ontario, Canada.

Fraser Sugden, Social Science Researcher at the International Water Management Institute (IWMI), co-leads (with the University of Southern Queensland) a four-year ACIAR-funded project on irrigation access for marginal and tenant farmers in Nepal and India, which will pilot models of collective farming. He specializes in the political economy of agriculture and water management. His PhD on land relations in lowland Nepal was completed in 2009 from the University of Edinburgh. He then joined the University of Stirling as a Postdoctoral Research Fellow for a project on highland aquatic resources in upland Asia (HighARCS), before joining IWMI. He is a geographer by training, with research interests in rural class relations, land tenure, gender and migration, and their role in mediating access to water and other land resources. More recently his research has focused on 'agrarian stress' in South Asia due to climate change and its intersection with local political-economic structures on the one hand, and the economic pressures of globalization on the other. He has worked for over a decade in the eastern Gangetic Plains, with a focus on Nepal's Terai-Madhesh, Bihar and West Bengal. He has been engaged in field research in Nepal's eastern and far-western Himalaya, China, Vietnam and Bangladesh.

Diana Suhardiman is Senior Researcher at the International Water Management Institute (IWMI), Southeast Asia Regional Office, based in Vientiane, Lao PDR. Prior to joining IWMI, she received a PhD at Wageningen University, the Netherlands, where her research focused on the role of government bureaucracies in shaping water policy formulation and implementation in Indonesia. She has over 10 years of experience in natural resources governance, with particular focus on water governance in Asia. Her research focuses on multilevel policy and institutional analysis at the intersection of land, water, environment and energy in South and Southeast Asia (Indonesia, Laos, Cambodia, Vietnam, Thailand, Myanmar, China Yunnan Province and Nepal). She leads research on environmental policy in the Mekong region, focusing on forest conservation within the wider policy context of land use allocation, watersheds protection and poverty reduction. Her most recent research includes the political economy of water governance in the Salween River Basin and the role of foreign direct investment in relation to natural resource management in Laos and Cambodia.

Vinod Tare is a professor of Environmental Engineering and Management at the Indian Institute of Technology (IIT), Kanpur, India. He is a civil engineer, obtained his Masters and Doctoral degrees in Environmental Engineering from IIT Kanpur, and subsequently did postdoctoral research at the Illinois Institute of Technology, Chicago. His research areas include processes for natural resource conservation and regeneration, physicochemical, biological and ecological processes, water and wastewater treatment, and modelling and simulation of environmental systems. He was the Chairman of the India Water Impact Summit 2013 held at the India Habitat Centre, New Delhi and the International Conference on Water-harvesting, Storage and Conservation (2009). He is a consultant to many government organizations, NGOs, industries and institutions. He has developed an environment-friendly passenger coach toilet system being deployed for Indian Railways, communities, house boats and Kumbh Mela, amongst others. He led a consortium of seven IITs for the preparation of the Ganga River Basin Environment Management Plan to be submitted to the National Ganga River Basin Authority chaired by the Prime Minister. He has written a number of journal articles including primary sewage sludge, source-separated human faeces for nutrient recycling, post-mortem examination and analysis of anaerobic filters and removal of heavy metals from wastewaters.

Bhuwan Thapa is a PhD student in the School of Geography and Development at the University of Arizona. His PhD research focuses on climate change adaptation in water resources management and agriculture in the Gandaki Basin, a tributary of the Ganges, in Nepal. He is a recipient of a doctoral research fellowship from the Himalayan Adaptation, Water and Resilience (HI-AWARE) research initiative of the International Centre for Integrated Mountain Development (ICIMOD). He has five years of research and professional experience in Nepal and the United States, working for various national and international agencies such as the Japan Water Agency, the Danish Hydraulic Institute, the World Wildlife Fund-Nepal, WaterAid-Nepal, the Center for Clean Air Policy (USA) and A4 Scientific (USA). He has worked on diverse topics related to water resources management and planning, including river basin planning, water and sanitation, decentralized wastewater systems, climate change and modelling, community forestry and environmental impact assessment.

Paul M. Thompson is Senior Research Fellow at Middlesex University in the UK. He is a social scientist specializing in the management of natural resources and has worked largely in Bangladesh since the late 1980s to the present time. Since 1996 he has been involved in development and research on community-based co-management of fisheries, wetlands and floodplains in Bangladesh, with a focus on institutions and sustainability. He also has a keen interest in the birds of Bangladesh and their conservation. He has a PhD in Development Studies from Middlesex Polytechnic in the UK.

Anwar Zahid is Deputy Director of Ground Water Hydrology, Bangladesh Water Development Board (BWDB), Dhaka. He did his PhD in Hydrogeology and Environmental Geology from the University of Dhaka, Dhaka, Bangladesh in 2009, and MSc in Tropical Hydrogeology at Eberhard-Karls University of Tuebingen, Tuebingen, Germany in 2004. At BWDB, he leads qualitative and quantitative research for the optimum utilization of water resources and preparation of hydrogeological, geochemical, geotechnical and sedimentological maps, reports and numerical models, and projects to monitor and assess groundwater situations across the country. His research areas include integrated water resources management, hydraulic and chemical characterization of aquifers, water contamination, coastal hydrogeology and salt water intrusion, climate change impacts on water resources and socio-economic conditions, groundwater modelling, river morphology, flood management, water and sanitation technologies. He has written three books and a number of peer-reviewed articles for national and international journals.

COVER AND CHAPTER PHOTOS

Cover photos

Top: Arun valley on the border between Bhojpur and Sankhuswabha district, eastern Nepal, Fraser Sugden (IWMI), September 2010.

Middle left: Frozen river at 4,800m descending from the glacial Lhonak Valley in the Kanchanjunga region, Taplejung district, Nepal. Fraser Sugden (IWMI), November 2007.

Middle centre: Generating energy using solar panels to pump water for agriculture. Sukhdev Vishwakarma and his daughter Meenu, both farm workers, use water pumped from a solar water pump at the farms of Gurinder Singh, a farmer with a land-holding of 80 acres in Jagadhri, India. Prashanth Vishwanathan (IWMI), 13 November 2014.

Middle right: General view of the Ganges River at Varanasi, India. Neil Palmer (formerly IWMI), October 2014.

Bottom: Paddy cultivation in Rajarhat, located east of Kolkata, India. The skyscrapers of the modern township of Newtown can be seen in the background. Chhandak Pradhan (IWMI), 31 January 2012.

Chapter photos

Foreword: Neil Palmer (formerly IWMI).

1 Nepal. Fraser Sugden (IWMI).
2 Ganges. Neil Palmer (formerly IWMI).
3 Nepal. Fraser Sugden (IWMI).
4 India. John Appelby (IWMI), 2014.
5 Mighty Ganges – when it floods. Ramesh Iyanswamy/Flickr, 19 September 2010.
6 Farmer at Safai Distributary, Ganges Basin. Nitasha Nair (IWMI).
7 Kali Gandaki hydropower project in Nepal. Asian Development Bank. Flickr.
8 24-year-old Mohan Das works on a sprinkling system energized through a submerged solar pump at a private farm in Klashar, India. Prashanth Vishwanathan (IWMI), 2014.
9 Ganges in Varanasi area, India. Neil Palmer (formerly IWMI).
10 Trishuli River in Nepal which flows into the Ganges. Nitasha Nair (IWMI).
11 A view in Nepal. Fraser Sugden (IWMI).

12 Women extracting water from the riverbed, Gaya, Bihar, India. Prasanta Biswas (ICIMOD).
13 Neil Palmer (formerly IWMI).
14 A participatory training for women farmers to voice their experiences of vulnerability to policy-makers, Dhanusa district, Nepal. Pawan Kumar, November 2012.
15 Preparing soil to plant paddy, Nepal. Jim Holmes (IWMI), June 2013.
16 A woman scatters fertilizer on newly planted rice seedlings, Chobar, Kathmandu Valley, Nepal. Jim Holmes (IWMI), 2013.
17 A view in Bangladesh. Neil Palmer (formerly IWMI).
18 Crossing the river, Bangladesh. Neil Palmer (formerly IWMI), November 2014.

FOREWORD

The Ganga: between faith and reality

The Ganga – one of South Asia's largest rivers – is also undoubtedly the most revered river in the world. Its spiritual and cultural significance remains undiminished in the hearts and minds of people since time immemorial. Even today, many refer to it as 'Ganga ji', adding the suffix as a mark of respect. Ma, mythical, majestic, mighty – are words commonly used in conversations to describe Ganga. They derive from its multiple identities, forms, size and above all its powerful hold on the human mind across ages.

Jawaharlal Nehru, India's much admired first Prime Minister, wrote in his will and testament: 'The Ganga beloved of her people, round which are intertwined her racial memories, her hopes and fears, her songs of triumph, her victories and her defeats. She has been a symbol of India's age-long culture and civilization, ever-changing, ever-flowing and ever the same Ganga.' A statue of Ganga adorns the royal palace at Mul Chowk in Patan, Nepal, which was built by the Malla kings in the 1600s.

Most Hindus also worship the River Ganga as a goddess. She is the consort of Lord Vishnu, the progenitor in the Hindu trinity along with Brahma the creator and Shiva the destroyer. For devotees, the river is a physical manifestation of the goddess. This belief endows her waters with mythical powers. According to Hindu scriptures a mere drop of *Gangajal* (Ganga water) or the sight or even remembrance of Ganga can absolve its faithful of all sins.

This belief derives from the popular myth of Ganga narrated during childhood over millennia to most of the Basin population. According to the story, the goddess Ganga descended to Earth in response to the prayers and penances of Prince Bhagirath to liberate the souls of his ancestors by sweeping over their ashes. This story is so strongly imprinted in the Hindu mind that to this day Hindus all over the world desire that their ashes be immersed in the Ganga so that they can attain *moksha* or salvation.

Ganga is known by over 100 names based on her different forms and legends of origins and powers. Some of the more common names are *Jahnavi* (daughter of the sage Jahnu), *Vishnupadi* (emerging from the foot of Lord Vishnu), *Trilokapathagamini* (one who traverses the three worlds, Heaven, Earth and the netherworld), *Mokshadaayani* (provider of salvation), *Paapharini* or *Paapvinaashini* (sin-cleansing) and *Sursari* (river of the gods) among others. Each name and legend creates a different image in the mind of the listener.

All along the course of the River Ganga and many of its tributaries every day, in fair weather or foul, thousands of devotees can be seen worshipping at its banks, performing numerous rituals offering obeisance to the river, the sun and the Almighty. The faithful turn to the rising sun at dawn to perform the *aachaman* (the ritual purification sip before praying); having faith that *Gangajal* is pure, they take dips or bathe in the river even at Gangotri where the water is icy cold. Hindus cremate the dead on the river banks followed by the ritual purifying bath. Many devotees offer flowers and garlands to the river at the banks while some ride in boats to the middle of the river. At many pilgrim towns the main worship ritual is the evening *aarti* (prayer) when the faithful gather on the river banks to pray *en masse*. At the end they place colourful flowers and *diyas* (small earthen lamps) on leaves and release them to float on the river water. Ganga Aarti (prayer) at Haridwar, Prayag and Varanasi is the most mesmerizing. All 6,000 rivers of Nepal also eventually flow into the Ganga so are a part of the Ganges Basin. Many rivers such as the Bagmati, Kali, Gandaki etc. are considered holy by both Hindus and Buddhists. There are also many temples and shrines along the banks of the rivers such as the Pashupatinath Temple, which is an important pilgrimage site for people from both India and Nepal.

While the Ganga is worshipped by most Hindus and Buddhists, people of other faiths also revere it. Muslim artisans make *kaanwars* for millions of Hindu devotees who gather *Gangajal* at Haridwar and other towns on the banks of the river during the annual *kaanwar yatraa* festival. Many of these artisans also revere the river. In recent years hundreds of burqa-clad Muslim women in Bhagalpur have begun taking part in the *kaanwar yatraa*. In some districts of Bengal, Hindu and Muslim boatmen are known to pray to Ma Ganga. The third Sikh Guru, Guru Amar Dass, was fascinated by the Ganga. He visited Haridwar regularly during his lifetime and performed prayers on its banks. The might of the Ganga also finds reference in Buddhist narratives.

A variety of festivals sustain the legends of River Ganga and people's faith. Perhaps the most famous festival is the Kumbh Mela. It is held every four years by rotation at Haridwar and Allahabad (both on the Ganga), Nashik (Godavari) and Ujjain (Kshipra). Bathing in these rivers during the festival is said to cleanse a person of all sins. In recent years the festival has attracted tens of millions of devotees from all over the world, making the Kumbh Melas the largest human gatherings on Earth. Once in six years, an Ardh Kumbh (half Kumbh) Mela is held between the two Kumbhs at Allahabad and Haridwar.

Every year in the month of *Jyeshtha* (May–June) the Ganga Dussehra, observed all along the river's course, celebrates the descent of Ganga to Earth. Twice every year there is the *kaanwar yaatra*, a small one around *Shivrartri* (around March) and then the major one during the monsoon month of *Saawan* (July–August). During the *kaanwar* festival Ganga worshippers walk back to their homes carrying *Gangajal*, without allowing the pot to touch the ground, to anoint a local Shiva *lingam*. In recent years the number of *kaanwariyaas* visiting Haridwar during *Saawan* has grown to well over a million. Thus these festivals add substantially to the local economies. They have also helped make Ganga a cradle for several important pilgrimage towns and locations.

By extolling the forms, virtues and powers of River Ganga, poets, singers, philosophers and writers have kept alive its pre-eminence for millennia. While little-mentioned in the early Rig Veda, the Ganga's importance is highlighted in the later three Vedas. In ancient times Valmiki, sometimes referred to as India's first poet and author of the Ramayana epic, describes the Ganga as full of whirlpools at places, sometimes flowing gently and sometimes braided. Its banks are celestial playgrounds frequented by gods (*devaas*), demons (*daanavas*) and celestial singers (*gaandharvas*) among others. Its body is home to cranes, swans, other birds and lotuses. The poet Kalidasa has also written in praise of the Ganga.

In the medieval period Tulsidas, the sixteenth century poet-saint, lived most of his life on the banks of the Ganga in Varanasi. Here he translated Valmiki's Ramayana from Sanskrit into the vernacular *Ramacharitmanas* and composed a number of hymns. Combining his understanding of the Ganga and the Ramayana, Tulsidas intimates that the Ganga represents devotion to Rama. Almost a century earlier, Kabir, the great weaver-poet, had also lived in Varanasi. In his *dohas* (couplets) Kabir often referred to the purity of *Gangajal*. Ganga features prominently in several works of Nobel laureate Rabindranath Tagore. He considered the Ganga as a symbol of India, uniting several centuries of Indian thought and culture. More recently, River Ganga is a setting for the writings of celebrated authors like V.S. Naipaul and Amitav Ghosh. The latter often takes a critical view of the Indians' faith in the Ganga as well as the contradictions between the faith and its practice. Many Bengali authors, poets and musicians have also celebrated the Padma. Traditional *Bhatiyali* songs and *baul* singers celebrate the Ganga and the Padma besides other rivers in West Bengal and Bangladesh. In recent decades maestros like Bhupen Hazarika, S.D. Burman and Salil Choudhary have been inspired by the music of the rivers, creating several popular compositions. The Great Poet of Nepal, Laxmi Prasad Devkota, was cremated at the Ghat of Bagmati River in Pashupatinath Temple.

In the past 50 years, a very limited conception of economic development focusing on massive water extraction for chemicals-based irrigated agriculture, power generation, industries and urbanization has become a growing threat to the very existence of River Ganga – its sanctity, purity and integrity. Several commentators have highlighted the paradox of the rapidly burgeoning number of devotees and their displays of worship on one hand and the total lack of concern for the well-being of the river on the other.

The establishment of the National Ganga River Basin Authority and National Mission for Clean Ganga in 2009 and more recently *Namami Gange* by the Government of India represents recognition of River Ganga as India's civilizational identity and as an ecosystem. It has taken a few initial steps to halt some of the damage done. But piecemeal efforts focusing mainly on cleaning up the river will not be enough. The goals of economic development will have to shift from consumption-driven growth to creating a society that cares for its natural and human resources. The first step has to be the restoration of River Ganga's flows all along its length so that it can flow from its origin to the sea in a manner befitting the majestic goddess that her devotees believe her to be.

This book is a comprehensive treatment of the important aspects of the river, its basin and the millions of people inhabiting it. The scientific rigour and treatment of the varied subjects by the eminent researchers and authors from the three riparian countries – India, Nepal and Bangladesh – and those with international experience make the chapters excellent pieces for understanding and appreciation of the complex issues related to this unique river system. I consider this book as a humble offering to the service of *Ma Gange* and all those interested in its divinity, river science and technology, irrigation and hydropower, people, policies and governance, and transboundary cooperation, and more importantly restoring its health and glory, will find it very useful.

Ravi Chopra
People's Science Institute
Dehra Doon, Uttarakhand, India

ACKNOWLEDGEMENTS

First and foremost, the editors are grateful to all the individual chapter authors for their contributions to the book that, we hope, will add to the knowledge base and also contribute towards cooperation in water resources management in the Ganges Basin. We gratefully acknowledge the financial support provided by the CGIAR Research Program on Water, Land and Ecosystems (WLE). We would like to thank Mrs Upamali Surangika and Mr Sumith Fernando (both of IWMI, Colombo, Sri Lanka) for very committed, significant and excellent administrative and graphic support throughout the course of the book project; Ms Beena Kharel (of IWMI-Nepal Office in Kathmandu) for essential help with author coordination; and Mr Salman Siddiqui (IWMI-Colombo, Sri Lanka) for efficient GIS support. We also thank anonymous reviewers of individual chapters.

ACRONYMS AND ABBREVIATIONS

ADB/N	Agricultural Development Bank, Nepal
AHEC	Alternate Hydro Energy Centre
AISMR	All-India Summer Monsoon Rainfall
APP	Agriculture Perspective Plan
AR4	Fourth Assessment Report
AR5	Fifth Assessment Report
AWS	Automated Weather Stations
BCM	Billion Cubic Metres
Bgl	below ground level
BHC	Butylated Hydroxytoluene
BIWTA	Bangladesh Inland Water Transport Authority
BMDB	Bangladesh Meteorological Department Board
BOD5	Biological Oxygen Demand after 5 days
BOOT	Build, Own, Operate and Transfer
BPDB	Bangladesh Power Development Board
BWDB	Bangladesh Water Development Board
BWFMS	Bangladesh Flood and Water Management Strategy
CA	Constituent Assembly
$Ca\text{-}HCO_3$	Calcium-bicarbonate
$Ca\text{-}Mg\text{-}HCO_3$	Calcium-magnesium-bicarbonate
$Ca\text{-}Na\text{-}HCO_3$	Calcium-sodium-bicarbonate
CEA	Central Electricity Authority
CEGIS	Centre for Environment and Geographic Information Services
CGWA	Central Ground Water Authority
CGWB	Central Ground Water Boards
CGWG	Central Ground Water Board
CMIP5	Coupled Model Intercomparison Project 5
CMORPH	Climate Prediction Center Morphing
CPC	Climate Prediction Centre
CPCB	Central Pollution Control Board
CRU	Climate Research Unit

CSE	Center for Science and Environment
CWC	Central Water Commission
DO	Dissolved Oxygen
DDC	District Development Committee
DDT	Dichlorodiphenyltrichloroethane
DFO	Dartmouth Flood Observatory
DHM	Department of Hydrology and Meteorology
DL	Danger Level
DMC	Disaster Management Centre
DoA	Department of Agriculture
DoE	Department of Environment
DoED	Department of Electricity Development
DoF	Department of Forests
DOF	Department of Fisheries
DoLIDAR	Department of Local Infrastructure Development and Agricultural Roads
DoNPWC	Department of National Parks and Wildlife
DoRH	Department of Roads and Highways
DoSC	Department of Soil Conservation
DPR	Detailed Project Report
DPSIR	Drivers-Pressures-State-Impacts-Responses
DVC	Damodar Valley Corporation
DWIDP	Department of Water Induced Disaster Prevention
DWRC	District Water Resources Committee
DWSS	Department of Water Supply and Sanitation
EDC	Electricity Development Center
EIA	Environmental Impact Assessment
ENSO	El Niño Southern Oscillation
EPADC	East Pakistan Agriculture Development Corporation
EPC	Environment Protection Council
EPWAPDA	East Pakistan Water and Power Development Authority
ETCF	Electricity Tariff Fixation Commission
ETP	Industrial Effluent Treatment Plant
EYC	Eastern Yamuna Canal
FAO	Food and Agriculture Organization of the United Nations
FEWS	Flood Early Warning System
FFWC	Flood Forecasting and Warning Centre
FMIS	Farmer Managed Irrigation Systems
FMP	Flood Management Programme
FRL	Full Reservoir Level
GAP	Ganga Action Plan
GBM	Ganga–Brahmaputra–Meghna
GBSA	Ganges Strategic Basin Assessment
GCM	General Circulation Model
GDB	Groundwater Development Board
GEC	Groundwater Estimation Committee (India)
GFCB	Ganga Flood Control Board
GFCC	Ganga Flood Control Commission
GoB	Government of Bangladesh

GoI	Government of India
GPM	Global Precipitation Mission
GRBEMP	Ganga River Basin Environment Management Plan
GW	Gigawatt
GWh	Gigawatt Hour
GWRDB	Groundwater Resources Development Board (Nepal)
HCH	Hexachlorocyclohexane
HIDC Ltd	Hydroelectricity Investment and Development Company Ltd
HP	Hydropower
I&D	Interception and Diversion
IBN	Investment Board Nepal
ICIMOD	International Centre for Integrated Mountain Development
IEE	Initial Environmental Examination
IIT	Indian Institute of Technology
IITC	Consortium of Indian Institutes of Technology
IITM	Indian Institute of Tropical Meteorology
IIT R	IIT Roorkee
IMD	India Meteorological Department
INR	Indian Rupees
IPB	Industrial Promotion Board
IPCC	Intergovernmental Panel on Climate Change
ISF	Irrigation Service Fees
ISM	Irrigation Service Markets
ISP	Irrigation Service Providers
ISRO	Indian Space Research Organization
IWM	Institute of Water Modeling
IWRM	Integrated Water Resources Management
JAXA	Japan Exploration Agency
JRC	Joint Research Centre
JRC	Joint Rivers Commission
K	Hydraulic Conductivity
KW	Kilowatt
LGC	Lower Ganga Canal
LGED	Local Government Engineering Department
Lpm	Litres per Minute
LTA	Long Term Average
MCM	Million Cubic Metres
MGC	Madhya Ganga Canal
$Mg\text{-}Na\text{-}HCO_3$	magnesium-sodium-bicarbonate
Mha	Million Hectares
MoAC	Ministry of Agriculture and Cooperatives
MODIS	Moderate Resolution Imaging Spectroradiometer
MoE	Ministry of Energy
MoEF	Ministry of Environment and Forests
MoEF	Ministry of Environment, Forest and Climate Change
MoFALD	Ministry of Federal Affairs and Local Development
MoFSC	Ministry of Forest and Soil Conservation
MoI	Ministry of Industry

MoIRR	Ministry of Irrigation
MOM	Maintenance, Operation and Management
MoSTE	Ministry of Science, Technology and Environment
MoUA	Ministry of Urban Affairs
MoWR	Ministry of Water Resources
MPO	Master Plan Organization
Msl	Mean sea level
MSP	Minimum Support Price
MU	Million Units of Electricity
MW	Megawatt
NASA	National Aeronautics and Space Administration
NDC	National Development Council
NDVI	Normalized Difference Vegetation Index
NDWI	Normalized Difference Water Index
NEA	Nepal Electricity Authority
NGO	Non-Governmental Organization
NGRBA	National Ganga River Basin Authority
NMCG	National Mission for Clean Ganga
NPC	National Planning Commission
NWMP	National Water Management Plan (Bangladesh)
NWP	National Water Plan
NWRDC	National Water Resources Development Council
NWRSP	National Water Resources Policy
NWSC	Nepal Water Supply Corporation
O&M	Operation and Maintenance
PERSIANN	Precipitation Estimation from Remotely Sensed Imagery Using Artificial Neural Networks
PHED	Public Health Engineering Department (India)
PPP	Public–Private Partnership
PTA	Electrical Power Trade, Cross-Border Transmission Interconnection and Grid Connectivity Agreement
PV	Photovoltaic
R&R	Rehabilitation and Resettlement
RHWL	Recorded Highest Water Level
RIMES	Regional Integrated Multi-Hazard Early Warning System
RRI	River Research Institute
RRR	Resource Recovery and Reuse
SAARC	South Asian Association for Regional Cooperation
SANDRP	South Asian Network of Dams, Rivers and People
SAR	Synthetic Aperture Radar
SEBON	Securities Board of Nepal
SOD	Stage of Development
SPCB	State Pollution Control Board (India)
SPMG	State Program Management Group
SPV	Special Purpose Vehicle
STP	Sewage Treatment Plant
STW	Shallow Tube Wells
TRMM	Tropical Rainfall Measuring Mission

UGC	Upper Ganga Canal
ULB	Urban Local Body
UNEP	United Nations Environment Programme
USAID	United States Agency for International Development
VDC	Village Development Committee
WARPO	Water Resource Planning Organization
WASA	Water and Sewerage Authority
WCD	World Commission on Dams
WCRP	World Climate Research Programme
WEC	Water and Energy Commission
WECS	Water and Energy Commission Secretariat
WHO	World Health Organization of the United Nations
WMO	World Meteorological Organization
WRD	Water Resources Department
WRS	Water Resources Strategy Nepal
WSS	Water Supply and Sanitation
WSTFC	Water Supply Tariff Fixation Commission
WUA	Water User Association
WYC	Western Yamuna Canal

PART I

Resources and uses

1

Introduction

Luna Bharati, Bharat Sharma and Vladimir Smakhtin

The Ganges or Ganga, as it is known to most South Asians, and its tributaries have diverse meaning to the basin inhabitants, incorporating the mundane to the most profound. For farmers, it is a source of irrigation, for cities, it is a source of domestic and industrial water as well as hydropower. The river system is also a drainage for waste. The basin supports rich aquatic and terrestrial ecosystems, with rare and endemic species and fascinating landscapes. Furthermore, according to ancient scriptures, the river descends when King Bhagiratha summons a goddess to wash away the sins of his ancestors, which is still being followed by millions of Hindus who bathe in the holy waters and immerse the ashes of their dead for the river to carry them to heaven. In this book, we have made an attempt to comprehensively cover the various aspects of the Ganges/Ganga Basin from these vast and different perspectives. The Ganges is a trans-boundary basin shared among primarily Nepal, India and Bangladesh, with an estimated 4 per cent of the catchment area in China.

In an average hydrological year, around 1,200 billion m^3 of precipitation falls in the basin. Of this, over 500 billion m^3 becomes streamflow with the rest directly recharging groundwater or returned to the atmosphere through evapotranspiration. Availability of abundant water resources and fertile valleys and plains in the Ganges Basin supported a large agriculture-based civilization in India, Nepal and Bangladesh. With a catchment area of nearly 1.1 million km^2, the basin is the most populated in the world, being a home to over 500 million people. Yet, despite the relative abundance of natural resources, the basin still faces acute poverty, with average GDP per capita under US$2 per day and poverty rates around 30 per cent. The hydrology of the basin is primarily monsoon-driven and about 85 per cent of rainfall falls during the months of June–September. The temporal variability of rainfall and runoff is hence very high, and the problem of excess water during the monsoon, and water scarcity during the dry season, affects all aspects of life throughout the basin. Spatially, it is very heterogeneous in terms of geography, availability and access to resources including land, water and energy, and levels of socio-economic development. Large differences in the above are seen between its hills/mountains, plains and delta areas.

The potential of the hills/mountains in terms of water, energy and agriculture have not yet been sufficiently developed. Water availability in the mountains is the highest in the basin. Nepal alone has an estimated mean annual runoff of 224 billion m^3, and per capita water availability of 9,000 m^3. Yet, water infrastructure to store and manage the monsoon water for irrigation as well as hydropower is underdeveloped. Most of the hill and mountain systems in the basin are still practising low productive rain-fed subsistence farming. The green revolution, which brought change in the plains was not able to penetrate the hills and mountains because high input technologies were not as feasible here. The green revolution also mainly focused on rice, wheat and maize and overlooked many of the existing cropping systems in the hills, which could have benefited from research leading to increases in efficiency and productivity. On the other hand, the hills and mountain systems, which encompass forests, grasslands, wetlands and high-value horticulture, are also providing ecosystem services to low-land areas through freshwater that is captured, stored and purified in mountain regions. Mountain regions are hotspots of biodiversity; and from a societal point of view, mountains are of global significance as key destinations for tourist, spiritual and recreation activities. Therefore, the challenge is to identify areas where traditional subsistence systems can be transformed into market-oriented agricultural ones through diversification with high-value crops (vegetables, flowers, fruits) and agro-forestry systems – to make hills and mountains eventually more productive, profitable, resilient and attractive to out-migrating youth. Similarly, the mountains could also power the much needed economic growth in the basin through exporting hydropower. At the same time, the challenge is also to manage these upland systems without drastically impacting the ecological services that they provide.

The most profound use of water happens for agriculture, industry, domestic supplies and other livelihood options in the vast plains of the Ganges. The region is extensively cultivated with rice, wheat, maize and millets, sugarcane, pulses, oilseeds and a variety of flowers, vegetables and fruits; along with fisheries, agro-forestry systems and livestock. With small pockets of high productivity in the north-west region, the plains may generally be described as a low productivity-high potential region, where with small interventions, primary productivity of several commodities and enterprises can be substantially improved to alleviate the pervasive rural poverty.

Underdeveloped water infrastructure and poor water and benefit-sharing mechanisms among the riparian countries make the poor population excessively vulnerable to natural water-related disasters, such as recurrent, often catastrophic, floods and extensive and often extreme droughts. In spite of the large hydropower potential in the hills and mountains, the plains region is seriously deficient in affordable energy. The prolific groundwater aquifers, which can potentially kick-start the second green revolution in the eastern Gangetic Plains and terai, remain unexploited. And yet, arsenic presence in groundwater threatens the health and life of the poor. With appropriate scientific backstopping, efficient energy policies including use of solar and bio-energy, agricultural productivity can be significantly improved through integrated options along with crop diversification with a focus on high value crops. Coupled with proper attention to governance and social equity issues, this can help the poor to profit from the region's fertile alluvial soils and generous water endowments. Alarmingly poor water quality in large stretches of the main river and its tributaries, due to unregulated and largely untreated domestic and industrial effluents from large cities, towns and villages discharging directly into streams, is a matter of serious concern. Large recent initiatives like the National Mission for Clean Ganga and the *Namami Gange* have been initiated by India and the World Bank to restore the rivers' quality. But overall, the matter needs much more attention throughout the basin.

The Ganges *delta*, shared by India and Bangladesh and also known as the *Sundarbans*, is the most active, fragile and the world's largest and most densely populated deltaic system. The region witnesses strong and complex freshwater–sea brackish water interface modified by the construction of numerous embankments to create favourable conditions for cultivation and habitation in the form of small island polders. Unfortunately, these are frequently invaded by surges and storms in the form of cyclones and climate change-induced rise in sea levels threatening the very existence of the coastal regions. As such man and nature remain in perpetual struggle to retain dominance. There is a continuing conflict among the freshwater rice/crop farmers versus the more lucrative brackish water aquaculture farmers. The farming system is characterized by two key trends that serve to undermine significantly the resilience of farmers and the area's agro-ecosystems. The first is frequent flooding, with well-publicized devastating consequences; while the second is saline intrusion, which can cause widespread crop damage, and contaminate groundwater supplies for both drinking and agricultural use. Farmers' ingenuity to support year-round cultivation of rice and other crops in ultra-small farm plots with the support of science in developing flood and salinity-resistant rice varieties has helped in improving the productivity and livelihoods. Commercialization and global demand for the brackish and freshwater prawn culture has provided new opportunities to the fish farmers, but with larger and sometimes irretrievable environmental consequences. Lower riparians also complain about the existing transboundary water-sharing mechanisms and declining freshwater flows posing additional challenges, and call for robust treaties and innovative and sustainable production systems to enhance the resilience and improved livelihoods in the delta.

As clear from the above, many serious challenges face the basin. Thousands of pages have been written about various aspects of water resources and overall natural resources management in the basin, and billions of USD invested to resolve them. And yet, challenges remain, and some get worse. This book is an attempt to start a basin-wide dialogue between policy makers, scientists and practitioners from three riparian countries, and to take stock of what we know about some of these challenges. It is also an attempt to solicit views, and sometimes quite contrasting views, on how to resolve these challenges. In the book, in order to achieve a balanced view, most chapters have been jointly written by authors from Nepal, India and Bangladesh or by international experts who have experience of or worked in all three countries. Most of the existing literature does not offer this sort of multi-country and multidisciplinary approach to highlight the strengths and understand the fine nuances of the problems facing the basin and explore potential solutions. The 17 chapters cover bio-physical, socio-economic and environmental aspects of the basin, each chapter providing several key messages that, we hope, will facilitate future discussions on how the basin should be developed in the decades to come. Some main messages from the book are:

- The annual surface water availability in the Ganges Basin is high, but there is tremendous variability in the spatial and temporal distribution of water supplies. The temporal variability is increasing as we speak, with changing climate. Future climate projections are still uncertain, making adaptation planning difficult. High levels of poverty and lack of infrastructure make the basin vulnerable to changing climate. The latter affects all sectors and aspects of life in basin countries – whether it is hills/mountains, plains or delta. Climate change affects the main biophysical characteristic feature of the basin – the monsoon regime – and leads to increasingly frequent and more extreme floods and droughts that cause massive human and economic losses. Managing this variability by a variety of institutional, political and structural solutions will be the major area of investment for riparian governments and development donors.
- Groundwater plays a critical role in supporting the lives and livelihoods of both rural and urban populations in the Ganges Basin. Technical support systems and regulatory frameworks to enable sustainable allocation and use of the groundwater resource and accounting for ecosystem service requirements are largely in place in India and Bangladesh. However, implementation of these policies and regulations to manage the resources sustainably remains a major long-term challenge. In Nepal, except in urban centres, groundwater is an underutilized resource. In the mountain areas, spring sources are key to providing water during the dry season but are not adequately understood or managed. Growing water scarcity requires more attention to accelerated recharge technologies – beyond natural rates – in order to capture part of the monsoon and provide more water for irrigation from aquifers. The concept of a Ganges Water Machine, although not a panacea to basin problems, is probably more relevant now than in the past and needs to be explored in detail and invested into in certain parts of the basin.
- Although agriculture is highly varied between hills, plains and delta, the agricultural productivity is low-to-medium throughout the basin. It is estimated that around 25 per cent of the basin population is income poor, but 75 per cent are multidimensional poor, i.e. deprived of the basic human requirements of education, health and assets required for decent living. Rural poverty, especially in the eastern Gangetic region, may be alleviated through improved regulation of floods and droughts, availability of affordable rural power and intensification and diversification of the agricultural enterprise. However, efforts to raise agricultural productivity alone might not benefit the chronically poor. Male out-migration represents a critical transformation in the Ganges Basin over the last two decades. Besides some positive benefits of migration, it is actually affecting local agricultural productivity as well as increasing work burden and economic stress for women.
- Despite high hydropower potential, the riparian countries experience energy shortages, which constrain economic and social growth. Regional hydropower development is hampered by various issues which range from land acquisition problems, administrative lapses, inadequate information and financing to transboundary disputes over water and energy-sharing modalities. Developing hydropower in the basin is synonymous with development at large, and the countries should really make a renewed effort to come to terms with this and engage in a constructive dialogue on hydropower development. It does require a political

will at the highest level, while all technical issues may be resolved. Hydropower development may also be seen in a larger context of managing the water–energy–food nexus, as countries' priorities associated with the hydropower development process may deviate from hydropower development per se. Nexus could be an efficient and effective model leading to water, food and energy security in the basin.

- River pollution and groundwater contamination, arsenic in particular, have grown into huge issues with time, but effective systems to deal with these have not been put in place. The understanding and management of the impact of pollution on human health and biodiversity is also lacking. Ganga pollution needs to be addressed at the basin-wide scale, and dilution of river water is not an option: pollution has to be treated at the source. Technological interventions and policies that address these need to be put in place and reinforced.
- The basin sustains rich freshwater biodiversity and provides a range of ecosystems services. Critical biodiversity hotspots need to be identified and legally protected from anthropogenic pressures, based on irreplaceability, species assemblage and connectivity considerations. Understanding of the need for environmental flow allocations in basin management has started to grow in riparian countries. It is timely to make these the integral part of management practices and policies.
- Water policies and institutions throughout the basin are fragmented, sectoral and supply-side oriented, which has led to poor coordination for water resources planning and management across levels (local to transboundary). The paradigm needs to change from 'sharing the water resources' to 'sharing the potential benefits' and facing future challenges in a spirit of camaraderie.
- The countries need to open their water-related data archives to each other and ensure transparent and full extent of water data flow and sharing across country boundaries. This is a very focused and feasible action that can be achieved in the nearest future. This may revolutionize water management in the basin. Without this, it is essentially impossible to conduct reliable and defendable water resources assessments and development plans, design flood protection measures and alert systems or assess required environmental water needs.
- Achieving improved and integrated water management and transboundary collaboration in the basin remains the major development challenge. Transboundary collaboration is heavily constrained by a general feeling of mistrust, lack of political vision and goodwill by the riparian countries. If this is not overcome through transparent political dialogue in the nearest future, the challenges facing riparian countries and related to water, hydropower and food security will be exacerbated dramatically.

This book is being published at a time when there is strong national and global interest in improving the health, freshwater environment and benefits from this unique river system providing sustenance to millions of poor people spread across the riparian countries. We are hopeful that this neutral, candid and science-based book will be of interest to policy makers, the private sector, the donor community and the general public who have great interest in and respect for the Ganga.

2

Surface water resources

Sharad Kumar Jain, Marc A. Jeuland, Luna Bharati and Zahirul Haque Khan

Key messages

- The Ganga is an incredibly important resource for over 600 million people living in South Asia, and delivers a wide variety of economic benefits across the riparian countries that include irrigation, domestic and industrial water supply, hydropower production, cultural and religious services, pollution control, navigation and ecosystem benefits.
- The annual surface water availability in the Ganga Basin is high, as in an average hydrological year around 1,200 billion m^3 of precipitation falls in the basin. Of this, around 600 billion m^3 becomes streamflow with the rest directly recharging groundwater or returned to the atmosphere through evapotranspiration (World Bank, 2013). There is, however,

tremendous variability in the spatial and temporal distribution of water supplies. Over 70 per cent of the rainfall is during the monsoon months. This variability is likely to increase in the future due to climate change. It also leads to recurrent challenges with floods and droughts, the solution for which calls for a variety of institutional, management and infrastructure investments.
- The current storage capacity in large infrastructure is 10 per cent of total average annual flow (World Bank, 2013). Storage investments and enhanced management of excess monsoon water would help the riparian countries to better meet water requirements during the dry season, and to address water scarcity.
- Many parts of the basin have well-developed surface irrigation networks, but these remain underutilized due to insufficient storage in the basin. The possibility of enhanced aquifer recharge to augment the limited surface water storage potential in the basin must be explored.
- Achieving improved water management in the Ganga Basin is heavily constrained by the inconsistent monitoring and sharing of hydrological and other data.

Introduction

The Ganga Basin is one of the largest and most populous river basins in Asia with a population of more than 655 million. Population density is high with an average 551 people per km^2 (World Bank, 2013). The basin is bounded on the north by the Himalayas, on the west – by the Aravalli ridge separating it from the Indus basin, on the south – by the Vindhyas and Chhotanagpur plateaus, and on the east – by the Brahmaputra ridge. It is distributed between India (79 per cent), Nepal (13 per cent), Bangladesh (4 per cent) and China (4 per cent) (Figure 2.1). It covers a total area of 1,086,000 km^2 (Jain *et al.*, 2007). Despite Nepal and Bangladesh representing small fractions of the basin, all of Nepal and over 25 per cent of Bangladesh lie within the basin.

The Ganga is considered to take its source from the headwaters of the Bhagirathi River which rises from the Gangotri glacier at an elevation of about 4,255 m above sea level, in the Uttarakhand State, India. The main stem of the Ganga, however, is formed from the confluence of the Bhagirathi and Alaknanda (another major stream in the area) at Devprayag. Flowing downhill, the river enters the Gangetic Plains near Haridwar. The Yamuna River is an important tributary that joins the Ganga on the right bank at Allahabad. Downstream of Allahabad, the Ganga River flows in an eastward direction and meets a number of major tributaries, including the Karnali (Ghaghra in India), the Gandaki, the Bagmati, the Kos(h)i, the Tons, the Son and the Damodar. Further downstream, the Farakka barrage, built in 1975 in the Indian state of West Bengal, regulates the flow. At Farakka, some of the water from the main river is diverted south into the Hooghly River (Bhagirathi) through a feeder canal designed to keep the latter relatively silt-free, to maintain adequate depths of flow for navigation to Kolkata, and to keep the downstream Diamond Harbour functional (Figure 2.1).

The Ganga River enters Bangladesh downstream of the Farakka barrage. Around 40 km downstream of Farakka, the river splits into two arms. The Bhagirathi River (right arm) flows south and enters the Bay of Bengal downstream of Kolkata city (India). The Padma (left arm) turns eastwards and enters Bangladesh where it meets the Brahmaputra River. The combined river is known as the Padma and is joined by the Upper Meghna River downstream. Thereafter, the river is known as the Lower Meghna, and flows into the Bay of Bengal through the Sundarbans deltaic region (named after the Sundari trees found in the area). This delta is 400 km wide and is one of the largest in the world. The total length of the Ganga River from the source at Gangotri to its entry into the Bay of Bengal (measured along the Bhagirathi and the Hooghly) is 2,525 km (Figure 2.1).

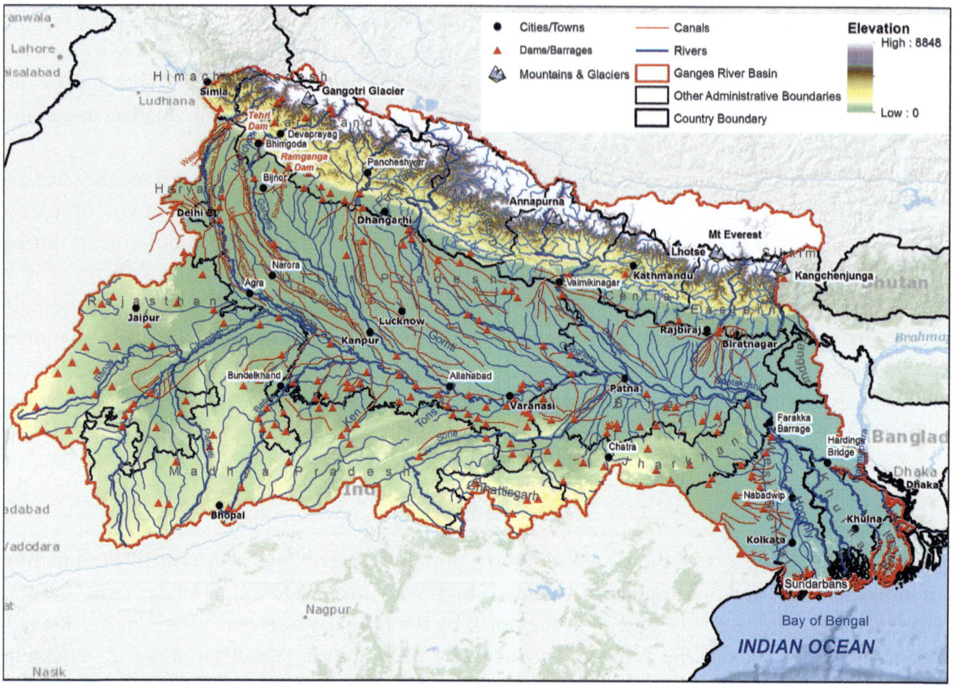

Figure 2.1 Ganges Basin map showing existing dams, barrages and major canals

Source: Ganga River Basin Management Plan – GeoSpatial Database (http://gisserver.civil.iitd.ac.in/grbmp/download dataset.aspx)

With their powerful flows emerging from the uplifted Himalaya and flowing swiftly over the flat lands of the Gangetic Plains, the highly dynamic Ganga and its large Himalayan tributaries have long been unstable. Upstream in the Indian state of Bihar and in the Terai in Nepal, the Kosi has periodically and dramatically shifted course, displacing millions of people (Dixit, 2009; Singh *et al*., 2009). Meanwhile geomorphological research and early maps of the region show that the downstream outlet of the Ganga River has steadily shifted eastward over the last several thousand years (Snedaker, 1991; Allison, 1998). The implications of this river migration have been great, first leading to efforts to save the port of Kolkata in the middle of the twentieth century (Crow *et al*., 1995), and today resulting in a steady drying of areas in western distributaries such as the Gorai, Mathabanga causing environmental problems, as river flows have increasingly turned towards the Brahmaputra River. Even after the merging of the two great rivers in Bangladesh, the powerful eastward combined flow of the Ganga and Brahmaputra continues to propel the river channel towards the east (Figure 2.1).

The Ganga River drains a basin of extraordinary variation in altitude, climate, land use and cropping patterns. Basin elevations range from 8,848 m to sea level. Topographic contrasts in the basin are the largest compared to any other basin on earth and can be appreciated by the fact that within a distance of nearly 200 km, the elevation plunges from the highest mountain in the world at 8,848 m (Mt Everest peak in Nepal) to about 100 m (the elevation of the flat Ganga Plains in India). The Himalayan region of the basin contains 9 of the 14 highest peaks in the world (all over 8,000 m in elevation), including Kangchenjunga, Lhotse, Annapurna and others (Figure 2.1).

Hydro-meteorological observational networks and data sources

Nepal

In Nepal, all hydrological and meteorological activities are monitored by the Department of Hydrology and Meteorology (DHM). The scope of DHM's work includes the monitoring of river hydrology, climate, agro-meteorology, sediment, air quality, water quality, limnology, snow hydrology, glaciology, and wind and solar energy. The government of Nepal first started monitoring hydrological and meteorological variables in 1962. The DHM lists a total of 51 hydrology and 282 meteorological stations, mostly in the hills and plains of Nepal, and all data are available for purchase (Figure 2.2) (www.dhm.gov.np). Not all of the stations are functional, however, and some have unstable river discharge relationships; as such, the data series from different stations are often incomplete and of variable quality. In addition, data coverage in the high mountains above 2,500 m is poor.

India

The India Meteorological Department (IMD) collects data on meteorological variables from a countrywide network of stations. In the Ganga alone, the IMD maintains 703 district monitoring stations that measure precipitation, temperature, humidity, wind speed, potential evaporation and other meteorological variables, all of which can be purchased. In addition, the Indian Space Research Organization (ISRO) maintains a network of 133 automated weather stations (AWS).

The Central Water Commission (CWC) measures stream flows on all major and interstate rivers. The CWC has established river gauging sites on the main stem of the Ganga and its tributaries; 100 of these measure water levels and 179 stations record discharge (Figure 2.2). Sediment and water quality parameters are also measured at 81 and 105 of these CWC stations, respectively (http://gisserver.civil.iitd.ac.in/grbmp/iitg.htm). In addition to these sites, there are six CWC snow gauging stations in the Ganga Basin; three of these are located in Uttarakhand and the other three are in Himachal Pradesh states. Besides the CWC stations, river flow is also measured at other locations by governmental organizations within each of the states in the basin (CWC, 2012). Despite the availability of flow data in principle, access to streamflow data in India, and in the Ganga Basin in particular, remains restricted, and this restriction is the major obstacle for detailed hydrological research, verification of its results, river basin planning and development, and larger transboundary cooperation.

Bangladesh

The Bangladesh Meteorological Department is responsible for maintaining a network of surface and upper air observatories, radar and satellite stations, agrometeorological observatories, geomagnetic and seismological observatories and meteorological telecommunication systems of Bangladesh (www.bmd.gov.bd/). The Bangladesh Water Development Board, under the Ministry of Water Resources, maintains a database from a network of rainfall and river gauging stations. Figure 2.2 shows the spatial distribution of 35 river gauging stations which fall within the Ganges Basin in Bangladesh (www.ffwc.gov.bd/).

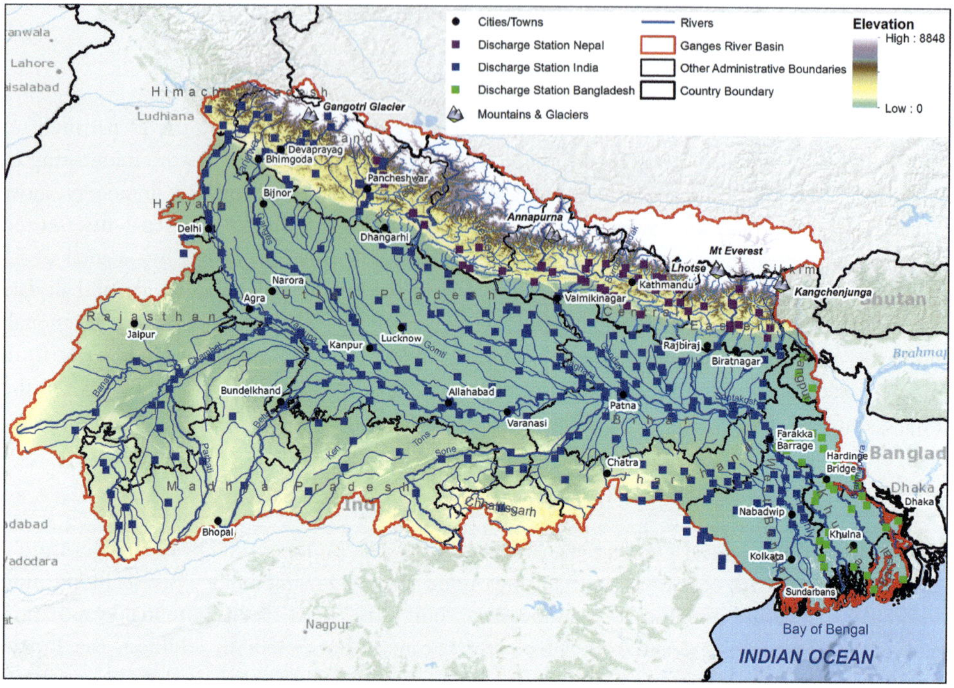

Figure 2.2 Ganges Basin map showing locations of the river network and discharge stations in Nepal, India and Bangladesh

Source: India stations: www.cwc.nic.in/main/downloads/Hydrological%20network%20details%20of%20CWC.pdf; Nepal stations: www.dhm.gov.np/hydrological-station/; Bangladesh stations: www.ffwc.gov.bd/?id=dis

Other data sources

Overall, the observed data on major components of the hydrological cycle in the basin remains scarce, and some types of data (discharge) may not be easily accessible. New technologies (e.g. the Tropical Rainfall Measuring Mission – TRMM, NCEP/NCAR reanalysis project – www.esrl.noaa.gov/psd/data/reanalysis/reanalysis.shtml) that rely on remote sensing are increasingly being employed to enhance data availability. In the future, these sources of data may decrease dependence on land-based, government-owned data. Rolling out these data into practice at large scales, however, takes time. Global data repositories such as the Global Runoff Data Base (www.bafg.de/GRDC/EN/01_GRDC/13_dtbse/database_node.html) also provide open access to the world's river discharge data, but they are, of course, also dependent on the individual countries' willingness to share their data. Crowdsourcing and community sourcing are being discussed as informal ways to increase data coverage and availability. Yet, these have issues of quality control and continuity.

Hydrological modelling and use of water resources assessment tools

Hydrological and water systems models are important tools for water resource assessments. In the Ganges, there has until recently been relatively limited modelling at basin scale (Johnson and Smakhtin, 2014). Inaccessibility to flow data hampers models' calibration and validation and makes the model result uncertain. Although data from stations in Nepal and Bangladesh can be purchased, the Indian government has not released flow data publicly since 1973 as it considers hydrological data for the Ganges to be a matter of national security. The few models that have been developed for the Ganges also emphasized the limitations imposed by the lack of data (de Condappa *et al.*, 2009; Eastham *et al.*, 2010). To promote an evidence-based dialogue between basin countries of the opportunities and risks of water resources development in the Ganges, the World Bank (2013) commissioned the development of a suite of models to give comprehensive coverage of the basin. The Ganges Strategic Basin Assessment (GBSA) included: water systems modelling, economic modelling of distribution of benefits and tradeoffs from water resources development, and social analysis of impact of water variability (World Bank, 2012; Johnson and Smakhtin, 2014). The studies carried out by the World Bank have, however, been criticized by the officials from Nepal and Bangladesh for having an India bias (Pun, 2013). The World Bank has also stressed the lack of data from India as a major constraint in basin-wide assessments. Models have, however, been developed at sub-basin scale where there is also data available for calibration and validation (e.g. Gourdji *et al.*, 2008; Biancamaria *et al.*, 2011; Bharati *et al.*, 2011; Bharati *et al.*, 2016).

Important climatic and hydrological characteristics of the basin

The water availability in the Ganges Basin is high, as in an average hydrological year around 1,200 billion m^3 (BCM) of precipitation falls in the basin. Of this, around 600 BCM becomes streamflow, with the rest directly recharging groundwater or returned to the atmosphere through evapotranspiration (World Bank, 2013). There is, however, tremendous variability in the spatial and temporal distribution of water supplies. The South Asian monsoon largely defines the climate and hydrology of the Ganga system. Each year from June to August, the two arms of the monsoon sweep up the coasts of the Indian sub-continent until they run into the towering mountains of the Hindu Kush–Himalaya range. In the process, they deliver 70–80 per cent of annual rainfall to the region (Trenberth *et al.*, 2007; World Bank, 2013; Bharati *et al.*, 2016; ICIMOD, 2014).

Unlike other nearby river systems like the Indus, glaciers have been shown to contribute a relatively small share (just 2 per cent) of the total river flow in the basin, and much of the glacier melt occurs during the early monsoon when rainfall is already quite heavy (Alford and Armstrong, 2010). There is greater uncertainty about the timing and overall contribution of snowmelt. Snow and ice melt together have been estimated to contribute about 9 per cent (almost 2/3 from snow) of total annual flows in the Ganga (Immerzeel *et al.*, 2010), though they contribute much more in some tributaries (e.g. 30 per cent of annual flow in the Buri Gandak Basin), and in small upstream catchments (see, for example, Brown *et al.*, 2014). Importantly, glaciers and snow provide valuable water storage that helps to ensure the perennial flow of the Himalayan tributaries, and enhances dry season flows in the Ganga. Estimates of their contribution to the low flows at Farakka in March–May vary greatly, ranging between 12 and 38 per cent (Siderius *et al.*, 2013). Groundwater contributions to base flows are also significant; the overall contribution of these to upstream base flows is thought to be 60 per cent (Andermann *et al.*, 2012).

Figure 2.3 shows a simplified diagram of the Ganga River and its main tributaries, along with their long-term average annual flows. According to the assessment by the CWC, the average surface water potential of the Ganga Basin at Farakka is 525 BCM (it consists of internal renewable surface and ground water plus inflow from upstream countries). The tributaries or sub-catchments which contribute the largest amount of water per annum are the Ghaghara including the Gomti (102 BCM), the Koshi system (81.8 BCM), the Gandak and Buri Gandak (59.3 BCM), the Yamuna (93.0 BCM), and the Son and East of Son (31.8 BCM). Therefore, nearly

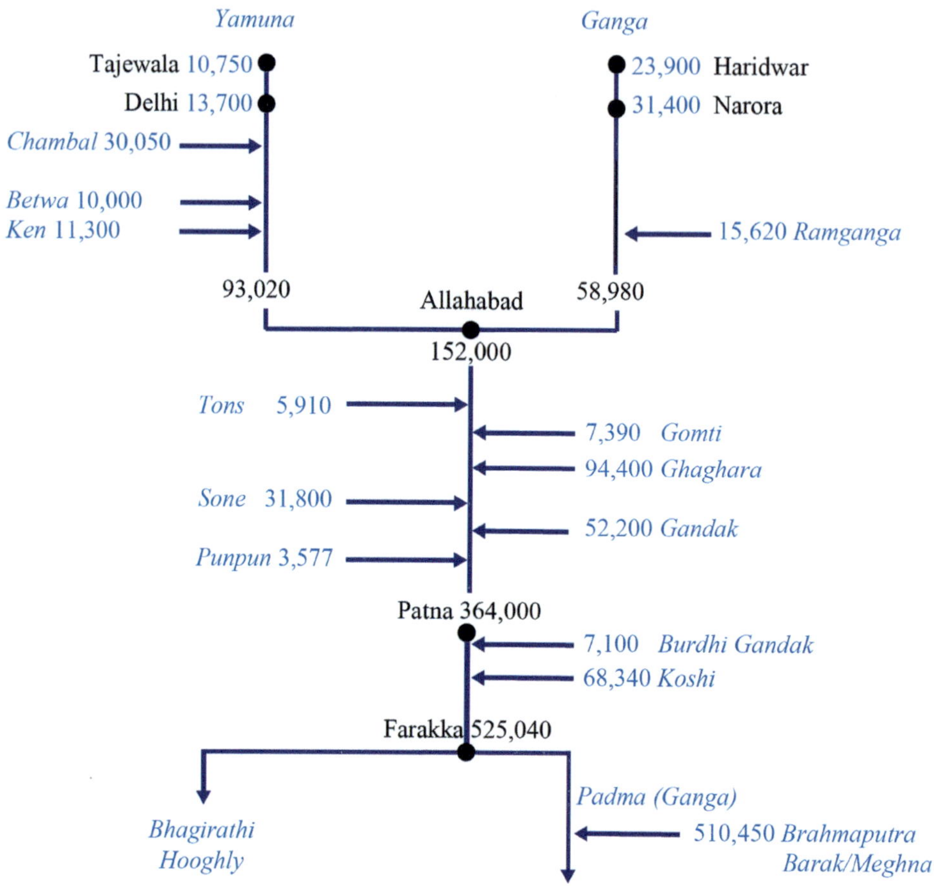

Figure 2.3 Line diagram of the Ganga and its major tributaries; numbers are average annual flows (MCM). Modified from Jain *et al.* (2007)

60 per cent of the water flowing in the Ganga comes from areas north of its course. Southern streams contribute the remaining 40 per cent (Jain *et al.*, 2007). Thus, hydrologically, the Himalayan rivers are of more importance than the southern streams of the basin.

Figure 2.4 shows historical flows at Farakka, located on the India–Bangladesh border. In correspondence with the rainfall pattern, river flows observed along the Ganges mainstem rise from very low levels in January–May to a sharp peak in July that then decrease through October. Low flows at the Farraka barrage during this pre-monsoon period (January-May) comprise just 6 per cent of total annual flow. In addition to annual variability, there is also high inter-annual variability in this system. In the maximum flow year, peak flow is more than twice the peak flow in the year of minimum flow. During the historical low-flow year (1991), just 153 BCM flowed through Farakka (38 per cent lower than average), while the historical high of 369 BCM in 1999 had 49 per cent higher flow than the average. Monthly variation across years can be even higher (Figure 2.4).

The Bangladesh Water Development Board maintains another discharge gauging station in the Ganges at Hardinge Bridge, downstream from Farakka. The minimum, average and

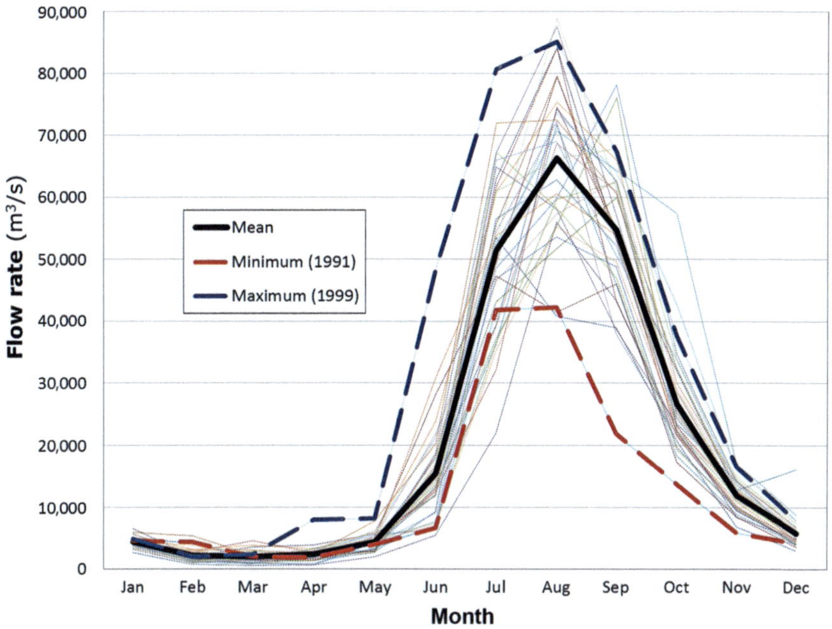

Figure 2.4 Historical flows at Farakka (India–Bangladesh border), for 1969–2001. Flows in individual years are shown by thin coloured lines; historical low (1991) and high years (1999) are shown by dotted red and blue lines; average flows are shown by a solid black line. As shown, roughly 80 per cent of annual flows through Farakka come during July–October

Source: World Bank (2013), using data from the Climate Research Unit (CRU), University of East Anglia

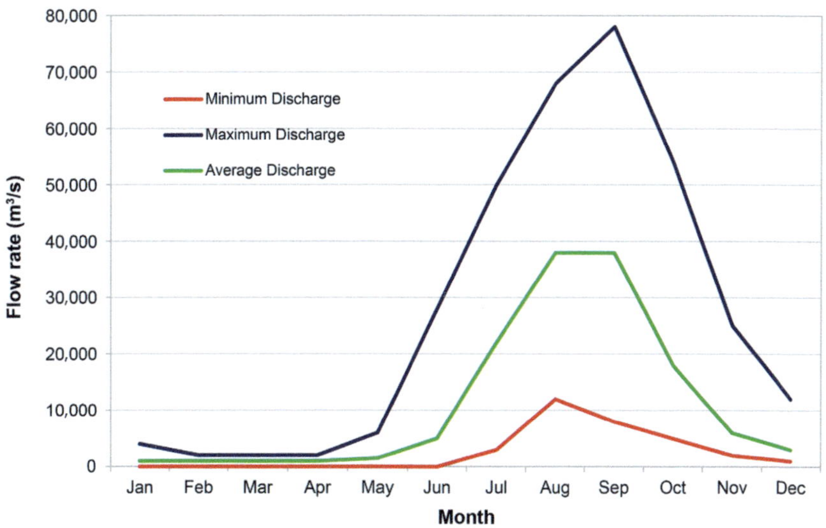

Figure 2.5 Historical flows at Hardinge Bridge (Ganges River, Bangladesh) for 1970–2013

Source: IWM (2014)

maximum discharges from 1970 to 2013 are shown in Figure 2.5. A comparison of maximum, mean and minimum flows at the two nearby locations given in Figures 2.4 and 2.5 show large differences. This may be partly due to the different time periods in the two graphs but it also serves to highlight some of the difficulties in reconciling data for this large transboundary river.

Average annual temperatures are high throughout the basin except in high elevation areas of the Himalaya (Figure 2.6) and reach their maximum in the pre-monsoon months (April–June). Mean annual precipitation varies significantly throughout the basin: rainfall is highest in the eastern Himalayan belt and in the delta areas (>2000 mm), and lowest in the desert of Rajasthan in the west (<250 mm) (Figure 2.6).

The resulting irregularity in the natural cycle of inundation and drying that emerges from the seasonal hydrology and this high interannual and spatial variation has both beneficial and destructive aspects. High flows typically lead to increased crop production during the winter (*rabi*) planting season (Yu *et al.*, 2010), and short periods of inundation help to mitigate land subsidence and coastal erosion by moving large amounts of silt downstream (ESA, 2011). Indeed, the Ganga remains one of the three most sediment-laden river systems in the world, and together with the Brahmaputra carries over 1 billion tons of silt to the Ganga–Brahmaputra–Meghna (GBM) delta every year (Milliman and Ren, 1995; Kuehl *et al.*, 2005). Estimates of the amount of transported sediment vary from 329 to 794 million tonnes/year (Subramanian and Ramanathan, 1996; Wasson, 2003). Migration of the tributaries draining the eastern part of the basin has resulted in conspicuous back-swamp and meander bolt deposits, and sedimentological features play a dominant role in the hydrodynamics of the delta region. The powerful and silt-laden flows of the Ganga and its tributaries helped over the ages to form the large, flat and fertile Terai (marshy land at the foot of the Himalayas) in Nepal, and the plains of North India and the Bangladesh Delta. In these regions, the combination of abundant water resources, high-quality soil and a suitable climate for agriculture gave rise to a highly developed civilization that was centred on agricultural production.

The downside of climate variability manifests itself in extreme floods and droughts, and in the substantial uncertainty that accompanies the timing of the monsoon onset. Floods are common and frequent throughout the basin with Ganga itself and its major tributaries (Yamuna, Sone, Ghagra, Gandak, Kosi and Mahananda). Floods take a significant toll on lives and livelihoods, damage infrastructure, homes and other assets, and generally limit investment and disrupt economic activities (Jain and Sinha, 2003; Dasgupta *et al.*, 2011; Mirza, 2011). Flood-affected areas in Bangladesh during 1954–2004 were estimated to have a median of over 30,000 km^2. In the Indian state of Bihar, over 20 million people were affected in 1987, 2004 and 2007, and flood-related deaths occur almost every year (World Bank, 2013). In Nepal, floods account for 90 per cent of the economic cost of natural disasters (Nepal Ministry of Home Affairs, 2010). Potential upstream dams alone may not be able to control lowland flooding, and hence a focus on non-structural measures (e.g. investments in forecasting and warning systems) is critical. The underground aquifers of the basin could also provide space for water storage to partially tame the floods. Similarly, many parts of the basin are prone to droughts. A more comprehensive look at floods and droughts in the Ganga is presented in Chapter 5 (this volume).

Water use and infrastructure

Ganga flow is used to irrigate around 39 million ha of agricultural land (Thenkabail *et al.*, 2005), generate hydropower, provide water for the domestic, cultural and spiritual practices of millions of people and maintain ecosystems. Agriculture dominates both surface and groundwater use in the basin, and accounts for roughly 90 per cent of all of its consumptive water abstractions

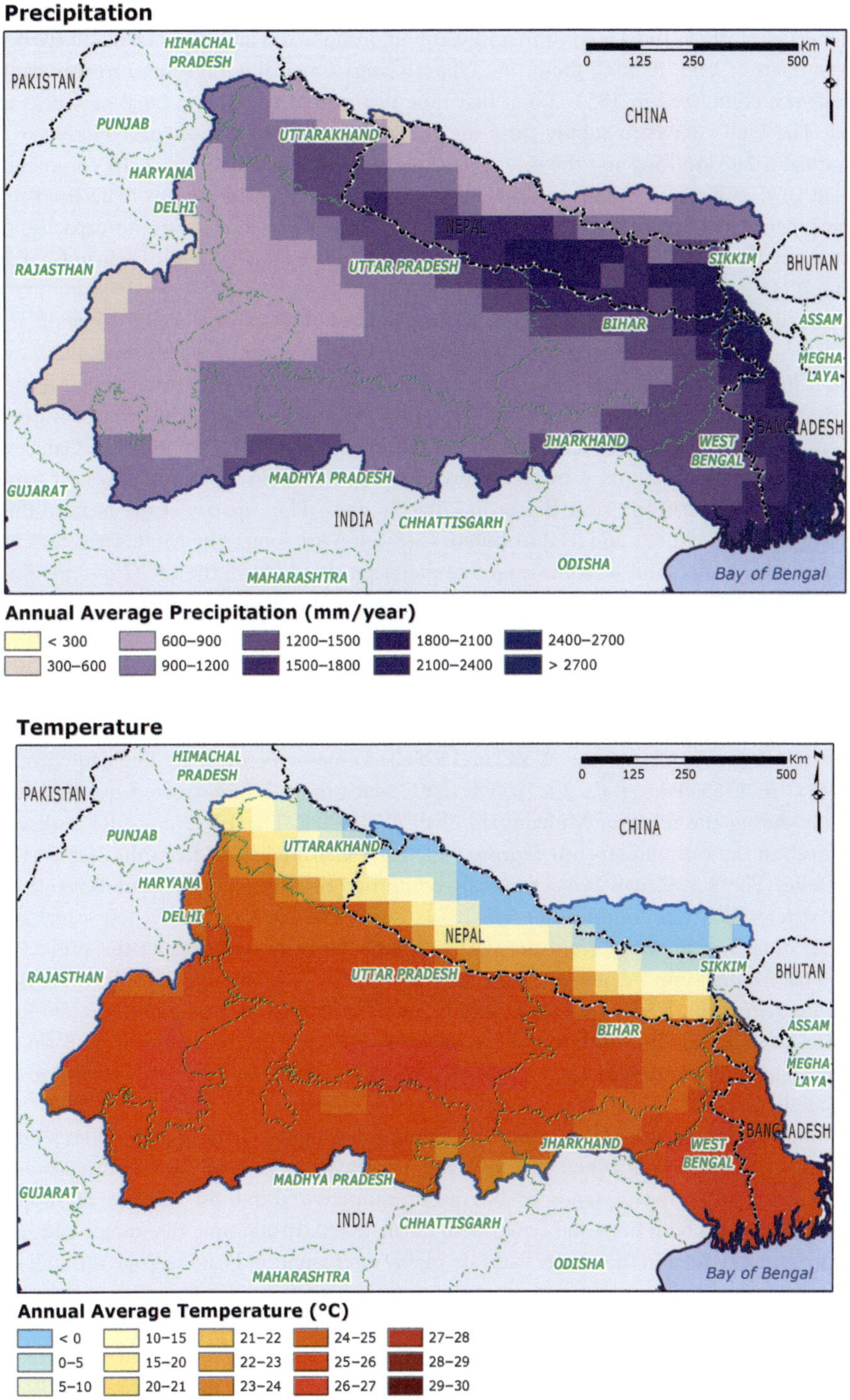

Figure 2.6 Historical average precipitation and temperature distribution over the Ganges Basin

Source: CRU TS 3.0 1901–2006; reproduced by permission; from Jeuland *et al.* (2013)

(World Bank, 2013). A vast system of canals constructed in the upper and middle stretches of the Ganga – mostly in India – continues to ward off drought and famine by facilitating irrigation of large tracts of land. Among these, the Upper Ganga Canal (UGC) system in western Uttar Pradesh was completed in 1854 and at that time was one of the largest canal networks in the world. The UGC draws its supply from the Ganga River at Bhimgoda (near Haridwar). The main canal is 292 km long and the discharge capacity at its head is 297 m^3/s. Its command area is about 0.92 million ha, falling in dry sub-humid to moist humid regions. The head reaches of the canal were remodelled during the 1990s to increase its flow-carrying capacity. Some water from the UGC is also diverted to meet the urban water supply demands of Delhi (Figure 2.1).

Downstream of the UGC, the Lower Ganga Canal (LGC) system that takes off from a weir across the Ganga at Narora was completed in 1879 (Figure 2.1). The main LGC canal is 100 km long, has a discharge capacity of 156 m^3/s, and irrigates 0.5 million ha. Together, the combined UGC–LGC canal systems are used to withdraw about 450 m^3/s of water from the Ganga, which represents a large fraction of the lean season flow. The Madhya Ganga Canal (MGC), a seasonal canal that is primarily used during higher flow periods, takes off from the Ganga at the Raoli barrage, near Bijnor in Uttar Pradesh. The capacity at the head of this 115 km-long canal is 234 m^3/s and its distributaries are 1,466 km long. The MGC irrigates 114,000 ha (mostly rice paddy) and serves to augment the water supply from the UGC system. Another canal in the system, the East Ganga Canal Project, uses water from the Ganga River during the monsoon that is diverted from the existing Bhimgoda barrage at Haridwar to provide irrigation to 105,000 ha (again mostly paddy crop).

Further to the west, the Yamuna canal system is one of the ancient canal systems of India (Figure 2.1). At Tajewala, the Yamuna water is diverted for irrigation by the Western and Eastern Yamuna Canals (the WYC and EYC). Emperor Feroz Shah Tughlaq constructed the WYC in 1355 (Jain *et al.*, 2007). The EYC was originally constructed by the Mughals, probably during the reign of Muhammad Shah (1719–48). The EYC was later realigned by the British in the early nineteenth century. The Agra Canal takes off from the Yamuna River near Delhi. The Son Canals Project initially comprised of a weir on the Sone River at Dehri (Bihar state) which was commissioned in 1874. Two canals, namely the Eastern canal and the Western canal system, were constructed. The cultivable command area of the project is 0.7 million ha. Another important system, the Sarda Canal, takes off from the Sarda River at Banbasa (Uttar Pradesh state). It was completed in 1926 and irrigates an area of about 800,000 ha. The length of canal along with distributaries is 12,368 km. Finally, in the south of the basin and in the delta region, there are important irrigation canals along the Tons River, as well as throughout Bangladesh (Figure 2.1).

Groundwater from the vast aquifers underlying the Gangetic Plain also provides a critically important supply reservoir when surface water resources are lacking, and land irrigated with at least some groundwater tends to be most productive and most capable of supporting double cropping, in both India and Bangladesh (Shah, 2008). In addition, the surface and groundwater irrigation systems in the Ganga Plain are highly interconnected, and seepage through surface irrigation canals recharges aquifers.

Total storage capacity in the basin is still only about 55 BCM, and active storage is 36 BCM, less than 10 per cent of annual flow. This is among the lowest storage ratios among major rivers of the world. For example, many 'developed' rivers have storage to annual flow ratios of 100–200 per cent. This is because of the limited active surface water storage capacity in the system due to several features of the basin and its existing dams (World Bank, 2013). First, most reservoirs (only the most notable ones are discussed below – Figure 2.1) are modest in size; only five are more than 100 m tall. Second, with the possible exception of the Upper Ganga Basin in India (Xu and Grumbine, 2014), many of the more than 1,000 Ganga dams are located in relatively flat portions of the basin and have very low volume to area ratios. Third, some of the largest tributaries originating in Nepal (e.g. the Ghagara, the Gandak and the Kosi) have almost no storage (Bandyopadhyay, 2009; Wu *et al.*, 2013). Thus, there is limited existing capacity to regulate upstream flows and to generate storage-backed hydropower (relative to energy generation potential), reduce flooding, and/or augment low flows for irrigation or ecosystem needs in the delta (Sadoff *et al.*, 2013). It has, however, also been argued by some eminent scholars that Himalaya Ganga does not have enough storage sites in the deep gorges to regulate the Ganga floods (Gyawali, 2003). Neutral and un-biased bio-physical assessments, which are accepted by all riparian countries on the potential of developing large storage dams in the basin, have not as yet been carried out. As the Ganges Basin is also in a seismic active zone, the safety issue regarding such large infrastructure has also not been adequately analysed.

The Tehri dam is 261 m high and is the largest storage project in the basin today (Figure 2.1). Tehri is a multipurpose dam that is used for hydropower, irrigation and water supply, flood control, fisheries and tourism. The dam traps a catchment area of 7,511 km^2, including 2,323 km^2 of snow-covered area. This catchment varies in elevation from 7,140 m to 600 m. The reservoir has a live and gross storage capacity of 2,613 and 3,540 million m^3. Together with a pumped storage project located at Koteshwar, the entire Tehri Hydropower Complex would have an installed capacity of 2,400 MW, making it the largest hydropower project in the Ganga Basin. The annual irrigation potential from the reservoir is 0.27 million ha, and the reservoir is also operated to provide water to meet Delhi's domestic water requirements. The Tehri dam has helped to moderate high flows in the Ganga, most notably during the June 2013 floods.

Another major dam in the Indian Himalaya is the Ramganga dam (Figure 2.1) in Uttar Pradesh, with a live storage capacity of 2,190 MCM. It is a multipurpose project that produces hydropower and provides flood-control benefits. Many other significant storage dams are located on the right bank tributaries or distributaries. For example, downstream in West Bengal, the Damodar Valley Corporation (DVC) was set up in 1948 on the Damodar River to regulate flows and reduce floods, based on the idea of the Tennessee Valley Authority in the United States. By now, five DVC dams are jointly operated for flood control, municipal and industrial water supply, irrigation and power generation. A somewhat similar system can be found in a cascade of three storage reservoirs (Gandhisagar, Jawaharsagar and Rana Pratapsagar) constructed in the Chambal (a tributary of the Yamuna) Basin, which together provide a live storage of 8,500 MCM. Despite the relative successes of the DVC and Chambal systems, there are few other examples of coordinated water control infrastructures in India.

There is significant potential for hydropower and multipurpose dam development in the Nepalese tributaries to the Ganga. The Koshi, the Karnali (Chisapani) and the Pancheshwar are three proposed mega-projects that would be located in Nepal, that would together multiply Nepal's current total installed capacity of electricity by a factor of 30. Recently, discussions on developing some of these and other projects in Nepal have begun to gain momentum. There are plans to phase the Karnali mega-project with an upstream run-of-the-river hydroelectric project and storage development on two tributaries of the Karnali. These would supplement an existing small-scale Karnali dam (a 15 MW dam built in 1985) and would generate 5,400 MW of electricity. The version of the Karnali project currently being considered would be a completely Nepalese project. Meanwhile, the Pancheshwar project on the Mahakali (a western boundary river between India and Nepal) River is of interest to both India and Nepal. The project can generate 2,000 MW and provide irrigation benefits. This joint project has been under investigation for decades. Finally, the controversial Kosi high dam was first proposed in the 1950s, and consists of a 239 m high dam located at Barakshetra in Nepal, plus a downstream regulating barrage at Chatra, also in Nepal. This huge dam would help attenuate the flood peak of the Kosi and would generate 1,800 MW of power and irrigate large tracts of land in Nepal and Bihar, through the Kosi canal system as well as other new ones. The Gandak project at Valmikinagar intercepts water of a catchment of 37,410 km^2 (partly in Nepal and partly in India). The Gandak barrage, over the Gandak River, is to provide irrigation to Nepal, Uttar Pradesh and Bihar. A hydroelectric generating station with an installed capacity of 15 MW has also been constructed. The collection of upstream dams in Nepal could nearly double the low flows downstream in the dry season, but the resultant economic value of this flow augmentation is unclear given current water-use practices.

Besides these infrastructure-related projects and uses, the waters of the Ganga also provide very important social, religious and ecological services (Hosterman *et al.*, 2012). Many upstream wetlands in the basin are considered sacred by nearby inhabitants and the mainstem of the Ganga is famous for its significance to Hindus. These cultural values, as well as other ecosystem services (e.g. fish and timber resources), are increasingly threatened by land conversion, water pollution by industrial and domestic discharges, invasive species, and intense and growing demand. Furthermore, flow regulation by water infrastructure projects has affected the hydrology of the Ganga Basin. For example, due to diversion of water in the UGC system, the flows in the Ganga River downstream of Haridwar are considerably reduced during the lean season. The unique ecological habitat of the Sunderbans mangrove swamps is increasingly under threat from the dual threats of hydrological changes related to human development and sea level rise from climate change. In addition, intense groundwater pumping is lowering water tables in the basin which is contributing to reduced lean season flows. Vertical fragmentation of habitat and ecosystems due to dams is also affecting the aquatic species in the river as well as riverine and riparian ecology.

Future challenges

The Ganga Basin presents a unique set of water management problems. The basin presents great opportunities but also major challenges for effective water management. These major challenges, discussed in this chapter, span lack of data availability, hydrological variability and lack of transparent and neutral basin-wide studies on resource assessment and development, which is accepted by all riparian countries.

Although considerable amounts of data are collected by the three major Ganga countries, the basin can be classified as data-deficient due to lack of sufficient and transparent data collection, and in general lack of data sharing. As noted above, many stations in the countries' hydrometeorological monitoring systems are not fully functional. In addition, much data (e.g. on hydrological flows in India) are not available in the public domain. These data-related shortcomings hinder serious efforts at basin-level planning, and lead to inconsistencies in the different riparians' conceptions of how the basin works.

Flow variability is a major challenge in this basin. The water availability in the Ganga is quite high, but the intra-annual variation in flows leads to periods of scarcity during the dry season and flooding problems during the monsoon. Meanwhile, the potential impacts from future climate change (discussed further in Chapter 4, this volume) are increasing the complexity of water management challenges by adding a layer of uncertainty to existing concerns, and by raising the prospect of increases in the frequency and extent of extreme events (e.g. floods and droughts).

The long-term objective to sustainably manage this variability for the benefit of all co-basin countries calls for an appropriate mix of new investments in storage and more efficient use of monsoon flows from the Ganga and its tributaries, a large part of which currently flows to the Bay of Bengal, inflicting flood damage along its way. Enhanced storage would enable irrigation of larger areas and higher lean season flows which are critical for water quality and river health, significant new power generation, maintaining of navigation during the dry season and would help buffer against the effects of climate change and sea level rise. Navigation has not been a historical focus for riparians of the Ganga, but there is growing interest in India, Nepal and Bangladesh in making investments that would allow for year-round navigation. Effective utilization of flows for irrigation throughout the year could also help to reduce water logging. More neutral and transparent basin-scale studies are needed, however, to assess water resources development considering economic costs and benefits, bio-physical constraints as well as environmental and social issues.

Finally, the basin as a whole, and India and Bangladesh in particular, are currently experiencing strong economic and income growth, which will have substantial implications for future water and food requirements. In all countries, population growth and industrialization are occurring at a rapid pace, increasing water demand and pollution. Land use, cropping, food choices and water-use patterns are also changing, partly as responses to these changing demographic and consumption patterns, and partly as responses to changing investment. Some predictions therefore suggest that food production in the basin is likely to soon fall short of consumption (Sharma *et al.*, 2008). Changes in population, economic development, urbanization and the concomitant patterns of consumption will thus have important implications for basin water management.

References

Alford, D. and Armstrong, R. (2010). The role of glaciers in stream flow from the Nepal Himalaya. *The Cryosphere Discuss* 4: 469–94.

Allison, M. (1998). Historical changes in the Ganges-Brahmaputra delta front. *Journal of Coastal Research* 14(4): 1269–1275.

Amarasinghe, U.A., Shah, T. and McCornick, P.G. (2008). *Seeking Calm Water: Exploring Policy Options for India's Water Future.* Natural Resources Forum, Wiley Online Library.

Andermann, C., Longuevergne, L., Bonnet, S., Crave, A., Davy, P. and Gloaguen, R. (2012). Impact of transient groundwater storage on the discharge of Himalayan rivers. *Nature Geoscience* 5(2): 127–32.

Bandyopadhyay, J. (2009). *Water, Ecosystems and Society: A Confluence of Disciplines.* Delhi: Sage Publications.

Bharati, L. and Jayakody, P. (2011). Hydrologic impacts of inflow and land-use changes in the Gorai River Catchment, Bangladesh. *Water International* 36(3): 340–56.

Bharati, L., Gurung, P., Maharjan, L. and Bhattarai, U. (2016). Past and future variability in the hydrological regime of the Koshi Basin, Nepal. *Hydrological Sciences Journal* 61(1): 79–93.

Bharati, L., Lacombe, G., Gurung, P., Jayakody, P., Hoanh, C.T. and Smakhtin, V. (2011). *The Impacts of Water Infrastructure and Climate Change on the Hydrology of the Upper Ganges River Basin.* Colombo, Sri Lanka: International Water Management Institute (IWMI), p. 28.

Brown, M., Racoviteanu, A., Tarboton, D., Gupta, A.S., Nigro, J., Policelli, F., Habib, S., Tokay, M., Shrestha, M. and Bajracharya, S. (2014). An integrated modeling system for estimating glacier and snow melt driven streamflow from remote sensing and earth system data products in the Himalayas. *Journal of Hydrology* 519: 1859–1869.

BWDB (2010). *Ganges Barrage Study.* Dhaka: Bangladesh Water Development Board.

Cai, X. and Sharma, B.R. (2010). Integrating remote sensing, census and weather data for an assessment of rice yield, water consumption and water productivity in the Indo-Gangetic river basin. *Agricultural Water Management* 97(2): 309–16.

Crow, B., Lindquist, A. and Wilson, D. (1995). *Sharing the Ganges: The Politics and Technology of River Development.* New Delhi and Thousand Oaks, CA: Sage.

Dasgupta, S., Huq, M., Khan, Z.H., Masud, M.S., Ahmed, M., Mukherjee, N. and Pandey, K. (2011). Climate proofing infrastructure in Bangladesh: The incremental cost of limiting future flood damage. *The Journal of Environment & Development* 20(2): 167.

de Condappa, D., Chaponnière, A. and Lemoalle, J. (2009). A decision-support tool for water allocation in the Volta Basin. *Water International* 34(1): 71–87.

Dixit, A. (2009). Kosi embankment breach in Nepal: Need for a paradigm shift in responding to floods. *Economic & Political Weekly* 44(6): 70–8.

Eastham, J., Kirby, M., Mainuddin, M. and Thomas, M. (2010). *Water Use Accounts in CPWF Basins: Simple Water-Use Accounting of the Ganges Basin.* CPWF Working Paper: Basin Focal Paper series BFP05. The CGIAR Challenge Programme on Water and Food, Colombo.

ESA (2011). *ESA and World Bank Move Towards Closer Collaboration, European Space Agency and World Bank.* Available at: www.esa.int/esaCP/SEMOJK9K73G_index_0.html. Accessed on: 17 September 2015.

Gourdji, S.M., Knowlton, C., Platt, K. and Wiley, M.J. (2008). 'Modelling the interlinking of the Ganges River: Simulated changes in flow'. In M.M.Q. Mizra, A.U. Ahmed and Q.K. Ahmad (eds). *Interlinking of Rivers in India: Issues and Concerns.* London: CRC Press, Taylor and Francis, pp. 107–28.

Gyawali, D. (2003). *Rivers, Technology and Society.* London: Zed Books.

Hosterman, H.R., McCornick, P.G., Kistin, E.J., Sharma, B. and Bharati, L. (2012). Freshwater, climate change and adaptation in the Ganges River Basin. *Water Policy* 14(1): 67.

ICIMOD (2014). *Research Insights on Climate and Water in the Hindu Kush Himalayas.* Kathmandu, Nepal: International Centre for Integrated Mountain Development.

Immerzeel, W.W., Van Beek, L.P.H. and Bierkens, M.F.P. (2010). Climate change will affect the Asian water towers. *Science* 328(5984): 1382.

India-WRIS (2014). *Ganga Basin.* Available at: www.india-wris.nrsc.gov.in/. New Delhi, India-WRIS, Ministry of Water Resources, Government of India. Accessed on: 17 September 2015.

IWM (2014). Salinity Intrusion in Coastal Zone of Bangladesh due to Climate Change.

Jain, S.K., Agarwal, P.K. and Singh, V.P. (2007). *Hydrology and Water Resources of India.* Netherlands: Springer.

Jain, V. and Sinha, R. (2003). Geomorphological manifestations of the flood hazard: A remote sensing based approach. *Geocarto International* 18(4): 51–60.

Jeuland, M., Harshadeep, N., Escurra, J., Blackmore, D. and Sadoff, C. (2013). Implications of climate change for water resources development in the Ganges basin. *Water Policy* 15: 26–50.

Johnston, R. and Smakhtin, V. (2014). Hydrological modelling of large river basins: How much is enough? *Water Resources Management* 28(10): 2695–2730.

Kuehl, S.A., Allison, M.A., Goodbred, S.L. and Kudrass, H. (2005). The Ganges-Brahmaputra delta. *Special Publication-SEPM (Society for Sedimentary Geology)* 83: 413–34.

Milliman, J.D. and Ren, M. (1995). 'River flux to the sea: Impact of human intervention on river systems and adjacent coastal areas'. In D. Eisma (ed.). *Climate Change: Impact on Coastal Habitation.* Boca Raton, FL: Lewis, pp. 57–83.

Mirza, M.M.Q. (2011). Climate change, flooding in South Asia and implications. *Regional Environmental Change* 11(1): 95–107.

Mittal, N., Mishra, A., Singh, R. and Kumar, P. (2014). Assessing future changes in seasonal climatic extremes in the Ganges river basin using an ensemble of regional climate models. *Climatic Change* 123(2): 273–86.

Moench, M. and Dixit, A. (2007). *Adaptive Capacity and Livelihood Resilience: Adaptive Strategies for Responding to Floods and Droughts in South Asia.* Kathmandu: Institute for Social and Environmental Transition.

Nepal Ministry of Home Affairs (2010). *Economic Survey.* Kathmandu, Nepal.

Papa, F., Frappart, F., Malbeteau, Y., Shamsudduha, M., Vuruputur, V., Sekhar, M., Ramillien, G., Prigent, C., Aires, F. and Pandey, R.K. (2015). Satellite-derived surface and sub-surface water storage in the Ganges–Brahmaputra River Basin. *Journal of Hydrology: Regional Studies* 4: 15–35.

Parua, P.K. (2010). *The Ganga.* Water Science and Technology Library 64, The Netherlands: Springer.

Pun, S.B. (2013).World Bank's 2013 Ganges strategic basin assessment: A view from Nepal. *Hydro Nepal: Journal of Water, Energy and Environment* 12: 6–12.

Sadoff, C., Rao, H.N., Blackmore, D., Wu, X., Jeuland, M. and Lee, S. (2013). Ten fundamental questions for water resources development in the Ganges: Myths and realities. *Water Policy* 15: 147–64.

Shah, T. (2008). *Taming the Anarchy: Groundwater Governance in South Asia.* Washington, DC: Earthscan.

Sharma, B.R., Amarasinghe, U.A. and Sikka, A. (2008). *Indo-Gangetic River Basins: Summary Situation Analysis.* Available at: cpwfbfp.pbworks.com/f/IGB_situation_analysis.pdf. Accessed on: 17 September 2015.

Siderius, C., Biemans, H., Wiltshire, A., Rao, S., Franssen, W., Kumar, P., Gosain, A., van Vliet, M. and Collins, D. (2013). Snowmelt contributions to discharge of the Ganges. *Science of the Total Environment* 468: S93–S101.

Singh, P., Ghose, N., Chaudhary, N. and Hansda, R. (2009). *Life in the Shadow of Embankments – Turning Lost Lands Into Assets in the Koshi Basin of Bihar, India.* Kathmandu: International Centre for Integrated Mountain Development.

Snedaker, C. (1991). 'Notes on the Sundarbans with emphasis on geology, hydrology and forestry'. In J. Seidensticker, R. Kurin and A. Townsend. *The Commons in South Asia: Societal Pressures and Environmental Integrity in the Sundarbans.* Washington, DC: The International Center, Smithsonian Institution.

Subramanian, V. and Ramanathan, A. (1996). Nature of sediment load in the Ganges-Brahmaputra river systems in India. *Sea-Level Rise and Coastal Subsidence.* Netherlands, Springer: 151–68.

Thenkabail, P.S. Schull, M. and Turral, H. (2005). Ganges and Indus river basin land use/land cover (LULC) and irrigated area mapping using continuous streams of MODIS data. *Remote Sensing of Environment* 95: 317–41.

Trenberth, K., Jones, P., Ambenje, P., Bojariu, R., Easterling, D., Klein, A., Tank, D., Parker, F., Rahimzadeh, J., Renwick, M.S., Rusticucci, B. and Zhai, P. (2007). 'Observations: Surface and atmospheric climate change'. In S. Solomon, D. Qin, M. Manning, Z. Chen, M. Marquis, K.B. Avery, M. Tignor and H.L. Miller (eds). *Climate Change 2007: The Physical Science Basis. Contribution of Working Group I to the Fourth Assessment Report of the Intergovernmental Panel on Climate Change.* Cambridge: Cambridge University Press.

Wasson, R. (2003). A sediment budget for the Ganga–Brahmaputra catchment. *Current Science* 84(8): 1041–1047.

World Bank (2013). *Ganges Strategic Basin Assessment: A Regional Perspective on Opportunities and Risks.* Washington, DC: The World Bank.

Wu, X., Jeuland, M., Sadoff, C. and Whittington, D. (2013). Interdependence in water resource development in the Ganges: An economic analysis. *Water Policy* 15(S1): 89–108.

Xu, J. and Grumbine, R.E. (2014). Building ecosystem resilience for climate change adaptation in the Asian highlands. *Wiley Interdisciplinary Reviews: Climate Change* 5(6): 709–18.

Yu, W., Alam, M., Hassan, A., Khan, A.S., Ruane, A., Rosenzweig, C., D. Major, D. and Thurlow, J. (2010). *Climate Change Risks and Food Security in Bangladesh.* Washington, DC: Earthscan.

3
Groundwater resources

Dipankar Saha, Anwar Zahid, Surendra Raj Shrestha and Paul Pavelic

Key messages

- Groundwater plays a critical role in supporting the lives and livelihoods of both rural and urban populations in the Ganges Basin. Groundwater resources are ubiquitous in most regions and available in large volumes and of a quality that is generally good or sufficient for its intended use (except for the lower Ganges Delta and parts of the middle Gangetic Plains). This has contributed immensely to poverty alleviation, food security, livelihood and wider socio-economic development within the basin. Groundwater now represents a well-recognized resource shared amongst the basin countries, although transboundary cooperation is limited.

- Broad generalizations for such a vast and complex basin, with its dramatic variations in terrain, hydrogeology, aquifer replenishment, groundwater use and quality, is not possible although distinct trends have emerged. In parts of the basin significant and growing problems associated with groundwater overexploitation have emerged (e.g. 35 per cent of the administrative blocks in four Indian states are overexploited). In parts, problems are also associated with arsenic, fluoride, saline water intrusion and other natural as well as anthropogenic forms of pollution that create great challenges for providing safe drinking water supplies (e.g. 62 and 23 districts in India and Bangladesh respectively are affected by arsenic contamination).
- Technical support systems and regulatory frameworks to enable sustainable allocation and use of the groundwater resource and accounting for ecosystem service requirements are largely in place. However, implementation of these policies and regulations to manage the resources sustainably remains a major long-term challenge.

Introduction

Groundwater is an essential input for human survival and prosperity in river basins which are the cradles of civilization and this particularly holds true for the Ganges Basin (Shah, 2009; World Bank, 2010). The highly variable monsoon rains, followed by a protracted dry season, and occasional drought, are the major drivers for groundwater development in urban and rural areas (Moench and Dixit, 2004). It is broadly accepted that groundwater can provide some degree of buffering to climate threats by supplementing surface water supplies and increasing water security.

The riparian countries of the Ganges (Ganga) Basin have abundant water resources, starting from Nepal where abundant rainfall and large glaciers make it one of the most water-rich countries in the world. Parts of the basin in India and Bangladesh too have high water endowments. Due to growing population, rapid economic growth, technological improvements and availability of potential aquifer systems, groundwater use continues to increase in the Ganga Basin. In particular, groundwater resources located in the Gangetic Plains are experiencing heavy extraction (e.g. Patna, the capital of Bihar state in India). Rural areas on alluvial terrain are also experiencing a rapid rise in groundwater-based irrigation demand (Saha *et al.*, 2009; Saha *et al.*, 2013). It is increasingly acknowledged that effective groundwater management is vitally important to provide sustainable access to groundwater of suitable quality for the 8 per cent of the world's population which reside within this basin.

Surface and groundwater resources within the basin are highly connected. Floods recharge the aquifer, and in the dry season, groundwater provides baseflow which is an important source of environmental flows in rivers. The contribution of aquifers to maintain environmental flows is not adequately quantified. However, it is recognized that groundwater development will reduce baseflow, and therefore there is a possibility that the environmental demand may be compromised. Surface water quality is compromised by diffuse and point sources of contaminants, and the extent to which groundwater dilutes contaminant concentrations unknown. Additionally, arsenic and fluoride contamination due to geogenic factors are also posing challenges to aquifer management.

A vast amount of data, information and knowledge at various spatial and temporal scales within the basin have been generated. This chapter attempts to compile and synthesize the current understanding of the groundwater resources. In so doing, it attempts to identify and highlight the most critical knowledge gaps which hinder sustainable groundwater management in the basin.

Regional hydrogeological framework

The hydrogeology of the Ganges Basin may be divided into four major regions: (i) Himalayan terrain – which comprises consolidated sedimentary and crystalline rock forming hills and mountain chains; (ii) Gangetic Plains – a vast alluvial track extending across the inner lowlands of the basin; (iii) southern peninsular region – comprised of lava flows and consolidated sedimentary rocks nestled between the Gangetic Plain and Indian Peninsula Shield; and (iv) deltaic plain – fluvio-deltaic sedimentary plain situated at the downstream end of the Ganga (Figure 3.1; Table 3.1). Details on each region are presented below.

Himalayan terrain

The aquifer systems of the Himalayan terrain, which encompass south-eastern Himachal Pradesh, Uttarakhand, West Bengal and north-western India and Nepal, are highly diverse due to extreme contrasts in geology, geomorphology and physiography. Intermontane valleys such as Kathmandu and Dang in Middle and Upper Himalayas in Nepal or Doon Valley (Dehradun) in the Lower Himalayas in India represent isolated aquifer systems with generally good groundwater potential.

The volume, depth and quality of groundwater in most parts of the Himalayan region varies from place to place. Except for the granite, gneiss and meta-sediments of higher metamorphic grade in the Middle and Upper Himalayas, other rock types have moderate to high potential for groundwater development. Apart from the unconsolidated sediments of the Terai bordering the Siwaliks and distributed over the India-Nepal border area, karstified and fractured carbonate rocks of the Middle Himalayas are considered to be productive aquifers. The unconsolidated deposits of Kathmandu and Surkehet (Birendranagar area) Valleys, the Doon Valley in Uttarakhand in India, Siwalik rocks, non-karstic but fractured carbonate rocks in the Lower Himalayas are interpreted as moderately productive aquifers.

The groundwater potential of the Kathmandu Valley is moderate to high where transmissivity ranges between 200 and 7,000 m^2/day in the northern part. Recharge takes place mainly along the foothills, though at a low rate. The potential for groundwater development is limited in the Lower Himalayan Series and the crystalline complexes, because they contain a substantial percentage of partly fractured hard rocks. Within this group, the best aquifers are carbonate rocks, whose occurrence is mostly local. The higher crystalline complexes feature high-grade metamorphism and do not contain major groundwater resources, though some permeable rocks are found in the upper part of the succession. Granites and gneisses have been classified as being poor aquifers because of limited secondary porosity. The Tethys Group contains some good aquifers of substantial thickness, mainly limestone and quartzites, but for a major part of the succession, they are impervious. As detailed maps are available for only the outcrop areas, the unit as a whole must be classified as a single unit with local groundwater.

In the Siwalik range of the Outer Himalayas, there are a number of longitudinal valleys known as Duns. One of the largest is the Doon Valley in India which is a synclinal depression between the Lower Himalayan mountains in the north and the Siwalik Hills in the south, representing a typical intermontane valley. The valley is principally underlain by the rocks of the Siwalik Group, which lies under post Siwalik poorly consolidated gravel, sand and boulders with clay bands. The thickness of this unconsolidated sequence is variable from 50 to 600 m. The springs are the potential source of water in the Doon Valley where discharge varies from 3 to 4.6×10^3 m^3/day. The depth to water table varies considerably from 10 to 100 m below ground level (bgl). Perched conditions of the shallow water table of local extent are also observed. Bore wells drilled by the Central Ground Water Board (CGWB) discharge between 360 and 4,600 m^3/day.

Gangetic Plains

The Gangetic Plains extend from the Aravali-Delhi Ridge in the west to the Rajmahal Hills in the east, and the Himalayan foothills (Siwalik Hills) in the north to the Indian peninsular plateau like Bundelkhand-Vindhyan-Hazaribagh in the south, covering an area of about 250,000 km^2 (Singh, 1996). The length of the plains is about 1,000 km, while the width ranges between 200 and 450 km, being wider in the west and narrower in the east (Sinha and Friend, 1994). The Gangetic Plains are marked by an irregular southern boundary, and outcrops of basement rocks protrude out of the alluvium at many places. The plains feature limited variation in topography and appear as a flat alluvial plain with a variation in altitude along the major axis from 175 m near Mathura in the west to 25 m above mean sea level at Ajmabad in the east.

The Gangetic Plains are divided into four broad units representing different hydrogeological units. (i) Bhabher Belt, 10–30 km wide along the Himalayan base in Himachal Pradesh and Uttarakhand states in India, and then entering Nepal and again observed in the northern part of West Bengal in India. (ii) Terai Belt, 10–50 km wide, occurring along the southern margin of the Bhabher, running along the states of Himachal Pradesh, Uttarakhand, then entering into Nepal and also delineated in Bihar, Uttar Pradesh and West Bengal. (iii) Central Alluvial Plain, representing the major part of the Gangetic Plains covering the area between the Terai Belt and the axial course of the Ganga and its tributary, the Yamuna. This unit extends further southward up to about 20 km, depending upon the basement configuration. (iv) Marginal Alluvial Plain, lying between the Central Alluvial Plain and the exposed Precambrian/Proterozoic rocks (Singh, 1996).

Bhabher Belt

The Bhabher Belt is situated along the foothills of the Siwaliks, consisting of poorly sorted coarse sediments (boulder, cobble and pebbles) of alluvial and colluvial origin. This is a major recharge area of the Terai Plain as well as the Central Alluvial Plain. Based on the drilling of the CGWB in India, the composition of the Bhabher Belt has been found to be 60 per cent gravel, 34 per cent sand and 16 per cent mud with thicker horizons of gravel in the lower part. This unit is marked with high seasonal water level fluctuation, in the order of 10–12 m and even more at places with abundant rainfall (~1,700 mm/yr). Darjeeling and Jalpaiguri districts in West Bengal indicate yields of up to 68 m^3/hr from the granular zone within 150 m bgl, whereas in Uttar Pradesh the discharge has been reported to vary between 100 to 300 m^3/hr.

Terai Belt

The width of the Terai Belt varies from 10 to 50 km and is characterized by extensive development of swamps, water bodies and thin rivers. The deposits of the Terai Belt have been formed under fluvial processes that are still underway. The sediments consist of thick layers of clastic deposits of Quaternary age, marked with more fine grained sediments than the Bhabher Belt. Studies taken up in Nepal clearly indicate that the thickness of the finer sediments increase towards the south (Figure 3.2a). Analyses of lithologs from CGWB bore wells indicate 24 per cent gravel, 36 per cent sand and 38 per cent clay/mud. The Terai Belt features some good aquifer layers, which can be clubbed into: (i) shallow unconfined aquifers of 30 to 50 m thickness; and (ii) deep confined aquifers, typically between 50 and 200 m bgl which may extend further to 450 m bgl or more (Mathur and Kohli, 1963). The shallow aquifers are widely used for drinking and irrigation. In the Terai region of Nepal, 20–50 per cent of the screenable sand layers are limited to a depth of 46 m.

CGWB has reported that in the Terai Belt of India the discharge of wells varies from 100 to 200 m^3/hr. Pumping test analysis shows transmissivity in the Terai area of Nepal to range between 10 and 10,000 m^2/d. Water level is shallow and the annual water level fluctuation is much less in comparison to the Bhabher Belt. Towards the south, water level comes closer to the land surface and artesian flow conditions exist along the southern part of the Terai Belt. The general direction of groundwater flow in the Nepal Terai follows the surface drainage pattern from north to south.

Auto flow conditions

Within the Terai Belt and in the northern part of the Central Alluvial Plain, the southerly dipping aquifers, sandwiched between the clay aquitards, result in artesian conditions in topographically favourable areas. They are observed along the northern part of the Central Alluvial Plain in the districts of Uttarakhand, Uttar Pradesh and Bihar bordering with Nepal. In Uttar Pradesh, depths of artesian wells range from 154 to 477 m bgl, and the hydraulic head varies from 1.0 to 4.90 m above mean sea level (msl) with discharge between 5.4 and 198 m^3/hr.

Central Alluvial Plain

The Central Alluvial Plain spans a vast area characterized by predominantly fluvial as well as locally lacustrine and aeolian deposits (Figure 3.2b). It covers vast tracts of Bihar and Uttar Pradesh, the eastern part of Haryana and Rajasthan, northern tips of Madhya Pradesh and Jharkhand and a major part of West Bengal.

The Central Alluvial Plain hosts large amounts of sediments from the Himalayas and less from the Precambrian craton during the Quaternary Period. The sediments are comprised of Recent and Holocene alluvium and older alluvium of Pleistocene age mainly covering the states located to the east of Delhi. In parts of Haryana and Rajasthan, the aeolian deposits of Quaternary age are also included in this category.

The thickness of unconsolidated deposits is a few metres along the basin boundary, but in excess of 1,000 m at places in the northern part of the plains. The Quaternary alluvium often grades to unconsolidated Upper Tertiary deposits. Together, unconsolidated deposits extend up to 4,000 m below in some basement depressions such as the Gandak and Sarda. The deposits are made up of alternating layers of sand, silt, clay and sandy clay with frequent change of sedimentary facies. The sand layers, with occasional gravel beds, form potential aquifers, which are interlayered with relatively impermeable aquitard (clay, sandy-clay) layers, forming a multi-aquifer system. The aquifers are hydraulically connected at a regional scale, semi-confined and confined conditions with vertical hydraulic head gradients present at sub-regional or local scale (Saha and Shukla, 2013). In the middle Ganga Plain near Patna, three aquifers to depths of 300 m were reported by Saha *et al.* (2009). Khanna (1992) has reported five distinct aquifers within 400 m bgl in Uttar Pradesh. CGWB has identified four aquifer systems: (i) first aquifer 0–150 m; (ii) second aquifer 160–210 m; (iii) third aquifer 250–360 m; and (iv) fourth aquifer 380–600 m.

In Bhojpur district of Bihar, on the confluence of the Sone and the Ganga, two aquifer systems have been identified to a depth of 300 m bgl by Saha *et al.* (2010). The deeper aquifer which lies below 120 m bgl has transmissivity between 5,163 and 6,974 m^2/day with a hydraulic conductivity (K) of 64.9 to 82.0 m/day. The semi-confined nature of the deeper aquifer is confirmed by the storativity values of 1.3×10^{-3} to 6.9×10^{-4}.

Alluvial fans

Bordering the Siwaliks, the Terai Belt and the adjoining Central Alluvial Plain are the large alluvial fans, often referred to as 'mega fans' which intrude 2 to 50 km within the Central Alluvial Plain. The mega fans, from east to west, include Tista, Kosi, Gandak, Sarda and Yamuna-Ganga. These mega fans generally hold thick deposits of medium to coarse sand and offer good potential aquifers. For example, the Kosi fan covering 10,200 km^2 in Bihar with its tip in Nepal has uninterrupted medium to coarse sand layers up to 100 m bgl (Figure 3.2c). The groundwater level is within 2–5 m bgl. Shallow tube wells within 25 m depth tapping 6 to 8 m of aquifers can yield 50 m^3/hr with drawdowns of <2 m. In the lower reaches, the potential is less, where shallow tube wells within 50 m depth tapping 12 to 16 m aquifers can yield 50 m^3/hr with a drawdown of <3 m.

Marginal Alluvial Plain

This part of the Gangetic Plains lies between the Central Alluvial Plain and the exposed hard rocks in the Indian Peninsular Shield. This unit is observed in the southern part of Uttar Pradesh and Bihar, northern-most tips of Madhya Pradesh and Jharkhand and western part of West Bengal. The alluvial deposits are comparatively thin, generally within 100 m, deposited on a northerly sloping basement (in West Bengal on an easterly sloping basement). The potential of these aquifers are low in comparison to the Central Alluvial Plain but increases with distance from exposed hard rocks.

In southern Bihar where the marginal alluvial plain borders the peninsular uplands, the thickness of sediments increases from a couple of metres to >600 m within a distance of 45 km. Aquifer potentiality also increases as reflected in an increase in K from <25 m/day to >150 m/day, while the transmissivity increases from 75 to >10,000 m^2/day (Saha *et al.*, 2007).

Southern peninsular region

Large parts of the Ganga Basin are underlain by consolidated and semi-consolidated rocks of Pre-Pleistocene Age, where the topography is hilly and undulating, with a flat-topped plateau at places forming narrow valleys. They are along the right-hand tributaries of the Ganga such as the Chambal, Betwa and Sone. The groundwater potential of these formations is low. The most prominent aquifer systems in this group can broadly be divided into two: (i) consolidated formation, where the groundwater occurs in the top slice of the weathered zone as well as within secondary porosity zones associated with fractures and joints; and (ii) semi-consolidated formation where groundwater occurs within the primary as well as in the secondary porosity zones.

Consolidated formation

The formations are classified into four broad groups: the compact sedimentaries and meta-sedimentaries, effusive rocks, intrusive rocks and basal crystallines. The southern and western parts are occupied by the rocks belonging to Archaean, Proterozoic and Vindhayan groups which are often tectonically disturbed. The Archaean group of rocks belongs to Aravalli Supergroup, Delhi Group (compact sedimentaries and meta-sediments), Bundelkhand Gneisses (basal crystallines), schists and meta-sedimentaries and metabasics in the southern and western parts, compact sedimentaries and metasedimentaries of Bijawars, Iron Ore Group of rocks and Chottanagpur gneiss granulite complex (basal crystallines) in the southern part of the eastern half. Tectonically undisturbed but compact sediments of the Vindhayan Supergroup also cover large parts along the south of the basin. Effusive volcanic rocks known as the Deccan Trap of Upper Cretaceous to Eocene age cover large areas of Madhya Pradesh and Rajasthan.

The aquifers in the consolidated rock areas are made up of the weathered zone (20–35 m thickness in general) and the underlying fractures/joints. Generally, two to five sets of fractures are encountered within depths of 200 m (the depth to which CGWB investigations are generally restricted). Dug wells are constructed in the weathered zones, but bore wells tap the fracture zones. In metamorphic and gneissic rocks, the dug wells (depth 25–30 m) exhibit specific capacities of 20 to 100 lpm/m, while the permeability of the weathered zone is generally <1 m/d. The fractures are anisotropically distributed, which poses a challenge for targeting sites for groundwater development. Yield of bore wells tapping the secondary porosity generally remains between 3.5 and 11 m^3/hr, but in suitable locations it goes up to 30 m^3/hr. Groundwater potential of compact sedimentaries like quartzites and sandstones are lower than the crystallines. Limestone, on the other hand, has better yield prospects with yields up to 36 m^3/hr with hydraulic conductivity in the range of 5–15 m/day. The sandstones and other compact sedimentaries generally have yields in the range of 0.5 to 11 m^3/hr with maximum yield sometimes reaching up to 25 m^3/hr. Specific yield in consolidated formations is around 2 per cent, while transmissivity is within 60 m^2/day. Groundwater occurs under unconfined to semi-confined conditions within the secondary porosity zones. In the Deccan Traps groundwater potential varies distinctly with depth. The lower vesicular part is the most productive unit characterized by multilayered flows, with yields ranging from 3.5 to 18 m^3/hr for a drawdown of 3–6 m. The specific capacity of dug wells tapping this unit varies from 50 to 150 lpm/m. The hydraulic conductivity is in the range of 5–15 m/day with transmissivity values up to 250 m^2/d.

Semi-consolidated formation

The formations of the Paleozoic-Mesozoic and Cenozoic age, extending from Carboniferous to Pliocene, belong to this group. The famous Gondwana Supergroup of the late Palaeozoic-Mesozpoic age deposited in rift valleys belongs to this group. The Gondwana sediments consisting of shales, claystones, sandstones, limestones and coal seams are found in Chhattisgarh, Jharkhand and West Bengal. Groundwater exploration in the Gondwanas reveals potential aquifers with yields between 3.5 and 20 m^3/hr for a drawdown of 6–10 m. The specific capacity remains in the range of 75–250 lpm/m. Hydraulic conductivity is up to 150 m/day, while the transmissivity varies from 200 to 400 m^2/day. Storativity values indicate semi-confined to confined conditions (0.25×10^{-4} to 5×10^{-4}).

Deltaic region

The Lower Ganga Plain merges with the vast deltaic Bengal Basin of West Bengal in India and Bangladesh. The area is underlain by a thick pile of admixtures of unconsolidated sand and clay of Quaternary Age. The sand layers make good to excellent aquifers. Multi-layered aquifer systems are separated by clay and sandy clay layers with separate flow regimes at the sub-regional scale (Mukherjee *et al.* 2009). In Bangladesh, the aquifer configuration is a three-tier arrangement (Figure 3.2d). The first aquifer, the upper Holocene aquifer, may extend to depths of 50 to over 100 m below ground, and lies exposed to the surface or places below a clay and silt layer of variable thickness. The aquifer sediments are composed of fine sand with clay lenses. Age of water from this aquifer dates to about 40–135 years (Aggarwal *et al.*, 2000; Zahid *et al.*, 2009). The second aquifer (main aquifer) from the Mid-Holocene extends down to 250–350 m and is generally underlain and overlain by silty clay beds, and composed mainly of fine to very fine

Table 3.1 Aquifer characteristics in the Ganga Basin

Sr. no.	*Country*	*Region*	*Terrain*	*Hydrogeological environment*	*Depth of investigation*	*Aquifer type and characteristics*
1a	India	Pre-Mesozoic consolidated rocks	Hilly and undulating at places, plateau	Groundwater in weathered zone and underlying fractures/joints	200 m in general	Low potential aquifers. Mainly unconfined and locally semi-confined
1b		Upper Cretaceous basalt flows	Flat to rolling topography, hilly and undulating at places	Groundwater in weathered zone and fractures and vesicular zones in successive flows	200 m in general	Low to moderate potential aquifers. Unconfined to semi-confined and locally confined aquifers
1c		Tertiary sediments and laterites	Flat to undulating and occasionally hilly	Moderate to good potential aquifers made up of primary porosity	300 m in general	Moderate to good potentiality. Unconfined to locally semi-confined
1d		Quaternary alluvium and aeolian deposits and narrow alluvial belt along the major peninsular tributaries	Flat to gently sloping land, part of the Gangetic Plains	Multi-layered deposits of sand and clay/sandy clay	300 m in general and >450 m and more at places, <150 m in marginal alluvial areas	Good to excellent aquifer system. Semi-confined to confined
2	Nepal[a]	Quaternary alluvial, fluvial	Gangetic Plains	Alluvial plains, multi-layered deposits of gravel, sand and clay becoming finer to the south	150 m in general, often <50 m	Shallow: unconfined to semi-confined. Deep: confined to leaky confined
3	Bangladesh	Fluvio-deltaic	Deltaic and estuarine	Sands with silt and clay lenses	Generally 3 to 4 layers to depths of up to 350 m	Good to excellent aquifer systems. Shallow: unconfined to semi-confined. Deep: confined to leaky confined

Note: a. In addition to the Terai described in the table, within the Himalayan region, the karstified and fractured carbonate rocks of the Midland and Tethys group are also considered to be highly productive aquifers. Unconsolidated deposits of Kathmandu and Surkehet (Birendranagar area) Valleys, Siwalik rocks, non-karstic but fractured carbonate rocks in lower Himalaya, Crystalline Complex and Tethys group offer moderately productive aquifers sand, occasionally inter-bedded with clay lenses. This aquifer is under semi-confined/leaky conditions with water, aged between 2,300 and 3,000 years. This second aquifer is predominantly used for production. The third aquifer (deep aquifer) is encountered at depths of 300–350 m, generally below a silty clay aquitard. This aquifer is composed mainly of grey to dark grey fine sand that in places alternates with thin silty clay or clay lenses. On a regional basis, aquifers at 350 m depth seem hydraulically connected. Limited abstraction of groundwater from any aquifer does not affect the others. Water from the third aquifer is dated 1,400–20,000 years, depending on the distribution of aquifer/aquitard sediments.

Horizontal hydraulic conductivity of the Deltaic region is between 2.2 and 6.0 m/day for the second aquifer and between 2.9 and 19.9 m/day for the third aquifer. Vertical hydraulic conductivity is considerably low. For the first and the upper part of the second aquifer in floodplain areas, K ranges from 11 to 56 m/day (BWDB-UNDP, 1982; BWDB, 1989, 1994; DPHE-DANIDA, 2001; BGS-DPHE, 2001). In most parts of Bangladesh the first unconfined aquifer has high potential with mean transmissivity as 1270±770 m^2/day and specific yield ranging from 0.01 to 0.2 with a national average of 0.06±0.04 (BWDB-UNDP, 1982; BWDB, 1994; Shamsudduha *et al.*, 2011). Transmissivity of the shallow aquifers in the Ganges floodplains (3,000–5,000 m^2/day) are generally higher than the finer textured deltaic aquifers (300–3,000 m^2/day). Deep aquifers (250–350 m) indicate higher potential in places (transmissivity from 769 to 3,224 m^2/day), whereas lower potential emerges in certain places (144 to 916 m^2/day). Storage co-efficient values of deep aquifers were estimated to be between 4.4×10^{-3} and 1.6×10^{-4}, indicating leaky confined to confined conditions.

Model simulations (Zahid *et al.*, 2014) show aquifers are recharged by vertical percolation and through flow from elevated eastern hilly areas. Pumping from the first aquifer and the upper part of the second aquifer retards vertical percolation in deeper aquifers, resulting in increased travel time in the lower part of the second and the upper part of the third aquifers. Under the

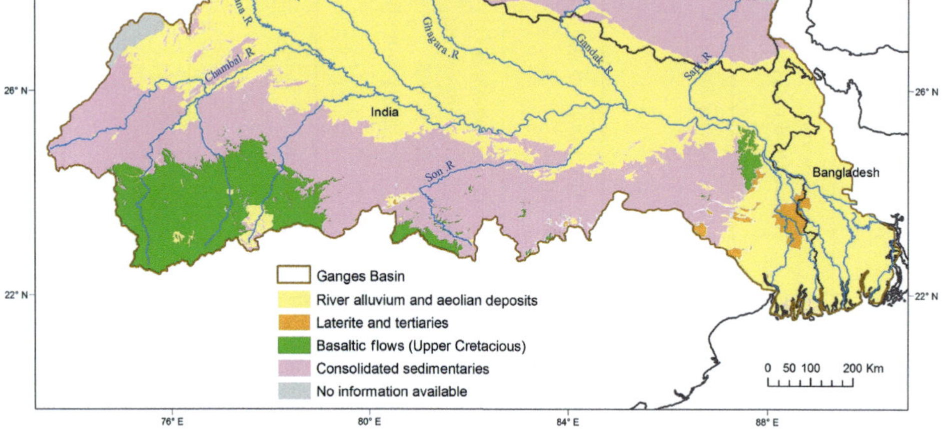

Figure 3.1 Distribution of major aquifers in the Ganga Basin

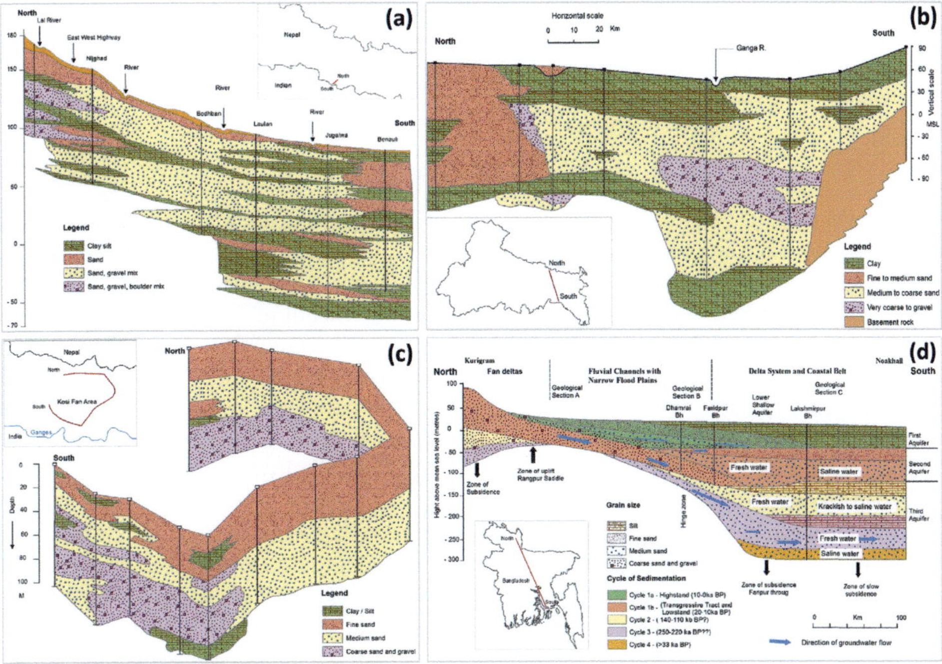

Figure 3.2 Selected geological cross-sections. Section (a) extends across the central area of the Nepal Terai (Bara); (b) Gangetic Plain; (c) Kosi Fan area; and (d) from north to south of Bangladesh (section (d) redrawn from BGS-DPHE, 2001)

current trend of groundwater development, the average travel time, i.e. age of water, for the upper and the lower parts of the first, second and upper part of the third aquifers at different geologic conditions are estimated between 37 and 234, 133 and 317, 832 and 2,485, 1,009 and 3,027 and 1,065 and 3,543 years respectively. Travel time of vertical recharge will be reduced in the first aquifer with increased irrigation pumping from the lower part of the first aquifer. Average travel time and length of flowlines will be increased in the second and third aquifers as huge pumping continues from the first aquifer. Groundwater level in the shallow aquifer varies mainly due to local-scale lithologic variations and groundwater pumping. The first and second aquifers are greatly influenced by irrigation abstraction. Under natural conditions the groundwater gradient is almost flat and horizontal flow is relatively small compared to vertical movement. Water from the first aquifer may move downward into the deeper fresh water zones through more permeable 'windows' within the aquitard, while groundwater in the third aquifer generally flows laterally.

Groundwater level behaviour and trends

Commensurate with the variations in elevation, terrain, hydrogeological conditions and groundwater use, the water levels vary considerably across the Ganga Basin. A strong temporal characteristic is observed due to the south-western monsoon, the main source of recharge. Shallowest water levels are observed around August (mid-monsoon); thereafter the water level steadily declines till the onset of the next monsoon. The declining part of the well hydrograph (August to pre-monsoon) can be divided into two segments. From August to November the rate of decline is slower as the extraction is mainly for domestic purposes and recharge is supplemented by seepage from the rivers, streams and other water bodies. The steeper part of the declining segment is observed between November and May, when millions of wells draw water from aquifers for winter crops and for domestic purposes. The water level variations in sub-regions of the basin are described below.

Nepal Terai

Water level measured from about 400 monitoring wells tapping shallow aquifers indicates that the level varies from artesian conditions to a depth of 16 m bgl, with an average of 4.6 m. The deepest level is encountered in Rupandehi district, lying in the western part. The levels representing the deeper granular zones indicate an average depth of 5.9 m, with a maximum of 30.7 m at Mahottari district in the central part.

Figure 3.3a shows the hydrographs for representative wells tapping the shallow aquifer over five years from January 2009 to December 2013. Across the Nepal Terai the average annual fluctuation in groundwater level is about 3 m. Groundwater levels are generally stable over the longer term and do not suggest a declining trend. However, the hydrographs possibly indicate that the variation in annual fluctuations is declining. This decrease in seasonal variation may be due to the change in recharge processes which in turn might be caused by climate change. In recent years the rainfall patterns have changed and although the average rainfall over the year remains the same, the number of rainy days has reduced and the rainfall intensity has increased. This has resulted in increased surface flow but less percolation to the water table.

India

In India, groundwater levels are monitored by the CGWB and state departments dealing with groundwater. The CGWB monitoring network includes about 2,400 wells, the majority of which tap shallow unconfined aquifers.

The depth to water level varies considerably in the Ganga Basin area of India, ranging from <1 m to more than 60 m. The deeper water levels are generally found in the western part of the basin, falling in the states of Delhi, Haryana and Rajasthan. Shallow water levels are observed in the eastern part of the basin, falling in the states of West Bengal, Bihar, Jharkhand, Chattisgarh and the eastern part of Uttar Pradesh (Figure 3.3b).

Groundwater is extensively used in Delhi, Uttar Pradesh, Bihar and West Bengal, which fall within the Ganga Basin. During the pre-monsoon of (May) 2015, the depth to water level is as deep as 62.2 m bgl. In about 40–45 per cent of the wells (n=2,342), the water level remains within 5 to 10 m bgl. In Bihar, Uttar Pradesh and West Bengal, in only 4 per cent of the wells, the groundwater level was deeper than 20 m bgl. In Delhi, only 15 per cent of the total wells (n=114) had groundwater levels within 20 m bgl, and 9 per cent of wells had groundwater at depths >40 m bgl. This indicates heavy groundwater extraction in Delhi state which is an urbanized state. During post-monsoon (November) 2014, depth to groundwater levels ranged between 0.2 and 63.6 m bgl. In Bihar, Uttar Pradesh and West Bengal 38–60 per cent of the measured wells (n=2,283) had groundwater at depths between 2 and 5 m. The number of wells falling within the 5–10 m bgl range has reduced to 15–28 per cent, indicating substantial rainfall recharge. In Delhi, however, the percentage of wells in the 10–20 m bgl range, as well as 20–40 m bgl range, remained almost the same as in the post-monsoon period. This points to two factors: (i) potentially less recharge due to extensive urbanization; and (ii) the extraction for domestic use continues throughout the year, even during the monsoon season.

Pre- and post-monsoon groundwater levels in 2015 reveal that 27–40 per cent of the wells in the states of Uttar Pradesh, Bihar and West Bengal had a rise of 0 to 2 m, while 2–5 per

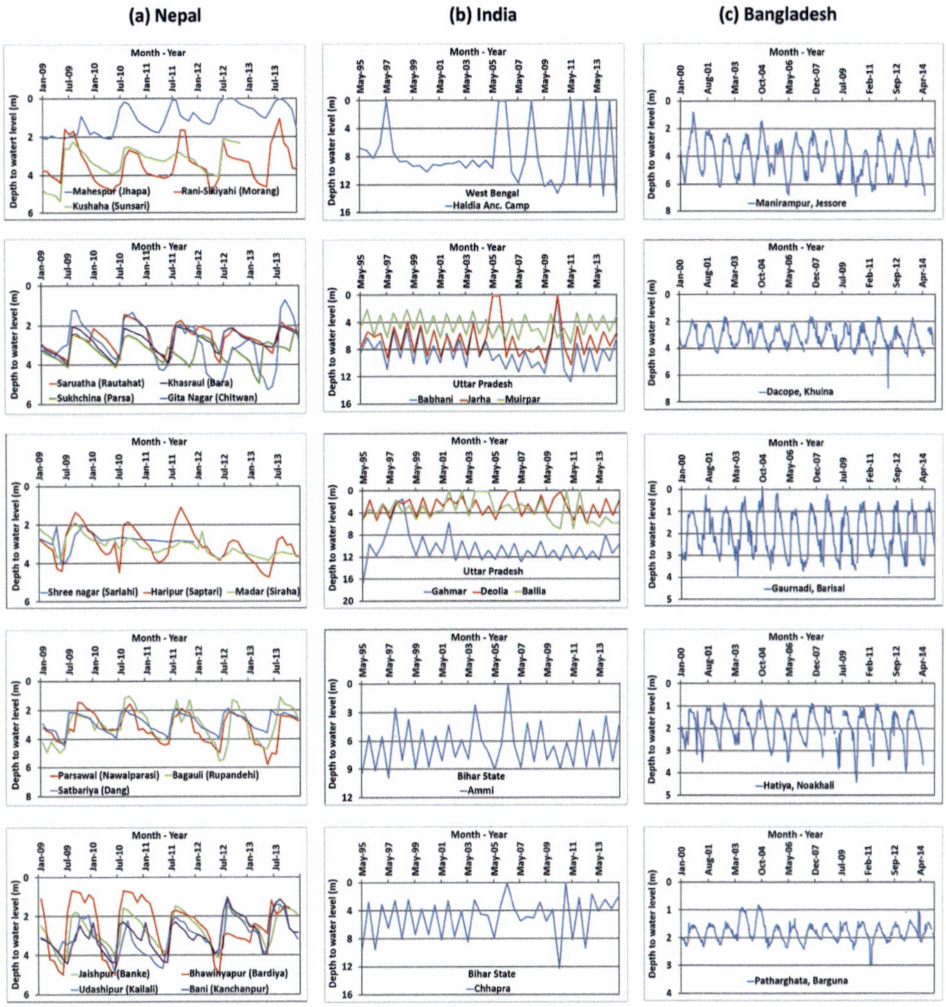

Figure 3.3 Well hydrographs from observation wells distributed throughout the basin. Plots (a–c) are representative of the Nepal (Terai), India and Bangladesh parts of the basin respectively

cent of cases had a rise of 2 to 4 m. A fall of 0 to 2 m has been reported in 38–61 per cent of the wells. In Delhi an overwhelming 73 per cent of the wells show a fall of 0–2 m bgl range during the same period.

Bangladesh

The water table fluctuation between pre- and post-monsoon is more in the central and northern part of the country where groundwater irrigation is extensive. This fluctuation is low or even nil near the coast as groundwater irrigation in the southern coast is negligible due to saline groundwater in the upper aquifers. Hydrographs show steepest changes when the irrigation season begins rather than at the start of the monsoon. During the monsoon, the hydrographs typically rise steadily until the groundwater is within 1–2 m bgl. This indicates a dynamic equilibrium between the water table, the extractive demands of deep rooted vegetation and linkages to surface water bodies (Ravenscroft, 2003).

Groundwater levels in the deeper part recover rapidly when seasonal pumping stops. This rapid recovery, parallel to water levels in the shallow aquifer, reveals that shallow and deeper aquifers are generally hydraulically well-connected.

Generally, groundwater withdrawal from the shallow aquifer for domestic use and irrigation is balanced with recharge from rain water and inflow from surrounding aquifers. Figure 3.3c reveals no permanent decline in groundwater level. In some areas the decline of groundwater levels below sea-level and/or tidal river water level create potential for the encroachment of saline water.

Shamsudduha *et al.* (2011) show that shallow groundwater levels are declining at a high rate in recent times (1985–2005) in Bangladesh. The decline is highest (up to 0.5 m/yr) in and around Dhaka City and the Barind Tract region, and lowest (up to 0.05 m/yr) in areas south of the River Ganges. In the well-buffered coastal areas, shallow groundwater levels indicate stable to slightly rising trends (0 to +0.1 m/yr).

Basin scale

At the basin scale, the pre-monsoon depths to water level range between 2 and 5 m bgl in the extreme north-eastern, north-central and coastal parts of the basin. Deeper groundwater levels (20–40 m bgl) occur in the states of West Bengal, Bihar and Uttar Pradesh. Deepest groundwater levels are observed in west and central parts of Haryana and Rajasthan. In the east-central part of the basin extending over the states of Bihar, Jharkhand and the southern parts of the state of

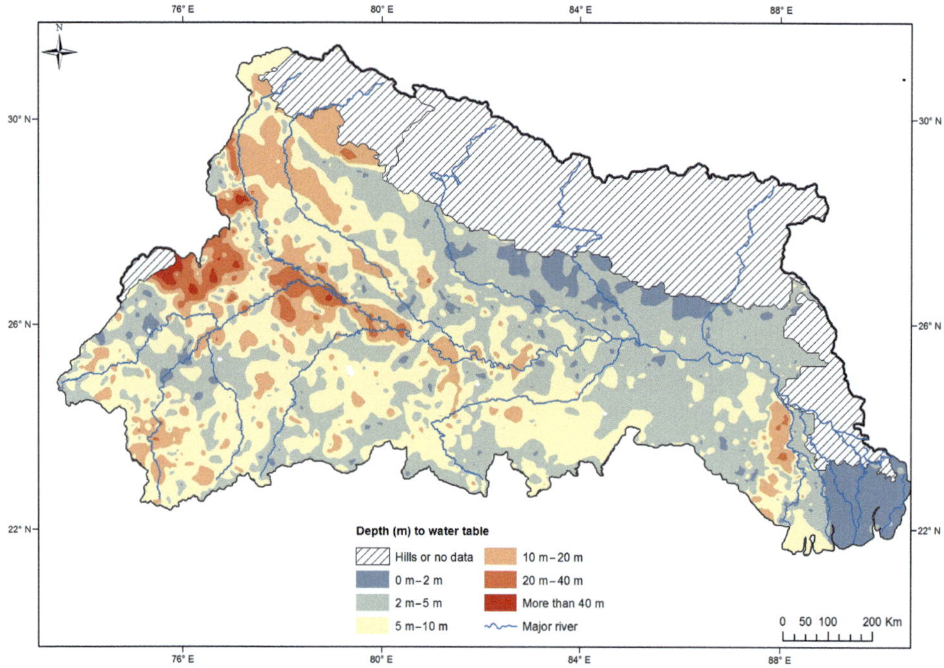

Figure 3.4 Post-monsoonal depth to water table across the Ganges Basin
Source: CGWB data used for India; BWDB (2013) data for Bangladesh

West Bengal, the depth to pre-monsoon water levels is in the range of 5–10 m bgl. The depth to water level is in the range of 10–20 m bgl in isolated patches of the southern part of the basin. In a major part of West Bengal, the water levels are within 10–20 m bgl consistently.

During the mid-monsoon season, the depth to water level shows perceptible changes from that of the pre-monsoon season, indicating the impact of monsoon recharge. During this season, the level largely remains in the depth range of 2–5 m bgl in the major part of the basin. In the northern part of Bihar and eastern Uttar Pradesh, the depth to water level is in the range of 0–2 m bgl. However, the west-central parts of the basin falling in Haryana and Rajasthan exhibit depth ranges akin to that during the pre-monsoon. Further, in this season too, the depth to water level remains in the depth range of 10–20 m bgl across much of West Bengal (Figure 3.4).

Groundwater recharge and resource utilization

Monsoon rainfall is the major source of groundwater recharge in the Ganga Basin. The other sources contributing recharge are return seepage from irrigation, seepage from canal and water bodies, infiltration from non-monsoon rainfall, etc. For the deeper semi-confined or confined aquifers, recharge may take place through: (i) leakage from the shallow unconfined aquifers through the aquitard layers in between; and (ii) lateral recharge through aquifers from the recharge areas.

Terai Plain (Nepal)

The Terai Plain gets recharge from rainfall and through-flow from the Bhabher zone. Lateral recharge is evident for the deeper aquifers in the Terai Belt, which are exposed in the Bhabher Belt. Precise demarcation of the major Terai and Bhabher recharge areas is unknown.

According to Duba (1982), the total annual dynamic recharge is 9,629 MCM, of which 3,114 MCM takes place in the Bhabher zone and the remaining 6,515 MCM in the Terai Plain. GDC (1987, 1994) estimated an average recharge of 600 mm/yr, of which about 460 mm/yr is contributed by direct percolation of rainfall and 140 mm/yr from flow through the aquifers from the Bhabher zone. These studies assume that about 75 per cent of annual recharge (i.e. 450 mm) can be safely exploited and thus the total annual volume of groundwater available for the abstraction would be 10,745 MCM for the entire Terai. Similarly, Electrowatt (1984) estimated around 465 mm/yr recharge in the Bhabher zone. In contrast, Tahal Consulting Engineers (1993) estimated over 1,100 mm/yr of annual recharge (i.e. 42 per cent of rainfall).

The 'Shallow Aquifer Investigation in the Terai' project (GWRDB/UNDP) estimated groundwater recharge by three different methods which resulted in a range of values with considerable contrasts. The most conservative estimate suggests annual recharge from the Terai to be 5,800 MCM, and applying Duba's assumption of upstream inputs, recharge is about 9,600 MCM. However, the most accepted recharge calculation is based on seasonal variations in water level, adjusted for the specific yield of the sediments, which derives a total annual groundwater recharge of 8,800 MCM (McNellis *et al.*, 1993). This is most generally accepted to be the annual recharge in the Terai and is reflected in Figure 3.5 and Table 3.2.

An estimated 1.05 million domestic shallow tube wells are in use for daily domestic water supplies, extracting groundwater mostly from unconfined aquifers in the Nepal Terai. Municipal and community water supply systems mostly tap groundwater from the deeper aquifers. As of 2013, there were about 1,000 deep and 112,401 shallow tube wells installed by the government agencies for irrigation purposes. The irrigation use is estimated to be 1,146 MCM/yr on the basis of the deep wells operating at the rate of 40 l/sec for 1,000 hr/yr and shallow tube wells operating at the rate of 10 l/sec for 200 hr/yr. Reliable estimates of the number of tube wells installed by various industries are not available and for this purpose it is assumed that about 115 MCM (25 per cent of domestic use) of groundwater is used by industry. Based on these assumptions, at present, only about 20 per cent of the available dynamic groundwater recharge is being utilized (Table 3.2). In 2011 and 1996 the corresponding figures were 16 per cent and 10 per cent respectively.

In recent decades there has been a boom in shallow tube wells caused by two factors: increase in rural electrification and introduction of cheap electric pumps. For the Terai farmers in Nepal, shallow tube wells are an attractive option because of good aquifers in many locations and higher cost of deep tube wells. The total area irrigated by groundwater in the Nepal Terai area suggests only about 42 per cent of the area has good potential for groundwater irrigation. There is still large scope to expand groundwater irrigation. Out of the total area of 2.64 Mha that is suitable for agriculture in Nepal, an estimated 0.73 Mha of Terai land can be potentially developed for irrigation by shallow tube wells. Similarly, 0.31 Mha have marginal potential for irrigation by shallow tube wells and a further 0.19 Mha show good potential for deep and medium deep tube wells (GDC, 1994).

India

In India, groundwater resource assessment units are blocks/talukas (the smallest unit in the administrative hierarchy) so that the estimated resources can be dove-tailed with policy decisions on groundwater resource development. The annual recharge versus the total draft for different societal uses (irrigation, domestic and industry combined) is expressed as Stage of Development (SOD), which forms the basic input for decision making on regulating groundwater extraction through direct and indirect means. The first national groundwater resource assessment was carried out in 1984, and detailed, block-wise resource assessment commenced in 2004. After the next assessment in 2009, the Government of India has decided to estimate the annually replenishable resource every two years.

Presently groundwater resources are estimated using GEC (1997). As per the latest 2011 resource estimate, 164.4 BCM is the net annual groundwater recharge within the Indian portion of the basin. Uttar Pradesh contributes more than any other state with 71.0 BCM, followed by Bihar at 29.2 BCM, West Bengal at 22.2 BCM and Madhya Pradesh at 21.3 BCM, with Delhi at 0.29 BCM contributing least, having an area of only 1,506 km^2. The total annual extraction from the Ganga Basin is 108.3 BCM, indicating an overall SOD of 65.9 per cent. However, this lumped SOD value does not reflect the wide spatial variations in the level of groundwater development across the basin. In areas under Chattisgarh and Jharkhand states, underlain predominantly by consolidated formations, the SOD is <37 per cent, whereas West Bengal and Bihar are underlain by thick, unconsolidated aquifer of high potential, and the SOD is ~45 per cent, pointing towards sufficient scope for further development of the resources. In Uttar Pradesh, the other state featuring a vast alluvial plain, the SOD on average is higher, at 72 per cent. However, a closer look within Uttar Pradesh reveals more intense draft in the western part. Further west, the states of Delhi and the parts of Haryana and Rajasthan that fall within the basin reveal significant overexploitation. The average SOD for these three states is 136 per cent, 125 per cent and 139 per cent respectively. Of the 533 administrative blocks that fall within the Ganges Basin for the states of Rajasthan, Haryana, Madhya Pradesh and Jharkhand, 188 of these (35 per cent) are classified as overexploited.

Groundwater is the major source for irrigation, which is reflected by the fact that about 90 per cent of total extraction is used for agriculture. The two states which show comparatively low groundwater use for irrigation are Delhi – the national capital of India uses around 36 per cent of total extraction – and Jharkhand, which uses 64 per cent of extraction for irrigation.

Bangladesh

Estimates of groundwater recharge at the national scale for the pre- (1975–1980) and post- (2002–2007) groundwater irrigation development periods, and long term (1985–2007), show that actual (net) recharge is higher in north-western (Dinajpur district) and western parts (Rajshahi district) than in southern (Khulna district) and eastern parts (except for Comilla district) (Shamsudduha *et al.*, 2011). Net recharge is high (300–600 mm/yr) along the Rivers Brahmaputra and Ganges and in north-western parts of the GBM Delta. In the south-eastern GBM Delta and Sylhet areas the net annual recharge is considerably lower (<150 mm/yr). Recharge has increased substantially (5–15 mm/yr) in north-western and western districts but has slightly decreased (–0.5 to –1 mm/yr) or remained unchanged in the rest of Bangladesh.

Net annual recharge in north-western and western parts of the country has substantially increased since the 1980s. Previous groundwater studies (e.g. BWDB-UNDP, 1982; MPO, 1987; WARPO, 2000) suggest that greater groundwater-fed irrigation will increase net recharge in areas where surface geology and soil properties are permeable and thereby favour recharge. Numerical modelling of regional groundwater flow suggests that actual (net) recharge increased from 70 mm/yr prior to widespread groundwater-fed irrigation to around 250 mm/yr (Michael and Voss, 2009). Net groundwater recharge can increase if available aquifer storage capacity increases due to pumping during dry seasons, increasing infiltration and reducing run-off. This finding conforms with the 'Ganges Water Machine' concept first described by Revelle and Lakshminarayana (1975). There are areas where recharge has increased due to fall in groundwater levels by 5–10 m as a result of groundwater pumping in dry seasons. Recovered groundwater levels were the same as at the end of the previous wet season. This rise in net recharge has occurred without an increase in annual rainfall. Caution must be taken to prevent saline water intrusion in coastal areas due to groundwater pumping for irrigation.

Increases in actual groundwater recharge are limited in areas where direct rainfall recharge is inhibited by low-permeable surface geology and net recharge has reached potential recharge. Net groundwater recharge along the River Ganges (Padma; 350–600 mm/yr) is much higher than that of the GBM Delta (<150 mm/yr). Sediments in floodplains of the major rivers are coarse grained and the storage capacity of adjacent aquifers is higher than those in the deltaic plains (Figure 3.5). Analysis of groundwater level and river-stage hydrographs (Shamsudduha *et al.*, 2011) reveals that water levels in almost all river channels rise above groundwater levels in adjacent aquifers during the monsoon season (May–September) and indirect recharge is restricted to lateral river-bank infiltration during the early monsoon time (April–June).

Over the past few decades groundwater has become the main source of irrigation water and one of the key determinants of Bangladesh's food security (Zahid *et al.*, 2009). This is largely due to its availability at shallow depths and accessibility at low costs through the dry season. During the late 1970s and early 1980s groundwater development was encouraged to avoid utilizing of contaminated surface water. An estimated 10 million tube wells now exist in the country. Groundwater-fed irrigation was introduced in the 1960s and has extended rapidly since then. The proportion of arable land irrigated by groundwater increased from <1 per cent in 1965 to approximately 78 per cent in 2007 (Shamsudduha *et al.*, 2011). By 2006, nearly 78 per cent of the irrigated rice fields were supplied by groundwater, of which approximately 80 per cent of the irrigation water derived from low-capacity (average discharge rate 10 l/s) shallow tube wells (depth <80 m); the rest was irrigated by high-capacity (average discharge rate 56 l/s) deep tube wells (depth >80 m) to produce Boro rice (BWDB-UNDP, 1982; BBS, 2009). During the so called pre-developed groundwater-fed irrigation period (1975–1980), shallow tube wells based irrigation covered an average area of 0.057 Mha and deep tube wells covered an average area of 0.14 Mha (Shamsudduha *et al.*, 2011). By the developed groundwater-fed irrigation period (2002–2007), the shallow well-supplied area had increased to 3.04 Mha and deep wells to 0.70 Mha. As a result, the abstraction of groundwater has also been increased and currently the estimated withdrawal by these irrigation wells is about 2,600 MCM/yr. About 30 per cent of the total groundwater recharge is being utilized (Table 3.2).

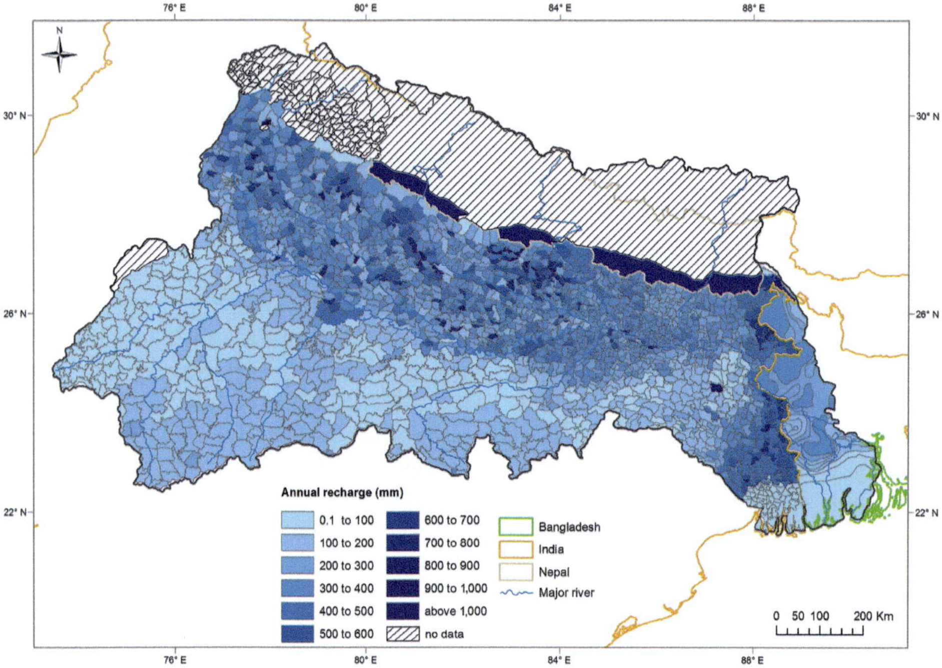

Figure 3.5 Distribution of annual groundwater recharge across the Ganges Basin

Sources: Adapted from McNellis *et al.* (1993) for Nepal; CGWB data for India; Shamsudduha *et al.* (2011) for Bangladesh

Table 3.2 Annual groundwater recharge and draft in the Ganga Basin

Sr. no.	*Country*	*Monsoon recharge (MCM/yr)*	*Non-monsoon recharge (MCM/yr)*	*Total annual draft (MCM/yr)*	*Stage of development*[a] *(%)*
1	India	164,390		108,260	65.9
2	Nepal (Terai)	7,040 (80% of total recharge)	1,760 (20% of total recharge)	1,723 (irrigation: 1,146; domestic: 462; industry: 115)	19.6
3	Bangladesh	6,500	<2,100	2,600 (mostly non-monsoon)	30

Note: a. Stage of development is defined as the total annual draft (abstraction) divided by total annual recharge, expressed as a percentage

Sources: CGWB data, 2013; McNellis *et al.* (1993); MPO (1987); CSIRO (2014)

Groundwater quality and suitability for societal uses

Generally, considering the hydrochemistry, the groundwater is Ca-HCO_3 and Ca-Mg-HCO_3 types, with localized enhanced concentration of sodium, making the water type as Ca-Na-HCO_3 or Mg-Na-HCO_3 (Saha *et al.*, 2009; Kumar *et al.*, 2006). The major groundwater quality concern in the Ganga is elevated arsenic concentration beyond the permissible limit of 10 μg/l. Furthermore, fluorides (limit for potable use is 1.5 mg/l), nitrates (limit for potable use is 45 mg/l), iron (potable limit 1 mg/l) and salinity (potable limit 2,000 mg/l) are causes for concern. The origin of arsenic, fluoride and iron is predominantly geogenic. The fluoride and iron are contributed by simple dissolution from rock-water interactions, while there are a range of views on mechanisms for the mobilization of arsenic in groundwater. The major groundwater pollution issues in the riparian countries are described below.

Nepal Terai

A major groundwater quality issue in the Nepal Terai region is arsenic contamination. Approximately 7 per cent of all tube wells tested are found to be beyond the permissible limit (50 μg/l) and 16 per cent of tube wells exhibited levels ranging from 11 to 50 μg/l (Table 3.3). The majority of the contaminated wells are less than 50 m deep (NASC and ENPHO, 2004). The maximum contaminated zones are found at depths from 11 to 30 m bgl. Other potential natural groundwater quality problems are high concentrations of dissolved methane and manganese. High levels of iron, nitrate and pesticides have also been reported. Groundwater quality of confined aquifers is higher than from unconfined aquifers and is particularly important for drinking water supplies.

India

Arsenic contamination has now been found to be spread along the entire length of the Ganga Plains. Approximately 50 million people are residing in the high-risk zone. The concentration of arsenic in groundwater exhibits high spatial variability, giving rise to patchiness in the distribution of contaminated wells. Depth-wise, the contamination is restricted to ~60 to 100 m bgl. The contaminated areas are confined within the Younger Flood Plain of the Ganga associated with the Holocene and Upper Pleistocene deposits. Arsenic contamination is not reported from the consolidated and semi-consolidated formations in the Ganga Basin. Five states falling in the Ganga Basin, Haryana, Uttar Pradesh, Bihar, Jharkhand and West Bengal (all alluvial late Quaternary deposits), are dotted with arsenic contaminated wells. Out of the five states, all four except for Haryana, where the entire alluvial area of the state falls within the Ganga Basin, contamination has been reported from 44 districts.

Fluoride contamination is also a widespread health hazard. The affected areas are located in consolidated and semi-consolidated formations; particularly vulnerable are the granitic areas. Fluoride contamination is also observed from alluvial areas, particularly in marginal alluvial areas and from acquifers in aeolian deposits. The marginal alluvial areas with thickness of less than 50 m or so are found to be vulnerable. Two interesting observations have been made in the pattern and extent of contamination, namely: (i) less spatial variability is observed in fluoride than arsenic; and (ii) there is no depth control on the concentration as observed in the case of arsenic in alluvial aquifers of the Ganga Basin. Several studies carried out by CGWB reveal that different fracture zones in hard rock areas down to depths of 200 m as well as the weathered zone are contaminated by fluoride. Fluoride contamination has been reported for more districts

Table 3.3 Ground water arsenic contamination (> 0.05 mg/L) in the Ganga Basin

Sr. no.	*Country*	*State/district where reported*	*Number of districts where reported*	*Number of blocks from where reported*[a]
1	India	States: 1. Haryana, 2. Uttar Pradesh, 3. Bihar, 4. Jharkhand, 5. West Bengal	1. Haryana – 18[b] 2. Uttar Pradesh – 20 3. Bihar – 15 4. Jharkhand – 1 5. West Bengal – 8	1 – 21[b] 2 – 71 3 – 57 4 – 3 5 – 79
2	Nepal[c]	Arsenic >0.05 mg/L: Dang, Chitwan, Sunsari, Nawalparasi Arsenic >0.01 mg/L: Rautahat, Nawalparasi	High vulnerability: 1 district; moderately high vulnerability: 2 districts; moderate vulnerability: 16 districts; low vulnerability: 1 district	
3	Bangladesh[d]	Arsenic >0.05 mg/L: Barguna, Patuakhali, Bhola, Jhalkathi, Pirojpur, Magura, Khulna, Rajbari, Jhenaidah, Kushtia, Barishal, Narail, Chuadanga, Jessore, Gopalganj, Madaripur, Noakhali, Satkhira, Faridpur, Shariatpur, Meherpur, Bagerhat and Lakshmipur	All (about 23) districts in the Ganges Basin more or less (countrywide 61 districts) Least affected: 4 Moderately affected: 10 Highly affected: 9	433 out of 496 Upazilla (sub-district) countrywide

Notes: a. The contaminated groundwater is reported at the district/block level; it may not be that the entire unit is contaminated. b. Out of the total number of districts in the state. c. Source: NASC and EPHO (2004). d. Source: BGS-DPHE (2001); GWTF (2002)

than arsenic primarily because it occurs both in hard rock and alluvial (marginal) areas. For example, in Delhi state, underlain primarily by hard rock aquifers, seven out of nine districts are affected by fluoride contamination.

Nitrate contamination is generally related to overuse of fertilizer as well as open defecation. This contamination (>45mg/l) has been reported from large areas of the basin cutting across the different hydrogeomorphic zones, rock types and depths to water level. Salinity contamination (>2,000 mg/l) is the result of both rock-water interactions and connate water, and is mainly confined to the western part of the Gangetic Plains covering the states of Uttar Pradesh (western part), Delhi, Rajasthan and Haryana. However, in the delta front areas of West Bengal in India and Bangladesh, the salinity increases are attributed to seawater ingress through aquifers due to over-pumping of coastal fresh water. The other reason is inundation by sea water in coastal areas during storms which percolate and increase the salinity of shallow aquifers.

Anthropogenic groundwater pollution varies from excessive pesticides which are related to its overuse in agriculture and horticulture. Heavy metal pollution is also noticed from localized places such as Kanpur in Uttar Pradesh which is related to industrial pollution. However, it is localized in comparison to the geogenic pollution which affects larger swaths of the landscape. Other major sources of heavy metals in groundwater stem from seepage of leachate from urban sewerage systems and landfills.

Bangladesh

Arsenic is the major groundwater contamination issue in Bangladesh. About three million tube wells, installed at shallow depths (10 to 50 m), discharge groundwater with arsenic concentrations more than the Bangladesh drinking water standard of 50 μg/l (Ahmed *et al.*, 2004; Zahid *et al.*, 2008). Between 28 and 35 million people in Bangladesh have been exposed to drinking water containing arsenic concentrations more than 0.05 mg/l and 46 to 57 million people have been exposed to 0.01 mg/l (BGS-DPHE, 2001). In the Ganges Basin, all districts have wells that do not conform to the Bangladesh drinking standard (Table 3.3).

Deep tube wells (100 to 300 m deep) have been installed in the main aquifer in recent years as an attempt to find groundwater without arsenic. Unfortunately, these wells often contain high concentrations of iron, manganese and salinity. There are concerns that large-scale groundwater abstraction from deep aquifers could eventually result in contamination of the deeper fresh, unpolluted groundwater by inducing relatively rapid downward flow from upper high-arsenic/saline layers.

Saltwater intrusion is occurring in some areas due to sea level rise and storm surges, especially in the low-topography regions of central Bangladesh. In the coastal part of Bangladesh, aquifers up to depths of 350 m have been affected by salinity. This process may explain in part why groundwater in much of the coastal area is brackish; in some parts of this area brackish water may also be relic saltwater deposited at the same time as the sediments.

In the coastal zone, salinity of the upper aquifers (down to 200–250 m depth) is extremely variable and changes abruptly over short distances. In most areas, the water is too saline for domestic and irrigation use due either to connate salts or estuarine flooding. In the deep aquifers (>250–300 m depth), fresh water occurs almost in the entire coastal area except some parts of the central coast. Seasonal variability of salinity is also noticed, and generally dry season salinity is higher than wet season salinity, where chloride concentrations vary between 30 and 60 mg/l. In the shallow groundwater, groundwater electrical conductivity (EC) values vary from 31 to 43,300 and 38.9 to 46,000 μS/cm in the dry and wet seasons respectively. At intermediate depth (100–200 m), it ranges from 309 to 37,300 μS/cm in the dry season and 109.3 to 34,100 μS/cm in the wet season. The deep aquifers (>250–300 m depth) show groundwater EC values between 231 and 37,000 μS/cm in the dry season and 72.3 and 10,340 μS/cm in the wet season (Zahid *et al.*, 2013).

High concentrations of dissolved iron are common in groundwater in many areas of Bangladesh (BGS-DPHE, 2001); however, there are no reported public health effects of high iron in drinking water. Groundwater is commonly treated for high iron (and manganese) concentrations by aeration in drinking and industrial water supplies for aesthetic reasons. Adverse neurological effects (i.e. reduced intellectual function) as well as infant mortality with high exposure to manganese through drinking water have recently been reported from long-term monitoring in central Bangladesh (Wasserman *et al.*, 2006; Hafeman *et al.*, 2007).

Policy, regulation and institutional arrangements for groundwater resource management

The highest volumes of groundwater extraction in the world occur in the Ganga Basin, where it plays a significant role in maintaining the economy, and standard of living. The policy and institutional mechanisms to support this varies significantly from country to country.

Terai (Nepal)

In the Terai area of Nepal the spurt in groundwater irrigation commenced in the 1990s as a result of the increase in rural electrification and introduction of cheap electric pumps. The Groundwater Resources Development Board (GWRDB), under the Ministry and Department of Irrigation, along with the Department of Water Supply, are key agencies responsible for groundwater management in Nepal. They are supported by the Agricultural Development Bank, Nepal (ADB/N), many NGOs, private companies, individual farmers and equipment suppliers (Table 3.4). The Groundwater Resources Development Board, which is tasked to regulate groundwater use, does not have any jurisdictional powers. Overlapping mandates of different groundwater institutions often lead to problems in regulating groundwater.

The Water Resources Act of 1992 and the Water Resources Regulations of 1993 provide the key water-related legal and regulatory basis for the country, but are formulated mainly for surface waters and offer no specific provision to regulate groundwater. An order created by the Committee for Underground Water in 1975, chaired by the Secretary of Ministry of Irrigation, mandates surveys, studies, innovates, supervises and monitors groundwater, as well as developing information systems and plans for its utilization. This order is insufficient to regulate use and address other groundwater issues. The National Water Plan 2005, together with the earlier National Water Resource Strategy 2002, recognizes the importance of groundwater resources for socio-economic development. Less priority is given to the regulatory framework for groundwater, with most attention given to surface water resources.

India

The government has no direct control over groundwater extraction which is primarily taking place through millions of private well owners – both in rural and urban areas. This is largely due to the absence of systematic registration of wells with attached rights for use as well as metering for abstraction.

Other forms of regulation, facilitated by provisions of institutional credit and subsidized energy, the major part of groundwater development post-1960s has been accomplished by private investments by millions of farmers. Groundwater rights are customarily attached to land ownership, hence its development is left to private initiatives. More than 80 per cent of total groundwater extraction in the Ganga Basin is used for agriculture. The 1960s were a turning point in India's agricultural development, which was closely tied to the expansion of groundwater-based irrigation. The 'Green Revolution' in India adopted new seeds and fertilizers, where the aquifers provided controlled supply of water at the will of the farmers. The impact of the revolution was largely confined to the north-western part of India, which extends over the alluvial part of the Ganges Basin, covering western Uttar Pradesh and Haryana. The revolution ensured that groundwater-based irrigation grew at a much faster rate than surface water-based irrigation (Briscoe and Malik, 2006). The huge increase in groundwater irrigation in Bangladesh started in the 1970s with the advent of cheap drilling techniques. In Uttar Pradesh (eastern part), Bihar and West Bengal, there is still scope for further groundwater development to boost food security. On the other hand, in the western part covering the states of Uttar Pradesh (western part), Haryana, Rajasthan, Delhi and hard-rock parts of Madhya Pradesh, the core groundwater challenge has shifted from development (i.e. additional extraction of water) to management. In the federal structure of India, constitutionally, groundwater remains a state responsibility. However, the central government is undertaking groundwater-related activities, particularly in the domain of monitoring, survey and assessment of recharge and extraction. The government's effort to tame groundwater extraction through restriction on credit and electricity connections had limited success.

The system for groundwater management falls within a complex multi-layered framework of constitutional and statutory provisions at the central and state levels. The central government circulated a groundwater bill in 1970, with a view to protecting the groundwater resources, taking measures against overexploitation and ensuring its equitable extraction. The bill was recirculated in 1992, 1996 and 2005 but had limited success. In the Ganga Basin only in Delhi and West Bengal is groundwater extraction regulated (mainly for industries) through State Government initiatives.

The Central Ground Water Authority (CGWA) has been constituted under the Environment Protection Act 1986 for the purpose of regulation and control of groundwater development and management in the country. It has issued notification to 162 areas in the country, prohibiting construction of new wells. A number of notified areas fall within the states of Delhi, Haryana, Rajasthan, western Uttar Pradesh and an isolated area in coastal West Bengal. Besides these notifications, CGWA is imposing conditions on the industries seeking to extract groundwater in over-exploited areas to invest in artificial recharge structures, particularly in the states where the State Government are not regulating groundwater.

There is also an urgent need for groundwater quality management. This is particularly true to save the aquifers from anthropogenic pollution. The Water (Prevention and Control of Pollution) Act of 1974 and the Environmental Protection Act of 1986 deal with the pollution of surface and groundwater in India. The Central Pollution Control Board and the State Pollution Control Boards are the agencies with responsibility to control groundwater pollution.

The Planning Commission of India, in its report on groundwater management and ownership in 2007, addressed issues related to sustainable groundwater management. It touched upon issues such as strategies of natural groundwater recharge and protection as a part of watershed development, protection of natural recharge areas, efficient well use, crop water management and participatory management of groundwater resources. The report emphasized the need to develop and adopt a framework of groundwater governance.

Bangladesh

The principal agency for groundwater survey, investigation and monitoring is the Bangladesh Water Development Board (BWDB) (Table 3.4). The role of the BWDB in groundwater management includes fundamental data collection and analysis, basic assessment and processes to understand the groundwater system, and development of analytical tools for water resource planning and monitoring. The Department of Public Health Engineering and the Bangladesh Agricultural Development Corporation also have monitoring activities targeting their respective mandates of rural potable water supply and irrigation.

The first major step towards water policy formulation commenced with the preparation of the 1964 Master Plan, which focused predominantly on flood control for agriculture, emphasizing large sector surface water interventions and overlooking the groundwater resources. The National Water Policy of 1999 (Ministry of Water Resources, 1999) provided the framework for formulating the National Water Management Plan (NWMP). It also supports private development of groundwater irrigation for promoting agricultural growth, with surface water usage where feasible. The main objective of the NWP is to improve water resource management, protect the environment and provide direction to all agencies working in the water sector for the achievement of specified objectives. The need of conjunctive use of surface and groundwater is highlighted in the NWP. Major water resource management issues addressed by the NWP

Table 3.4 Proposed institutional responsibilities for the development and management of groundwater resources in the Ganga Basin

Sr. no.	*Country*	*Institution*	*Responsibility*
1	India – State Governments	1 Water Resources Department/ Minor Water Resources Department 2 Public Health Engineering Department (PHED) 3 State Pollution Control Board (SPCB) 4 State Ground Water Authorities	1 Water level and quality monitoring. Groundwater resources development schemes for irrigation. Dissemination of information to the stakeholders 2 Groundwater-based drinking water supply to rural and urban areas. Treatment of contaminated water before supply 3 Identification of polluting industries and urban sewage. Monitoring of groundwater quality 4 Regulating groundwater extraction, particularly for industries. Making artificial recharge mandatory in critical areas
	India – Central Governments	1 Central Ground Water Board 2 Central Ground Water Authority 3 National Bank for Rural Development 4 Department of Drinking Water Supply 5 Central Pollution Control Board 6 Ministry of Rural Development	1 Monitoring and assessment of groundwater resources in collaboration with the State Goverment Department. Aquifer Mapping Programme with the help of the State Goverment. Monitoring of water levels and water quality Data dissemination, new technology inputs and providing guidelines. Technical support to the State Goverments 2 Regulating groundwater extraction, particularly for industries. Making artificial recharge mandatory in critical areas. Preparing bills to implement for groundwater regulation and participatory management 3 Evaluation of groundwater-based schemes. Providing credit to farmers through banks 4 Financing and facilitating rural drinking water supply in the country through PHED 5 Pollution monitoring of groundwater, identifying polluting industries and polluted aquifers/areas. Providing guidance to the SPCB 6 Renovation of tanks and water-harvesting structures to facilitate groundwater recharge, under various schemes

Table 3.4 (continued)

Sr. no.	*Country*	*Institution*	*Responsibility*
2	Nepal[a]	1 Groundwater Resources Development Board (GWRDB) 2 Department of Irrigation 3 Department of Water Supply and Sewerage 4 Agricultural Development Project, Janakpur 5 Agricultural Development Bank	1 Identification of areas of groundwater potential 2 Aquifer exploration (shallow and deep) 3 Develop technical capacity to support groundwater development and management 4 Monitor existing observation wells for water-level fluctuation, groundwater reserves and water quality 5 Regulation of groundwater use (GWRDB lack authorization) 6 Provide drinking water supply in the country
3	Bangladesh	1 Water Resources Planning Organization 2 Bangladesh Water Development Board 3 Department of Public Health Engineering 4 Bangladesh Agricultural Development Corporation	1 National level organization responsible for the planning and policy making of the water resource as a whole, including the groundwater resource 2 Overall water resources management including survey, investigation and monitoring of groundwater resources as well as fundamental data collection and analysis 3 Potable water supply and basic assessment and processes to understand the groundwater system to some extent 4 Planning and monitoring for irrigation water supply including development of technologies

Note: a. In Nepal, the Department of Water Supply and Sewerage is also one of the major institutions that constructs especially deep tube wells in the Terai. So far there is lack of information regarding how many tube wells they have constructed are: river basin management; water rights and allocation; public and private investment; water supply and sanitation; water and agriculture; water for environment; stakeholder participation; legislative framework etc. The NWP also defines major institutional reform and the role of the government, the private sector and civil society in the management of water. Along with the government's programme, NGOs are also involved in playing a more proactive role in transforming agriculture and the rural economy. Very recently the Bangladesh Water Act of 2013 has been enacted for integrated development, management, abstraction, distribution, use, protection and preservation of water resources.

Conclusions

Across the Ganges Basin, groundwater resources are ubiquitous in most regions and available in large volumes of a quality that is good or sufficient for its intended use. The two dominant hydrogeological formations in the basin are the unconsolidated alluvial/aeolian aquifers that occupy the inner plains and account for more than half of the basin area, as well as the consolidated sedimentary aquifers that flank the unconsolidated sediments to the north and south. Thus, the most productive aquifers in the basin are dominantly sedimentary aquifers with hard-rock aquifers having lower significance.

Developing access to groundwater resources is rather simple and cheap, and is usually driven by private investments, so does not entail large governmental coordination or support. The intrinsic advantages of groundwater has given rise to its accelerate development by making it a favoured grassroots-level source. Dependency upon and importance of groundwater for the lives, livelihoods and economic advancement of the people that reside within the Ganges Basin are of a scale that is not easily surpassed by other major basins of the world. This defining characteristic will only strengthen in coming years as the large population continues to grow and socio-economic development brings with it more water-demanding dietary requirements and expanding urbanization and industrialization. Rural communities rely heavily upon groundwater for multiple societal needs that contribute to poverty alleviation and livelihood improvements. Groundwater is also extensively utilized in urban centres located in valleys and flood plains with productive aquifers. A rapid expansion in groundwater use for irrigation has emerged in recent decades, driven in part by the increased affordability and freedoms that farmers can exercise with groundwater, and in part by the buffering provided against rainfall variability including seasonal and more prolonged droughts.

Groundwater resources represent a finite, renewable resource of generally good quality but which may be compromised by natural or anthropogenic pollution. This individual access of common pool resource, combined with its hidden nature, makes it challenging to manage. At present, there are a variety of challenges and threats from groundwater overexploitation in intensively irrigated areas such as Haryana, western Uttar Pradesh, northwest Bangladesh (Barindarea) and in major urbanized areas such as the Kathmandu Valley and Dhaka. Arsenic pollution alone has affected millions of people residing in the high-risk zones across the basin and is expected to worsen in coming decades unless major interventions are undertaken. The riparian countries face specific sets of issues and are at different stages in developing strategies for sustainable groundwater management.

Understanding the complexity of the hydrogeology and hydrogeochemistry of the basin is a key starting point for effective and sustainable groundwater management. The data, human resources capacity and institutional arrangements to support effective groundwater management are largely present in individual countries, although certainly not to a uniform extent and stage of advancement across the basin. Data and databases for individual countries exist, but an important constraint for this river basin which supports transboundary aquifers is the lack of a unified and comprehensive groundwater database that is readily accessible to all stakeholders. Each of the riparian countries has developed policies and strategic legal/regulatory frameworks to varying extents; however, implementing these policies remains a challenge.

The importance of sustainable management of groundwater is now broadly acknowledged within the Ganges Basin at the national and lower levels across a broad spectrum of stakeholders. We hope that the contribution provided through this chapter will help researchers, planners and policy makers to better achieve the overarching goals of sustainable and equitable groundwater development and management, recognizing the high diversity in biophysical and socio-economic settings that exists across the Ganges Basin.

Acknowledgements

We wish to thank the Chairman of the Central Ground Water Board, India for his support and guidance. Thanks are also extended to Mr Nir Shakya, Senior Hydrogeologist (retired), Department of Irrigation, Nepal for his help in preparing the Nepal portion of this chapter. Similarly, we are highly indebted to Mr Surendra Shah and Mr Surendra Maharjan, also from the Department of Irrigation, for preparing GIS maps and the lithological cross section for the Nepal Terai. The basin-level maps were kindly prepared by Mr Touleelor Sotoukee from the IMWI Vientiane Office.

References

Aggarwal, P.K., Basu, A.R., Kulkarni, K.M., Froehlich, K., Tarafdar, S.A., Ali, M., Ahmed, N., Hussain, A., Rahman, M. and Ahmed S.R. (2000). *A Report on Isotope Hydrology of Groundwater in Bangladesh: Implications for Characterization and Mitigation of Arsenic in Groundwater.* Work performed within an IAEA – TC Project (BGD/8/016).

Ahmed, K.M., Bhattacharya, P., Hasan, M.A., Akhter, S.H., Alam, S.M.M., Bhuyian, M.A.H., Imam, M.B., Khan, A.A. and Sracek O. (2004). Arsenic enrichment in groundwater of the alluvial aquifers in Bangladesh: An overview. *Applied Geochemistry* 19: 181–200.

BBS (Bangladesh Bureau of Statistics) (2009). *Annual Agricultural Statistics 2008.* Agricultural Wing, Bangladesh Bureau of Statistics, Dhaka.

BGS-DHPE (2001). 'Arsenic contamination of groundwater in Bangladesh'. In D.G. Kinniburgh and P.L. Smedley (eds). Vol. 2: Final Report. *British Geological Survey Report WC/00/19*, British Geological Survey, Keyworth, UK.

Briscoe, J. and Malik, R.P.S. (2006). *India's Water Economy: Bracing for a Turbulent Future.* New Delhi: Oxford University Press.

BWDB (1989). *Report on the Compilation of Aquifer Test Analysis Results as on June, 1988.* BWDB Water Supply Paper No. 502.

BWDB (1994). *Report on the Compilation of Aquifer Test Analysis Results as on June, 1993.* BWDB Water Supply Paper No. 534.

BWDB (2013). *Hydrogeological study and mathematical modelling to identify sites for installation of observation well nests, selection on model boundary, supervision of pumping test, slug test, assessment of different hydrogeological parameters collection and conduct chemical analysis of surface water and ground water*, Final Report (Main Volume, Package 3). Bangladesh Water Development Board and Institute of Water Modelling.

BWDB-UNDP (1982). *Groundwater Survey: The Hydrogeological Conditions of Bangladesh.* UNDP Technical Report DP/UN/BGD-74–009/1.

CSIRO (2014). *Bangladesh Integrated Water Resources Assessment Supplementary Report: Approximate Regional Water Balances.* Commonwealth Scientific and Industrial Research Organisation.

DPHE-DANIDA (2001). *Hydrogeology Summary Report.* Department of Public Health Engineering–Danish International Development Assistance Water Supply and Sanitation Components, Ministry of Foreign Affairs, Government of Bangladesh.

Duba, D. (1982). *Groundwater Resources in the Terai of Nepal.* Water and Energy Commission Secretariat Report.

Electrowatt (1984). *Eastern Terai Irrigation Project Nepal*, Volumes 1 to 4.

GDC (1987). Study of Groundwater Development Strategies for Irrigation in the Terai.

GDC (1994). Reassessment of Groundwater Development Strategy for irrigation in the Terai, 6 Volumes.

GEC (Ground Water Estimation Committee) (1997). *Ground Water Estimation Methodology – 1997.* Ministry of Water Resources, Government of India, New Delhi.

GWTF (2002). *Report of the Ground Water Task Force.* Ministry of Local Govt. Rural Dev. and Co-operatives, Govt. of the Peoples Republic of Bangladesh.

Hafeman, D., Factor-Litvak, P., Cheng, Z., van Geen, A. and Ahsan, H. (2007). Association between manganese exposure through drinking water and infant mortality in Bangladesh. *Environ Health Perspect.* 115: 1107–1112.

Khanna, S.P. (1992). 'Hydrogeology of the Central Ganga Plain, Uttar Pradesh'. In I.B. Singh (ed.). *Gangetic Plain Terra Incognita*, Geology Department, Lucknow University, pp. 23–7.

Kumar, K., Ramanathen, A.L., Rao, M.S. and Kumar, B. (2006). Identification and evaluation of hydrogeochemical processes in the roundwater environment of Delhi, India. *Environ Geol.* 50: 1025–1039.

McNellis, J., Tuladhar, R.M., Shrestha, R., Amatya, S.C., Shrestha I. and Kansakar D.R. (1993). *Ground Water in the Terai, Nepal.* UNDP/HMGN, Shallow Ground Water Investigations in Terai, Technical Report No. 33.

Mathur, L.P. and Kohli, G. (1963). Exploration and development of the oil resources of India. Proc. 6th World Petroleum Congress, 19–26 June, Frankfurt, Germany.

Michael, H.A. and Voss, C.I. (2009). Controls on groundwater flow in the Bengal Basin of India and Bangladesh: Regional modeling analysis. *Hydrogeology Journal* 17: 1561–1577.

Ministry of Water Resources (1999). National Water Policy, The Ministry of Water Resources, Government of the People's Republic of Bangladash, Dhaka, Bangladesh.

Moench, M. and Dixit, A. (eds) (2004). *Adaptive Capacity and Livelihood Resilience: Adaptive Strategies for Responding to Floods and Droughts in South Asia.* Published by The Institute for Social and Environmental Transition, International, Boulder, Colorado, USA. and the Institute for Social and Environmental Transition, Nepal, June, 2004.

MPO (1987). *The Groundwater Resource and its Availability for Development.* Technical Report No. 5, Ministry of Irrigation, Water Development and Flood Control, Dhaka.

Mukherjee, A., Fryar, A.E. and Thomas, W.A. (2009). Geologic, geomorphic and hydrologic framework and evolution of the Bengal basin, India and Bangladesh. *Journal of Asian Earth Sciences*, 34(3): 227–44.

NASC & ENPHO (2004). *The State of Arsenic in Nepal – 2003.* National Arsenic Steering Committee (Nepal) and Environment and Public Health Organization (Nepal).

Ravenscroft P. (2003). 'Overview of the hydrogeology of Bangladesh'. In A.A. Rahman and P. Ravenscroft P (eds). *Groundwater Resources and Development in Bangladesh: Background to the Arsenic Crisis, Agricultural Potential and the Environment.* Bangladesh Centre for Advanced Studies, University Press, Dhaka, pp. 43–86.

Revelle, R. and Lakshminarayana, V. (1975). The Ganges Water Machine. *Science*, 188: 611–16.

Saha, D. and Shukla, R.R. (2013). Genesis of arsenic-rich groundwater and search for alternative safe-aquifers in the Gangetic plains, India. *Water and Environment Research*, 85: 2254–2264.

Saha, D., Dhar, Y.R. and Vittala, S.S. (2009). Delineation of Groundwater Development Potential Zones in Parts of Marginal Ganga Alluvial Plain in South Bihar, Eastern India. *Environmental Monitoring and Assessment.*

Saha, D., Dwivedi, S.N. and Singh, R.K. (2013). Aquifer system response to intensive pumping in urban areas of the Gangetic plains, India: The case study of Patna Urban area. *Environmental Earth Science.*

Saha, D., Sahu, S. and Chandra. P.C. (2010). Arsenic-safe alternate aquifers and their hydraulic characteristics in contaminated areas of Middle Ganga Plain, Eastern India. *Environmental Monitoring and Assessment.*

Saha, D., Sarangam, S.S., Diwedi, S. and Bhartariya, K.G. (2009). Evaluation of hydrogeochemical processes in arsenic-contaminated alluvial aquifers in parts of Mid-Ganga Basin, Bihar, Eastern India. *Environmental Earth Sciences,* 61(4): 799–811.

Saha, D., Upadhyay, S., Dhar, Y.R. and Singh, R. (2007). The aquifer system and evaluation of its hydraulic parameters in parts of South Ganga Plain, Bihar, India. *Journal Geological Society of India*, 69: 1031–1041.

Shah, T. (2009). *Taming the Anarchy: Groundwater Governance in South Asia.* Washington, DC: Resources for the Future Press.

Shamsudduha, M., Taylor, R.G., Ahmed, K.M. and Zahid, A. (2011). The impact of intensive groundwater abstraction on recharge to a shallow regional aquifer system: Evidence from Bangladesh. *Hydrogeology Journal* 19: 901–16.

Singh, I.B. (1996). Geological evolution of the Ganga plain – An overview. *J. Paleontological Society of India*, 41: 99–137.

Sinha, R. and Friend, P.F. (1994). River systems and their sediment flux, Indo-Gangetic Plains, Northern Bihar. *India Sedimentology*, 41: 825–45.

Tahal Consulting Engineers (1993). *Bhairahawa Lumbini Groundwater Project Environmental Assessment Study*.

WARPO (2000). *Draft Development Strategy (DDS)*. Estimation of groundwater resources, Annex-C, Appendix 6, National Water Management Plan, Water Resources Planning Organization, Ministry of Water Resources.

Wasserman, G., Liu, X., Parvez, F., Ahsan, H., Levy, D., Factor-Litvak, P., *et al.* (2006). Water manganese exposure and children's intellectual function in Araihazar, Bangladesh. *Environ Health Perspect*. 114: 124–29.

World Bank (2010). *Deep Wells and Prudence: Towards Pragmatic Action for Addressing Groundwater Overexploitation in India*. Washington, DC: The World Bank.

Zahid, A., Haque, M.A., Islam, M.S. and Rashidul Hasan, M.A.F.M. (2009). 'Development of groundwater under lower Gangetic plain of Bangladesh: The impact of shallow tubewells on irrigation water availability, access, crop productivity and farmer's income'. In A. Mukherji, K.G. Villholth, B.R. Sharma and J. Wang (eds). *Groundwater Governance in the Indo-Gangetic and Yellow River Basins: Realities and Challenges*. IAH Selected Papers Series, Netherlands: CRC/Balkema Press, pp. 141–62.

Zahid, A., Hassan, M.Q. and Ahmed, K.M.U. (2014). Simulation of flowpaths and travel time of groundwater through arsenic contaminated zone in the multi-layered aquifer system of Bengal Basin. *Journal of Environmental Earth Sciences*.

Zahid, A., Hassan, M.Q., Balke, K.-D., Flegr, M. and Clark, D.W. (2008). Groundwater chemistry and occurrence of arsenic in the Meghna floodplain aquifer, southeastern Bangladesh. *Journal of Environmental Geology*, 54: 1247–1260.

Zahid, A., Jahan, K., Ali, M.H., Ahmed, N., Islam, M.K. and Rahman, A. (2013). 'Distribution of groundwater salinity and its seasonal variability in the coastal aquifers of Bengal Delta'. In A. Zahid, M.Q. Hassan, R. Islam, Q.A. Samad, M.S. Khan and R. Haque (eds). *Impact of Climate Change on Socio-economic Conditions of Bangladesh*. Published by German Academic Exchange Service, German Development Cooperation, Alexander von Humboldt Foundation and Alumni Association of German Universities in Bangladesh, pp. 170–93.

4

Climate change and the Ganges Basin

Marc A. Jeuland

Key messages

- Climate change is likely to have a diverse set of impacts on the highly variable climate of the Ganges Basin, but predicting the direction, magnitude and distribution of these impacts remains challenging due to uncertainties about the future trajectory of emissions as well as inconsistencies across the different General Circulation Models (GCMs) that are used to make projections of future climate conditions.
- While temperatures and the frequency of extreme precipitation events both seem likely to increase in the future, there is little agreement in the literature over future changes in average

precipitation; this lack of consensus is also reflected in the divergent predictions of future runoff and hydrological flows in the Ganges system.

- Vulnerability to climate change disruptions in this basin is highly related to lack of development, but must also be examined through the lens of important structural inequities that exist along gender, caste, ethnicity and class lines. The severity of climate change impacts will be determined by the spatial and temporal pattern of the physical changes that occur as well as the nature of these inequities.
- In light of the divergence in projections of future water availability in this system, the basin riparians would benefit from adaptation and resource management strategies that enhance their capacity to cope with variability (and therefore uncertainties over water availability), foster economic development and enhance livelihoods.
- Adaptation planning should consider the role of water control infrastructure, institutions and investments that facilitate conjunctive use of water resources and enhanced management of water-use tradeoffs, and targeted interventions that help to address specific impacts (e.g. development of better flood and drought warning systems).

Introduction

The Ganges Basin and its downstream delta, covering an area of roughly 1.2 million km^2 and home to more than 500 million people living in China, Nepal, India and Bangladesh, is the most populous river system in the world. People living in the region have long coped with the extreme temporal and spatial climatic variability that characterizes the South Asian monsoon, and continue to be highly exposed to its irregularities today (World Bank, 2013). Meanwhile, the population density of the Gangetic Plain and Delta regions of the basin is among the highest in the world, poverty remains widespread and environmental degradation is acute (Hamner *et al.*, 2006). Due to the confluence of geographic and climatic features (e.g. exposure to high rates of sea level rise and tropical storms), low development, and high population density and freshwater dependence, downstream portions of the basin have been listed among the world's most vulnerable to climate change (Ali, 1999). Meanwhile, remoteness, high slopes, thin soils and climatic variability also render its mountainous regions vulnerable to disturbance.

The major basin riparians – Nepal, India and Bangladesh (the fourth riparian, China, plays only a minor role in this basin, with less than 4 per cent of its area) – all appear on global lists of the most exposed countries to floods and droughts (Guha-Sapir *et al.*, 2015). Low development in the basin exacerbates vulnerability to such natural disasters. Floods result from the unique combination of extreme precipitation events from cyclones or other monsoon events and the very flat topography of the Gangetic Plains. As discussed in the chapter on surface waters, the powerful flows of the highly dynamic Ganges and its large Himalayan tributaries are also the source of a uniquely unstable hydrology (Allison, 1998).

In light of the existing challenges to living in this dynamic system and coping with its extreme climatological and hydrological variability, the added insult of climate change appears daunting. Increases in temperature, changes in precipitation and glacier melt, intensified monsoons, water-induced disasters, sea level rise (IPCC, 2013) – all associated with climate change – are expected to upset the delicate hydrological balance existing in this basin today, giving rise to new water management challenges. Though predictions remain highly speculative, many expect that climate change dynamics will exacerbate existing environmental pressures in the basin and threaten unique ecosystems (Sapkota *et al.*, 2013; Alongi, 2015), impose high costs on already fragile social systems (Mirza, 2011), test governance and institutions (Salman and Uprety, 2002) and possibly feed conflict (Wirsing and Jasparro, 2007).

Climate change predictions in the basin are, however, plagued by enormous uncertainties, about future emissions as well as their effects on regional climatology. These uncertainties are compounded by a profound lack of collection and sharing of historical data and of modelling skill for predicting changing monsoon patterns and resulting hydrology. To be sure, some progress is being made: the latest generation of CMIP5 models appears more reliable for understanding potential future changes in the region. Yet the Intergovernmental Panel on Climate Change (IPCC) continues to highlight the divergence in predictions of precipitation and the dynamics of the monsoon (IPCC, 2013). In addition, the coarse resolution of global models creates difficulties for projecting changes in the Himalaya (Whitehead, Barbour *et al.*, 2015), from which the majority of the runoff into the Ganges originates. Researchers have begun to make greater use of regional climate models, but their validity remains uncertain and partly depends on the uncertain boundary conditions drawn from the General Circulation Models (GCMs) on which they depend.

This chapter presents an overview of previous research related to the potential impact of climate change on the hydrology and economic water uses of the Ganges, and provides some ideas on how adaptation strategies for coping with climate change can be aligned with riparians' broader development needs. The next section summarizes previous work on climate change in the basin – as it pertains to temperature, precipitation, snow/glacier melt, hydrology (including extreme events), sea level rise, water-use and economic production, environmental services and climate vulnerability. The third section then offers suggestions aimed at integrating development and adaptation in this basin under climate change, and the fourth concludes.

Climate change impacts in the Ganges Basin

This section summarizes the literature on climate change impacts on temperature, precipitation, snow and glacier melt, hydrology, sea level rise, water-use and economic production, and other ecosystem services. In what follows, we discuss results from peer-reviewed articles, many of which relied on the IPCC's 4th Assessment Report (AR4), or on even earlier reports. We do, however, refer extensively to the relevant chapters and findings in the 5th Assessment Report (AR5), wherever possible.

Temperature change

Scientists universally agree that the climate is warming due to anthropogenic influences, though the rate of future temperature change remains uncertain and depends on the unpredictable future rate of climate-altering emissions (IPCC, 2013). Several studies have examined historic trends in the Ganges Basin and its wider region. Over the past 100 years, weather station data suggest that the broader region has warmed by about 0.5°C, and six of the 10 warmest years of the twentieth century in the nearby Brahmaputra Basin occurred after 1994 (Ahmed and Alam, 1999; Immerzeel, 2008). Recently published studies have similarly detected increases in temperature in Nepal and India in recent years (Mittal *et al.*, 2014); Bhutiyani *et al.* (2010) document particularly rapid increases (+1.4°C) in the north-west Himalaya.

Meanwhile, multi-model ensembles from the AR5 based on various scenarios of future emissions predict additional increases of 0.5–1.5°C by 2050 and 0.5–4°C by 2100 in the basin, with greater increases occurring during the winter and at higher elevation (Collins *et al.*, 2013; Kirtman *et al.*, 2013). These predictions vary considerably owing to the use of representative concentration pathways, which correspond to four scenarios ranging from very high to very low mitigation (for example, the two middle emissions scenarios correspond to a range of additional warming in the range of about 0.5–1.5°C by 2050 and 1–3°C by 2100, across 42 CMIP5 models). Overall, the range of predicted increases is not substantially different from that previously discussed in the IPCC's AR4 (IPCC, 2007), and AR4 models have been more extensively used in Ganges-specific studies (Agrawala *et al.*, 2003; Jeuland *et al.*, 2013; Masood *et al.*, 2015). A few analyses using regional climate models and AR4 scenarios notably suggested the possibility of larger increases (Moors *et al.*, 2011; Kulkarni *et al.*, 2013).

Precipitation change

In contrast to the consensus over the fact that temperatures will rise (if not about the precise rates of future change), there is low confidence over sub-regional changes in precipitation in South Asia and over the Ganges Basin (Moors *et al.*, 2011; Hijioka *et al.*, 2014). Historical analysis of trends is challenged by a lack of sufficient observational records, high inter-decadal variability and the influence of large-scale climatological phenomena such as the El Niño Southern Oscillation (ENSO) (Kirtman *et al.*, 2013). Nonetheless, seasonal mean rainfall in South Asia shows a declining trend and reveals more frequent monsoon deficits despite an increased prevalence of extreme rainfall events (IPCC, 2013).

Precipitation projections from climate models are divergent over the basin (Jeuland *et al.*, 2013), as shown in Figure 4.1 (using the AR4 models). Still, an increasing number of studies appear to suggest that climate change may induce greater mean precipitation over the near to medium term, due to potentially higher monsoon rainfall (Immerzeel *et al.*, 2013; Kulkarni *et al.*, 2013; Hijioka *et al.*, 2014; Masood *et al.*, 2015; Whitehead, Sarkar *et al.*, 2015). Interannual variability and the frequency of heavy rainfall events are also projected to increase, except in the arid western parts of the basin (Hijioka *et al.*, 2014; Mittal *et al.*, 2014). The multi-model ensemble mean in the CMIP5 points to decreased precipitation in the winter (Kirtman *et al.*, 2013), as do more limited scale or sub-basin assessments from other studies (Bharati, Gurung, Maharjan *et al.*, 2014; Bharati, Gurung, Smakhtin *et al.*, 2014).

Accelerated snow and glacier melt

The predicted rise in temperature across the basin will alter the timing and magnitude of contributions from snow and glacier melt to the Ganges. In the medium term, melting in the Himalaya will begin earlier in the year and persist longer (Rikiishi and Nakasato, 2006), increasing runoff. In the long term, the consistent contributions from meltwater will decrease as storage gradually declines and the zero degree isotherm pushes further upwards (Alford and Armstrong, 2010). Seasonal storage in the form of snow will decrease, and more variable rainfall will then become increasingly dominant (Miller *et al.*, 2012). In the AR5, Jimenez *et al.*, (2014) note that the loss of storage in snow and ice will be slower than once thought; for example, the multi-model mean loss of glaciers in the Himalaya in 2100 under the moderate RCP4.5 emissions scenario is 45 per cent.

Changes in snow and ice will also vary across space. In particular, moving from the east to the west across the Himalaya, monsoon rainfall decreases greatly, and winter precipitation increases. Summer rain (and snowfall at very high elevation) in the eastern catchments of the Ganges therefore contributes much more to total basin runoff than glacier melt does, whereas glacier melt is more important in the western Ganges (Rees and Collins, 2006). This precipitation gradient implies that perturbations to the rate of glacier melt will have spatially differentiated effects, and somewhat muted impacts at basin scale (Alford and Armstrong, 2010; Immerzeel *et al.*, 2010).

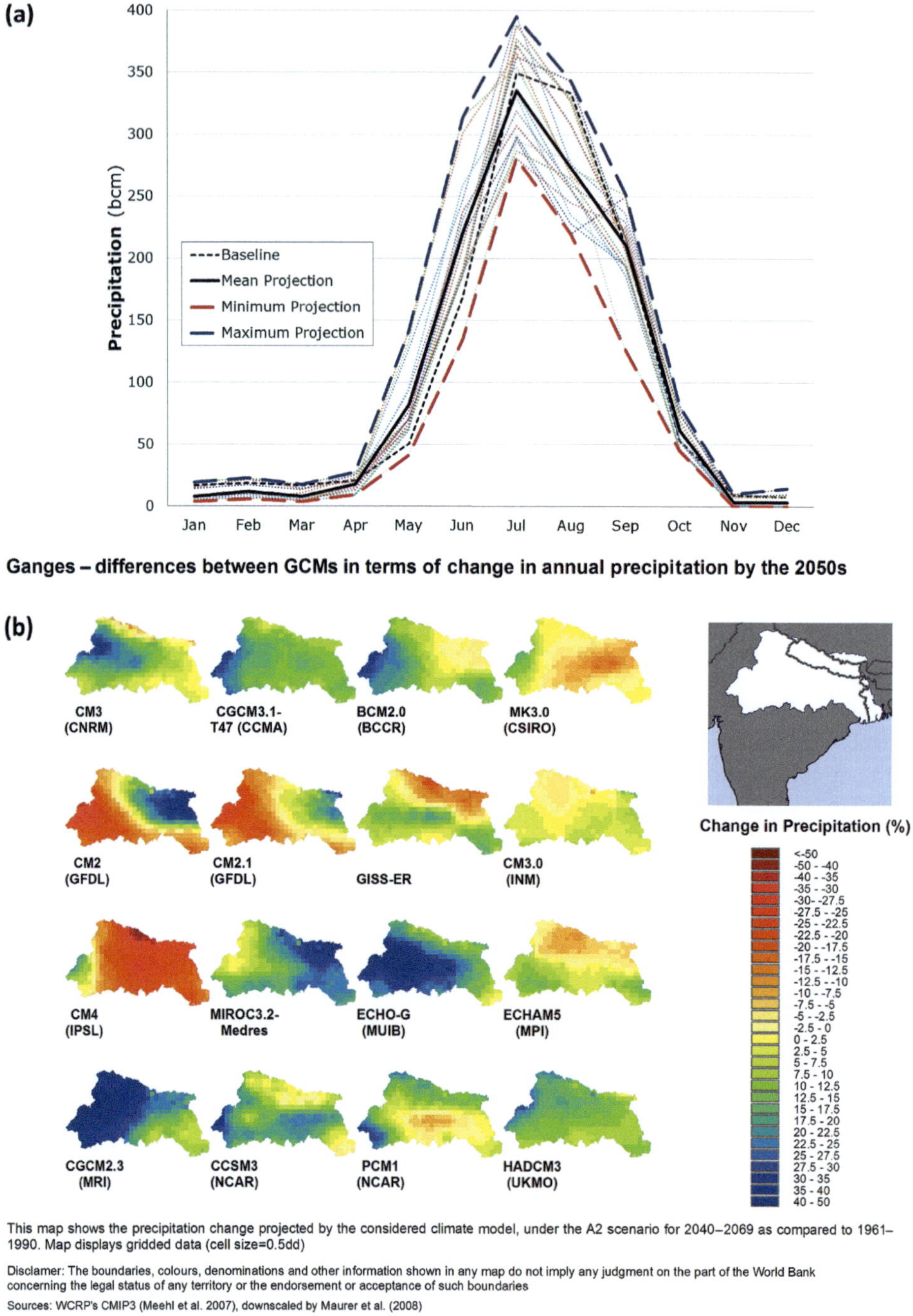

Figure 4.1 From Jeuland *et al*., (2013). Range of projected average precipitation in the Ganges Basin for the A2 emissions scenario: (a) by month, in 2050 (each GCM outcome indicated by the dotted lines); and (b) across space, for 2040–2069 relative to 1961–1990

Notes: The maxima and minima do not correspond to a single GCM

Sources: WCRP CMIP3 (Meehl *et al*., 2007), downscaled by Maurer and Hidalgo (2008)

Hydrological change

Given the nature of the climate effects discussed above, altered runoff into the Ganges and its tributaries will mostly be driven by changes in precipitation, except in sub-catchments that experience high levels of glacier melt. Perturbations in streamflows arising from climate change will in turn be affected by these evolving runoff patterns, increased evaporation losses, and by demand side drivers that relate to climate change. The latter, for example, might include increased irrigation demands (Jeuland, 2010), or effects on downstream demand for freshwater given salinity ingress (Dasgupta *et al.*, 2014).

It is therefore not surprising that studies conducted to date on historical changes in runoff and streamflow largely fail to find consistent evidence of changes (Mirza *et al.*, 2003; Shrestha and Shrestha, 2004; Miller *et al.*, 2012), particularly given the very high interannual variability of the system (Table 4.1). Similarly, predictions of future runoff and streamflow largely follow those for precipitation, with wide divergence across models and studies (Figure 4.2). Some recent studies suggest the possibility of decreased monsoon season runoff in the Brahmaputra and Ganges (Gupta *et al.*, 2010), while others predict that average and peak flows will increase and that extreme low flows will become less frequent (Gain *et al.*, 2011; Masood *et al.*, 2015; Whitehead, Barbour *et al.*, 2015). More systematic analyses across a range of GCMs instead highlight the lack of consistency in model predictions (Jeuland *et al.*, 2013) (Table 4.1 and Figure 4.2).

In addition to mean changes, hydrologists worry that higher temperatures will introduce greater uncertainty over the timing of water flows and increase the frequency of extreme drought and flood events, due to intensification of the water cycle (Trenberth *et al.*, 2003). Some research discusses findings of increased frequency of drought over much of India, and flood in West Bengal and Bangladesh (World Bank, 2013). These trends are consistent with predictions made with hydrological models that rely on climate projections (Hassan *et al.*, 2014; Tsarouchi and Buytaert, 2014; Whitehead, Barbour *et al.*, 2015). Increased variability and higher frequency of intense rainfall events would strengthen sediment flows and exacerbate existing maintenance challenges for embankments in the downstream system, and potentially increase flood risks.

Sea level rise

Sea level rise associated with increased temperatures will have important implications for flooding, storm-surge and salinity levels in the Ganges Delta, which has a very low elevation gradient (Wong *et al.*, 2014; Auerbach *et al.*, 2015). Studies suggest that the low-end projections of sea level rise (0.3 m) will increase the coastal area at high risk of flooding by roughly 15 per cent, and increase flood depths in severely affected zones by 30–40 per cent, after accounting for storm surge (World Bank, 2013). Without large-scale investments in coastal protection, a 1 metre rise would inundate 2,062 km^2 (1.5 per cent of the area of Bangladesh) and affect over 1.5 million people. Still, the delta may not suffer a loss in land area, since sediment deposits in some portions of the delta by the silt-laden rivers of the region have recently led to net accretion (Auerbach *et al.*, 2015). Indeed, the net effect of sea level rise and accretion will depend on the extent to which silt transport changes with upstream flow perturbations and alterations to embankment infrastructures (Jeuland *et al.*, 2013). Finally, some speculate that the frequency and intensity of tropical cyclones may increase with higher Indian Ocean temperatures (Mirza, 2011), though there is currently no documented evidence of such changes so far (Hijioka *et al.*, 2014).

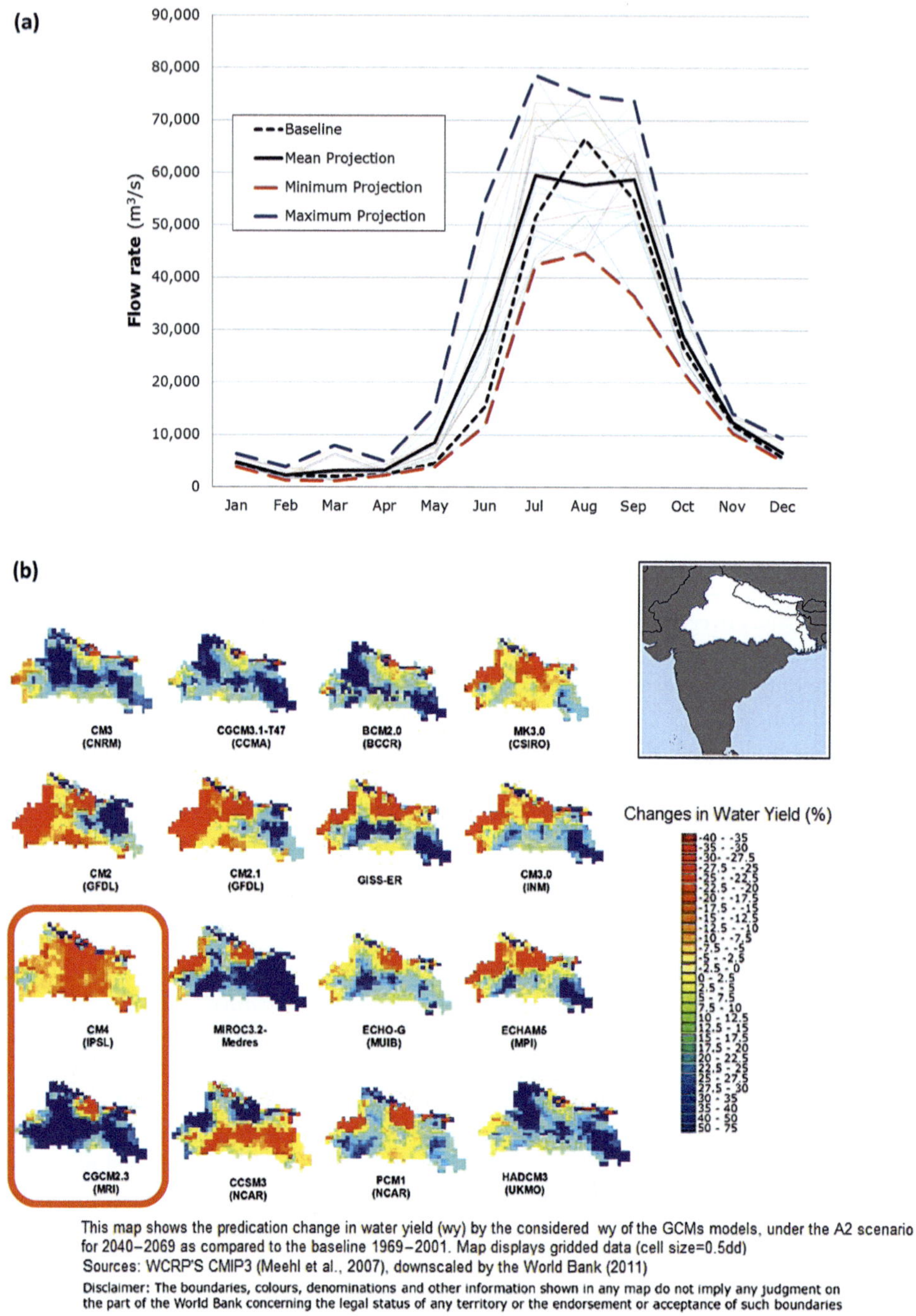

Figure 4.2 From Jeuland *et al.*, (2013): Range of projected: (a) average flow at Farakka in the Ganges Basin for the A2 emissions scenario by month, for 2050 (with each GCM outcome indicated by the dotted lines); and (b) change in runoff across space, for 2040–2069 relative to 1969–2001

Notes: The maxima and minima in (a) do not correspond to a single GCM

Sources: WCRP CMIP3 (Meehl *et al.*, 2007), downscaled by Maurer and Hidalgo (2008)

Table 4.1 Average historical and projected flows (in BCM/yr) at Farakka and in major tributaries of the Ganges across different climate GCMs (year 2050)

Indicator	*Ganges at Farakka*	*Yamuna (including Chambal)*	*Upper Ganga*	*Son*	*Ghagara (Kamali)*	*Gandak*	*Kosi*
Historical flows[1,2]	*1969–2001*	*Simulated: 1998–2007*	*Simulated: 1998–2007*	*Simulated: 1998–2007*	*Simulated: 1998–2007*	*Simulated: 1998–2007*	*Simulated: 1998–2007*
Average	493	101	60	42	109	69	74
Minimum	306 (1979)	54 (2002)	47 (1999)	19 (2004)	96 (2006)	51 (2006)	51 (2006)
Median	492 (1997)	104 (2003)	51 (2004)	49 (2000)	108 (1999)	72 (2005)	77 (1999)
Maximum	735 (1971)	141 (2005)	77 (2007)	73 (1999)	142 (2003)	97 (2001)	97 (2001)
Projected future flows in 2050[1,3]							
Ensemble mean	542	116	61	41	115	59	89
Minimum	448 (IPSL)	75 (GFDL0)	51 (GFDL0)	31 (GFDL0)	91 (ECHAM)	41 (GISS)	61 (IPSL)
Median	541 (NCAR CSM)	123 (MRI)	61 (NCAR CSM)	41 (ECHAM)	112 (NCAR PSM)	56 (MRI)	89 (CNRM)
Maximum	651 (MIROC)	138 (BCM)	75 (MRI)	50 (MIROC)	142 (UKMO)	79 (UKMO)	120 (MIROC)

Notes: Data from Jeuland *et al.*, (2013). 1 Flows reported here are virgin flows where the tributaries meet the Ganges mainstem, prior to withdrawals for irrigated agriculture or other purposes. 2 Minimum, median and maximum years are identified in parentheses; these are simulated for the tributaries since historical gauge data is not available. 3 Minimum, median, and maximum GCMs from the simulations are identified in parentheses

Implications for water-use and economic production

Hydropower generation from storage dams and run-of-the-river installations will clearly change with perturbations in streamflow, but the most clearly negative impacts of climate change will likely occur in the agricultural sector. Given that temperatures throughout the Gangetic Plain are already elevated during much of the growing season, rising temperatures are expected to threaten yields of rice, wheat, maize and sorghum (Hossain *et al.*, 2013; Hijioka *et al.*, 2014; Porter *et al.*, 2014), while increasing evaporation and crop water requirements (Jeuland *et al.*, 2013). Delays in the onset of the monsoon or increased variability in rainfall could similarly harm productivity (Hosterman *et al.*, 2012). Sea level rise threatens to flood fertile agricultural lands in Bangladesh (Wong *et al.*, 2014), and salinity ingress in Bangladesh could decrease *boro* rice yields by 10–30 per cent (Dasgupta *et al.*, 2014). The economic costs of these impacts on agriculture are hotly debated and wildly divergent, because they largely depend on assumptions about the efficacy of (and tradeoffs inherent in) adaptation to rising temperatures and changes in local water availability. Given the particular sensitivities apparent in South Asia (Lobell *et al.*, 2008), climate adaptation researchers have noted the need for interventions that both reduce climate vulnerability and advance economic development, to indirectly boost adaptive capacity (Ayers *et al.*, 2014).

Jeuland *et al.*, (2013) considered the sensitivity of hydropower generation and water productivity to projections of changes in surface water availability in the Ganges. The authors found that the value of energy production varied by about 30 per cent (or US$800 million/yr) across the range of AR4 GCMs in the moderate emissions A2 scenario, with existing infrastructure. Though the unit values for water productivity were highly uncertain, the authors calculated that the total value of irrigation water varied by about 20 per cent (or about US$800 million/yr) in a moderate productivity scenario, whereas the value of low flows to the delta varied by about US$1.3 billion/yr (Figure 4.3). The authors note that these changes do not include the effects of higher temperatures and changing water availability that is not included in 'blue water' flows, i.e. the water that flows through the basin surface water system.

Impacts on other ecosystem services

Much has been written about threats to water quality and the provision of ecological and cultural services by the waters of the Ganges (Singh and Singh, 2007). In the most highly populated portions of the river, the main concern is over elevated levels of various anthropogenic contaminants, from sources such as untreated wastewater, chemical or industrial discharges, and agricultural fertilizers or chemical runoff. Climate change will mainly affect concentrations of these contaminants through its uncertain link to perturbations in the magnitude of water flows. Increased monsoon flows, for example, would reduce average contamination levels in the river due to dilution, while reduced low flows would render dry season pollution more acute (Whitehead, Sarkar *et al.*, 2015). These effects, however, likely pale relative to those arising from other anthropogenic changes over the near and long term, e.g. from population growth, economic development and greater societal or governmental efforts to control pollution (e.g. improved implementation of the Ganga Action Plan).

In less populous regions such as the high elevations of the Himalaya or around unique coastal assets such as the downstream mangroves of the Sunderbans, climate change connections to hydrological services pose different threats. Sea level rise could irreversibly damage mangrove ecosystems, but differences in the rates of sediment accretion will influence the extent and speed of these impacts (McIvor *et al.*, 2013). The distribution of human settlements also increasingly constrains the potential for these ecosystems to migrate naturally as they have for millennia.

Many parts of the western Sunderbans are thus especially vulnerable to climate change because of the triple threats of temperature increases, land subsidence (which is largely independent of climate change) and sea level rise (Raha *et al.*, 2012; Rashid, 2014), but areas located closer to the main river channel are actually experiencing net accretion due to sediment deposition (Nicholls and Goodbred, 2004). Along with sea level rise, salinity concentrations are predicted to increase throughout the delta with negative consequences for high value species (Dasgupta *et al.*, 2014). Overall, the loss of mangroves threatens to reduce provision of important ecological services, protection from coastal erosion, storms and cyclones, and supplies of food, traditional medicines and renewable wood resources that are used extensively by local populations for fuel and construction (Hossain *et al.*, 2013; Alongi, 2015). Meanwhile, saline intrusion will threaten freshwater fisheries, particularly in areas suffering from land subsidence (Ali, 1999), while creating new opportunities for the development of brackish aquaculture (Dasgupta *et al.*, 2014).

There is considerably less work on the potential impacts from climate change on high mountain ecosystems in the Himalaya, which is rich in bio- and cultural diversity (Bharati, Gurung, Maharjan *et al.*, 2014; Bharati, Gurung, Smakhtin *et al.*, 2014; Xu and Grumbine, 2014). The lack of research likely stems from the tremendous uncertainties surrounding the micro-hydrologies of this region, and the inability of even high resolution climate models to yield reliable predictions of change. Some scientists have warned that accelerated temperature increases in the Himalaya and resulting changes in hydrology (from perturbations to patterns of glacier and snow melt), though minor at basin scale, could trigger significant local ecosystem regime shifts or 'landscape traps' (Lindenmayer *et al.*, 2011; Xu and Grumbine, 2014). Higher temperatures may encourage land conversion for agricultural purposes (Tsarouchi *et al.*, 2014), while reduced baseflows could affect fish migration and compromise provision of other important hydrological services, leading to rural outmigration (Sapkota *et al.*, 2013; Aryal *et al.*, 2014). These trends will of course interact with other drivers of change – related to development and land conversion, urbanization and infrastructure development.

Vulnerability to climate change

The previous discussion largely focused on the physical changes that climate change will bring to the basin, and to a somewhat lesser degree on the implications of those changes for water users and for economic production. Yet the Ganges Basin is also an incredibly diverse and complex social system, and social factors will influence the extent to which riparians are able to cope with climate change impacts. As Ganges water makes its way across heavily populated areas in Nepal, India and Bangladesh, it is used to generate hydropower, irrigate vast agricultural expanses, satisfy the domestic, cultural and spiritual practices of millions of people, nourish ecosystems, and for the dilution and treatment of pollutants and the transport of sediments. Many upstream wetlands in the basin are considered sacred by nearby inhabitants, and the mainstem of the Ganga is famous for its significance to Hindus. The effects of threats to these sociocultural values and those associated with the ecosystem services the river provides (e.g. fish and timber resources), whether from climate change, land conversion, water pollution by industrial and domestic discharges, invasive species or growing demand, will largely depend on different populations' vulnerabilities or differential exposures to them.

In the Ganges Basin, vulnerability is intricately connected to social structures such as gender, class, caste and ethnicity (Sugden *et al.*, 2014), and to the urban–rural divide in access to institutions and infrastructure. The region is known for its great gender and birth order inequities (Jayachandran and Pande, 2013), and for inequality in labour market participation and outcomes as well as control over income, property and other resources (Agarwal, 1998). Given that rural, lower caste or other ethnically marginalized households more heavily depend on natural resource-based livelihood activities such as collection of food, water and fuelwood, these groups are more heavily exposed to the effects of disruptions in these systems due to climate change (Kabeer, 2002; Sultana, 2009). Within households, women are often disproportionately involved in water and fuel collection activities. In addition, higher income and social status individuals or households have access to protections – alternative water sources, access to modern fuels and energy sources, secure land tenure and the ability to pay for food – whereas more vulnerable groups may not (Regmi and Fawcett, 1999). Gender norms also imply that men can often work in a broader labour market, and are more able to migrate in response to new opportunities, or when their livelihoods are threatened (Ahmad, 2012). Seasonal out-migration provides benefits in the form of remittances and perhaps reduced dependence on precarious subsistence activities, but it can leave those who are left behind – often women and children or the elderly – with weaker support structures (Zahur, 2009; Sugden *et al.*, 2014).

Rural and lower class groups, and women, also often suffer more when extreme events or natural disasters occur. Research has shown that marginalized populations have a lower ability to migrate; have fewer resources to invest in self-protection; have less access to or understanding of information, warnings and safe havens; and are at greater risk of gender-based violence or human trafficking during disruptive events (Ahmad, 2012; Dhungel and Ojha, 2012). Traditions and norms may require women to stay home in spite of calls for evacuation, increasing their exposure during disasters (Dasgupta *et al.*, 2010). Vulnerable groups also have a weaker voice in regional and national-level policy making and lower representation in institutions that help mitigate the effects of climate change or enhance adaptation (Neumayer and Plümper, 2007). These inequities may impede their ability to bargain over and receive disaster relief.

Development and adaptation options under climate change

As discussed above, the challenges posed by climate change are numerous. This section briefly considers a set of climate adaptation responses that might be considered alongside development objectives in the basin. We consider four broad categories of such responses, all framed within the context of economic development (Ayers *et al.*, 2014): infrastructure investment and operations; enhanced conjunctive management of ground and surface water resources; improved water allocation and demand management; and adaptation responses targeted to reducing specific vulnerabilities (namely promotion of agricultural technology adoption and more effective flood and disaster management).

Opportunities for development of water control infrastructure

Investment in new water control infrastructure can potentially enhance development in a variety of ways and alleviate some negative impacts of climate change, by providing storage to ensure against hydrological variability and seasonal water shortages, to help moderate extreme flows, and to generate carbon neutral hydropower to meet growing energy needs. Using a river basin optimization approach, Jeuland *et al.*, (2013) conducted an analysis of the benefits of construction of storage dams in Nepal, where water storage is currently limited compared to the Yamuna and Upper Ganga in northern India where dam density is considerably higher (Wu *et al.*, 2013; Xu and Grumbine, 2014). They found that the hydropower and irrigation benefits from a suite of the 23 largest potential projects would be substantial, reaching US$6–9 billion/yr from power generation and US$1.7–2.0 billion/yr from low flow augmentation. Nonetheless, the benefits vary considerably across climate models (Figure 4.3). Projects in the Kosi sub-basin (in the eastern

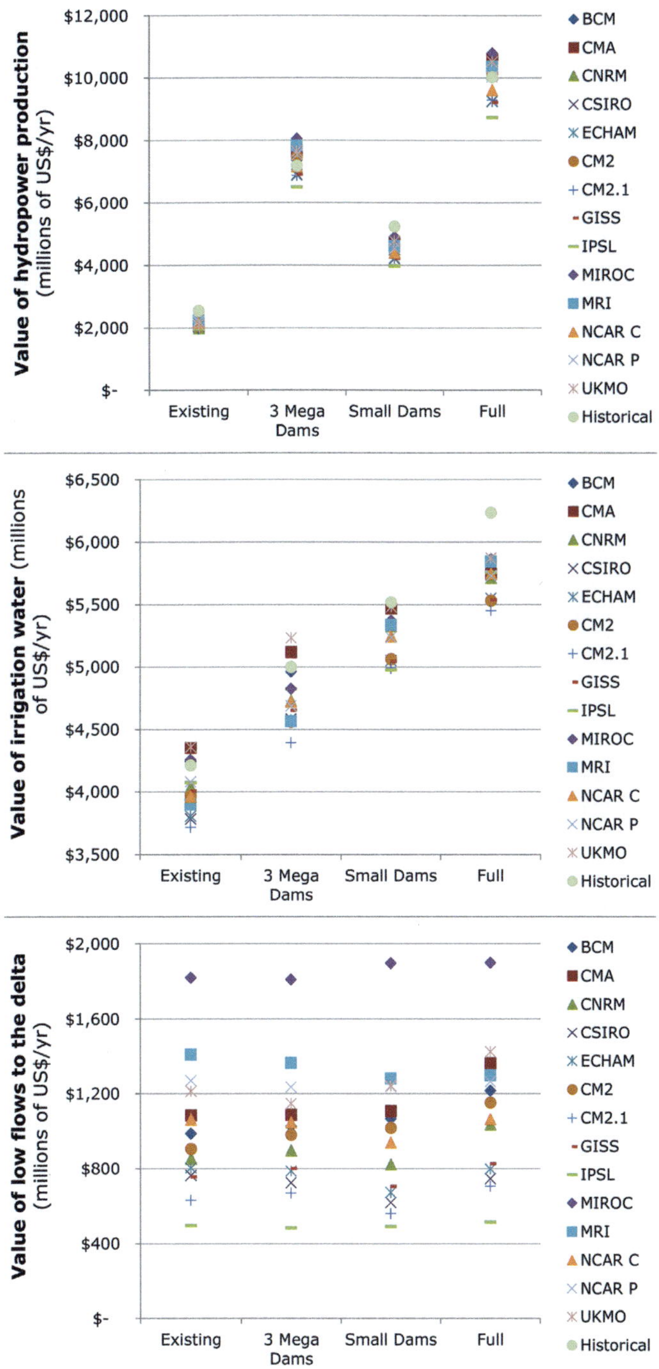

Figure 4.3 From Jeuland *et al*., (2013). Range of economic values (moderate value scenario) for optimal water allocations associated with: (top) hydropower production; (middle) irrigation abstractions; and (bottom) low flow allocations to Bangladesh, using the GCM projections for the A2 emissions scenario, for 2050. Unit values are: $0.1/kWhr (hydropower) and $0.05/m^3 (irrigation water and low flows)

Ganges) show a tendency across models towards increased hydropower under climate change due to increased flows. On the other hand, the energy production from projects in the western catchments, where flow decreases appear more likely, declines somewhat. Upstream storage projects generally help to enhance downstream water use in irrigation and low flows in Bangladesh by smoothing the temporal pattern of flows, but these benefits would likely be modest compared to those obtained from hydropower. Flood control benefits would meanwhile probably be low as a result of limited storage capacity (Sadoff *et al.*, 2013), and the range of projected changes in high flows across GCMs is much larger than the effects that the full set of these infrastructures has on them. Illustrative calculations nonetheless suggest that the suite of projects appears to deliver favourable economic returns (Jeuland *et al.*, 2013), though the analyses do not account for potentially high social and environmental costs (Xu and Grumbine, 2014), and therefore require additional study. In addition, managing water-use tradeoffs downstream of storage dams in a manner that would appropriately balance equity and efficiency would require strong transboundary institutions.

Besides hydropower dams or run-of-the-river facilities, the other water control infrastructures that receive a great deal of attention in the Ganges Basin are embankments. Thousands of kilometres of embankments have long protected low-lying and densely populated flood plains throughout the Gangetic Plain and Delta (Bandyopadhyay, 2009). Yet these remain controversial, for three main reasons. First, embankments disrupt natural processes of sediment deposition and lead to gradual heightening of embanked rivers over time and accelerated land subsidence in locations protected by them (Boyce, 1990; Gilvear, 1999). This increases the risks of catastrophic flooding from breaches (Mishra, 2008). Second, most severe flooding events in the basin are caused by local rainfall, high flows in small and flashy tributaries and embankment failures, and not by peak flows overtopping embankments (Brammer, 1990). Third, there is evidence that embankments increase risks for communities that are purportedly protected by them, because populations living behind their walls decrease their own self-protection and become less nimble for responding to extreme events (Singh *et al.*, 2009; Somanathan, 2013). The solutions to flood problems therefore have to lie in multi-faceted approaches (described further below) rather than hard infrastructure investments alone.

Irrigation: making better use of ground and surface water resources

Though the Ganges Basin is famous for its massive surface water irrigation schemes, especially in India, the reality is that groundwater is now tapped to meet an increasing proportion of farmers' water needs (Moors *et al.*, 2011). Irrigation with sub-surface water allows farmers to plant a dry season crop and provides water on demand in the event that the monsoon fails (Shah, 2008). Rates of groundwater recharge are moderate in the western parts of the basin but exceed 3 metres per annum throughout the eastern Ganges Plain (Zektser and Lorne, 2004). Use rates meanwhile exceed recharge in the most productive agricultural areas of the western Ganges, while remaining well below it in the vast zone stretching from eastern Uttar Pradesh to the delta (Rodell *et al.*, 2009). These imbalances contribute to extended periods of waterlogging of lands in the eastern parts of the basin, low productivity and increased flood risks (Cai *et al.*, 2010; Pandey *et al.*, 2010). The basic water balance that leads to waterlogging has not changed dramatically over the past half century, and seems unlikely to do so in the near to medium future, given the divergent predictions of future precipitation. Thus, the longstanding idea of engaging in more targeted recharge of aquifers where water tables are currently dropping (in the western basin) deserves consideration (Revelle and Lakshminarayana, 1975). Implementing this idea on a large scale would require better integration and strengthening of surface and groundwater management institutions. In addition, incentives could be better utilized to encourage groundwater abstractions in the eastern basin (where water is overly abundant but energy is lacking to pump it), while the cost of groundwater pumping should better reflect water scarcity in locations where this resource is being depleted (Sadoff *et al.*, 2013).

More effectively managing water-use tradeoffs

Wu *et al.*, (2013) showed that there are low tradeoffs – on the order of 5–10 per cent of power generation – between hydropower and consumptive water uses in the Ganges. This is because it is not economically feasible to supply water to upland locations facing scarcity (primarily located on or around flatter ridges) and because downstream locations primarily need water outside of the monsoon season. Storage dams help to alleviate this downstream scarcity by releasing stored water over the course of the year to meet energy demands. At the same time, dams hold back some peak flows and thus reduce flood risks when they are greatest. Nonetheless, storage potential is limited in the basin due to its extremely steep topography, which limits the amount of water that can be held back by even very tall dams.

The principal tradeoffs in the basin are therefore between farmers and populations located throughout the Gangetic Plain and Delta region, on the one hand, and local and downstream ecosystems (local and in the Sunderbans in the delta) as well as riparians located further downstream, on the other (Sapkota *et al.*, 2013; World Bank, 2013). In hydropolitical terms, these tradeoffs occur across state (within India) and national (between Nepal and India, or India and Bangladesh) boundaries (Crow *et al.*, 1995). They are likely to intensify with future population growth, economic development, warming temperatures and increasing demand for outflows into the Bay of Bengal that can help to offset sea level rise (Hijioka *et al.*, 2014; Whitehead, Sarkar *et al.*, 2015). In upstream reaches, farmers may be tempted to use abundant and cheap surface water inefficiently, for farming water intensive crops (Kaushal and Kansal, 2011). Downstream users currently only receive protection through treaties or agreements that curtail such upstream consumption. Basin stakeholders would benefit from development of institutions that allow greater flexibility and ability to allocate water to its most productive purposes, that foster diversification of economic activities in rural areas and thereby reduce vulnerability (Ellis, 2000), so long as basic needs are also prioritized. Indeed, a critical problem in the basin at this time is that the economic value of water in different locations and for different uses is largely unknown due to a dearth of research on this issue (World Bank, 2013).

Investing in targeted climate change adaptation

All of the strategies described above are both development-enhancing and potentially protective against the disruptions expected with climate change, but there is also a need for targeted adaptation. Effective strategies to forestall the costs of sea level rise and salinity intrusion downstream in the basin require maintenance of the flood pulse of the Ganges and a sustained focus on increasing the resilience of the build environment (Mallick and Rahman, 2013). Similarly, there are no simple solutions to the problem of flooding in the Ganges Plains, and a shift is needed from 'flood control' to 'flood management' that would entail a combination of structural and non-structural (institutional and social) interventions. Such an approach would place greater emphasis on regional forecasting and warning systems; work to enhance drainage and repair and strengthen embankments; and investing in 'soft' responses that would include disaster preparedness, land use zoning, provision of safe havens and flood insurance, and training and communications campaigns (World Bank, 2013).

In agriculture, higher temperatures and the possibility of less reliable monsoon precipitation should motivate experimentation with water-saving technologies and other means for increasing water productivity, e.g. demand management institutions (Cai *et al.*, 2010; Hijioka *et al.*, 2014). Successful experiences with such technologies and improvements would have more general positive impacts on food security and rural economic development. Finally, and more broadly, there is a need for policies that facilitate equitable climate change adaptation for both privileged and marginalized groups, and that take account of the particular vulnerability of the latter. Such policies should work to ensure women's rights to land and representation in the political system, allow for broader access to credit, and promote local and effective solutions that incorporate indigenous knowledge and facilitate collective action when it is needed (Sugden *et al.*, 2014).

Conclusions

This chapter asserts that climate change will bring complex new challenges to those living in the Ganges Basin. We reviewed a number of potentially disruptive changes, including temperature increases, uncertain and spatially variable disruptions to historical precipitation, and resulting changes in the basin's water cycle, which encompass perturbations in runoff, snow and glacier melt, and the effects of sea level rise. Such changes will all occur in the context of growing water demand, perhaps exacerbated by temperature change. These will have complex implications for energy production (from hydropower), irrigation productivity, water stress, storm surge vulnerability and livelihoods.

Nonetheless, it is noteworthy that the range of potential changes predicted by existing climate models are not outside of the experience of natural variability in the basin. Thus, a focus on development that builds resilience to stresses and helps to better manage current hydrological variability (whether attributable to climate change or not) is probably a good place to start when considering adaptation to the potential future climate change challenges of the Ganges. In this respect, a range of soft and hard infrastructure strategies – investment in upstream storage infrastructure, enhanced conjunctive use of ground and surface water, more efficient management of water tradeoffs, and a set of specific adaptation measures – that revolve around increasing the economic productivity of water and protecting against costly damages that are already incurred during extreme hydrological events, would be helpful. At present, the marginal productivity of water in the basin appears to be very low; investments that alters this situation of low development seems critical to the future well-being of the riparians of the basin.

References

Agarwal, B. (1998). Environmental management, equity and ecofeminism: Debating India's experience. *Journal of Peasant Studies* 25(4): 55–95.

Agrawala, S., Ota, T., Ahmed, A.U., Smith, J. and Aalst, M.V. (2003). *Development and Climate Change in Bangladesh: Focus on Coastal Flooding and the Sundarbans.* Paris: Organisation for Economic Co-operation and Development.

Ahmad, N. (2012). Gender and climate change in Bangladesh: The role of institutions in reducing gender gaps in adaptation program. *Social Development Paper.* Washington, DC: The World Bank: 126.

Ahmed, A.U. and Alam, M. (1999). 'Development of climate change scenarios with general circulation models'. In S. Huq, Z. Karim, M. Asaduzzaman and F. Mahtab (eds). *Vulnerability and Adaptation to Climate Change for Bangladesh.* Dordrecht: Kluwer Academic Publishers.

Alford, D. and Armstrong, R. (2010). The role of glaciers in stream flow from the Nepal Himalaya. *The Cryosphere Discuss* 4: 469–94.

Ali, A. (1999). Climate change impacts and adaptation assessment in Bangladesh. *Climate Research* 12(2–3): 109–16.

Allison, M. (1998). Historical changes in the Ganges-Brahmaputra delta front. *Journal of Coastal Research* 14(4): 1269–1275.

Alongi, D.M. (2015). The impact of climate change on mangrove forests. *Current Climate Change Reports* 1(1): 30–9.

Andermann, C., Longuevergne, L., Bonnet, S., Crave, A., Davy, P. and Gloaguen, R. (2012). Impact of transient groundwater storage on the discharge of Himalayan rivers. *Nature Geoscience* 5(2): 127–32.

Aryal, A., Brunton, D. and Raubenheimer, D. (2014). Impact of climate change on human–wildlife–ecosystem interactions in the Trans-Himalaya region of Nepal. *Theoretical and Applied Climatology* 115(3–4): 517–29.

Auerbach, L., Goodbred Jr, S., Mondal, D., Wilson, C., Ahmed, K., Roy, K., Steckler, M., Small, C., Gilligan, J. and Ackerly, B. (2015). Flood risk of natural and embanked landscapes on the Ganges-Brahmaputra tidal delta plain. *Nature Climate Change*.

Ayers, J.M., Huq, S., Faisal, A.M. and Hussain, S.T. (2014). Mainstreaming climate change adaptation into development: a case study of Bangladesh. *Wiley Interdisciplinary Reviews: Climate Change* 5(1): 37–51.

Bandyopadhyay, J. (2009). *Water, Ecosystems and Society: A Confluence of Disciplines*. Delhi: Sage Publications.

Bharati, L., Gurung, P., Maharjan, L. and Bhattarai, U. (2014). Past and future variability in the hydrological regime of the Koshi basin, Nepal. *Hydrological Sciences Journal*.

Bharati, L., Gurung, P., Smakhtin, V. and Jayakody, P. (2014). The Impact of Climate Change on Water Resources Development in the Koshi Basin, Nepal. *Mountain Research and Development* 34(2): 118–30.

Bhutiyani, M., Kale, V. and Pawar, N. (2010). Climate change and the precipitation variations in the northwestern Himalaya: 1866–2006. *International Journal of Climatology* 30(4): 535–48.

Boyce, J.K. (1990). Birth of a megaproject: Political economy of flood control in Bangladesh. *Environmental Management* 14(4): 419–28.

Brammer, H. (1990). Floods in Bangladesh: Geographical background to the 1987 and 1988 floods. *Geographical Journal*: 12–22.

Brown, M., Racoviteanu, A., Tarboton, D., Gupta, A.S., Nigro, J., Policelli, F., Habib, S., Tokay, M., Shrestha, M. and Bajracharya, S. (2014). An integrated modeling system for estimating glacier and snow melt driven streamflow from remote sensing and earth system data products in the Himalayas. *Journal of Hydrology* 519: 1859–1869.

Cai, X., Sharma, B.R., Matin, M.A., Sharma, D. and Gunasinghe, S. (2010). *An Assessment of Crop Water Productivity in the Indus and Ganges River Basins: Current Status and Scope for Improvement*. IWMI.

Collins, M., Knutti, R., Arblaster, J., Dufresne, J.-L., Fichefet, T., Friedlingstein, P., Gao, X., Gutowski, W.J., Johns, T., Krinner, G., Shongwe, M., Tebaldi, C., Weaver, A.J. and Wehner, M. (2013). 'Long-term climate change: Projections, commitments and irreversibility'. In T. F. Stocker, D. Qin, G.-K. Plattner, M. Tignor, S.K. Allen, J. Boschung, A. Nauels, Y. Xia, V. Bex and P.M. Midgley. *Climate Change 2013: The Physical Science Basis*. Contribution of Working Group I to the Fifth Assessment Report of the Intergovernmental Panel on Climate Change. Cambridge and New York: Cambridge University Press.

Crow, B., Lindquist, A. and Wilson, D. (1995). *Sharing the Ganges: The Politics and Technology of River Development*. New Delhi and Thousand Oaks, CA: Sage.

Dasgupta, S., Huq, M., Khan, Z.H., Ahmed, M.M.Z., Mukherjee, N., Khan, M.F. and Pandey, K.D. (2010). Vulnerability of Bangladesh to cyclones in a changing climate: Potential damages and adaptation cost. *World Bank Policy Research Working Paper Series*.

Dasgupta, S., Kamal, F.A., Khan, Z.H., Choudhury, S. and Nishat, A. (2014). River salinity and climate change: evidence from coastal Bangladesh. *World Bank Policy Research Working Paper* (6817).

Dhungel, R. and Ojha, R.N. (2012). Women's empowerment for disaster risk reduction and emergency response in Nepal. *Gender & Development* 20(2): 309–21.

Ellis, F. (2000). The determinants of rural livelihood diversification in developing countries. *Journal of Agricultural Economics* 51(2): 289–302.

Gain, A.K., Immerzeel, W.W., Sperna-Weiland, F.C. and Bierkens, M.F.P. (2011). Impact of climate change on the stream flow of the lower Brahmaputra: Trends in high and low flows based on discharge-weighted ensemble modelling. *Hydrology and Earth System Sciences Discussions* 8: 365–90.

Gilvear, D.J. (1999). Fluvial geomorphology and river engineering: Future roles utilizing a fluvial hydrosystems framework. *Geomorphology* 31(1): 229–45.

Guha-Sapir, D., Below, R. and Hoyois, P. (2015). *EM-DAT: International Disaster Database* – www.emdat.be. U.C.d. Louvain. Brussels, Belgium.

Gupta, P.K., Panigrahy, S. and Parihar, J.S. (2010). Modeling the spatiotemporal changes in the runoff of major basins in India under the future climate change scenario. *Bulletin of the National Natural Resources Management System*: 57–62.

Hamner, S., Tripathi, A., Mishra, R.K., Bouskill, N., Broadaway, S.C., Pyle, B.H. and Ford, T.E. (2006). The role of water-use patterns and sewage pollution in incidence of water-borne/enteric diseases along the Ganges river in Varanasi, India. *International Journal of Environmental Health Research* 16(2): 113–32.

Hassan, A., Wahid, S., Shrestha, M.L., Rashid, M.A., Ahmed, T., Mazumder, A., Sarker, M.H., Al Hossain, B.M.T., Mumu, S. and Sarker, M.H. (2014). Climate change and water availability in the Ganges-

Brahmaputra-Meghna basin: Impact on local crop production and policy directives. *Research Insights on Climate and Water in the Hindu Kush Himalayas*: 97.

Hijioka, Y., Lin, E., Pereira, J.J., Corlett, R.T., Cui, X., Insarov, G.E., Lasco, R.D., Lindgren, E. and Surjan, A. (2014). 'Asia'. In V. R. Barros, C.B. Field, D.J. Dokken, M.D. Mastrandrea, K.J. Mach, T.E. Bilir, M. Chatterjee, K.L. Ebi, Y.O. Estrada, R.C. Genova, B. Girma, E.S. Kissel, A.N. Levy, S. MacCracken, P.R. Mastrandrea and L.L. White. *Climate Change 2014: Impacts, Adaptation, and Vulnerability. Part B: Regional Aspects. Contribution of Working Group II to the Fifth Assessment Report of the Intergovernmental Panel on Climate Change*. Cambridge and New York: Cambridge University Press: 1327–1370.

Hossain, M.S., Hein, L., Rip, F.I. and Dearing, J.A. (2013). Integrating ecosystem services and climate change responses in coastal wetlands development plans for Bangladesh. *Mitigation and Adaptation Strategies for Global Change*: 1–21.

Hosterman, H.R., McCornick, P.G., Kistin, E.J., Sharma, B. and Bharati, L. (2012). Freshwater, climate change and adaptation in the Ganges River Basin. *Water Policy* 14(1): 67.

Immerzeel, W. (2008). Historical trends and future predictions of climate variability in the Brahmaputra basin. *International Journal of Climatology* 28(2): 243–54.

Immerzeel, W.W., Van Beek, L.P.H. and Bierkens, M.F.P. (2010). Climate change will affect the Asian water towers. *Science* 328(5984): 1382.

Immerzeel, W., Pellicciotti, F. and Bierkens, M. (2013). Rising river flows throughout the twenty-first century in two Himalayan glacierized watersheds. *Nature Geoscience* 6(9): 742–45.

IPCC (2007). *Climate Change 2007: The Physical Basis of Climate Change: Contribution of Working Group I to the Fourth Assessment Report of the Intergovernmental Panel on Climate Change*. Cambridge: Cambridge University Press.

IPCC (2013). *Climate change 2013: The physical science basis. Contribution of Working Group I to the Fifth Assessment Report of the Intergovernmental Panel on Climate Change*. Cambridge and New York: Cambridge University Press.

Jain, V. and Sinha, R. (2003). Geomorphological manifestations of the flood hazard: A remote sensing based approach. *Geocarto International* 18(4): 51–60.

Jayachandran, S. and Pande, R. (2013). *Why Are Indian Children Shorter Than African Children?* Chicago, Il: Northwestern University.

Jeuland, M. (2010). Economic implications of climate change for infrastructure planning in transboundary water systems: An example from the Blue Nile. *Water Resources Research* 46(11).

Jeuland, M., Harshadeep, N., Escurra, J., Blackmore, D. and Sadoff, C. (2013). Implications of climate change for water resources development in the Ganges basin. *Water Policy* 15: 26–50.

Jiménez Cisneros, B.E., Oki, T., Arnell, N.W., Benito, G., Cogley, J.G., Döll, P., Jiang, T. and Mwakalila, S.S. (2014). 'Freshwater Resources'. In C.B. Field, V.R. Barros, D.J. Dokken, K.J. Mach, M.D. Mastrandrea, T.E. Bilir, M. Chatterjee, K.L. Ebi, Y.O. Estrada, R.C. Genova, B. Girma, E.S. Kissel, A.N. Levy, S. MacCracken, P.R. Mastrandrea, and L.L. White. *Climate Change 2014: Impacts, Adaptation, and Vulnerability. Part A: Global and Sectoral Aspects. Contribution of Working Group II to the Fifth Assessment Report of the Intergovernmental Panel on Climate Change*. Cambridge and New York: Cambridge University Press.

Kabeer, N. (2002). Safety nets and opportunity ladders: Addressing vulnerability and enhancing productivity in South Asia. *Development Policy Review* 20(5): 589–614.

Kaushal, N. and Kansal, M.L. (2011). Overview of water allocation practices in Uttar Pradesh and Uttarakhand with a specific reference to future demands. *South Asia Water Studies Journal* 2(2): 28–43.

Kirtman, B., Power, S.B., Adedoyin, J.A., Boer, G.J., Bojariu, R., Camilloni, I., Doblas-Reyes, F.J., Fiore, A.M., Kimoto, M., Meehl, G.A., Prather, M., Sarr, A., Schär, C., Sutton, R., van Oldenborgh, G.J., Vecchi, G. and Wang, H.J. (2013). 'Near-term climate change: Projections and predictability'. In T.F. Stocker, D. Qin, G.-K. Plattner, M. Tignor, S.K. Allen, J. Boschung, A. Nauels, Y. Xia, V. Bex and P.M. Midgley. *Climate Change 2013: The Physical Science Basis. Contribution of Working Group I to the Fifth Assessment Report of the Intergovernmental Panel on Climate Change*. Cambridge and New York: Cambridge University Press.

Kulkarni, A., Patwardhan, S., Kumar, K.K., Ashok, K. and Krishnanm, R. (2013). Projected climate change in the Hindu Kush-Himalayan region by using the high-resolution regional climate model PRECIS. *Mountain Research and Development* 33(2): 142–51.

Lindenmayer, D.B., Hobbs, R.J., Likens, G.E., Krebs, C.J. and Banks, S.C. (2011). Newly discovered landscape traps produce regime shifts in wet forests. *Proceedings of the National Academy of Sciences* 108(38): 15887–15891.

Lobell, D.B., Burke, M.B., Tebaldi, C., Mastrandrea, M.D., Falcon, W.P. and Naylor, R.L. (2008). Prioritizing climate change adaptation needs for food security in 2030. *Science* 319(5863): 607–10.

McIvor, A., Spencer, T., Möller, I. and Spalding, M. (2013). The response of mangrove soil surface elevation to sea level rise. *Natural Coastal Protection Series: Report 3. Cambridge Coastal Research Unit Working Paper 42*, The Nature Conservancy and Wetlands International.

Mallick, F. and Rahman, A. (2013). Cyclone and tornado risk and reduction approaches in Bangladesh. *Disaster Risk Reduction Approaches in Bangladesh*, Springer: 91–102.

Masood, M., Yeh, P.-F., Hanasaki, N. and Takeuchi, K. (2015). Model study of the impacts of future climate change on the hydrology of Ganges–Brahmaputra–Meghna basin. *Hydrology and Earth System Sciences* 19(2): 747–70.

Maurer, E.P. and Hidalgo, H.G. (2008). Utility of daily vs. monthly large-scale climate data: An intercomparison of two statistical downscaling methods. *Hydrology and Earth System Sciences* 12: 551–63.

Meehl, G.A., Covey, C., Delworth, T., Latif, M., McAvaney, B., Mitchell, J.F.B., Stouffer, R.J. and Taylor, K.E. (2007). The WCRP CMIP3 multi-model dataset: A new era in climate change research. *Bulletin of the American Meteorological Society* 88(9): 1383.

Miller, J. D., Immerzeel, W.W. and Rees, G. (2012). Climate change impacts on glacier hydrology and river discharge in the Hindu Kush-Himalayas: A synthesis of the scientific basis. *Mountain Research and Development* 32(4): 461–67.

Mirza, M.M.Q. (2011). Climate change, flooding in South Asia and implications. *Regional Environmental Change* 11(1): 95–107.

Mirza, M.M.Q., Warrick, R.A and Ericksen, N.J. (2003). The implications of climate change on floods of the Ganges, Brahmaputra and Meghna rivers in Bangladesh. *Climatic Change* 57(3): 287–318.

Mishra, D. (2008). *Trapped Between the Devil and Deep Waters: The story of Bihar's Kosi River.* New Delhi, India, Peoples' Science Institute and South Asia Network on Dams, Rivers and People (SANDRP): 208.

Mittal, N., Mishra, A., Singh, R. and Kumar, P. (2014). Assessing future changes in seasonal climatic extremes in the Ganges river basin using an ensemble of regional climate models. *Climatic change* 123(2): 273–86.

Moench, M. and Dixit, A. (2007). *Adaptive Capacity and Livelihood Resilience: Adaptive Strategies for Responding to Floods and Droughts in South Asia.* Kathmandu, Institute for Social and Environmental Transition.

Moors, E. J., Groot, A., Biemans, H., van Scheltinga, C.T., Siderius, C., Stoffel, M., Huggel, C., Wiltshire, A., Mathison, C. and Ridley, J. (2011). Adaptation to changing water resources in the Ganges basin, northern India. *Environmental Science & Policy* 14(7): 758–69.

Neumayer, E. and Plümper, T. (2007). The gendered nature of natural disasters: The impact of catastrophic events on the gender gap in life expectancy, 1981–2002. *Annals of the Association of American Geographers* 97(3): 551–66.

Nicholls, R.J. and Goodbred, S. (2004). *Towards Integrated Assessment of the Ganges–Brahmaputra Delta.* Paper for Proc. 5th International Conference on Asian Marine Geology and 1st Annual Meeting of IGCP 475 Delta MAP and APN Mega-Deltas.

Pandey, A., Singh, S.K. and Nathawat, M. (2010). Waterlogging and flood hazards vulnerability and risk assessment in Indo Gangetic plain. *Natural Hazards* 55(2): 273–89.

Porter, J.R., Xie, L., Challinor, A.J., Cochrane, K., Howden, S.M., Iqbal, M.M., Lobell, D.B. and Travasso, M.I. (2014). 'Food security and food production systems'. In V.R. Barros, C.B. Field, D.J. Dokken, M.D. Mastrandrea, K.J. Mach, T.E. Bilir, M. Chatterjee, K.L. Ebi, Y.O. Estrada, R.C. Genova, B. Girma, E.S. Kissel, A.N. Levy, S. MacCracken, P.R. Mastrandrea, and L.L. White. *Climate Change 2014: Impacts, Adaptation, and Vulnerability. Part B: Regional Aspects. Contribution of Working Group II to the Fifth Assessment Report of the Intergovernmental Panel on Climate Change.* Cambridge and New York: Cambridge University Press: 485–533.

Raha, A., Das, S., Banerjee, K. and Mitra, A. (2012). Climate change impacts on Indian Sunderbans: A time series analysis (1924–2008). *Biodiversity and Conservation* 21(5): 1289–1307.

Rashid, T. (2014). *Holocene Sea-Level Scenarios in Bangladesh.* Singapore: Springer.

Rees, H.G. and Collins, D.N. (2006). Regional differences in response of flow in glacier fed Himalayan rivers to climatic warming. *Hydrological Processes* 20(10): 2157–2169.

Regmi, S.C. and Fawcett, B. (1999). Integrating gender needs into drinking-water projects in Nepal. *Gender & Development* 7(3): 62–72.

Revelle, R. and Lakshminarayana, V. (1975). The Ganges water machine. *Science* 188(4188): 611–16.

Rikiishi, K. and H. Nakasato, H. (2006). Height dependence of the tendency for reduction in seasonal snow cover in the Himalaya and the Tibetan Plateau region, 1966–2001. *Annals of Glaciology* 43(1): 369–77.

Rodell, M., Velicogna, I. and Famiglietti, J.S. (2009). Satellite-based estimates of groundwater depletion in India. *Nature* 460(7258): 999–1002.

Sadoff, C., Rao, H.N., Blackmore, D., Wu, X., Jeuland, M. and Lee, S. (2013). Ten fundamental questions for water resources development in the Ganges: Myths and realities. *Water Policy* 15: 147–64.

Salman, S.M. and Uprety, K. (2002). *Conflict and cooperation on South Asia's international rivers: A legal perspective.* World Bank Publications.

Sapkota, P., Bharati, L., Gurung, P., Kaushal, N. and Smakhtin, V. (2013). Environmentally sustainable management of water demands under changing climate conditions in the Upper Ganges Basin, India. *Hydrological Processes* 27(15): 2197–2208.

Shah, T. (2008). *Taming the Anarchy: Groundwater Governance in South Asia.* Washington, DC: Earthscan.

Shrestha, M.L. and Shrestha, A.B. (2004). *Recent Trends and Potential Climate Change Impacts on Glacier Retreat/Glacier Lakes in Nepal and Potential Adaptation.* Presentation by ML Shrestha, Director of the Department of Hydrology and Meteorology of Nepal, at the OECD Global Forum on Sustainable Development, Paris.

Siderius, C., Biemans, H., Wiltshire, A., Rao, S., Franssen, W., Kumar, P., Gosain, A., van Vliet, M. and Collins, D. (2013). Snowmelt contributions to discharge of the Ganges. *Science of the Total Environment* 468: S93–S101.

Singh, M. and Singh, A.K. (2007). Bibliography of environmental studies in natural characteristics and anthropogenic influences on the Ganga River. *Environmental Monitoring and Assessment* 129(1–3): 421–32.

Singh, P., Ghose, N., Chaudhary, N. and Hansda, R. (2009). *Life in the Shadow of Embankments: Turning Lost Lands into Assets in the Koshi Basin of Bihar, India.* Kathmandu, Nepal: ICIMOD.

Somanathan, E. (2013). Are embankments a good flood control strategy? A case study on the Kosi river. *Water Policy* 15: 75–88.

Sugden, F., d. Silva, S., Clement, F., Maskey-Amatya, N., Ramesh, V., Philip, A. and Bharati, L. (2014). A framework to understand gender and structural vulnerability to climate change in the Ganges River Basin: Lessons from Bangladesh, India and Nepal. *IWMI Working Paper* (159).

Sultana, F. (2009). Fluid lives: Subjectivities, gender and water in rural Bangladesh. *Gender, Place and Culture* 16(4): 427–44.

Trenberth, K.E., Dai, A., Rasmussen, R.M. and Parsons, D.B. (2003). The changing character of precipitation. *Bulletin of the American Meteorological Society* 84(9): 1205–1217.

Trenberth, K., Jones, P., Ambenje, P., Bojariu, R., Easterling, D., Klein Tank, A., Parker, D., Rahimzadeh, F., enwick, J., Rusticucci, M.S., Zhai, P. (2007). 'Observations: Surface and atmospheric climate change'. In S. Solomon, D. Qin, M. Manning, Z. Chen, M. Marquis, K.B. Avery, M. Tignor and H.L. Miller. *Climate Change 2007: The Physical Science Basis. Contribution of Working Group I to the Fourth Assessment Report of the Intergovernmental Panel on Climate Change.* Cambridge: Cambridge University Press.

Tsarouchi, G.-M. and Buytaert, W. (2014). *Comparing Land Use Change and Climate Change as Drivers of Hydrological Change in the Upper Ganges Basin, India.* EGU General Assembly Conference Abstracts.

Tsarouchi, G., Mijic, A., Moulds, S. and Buytaert, W. (2014). Historical and future land-cover changes in the Upper Ganges basin of India. *International Journal of Remote Sensing* 35(9): 3150–3176.

Whitehead, P., Barbour, E., Futter, M., Sarkar, S., Rodda, H., Caesar, J., Butterfield, D., Jin, L., Sinha, R. and Nicholls, R. (2015). Impacts of climate change and socio-economic scenarios on flow and water quality of the Ganges, Brahmaputra and Meghna (GBM) river systems: Low flow and flood statistics. *Environmental Science: Processes & Impacts.*

Whitehead, P., Sarkar, S., Jin, L., Futter, M., Caesar, J., Barbour, E., Butterfield, D., Sinha, R., Nicholls, R. and Hutton, C. (2015). Dynamic modeling of the Ganga river system: Impacts of future climate and socio-economic change on flows and nitrogen fluxes in India and Bangladesh. *Environmental Science: Processes & Impacts.*

Wirsing, R.G. and Jasparro, C. (2007). River rivalry: Water disputes, resource insecurity and diplomatic deadlock in South Asia. *Water Policy* 9(3): 231–51.

Wong, P.P., Losada, I.J., Gattuso, J.-P., Hinkel J., Khattabi, A., McInnes, K.L., Saito, Y. and Sallenger, A. (2014). 'Coastal systems and low-lying areas'. In V.R. Barros, C.B. Field, D.J. Dokken, M.D. Mastrandrea, K.J. Mach, T.E. Bilir, M. Chatterjee, K.L. Ebi, Y.O. Estrada, R.C. Genova, B. Girma, E.S. Kissel, A.N. Levy, S. MacCracken, P.R. Mastrandrea, and L.L. White. *Climate Change 2014: Impacts, Adaptation, and Vulnerability. Part B: Regional Aspects. Contribution of Working Group II to the Fifth Assessment Report of the Intergovernmental Panel on Climate Change.* Cambridge and New York: Cambridge University Press: pp. 361–409.

World Bank (2013). *Ganges Strategic Basin Assessment: A Regional Perspective on Opportunities and Risks.* Washington, DC: The World Bank.

Wu, X., Jeuland, M., Sadoff, C. and Whittington, D. (2013). Interdependence in water resource development in the Ganges: an economic analysis. *Water Policy* 15: 13–19.

Xu, J. and Grumbine, R.E. (2014). Building ecosystem resilience for climate change adaptation in the Asian highlands. *Wiley Interdisciplinary Reviews: Climate Change* 5(6): 709–18.

Yu, W., Alam, M., Hassan, A., Khan, A.S., Ruane, A., Rosenzweig, C., Major, D. and Thurlow, J. (2010). *Climate Change Risks and Food Security in Bangladesh.* Washington, DC: Earthscan.

Zahur, M. (2009). Climate Change, Migration and Gender: Reflection from Balochistan, Pakistan drought 1998–2002. Bonn, Gender CC/Gender Disaster Network. Available at: www.gendercc.net/fileadmin/inhalte/Dokumente/Actions/final-maira-original.pdf.

Zektser, I.S. and Lorne, E. (2004). Groundwater resources of the world and their use. *IhP Series on Groundwater,* UNESCO.

5

Managing variability

Floods and droughts

Giriraj Amarnath, A.K.M. Saiful Islam and Mandira S. Shrestha

Key messages

- The Ganges Basin countries routinely feature on lists of the most climate-vulnerable nations in the world. Increases in temperature, changes in precipitation and glacier melt, intensified monsoons, water-induced disasters, sea level rise – all the impacts of climate change – are expected to upset the delicate hydrological balance existing in the basin, giving rise to new water management challenges.
- Approximately 178 million people are exposed to floods and droughts. Floods affected 93 million people and droughts affected 85 million between 2000 and 2014. An estimated total of 1.7 million km^2 are regularly affected by floods and droughts in the Ganges Basin.
- Floods and droughts in the Ganges Basin are not yet evaluated systematically. Information on the scale of flooding and associated economic loss is also sparse. It is an urgent challenge to undertake a comprehensive analysis of historical floods and drought patterns, and to establish systematic monitoring of current extreme events. This will allow for better understanding of risk areas requiring investment and the development of adaptation strategies.

Introduction

Floods and droughts are major natural disasters that affect many regions around the world year after year, causing loss of life, damaging economies and affecting human health. More than one third of the world's land area is flood prone, affecting about 82 per cent of the world's population (Dilley *et al.*, 2005). Globally, between 1990 and 2015 about 2,375 floods and 393 droughts occurred, with an estimated affected population of 3.6 billion and economic losses of US$1.7 trillion (EM-DAT, 2015). In South Asia, the distribution of floods and droughts accounted for 516 extreme events, with an estimated economic loss of US$88 billion and causing 1.2 billion deaths. Over the same period in the Ganges Basin, (covering parts of China, Nepal, India and Bangladesh) floods and droughts affected 1 billion people and caused economic losses of US$58 billion (EM-DAT, 2015).

The Ganges Basin, covering an area of 1.2 million km^2, is the most populous in the world, home to more than 650 million people living in China, Nepal, India and Bangladesh. Poverty is widespread, and population density, particularly in the plains and delta, is extremely high (World Bank, 2013). People living in the basin have long coped with the extreme temporal climatic variability that manifests itself in floods and droughts. Nonetheless, they remain highly exposed to these extremes (World Bank, 2013). Both floods and droughts take a significant toll on lives and livelihoods, damage infrastructure, homes and other assets, and generally limit investment and disrupt economic activities (Amarnath *et al.*, 2012; Dasgupta *et al.*, 2011; Jain and Sinha, 2003; Smakhtin *et al.*, 2014).

Each year in Bangladesh about 26,000 km^2 (around 18 per cent of the country) is flooded, killing over 5,000 people and destroying more than seven million homes. During severe floods the affected areas may exceed 75 per cent of the country, as was seen in 1988. The floods have caused devastation in Bangladesh throughout history, especially during the years 1966, 1987, 1988 and 2007 (Hossain, 2015). In the Indian state of Bihar, people have been affected by floods every year since 1979, with over 20 million people affected in each of years 1987, 2004 and 2007. Flood-related deaths occurred in all but one year (World Bank, 2013). In Nepal, floods account for 90 per cent of the economic cost of natural disasters (Nepal Ministry of Home Affairs, 2010). Even larger areas of the basin are affected by drought, which hits vulnerable populations with limited or poor access to irrigation (Moench and Dixit, 2007; Jeuland *et al.*, 2013; World Bank, 2013).

The annual flooding in the Ganges is, at the same time, a characteristic feature of the local ecosystem and a determinant of human activities. For generations, local residents adapted rice varieties and cropping systems to seasonal water cycles influenced by the monsoon. The challenge in the Ganges, as in many other world regions with high temporal water resource variability, is how to reduce the negative impacts of floods and droughts while ensuring healthy ecosystems.

Historical characteristics of floods and droughts

Variability of Indian summer monsoon

Figure 5.1 shows the time-series evolution of AISMR (all-India summer monsoon rainfall) anomalies, expressed as per cent departures from its long-term mean, over more than a century. The index was developed by the Indian Institute of Tropical Meteorology (IITM), based on homogenous rainfall data sets of 306 rain gauges, to characterize variability in the country. It is interesting to note that there have been alternating periods extending to three to four decades

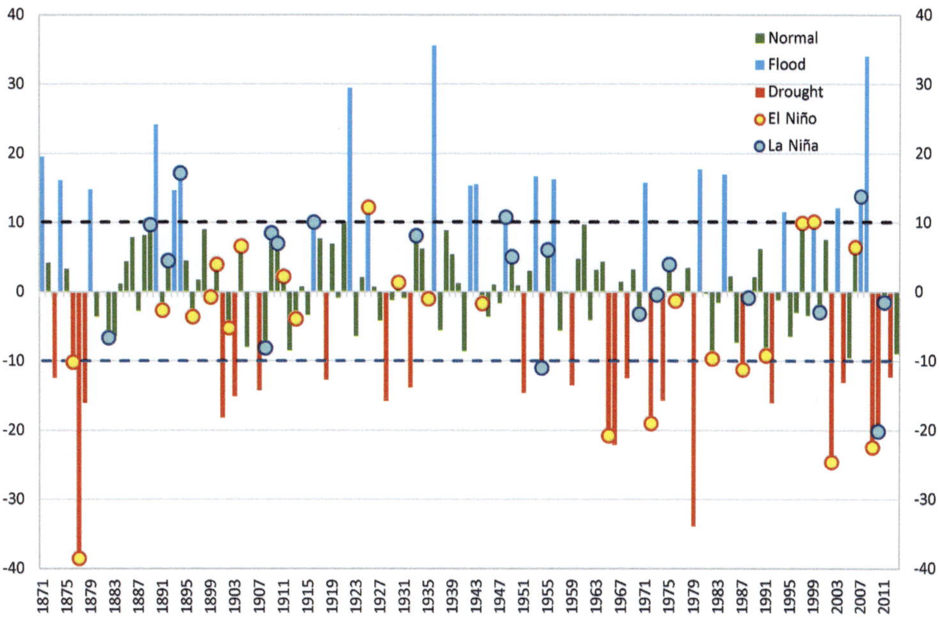

Figure 5.1 Variation of all-India summer monsoon rainfall anomalies during 1871–2012 and its association with El Niño/La Niña occurrences. The horizontal dashed lines delineate anomalies crossing the standard deviation

Source: Indian Institute of Tropical Meteorology

of more frequent floods and droughts and of higher rainfall. For example, the 44-year period between 1921 and 1964 witnessed just three drought years. During such periods, the monsoon was found to be less correlated with the El Niño Southern Oscillation (commonly called ENSO). During other periods like that of 1965–1987, which had as many as 10 drought years out of 23, the monsoon was found to be strongly linked to the ENSO (Parthasarathy *et al.*, 1991; Pervez and Henebry, 2015).

Occurrences of El Niño alone significantly reduce (88 per cent of the long-term average LTA) precipitation during the monsoon months in the western and south-eastern Ganges Basin. In contrast, occurrences of La Niña and co-occurrences of La Niña and negative IOD (Indian Ocean Dipole) events significantly enhance (110 per cent of LTA) precipitation. The above pattern implies that a regular water deficit is likely, especially in the Ganges Basin, with implications for the agriculture sector due to its reliance on consistent rainfall for successful production (Pervez and Henebry, 2015). Historically, major droughts occurred during El Niño and co-occurrences of El Niño and positive IOD events, while major flooding occurred during La Niña and co-occurrences of La Niña and negative IOD events in the basins. This observational analysis shall inform decisions that help minimize natural hazard risks and climate impacts on agriculture and water resources.

Flood characteristics

EM-DAT and Dartmouth Flood Observatory (DFO, 2013) are the two major global inventories used for quantifying the spatial and temporal distribution of floods and droughts in the Ganges

Table 5.1 Historical floods and droughts, occurrences between 1980 and 2015 reported by EM-DAT (2015). Since no information on events is reported at basin scale, riparian Ganges countries are used instead

Country	*Floods*				*Droughts*			
	Count	*Deaths*	*Total affected*	*Total damage ('000 $)*	*Count*	*Deaths*	*Total affected*	*Total damage ('000 $)*
Nepal	40	5,952	3,736,721	405,419	2	0	503,000	–
India	212	48,126	708,620,832	52,061,629	7	320	751,175,000	2,041,122
Bangladesh	74	12,669	250,438,897	11,388,800	3	0	25,000,000	–
Total	326	66,747	962,796,450	63,855,848	12	320	776,678,000	2,041,122

Basin. Extreme flood events totalled 326 (Table 5.1) of which India (212) ranks highest, followed by Bangladesh (74) and Nepal (40). There is little or no knowledge regarding Chinese floods, as the area is dominated by snow and glacier in the upstream. Analysing annual flood occurrences demonstrates the increase in flood numbers between 2006 and 2011 (53), compared to previous years (1985–2005) (Figure 5.2). Jonkman (2005) and Bradshaw *et al.* (2007) report on the increase in the global occurrence of floods. These are partially due to an increase in frequency of intense rainfall, known as cloud burst (IPCC, 2007; Verhaar *et al.*, 2010). This may result in an increase in various types of rainfall-related disasters like flash floods, landslides and landslide dammed outburst floods.

Densely populated countries in the Ganges Basin (i.e. India, Bangladesh) sustain repeated annual damage from floods, which is strongly linked to the onset of the monsoon. A notable observation is the disproportionate number of affected people between the two disaster types (Table 5.1). Occurrences of droughts are relatively low, but the affected population numbers are very similar to those for floods. Also, for droughts, economic losses are not well documented.

Monsoonal rain is the most frequently reported cause of flooding in the basin (44.6 per cent); heavy rain is the second most reported cause (31.2 per cent) followed by brief torrential rain (8.3 per cent) and tropical cyclones (4.5 per cent; Figure 5.3).

The analysis of flood duration suggests that the majority of floods in the Ganges are relatively short with an average duration of between three and five days. Flood durations of 20–60 days have occurred 20 times and there were seven floods of over 60 days during the period of analysis (Figure 5.2). The recent 2007–2008 floods in India and Bangladesh had the longest duration which is explained by abnormal heavy rains during the monsoon season. Flood events start increasing during June (18) and reach a peak in July (50) and August (42; Figure 5.2).

Figure 5.4 shows daily water level in the Ganges at Hardinge Bridge during the monsoon season (Islam *et al.*, 2010). The flood level of 1998 was the highest recorded (15.19 m), followed by 1988 which crossed the danger level of 14.25 m. Floods of 1987 and 1988 inundated a total area of 57,270 km^2 and 81,831 km^2 respectively. The other two flood events of 2007 and 2004 also caused enormous damage to properties and considerable loss of life.

The existing status of flood event observations is far from suitable in identifying flood risk areas. The situation has not improved significantly over the years. This suggests that alternative approaches for flood monitoring, for example, those based on satellite remote sensing data, are required.

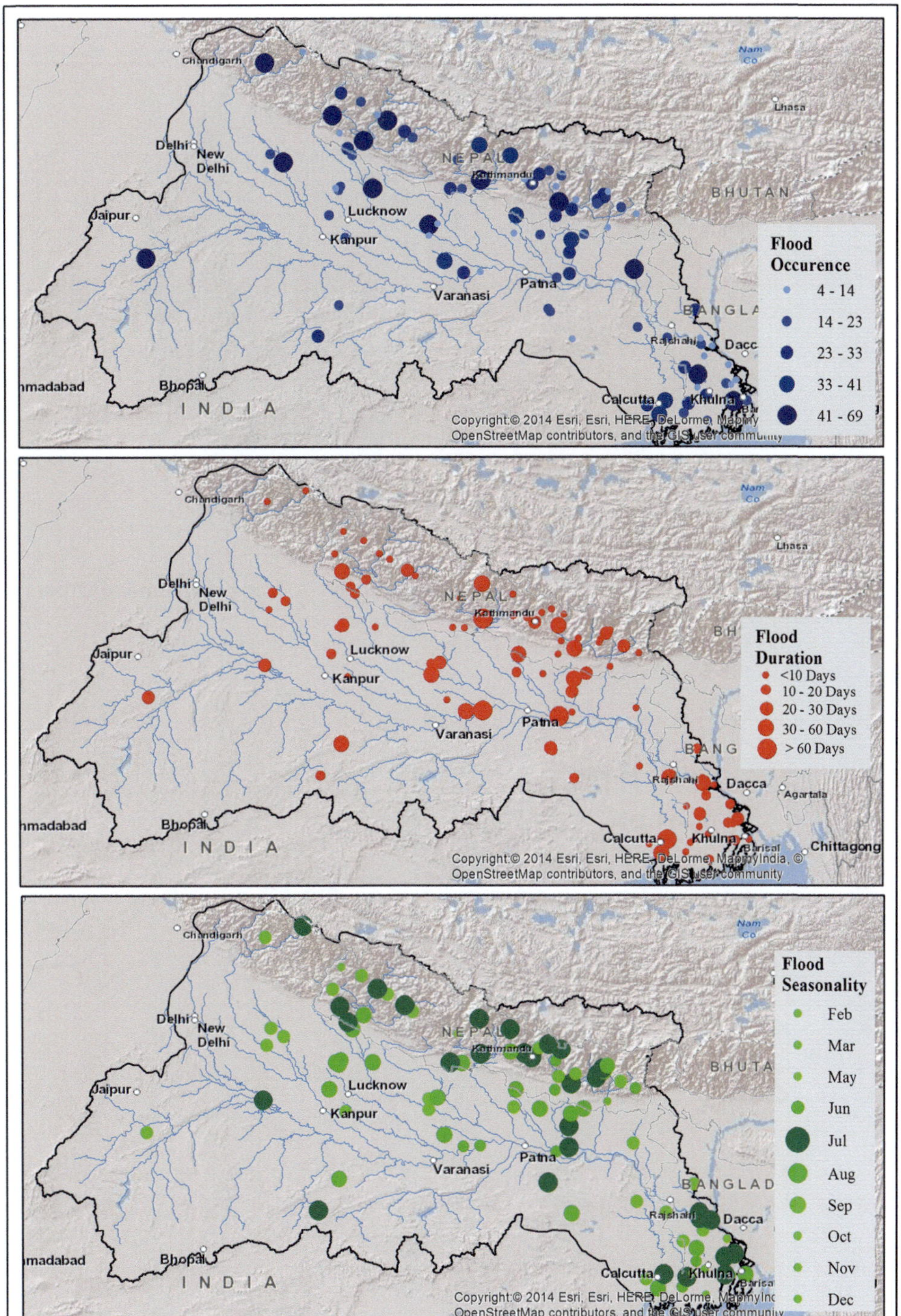

Figure 5.2 Spatial and temporal distribution of large flood events in the Ganges Basin between 1985 and 2011: (top) flood occurrence; (middle) flood duration; (bottom) flood seasonality

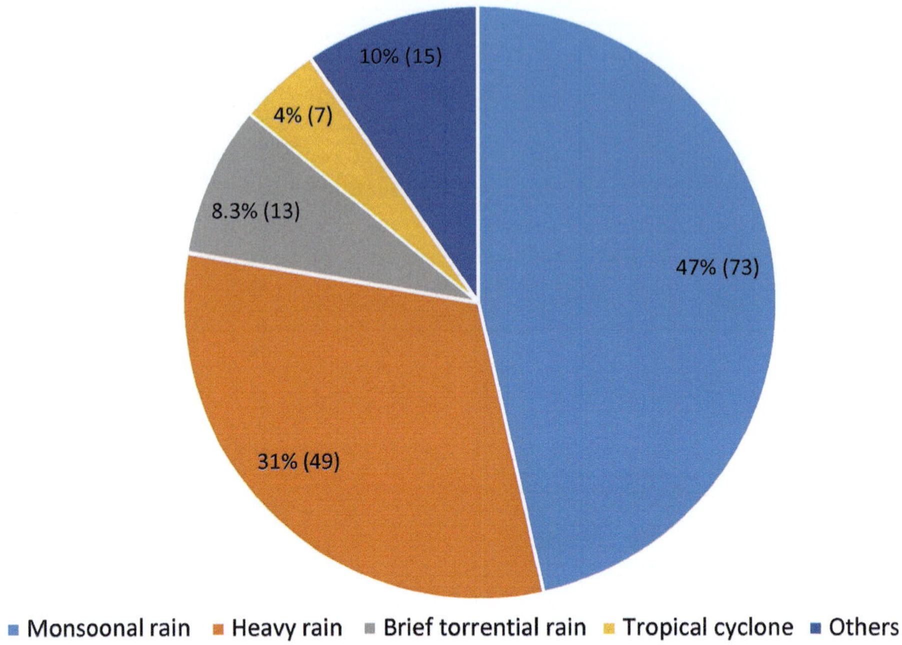

Figure 5.3 Causes of flooding. Data in parentheses show the number of flood events

Source: Dartmouth Flood Observatory (2012)

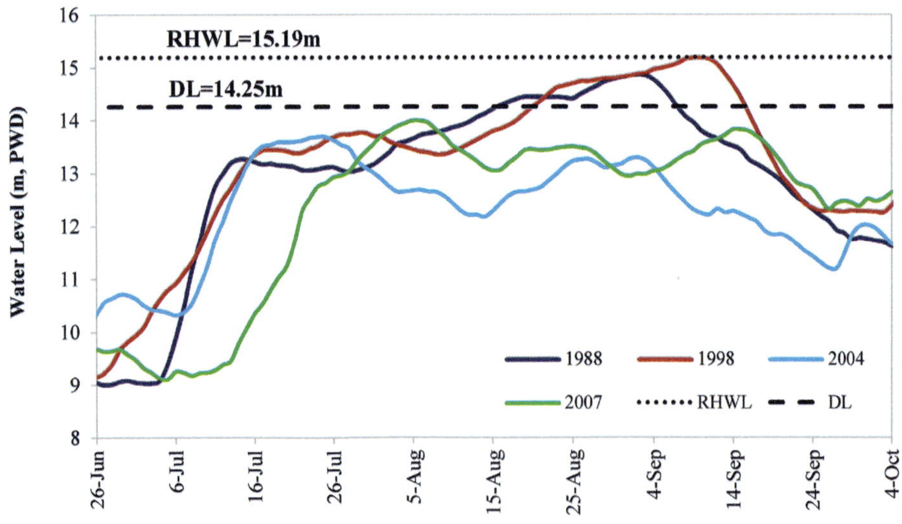

Figure 5.4 Water levels in Ganges at Hardinge Bridge during four major flood years: 1988, 1998, 2004 and 2007

Note: RHWL: recorded highest water level; DL: danger level (level above which it is likely that the flood may cause damage to nearby crops and settlements. In a river with no embankments, danger level is about annual average flood level. On this figure, the danger level is set for 14.25 m based on the historical data)

Drought characteristics

Droughts affect the food and water security of a region directly and indirectly. Meteorological droughts create soil moisture deficits (Vicente-Serrano *et al.*, 2010), affecting crop growth (Rojas *et al.*, 2011), which in turn hampers the overall quality and quantity of crops produced (Mishra and Cherkauer, 2010). Soil moisture depletion below the permanent wilting point results in crop failure. Prolonged meteorological droughts lead to hydrological droughts, which in turn result in reduced flow in streams and rivers (Shukla *et al.*, 2011), low groundwater table (Dahm *et al.*, 2003), and depleted lakes, ponds and reservoirs. Major river basins of South Asia face regular occurrences of droughts, largely driven by erratic monsoon conditions (Bollasina *et al.*, 2011; Mishra *et al.*, 2012).

The economic costs and humanitarian consequences of droughts can be enormous. Almost 50 per cent of the past drought occurrences were widespread and affected more than 40 per cent of the basin area. Seven major droughts (1877, 1899, 1918, 1965, 1972, 1987 and 2002) affected over 40 per cent of the territory of India (Figure 5.5). The 2002–2003 droughts affected more than half of India's crop area of approximately 1.8 million km^2 and led to a huge fall in crop production. During the recent drought of 2009, a total relief of US$4.6 billion was disbursed to 300 million affected people. In the eastern Ganges covering the states of Bihar and Uttar Pradesh with vast agricultural production, drought annual average loss accounted for US$162 million between 1990 and 2012. India accounts for the largest share (90 per cent) of the total drought-prone area in the Ganges Basin (EM-DAT, 2015; SAARC-DMC, 2010). Most of these drought-prone areas are rainfed. Thus, the additional loss in agricultural GDP during drought years is likely to be US$145 million. Total losses (including the production loss of US$856 million during drought years) thus exceed a billion dollars. Improved food availability and improved distribution systems, the integration of markets, better preparedness, and better

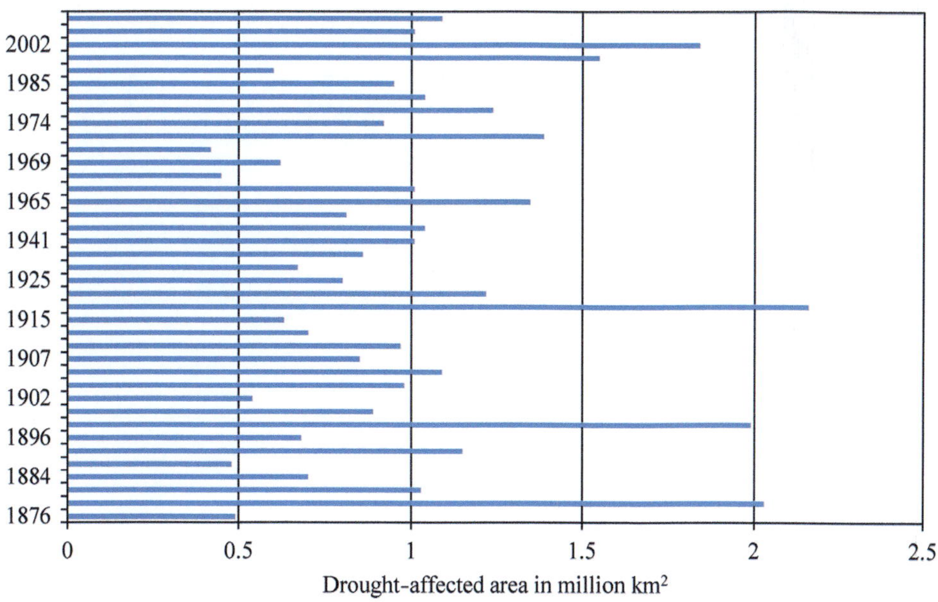

Figure 5.5 Drought years in India during 1876–2002

Source: Modified from Indian Agricultural Statistics Research Institute governance almost eliminated starvation deaths during the droughts between 1980 and 2015 (EM-DAT, 2015) However, post-drought consequences on human welfare are still substantial.

Monitoring of floods and droughts

Flood mapping and monitoring

Satellite remote sensing technology has emerged as one of the most important means of capturing catastrophic flood events, providing information in time and space and also continuously recording such events (Amarnath, 2013, 2014; Islam and Sado, 2000; Jain *et al.*, 2006). For large river basins, such as the Ganges, the satellite-based flood inundation maps can be an important input for assessing different stages of flood disasters. Unlike optical (visible/infrared) sensors, Synthetic Aperture Radar (SAR) sensors, because of their ability to operate 24 hours a day in almost all weather conditions, have emerged as one of the most important tools for providing reliable and near-real-time information on floods (Smith, 1997; Manjusree *et al.*, 2012; Amarnath and Rajah, 2015). They provide timely and cost-effective ways of gathering hydrological data in sparsely gauged basins, irrespective of the political or geophysical barriers (Brakenridge *et al.*, 2005; Hirpa *et al.*, 2012; Khan *et al.*, 2011a; Smith, 1997). One problem with optical satellite sensors (Figure 5.6) is their inability to 'see' through clouds. The land surface is obscured by clouds on most days when flooding occurs in the monsoon period. SAR, on the contrary, can detect flood inundation areas on cloudy days. The potential of SAR data for flood detection and mapping has been demonstrated by Manjusree *et al.* (2012), Amarnath and Rajah (2015), and others.

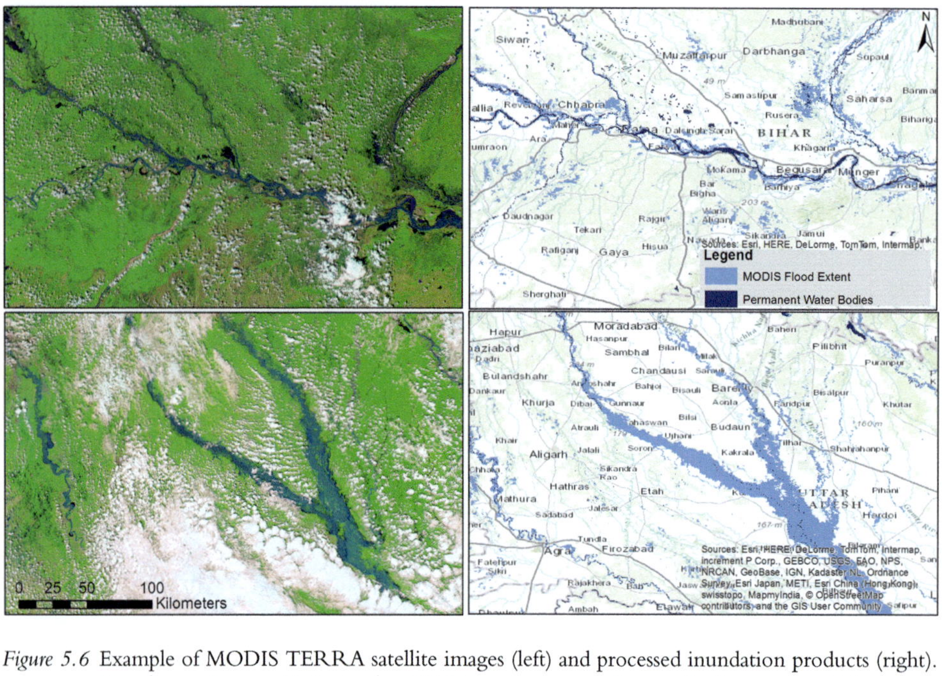

Figure 5.6 Example of MODIS TERRA satellite images (left) and processed inundation products (right). Top: lower Ganges, 30 September 2010. Bottom: upper Ganges, 24 September 2010

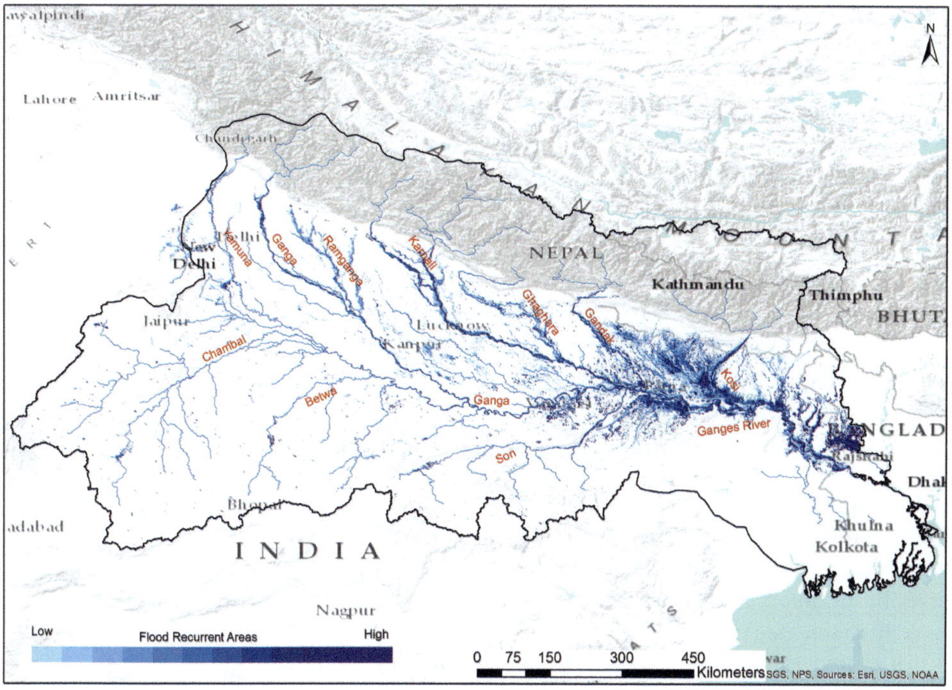

Figure 5.7 Flood map processed using 12 years of MODIS data (2000–2011) showing areas that are regularly flooded

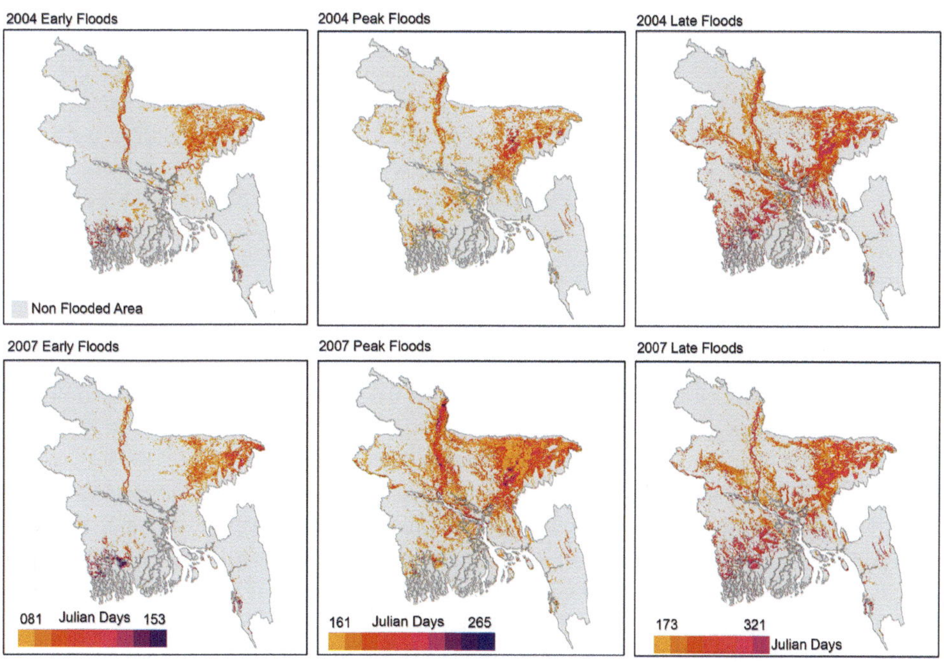

Figure 5.8 Spatial distribution of early floods (May–June), peak floods (July–September) and late floods (October–November) for 2004 and 2007 in Bangladesh

Overlaying multiple images of the inundated region is effective in detecting the maximum extent of the flooded area. This approach is also useful in identifying temporal changes in the simultaneous recession and expansion of flood cycles at a regional scale. Maximum flood inundation extent obtained from MODIS data (2000–2011) is illustrated in Figure 5.7. Figure 5.8 shows estimates of the start dates, end dates and duration of inundation cycle of 2004 and 2007. The spatial distribution of the start dates varies year by year. The start dates of the largest flood (2007) are mostly earlier than those of the medium-scale flood in 2004 (Figure 5.8). These early and late start dates of annual flooding appear to be in good agreement with the dates that the water level increased (Figure 5.4). The water level recorded in 2007 started to increase earlier in the season than that recorded in other years. There is a clear contrast in the estimated end dates of inundation cycles between large-scale floods (1988, 1998 and 2007) and medium floods (2004, 2007).

Building a flood early warning system

The forecasting of transboundary flooding in lower regions of the Ganges Basin, in particular Bihar and Bangladesh, remains notoriously difficult using conventional modelling approaches that rely on extensive and real-time *in-situ* data (Hossain *et al.*, 2014). Also, in transboundary river basins, information exchange and sharing of data between riparian countries is regarded as a first and essential step towards fostering cooperation and trust (Gerlak *et al.*, 2013; Khan *et al.*, 2011b). In the Ganges Basin, there are limited known examples of regional data and information exchanges, although bilateral processes already exist to a certain extent. Cooperation is ongoing between India and Nepal on sharing of information in the Koshi River and Burhi Gandaki dams (World Bank, 2013). More importantly since 1989, 'Flood Forecasting and Warning System on rivers common to India and Nepal' maintains 42 meteorological/ hydrometric sites in Nepalese territory (Nishat and Shams, 2013). Similar programmes exist between Bangladesh and India. The Joint River Commission between Bangladesh and India monitors river flows at Farakka in Bihar during January–April according to an agreement in the Ganges Treaty. However, this is just focused on one location. Additionally, there is a system of transmission, from India to Bangladesh, of flood forecasting data on major rivers like the Ganga, Teesta, Brahmaputra and Barak during the monsoon season (Nishat and Rahman, 2009).

The availability of technologically advanced geographic information systems, remotely sensed data, decision support systems, innovative analytical tools and web-based communication materials in the public domain could be the catalyst for bringing changes in the current level of transboundary information exchange in the Ganges Basin (Figure 5.9) (Khan *et al.*, 2011b; Nishat and Rahman, 2009; Shrestha, 2008; Shrestha *et al.*, 2011, 2015; Stimson Center, 2010; Gerlak *et al.*, 2013). The figure illustrates how geospatial datasets combined with hydrological models could provide 7–10 days' lead time in regional flood early warning systems. Successful implementation of a flood forecasting system requires a coupled rainfall-runoff model that uses mainly rainfall inputs either from *in situ* observations, satellite-based rainfall estimates such as Precipitation Estimation from Remotely Sensed Imagery Using Artificial Neural Networks (PERSIANN; Ashouri *et al.*, 2015), Tropical Rainfall Measuring Mission (TRMM)-based 3B42RT (Huffman *et al.*, 2010) and Global Precipitation Mission (GPM) and seasonal weather forecast datasets (e.g. European Centre for Medium-Range Weather Forecasts) to simulate river discharges with a lead time to predict flood inundation in downstream areas. Further, the early warning information can be rapidly disseminated to disaster risk management agencies to reduce economic losses and mitigate the number of injuries or deaths.

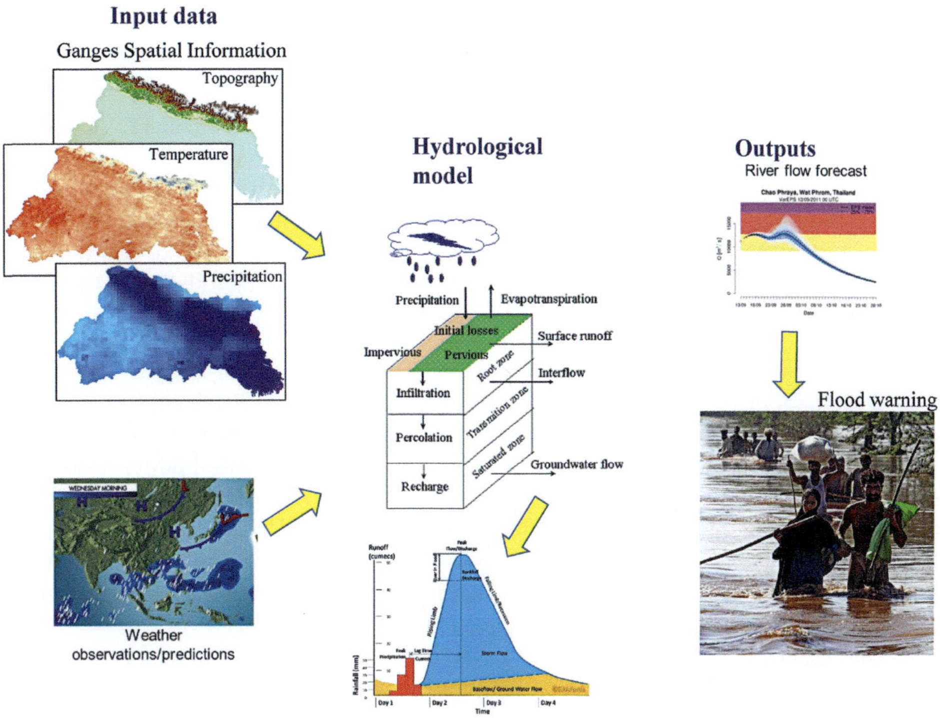

Figure 5.9 Example of a flood forecasting system by combining weather forecast inputs, geospatial data and hydrological models for streamflow predictions

Country-wide initiatives in flood early warning

There is continuing interest in developing management plans and appropriate response actions and communicating information about flood conditions and anticipated impacts. Basin countries are already using advanced prediction tools and in many cases integrating these into the management and decision making process. Selected initiatives at country and regional levels are listed below:

- The Nepal Department of Hydrology and Meteorology (DHM), under the Ministry of Environment, is mandated to monitor all the hydrological and meteorological activities in the country. DHM provides general and aviation weather forecasts, forecasting and early warning of hydrometeorological hazards. Flood alerts for major locations in Nepal are available on the website: http://hydrology.gov.np/new/bull3/index.php/hydrology/home/main.
- The Central Water Commission (CWC) of India, under the Ministry of Water Resources, is the national organization that provided early warning dissemination for the major rivers in India since 1958. At present, it operates a nationwide network of 945 hydrological observation stations and issues early warnings from 145 forecasting stations on major rivers and 27 inflow forecasting stations on major dams/barrages. It provides daily updates on flood forecast information which can be accessed online from: http://india-water.gov.in/ffs/.

- The Flood Forecasting and Warning Centre (FFWC) of Bangladesh was established in 1972 under the Bangladesh Water Development Board (BWDB). Its aim is to provide reliable and accurate flood forecasts with a lead time of 72 hours to help mitigate flood disasters and enhance the capacity of national agencies and communities. Close to 90 stations are being observed on a daily basis. These observations are also published on the FFWC website: www.ffwc.gov.bd/index.php.
- Apart from individual country initiatives, other international organizations including ICIMOD, RIMES, JRC and other development agencies support governments on flood forecasting and awareness initiatives. The Asian Development Bank, the European Commission, the World Meteorological Organization, USAID and the World Bank are also working to enhance flood forecasting capacity.

Despite considerable effort and success in improving flood early warning and communications, there is room for improvement. In response to the floods of 1998, 2007 and 2012, the National and Regional Cooperation of Ganges Forum was convened by governments in Nepal, India and Bangladesh. Along with its partners, the forum aimed to identify priority actions and opportunities for advancing flood preparedness and response. Two of these included improving the characterization and assessment of flood conditions, as well as the socioeconomic and environmental impacts of floods across temporal and spatial scales. A holistic approach towards flood management is needed that should include regional information and warning systems, coupled with a range of hard and soft, national and local investments.

Finally, significant climate change uncertainties remain in the basin. Current data and models give little compelling evidence of what the future holds. It appears that mean hydrological variability in the future will be similar to the pronounced variability seen in the basin today, but extremes may well be greater and could potentially create major new challenges for populations in the basin (Jeuland *et al.*, 2013; World Bank, 2013).

Drought prediction and monitoring

The detrimental impact of droughts can be mitigated by proactive planning such as monitoring drought conditions in real time. This is done in order to assess the conditions of precipitation, evapotranspiration, soil moisture, runoff and vegetation health, which can minimize risks to food and water security. Furthermore, monitoring can provide information on the intensity and spatial extent of droughts (Thenkabail *et al.*, 2004), which can assist in proactive planning and decision making.

Most of the global drought monitoring systems (AghaKouchak and Nakhjiri, 2012; Hao and AghaKouchak, 2013) operate at coarse spatial and temporal resolutions, which may not be sufficient for district-level decision making. Existing drought monitoring systems in the Ganges are limited to India and work only during the monsoon season (June–September [JJAS]), merely providing information on meteorological conditions. While near-real-time drought monitoring is challenging, remotely sensed data (Mu *et al.*, 2013; Huffman *et al.*, 2007) available in real time, and land surface hydrologic models (Rodell *et al.*, 2004; Sheffield and Wood, 2008) that can simulate hydrological variables, provide a basis for further advancement (Figure 5.10). However, such approaches have a limited spatial accuracy at which drought patterns can be mapped, because the indices are calculated from point-based meteorological measurements collected at weather stations. As a result, climate-based drought index maps depict broad-scale drought patterns and their accuracy is limited due to dependence on uneven and sparse weather station distributions. This limits drought planning and monitoring activities in areas not well covered by weather stations.

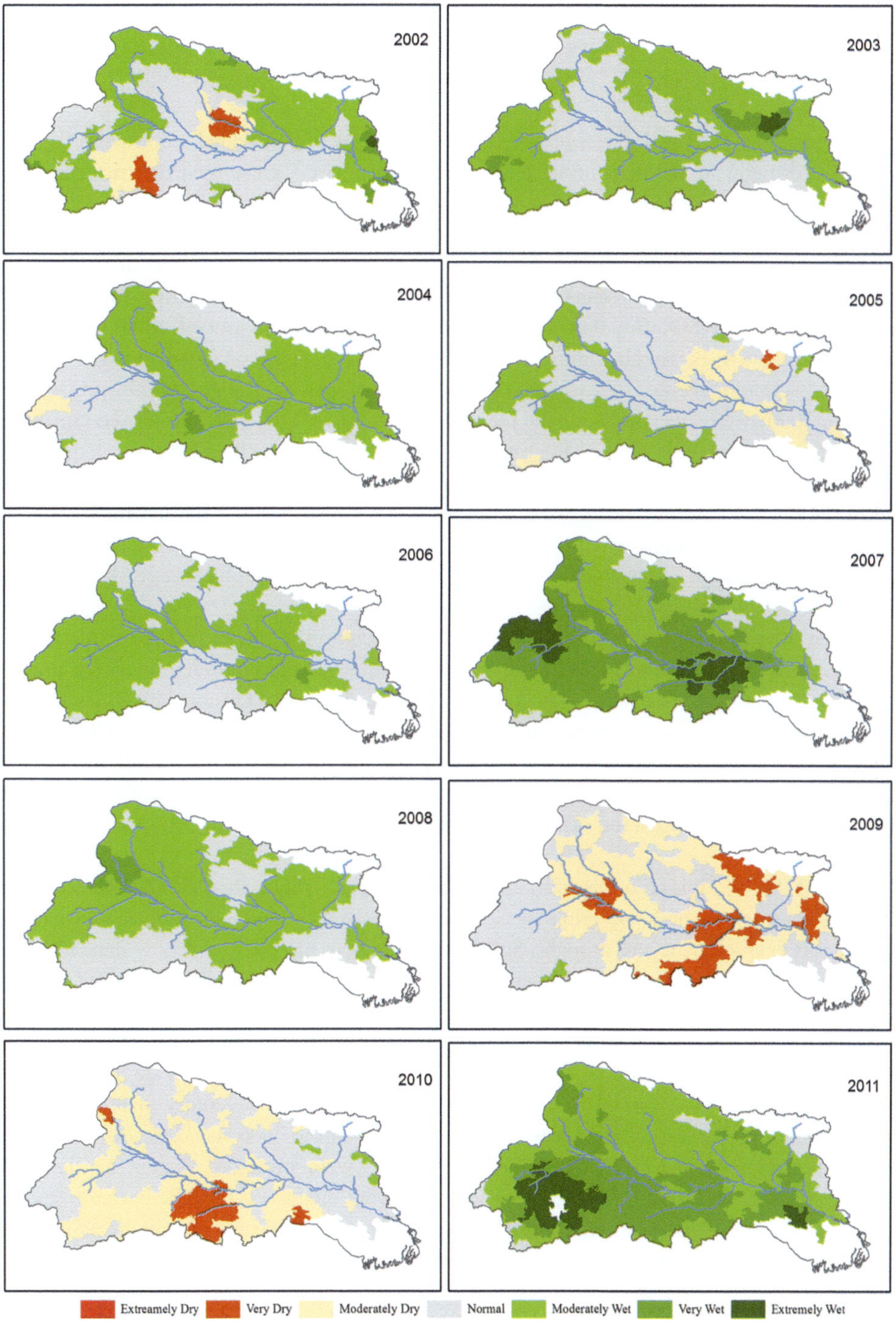

Figure 5.10 Standardized Precipitation Index (SPI) for 3 months' rainfall total calculated using satellite-based rainfall estimates between 2002 and 2011 for the months May–June–July. This index is negative for drought and positive for wet conditions. As the dry and wet conditions become more severe, the index becomes more negative or positive, respectively

Remotely sensed data resources (Mu *et al.*, 2013; Huffman *et al.*, 2007) available in real time can rapidly provide information on current drought situations. This can inform short-term emergency responses or crisis management, rather than long-term mitigation or risk management.

In India, the National Remote Sensing Centre (NRSC) provided information on the extent of drought for 13 agriculturally important states. The Centre used two main drought indices, namely NDVI and NDWI. The system was transferred to Mahalanobis National Crop Forecast Centre (MNCFC), Department of Agriculture and Cooperation, with a mandate to report on the agricultural situation for the entire country. The limitation of the drought system is that it mainly uses vegetation inputs to determine drought severity. In Bangladesh and Nepal SPARRSO, the Bangladesh Agricultural Research Council (BARC) and the Nepal Agriculture Research Council, World Food Programme prepared drought severity maps for selected events. In early 2003, IWMI developed a drought monitoring system covering south-west Asia combining the use of AVHRR and MODIS satellite datasets (Smakhtin *et al.*, 2006). However, neither the regional countries nor SAARC have maintained the drought portal.

An attempt is currently being made to develop a regional drought monitoring system for South Asia[1] by combining remote sensing data on vegetation, temperature, precipitation and soil moisture through a new monitoring tool, the Integrated Drought Severity Index (IDSI) (Figure 5.11).

Three major drought indices, namely the Vegetation Condition Index (VCI), the Temperature Condition Index (TCI) and the Precipitation Condition Index (PCI), are used to monitor drought from vegetation, soil and climate respectively. Data fusion techniques are used to improve the spatial characteristics and best correlating factors between VCI, TCI and PCI. In addition, a semi-automated tool for processing the information required by IDSI has been developed.[2] These developments and tools aim to support the governments of the region in formulating drought mitigation plans and in making decisions that reduce the harmful impacts of droughts.

From Table 5.2, it is clear that most events listed affected crop areas across the Ganges Basin. The growing season of *Kharif* crops in Central Ganges which begins in June and extends to October is exposed to severe drought. Since the *Kharif* season crops are directly dependent on the monsoonal rainfall, they are very sensitive to fluctuations in the amount of rainfall. During 2002, in *Kharif* approximately 320,000 km^2 were affected through three categories of the drought. In contrast, 2007 was more humid with drought reported over 53,000 km^2 only.

In Bangladesh, drought-prone areas for pre-*Kharif*, *Kharif* and *Rabi* seasons are mapped based on the Agro Ecological Zones (AEZ) database and land resource inventory map, in order to assess drought severity in different seasons. For example, for the 2006 and 2009 droughts, three different maps for pre-*Kharif*, *Kharif* and *Rabi* seasons were produced (Figure 5.12). Using these drought severity maps, drought-affected areas were quantified for different crop seasons (Table 5.3). Here it has been reported that every year 0.45 million ha of land are affected by very severe drought during the *Rabi* season, while 0.40 million ha and 0.34 million ha of land are affected during the pre-*Kharif* and *Kharif* seasons, respectively.

It is clear from the above sections that collecting drought impact information and measuring the economic impacts of drought are critical for understanding geographic or sectoral vulnerabilities to drought. Such understanding and information at basin, sub-national and regional scale, in turn, can be used to develop mitigation and management options that are intended to reduce the vulnerability to and impact of drought.

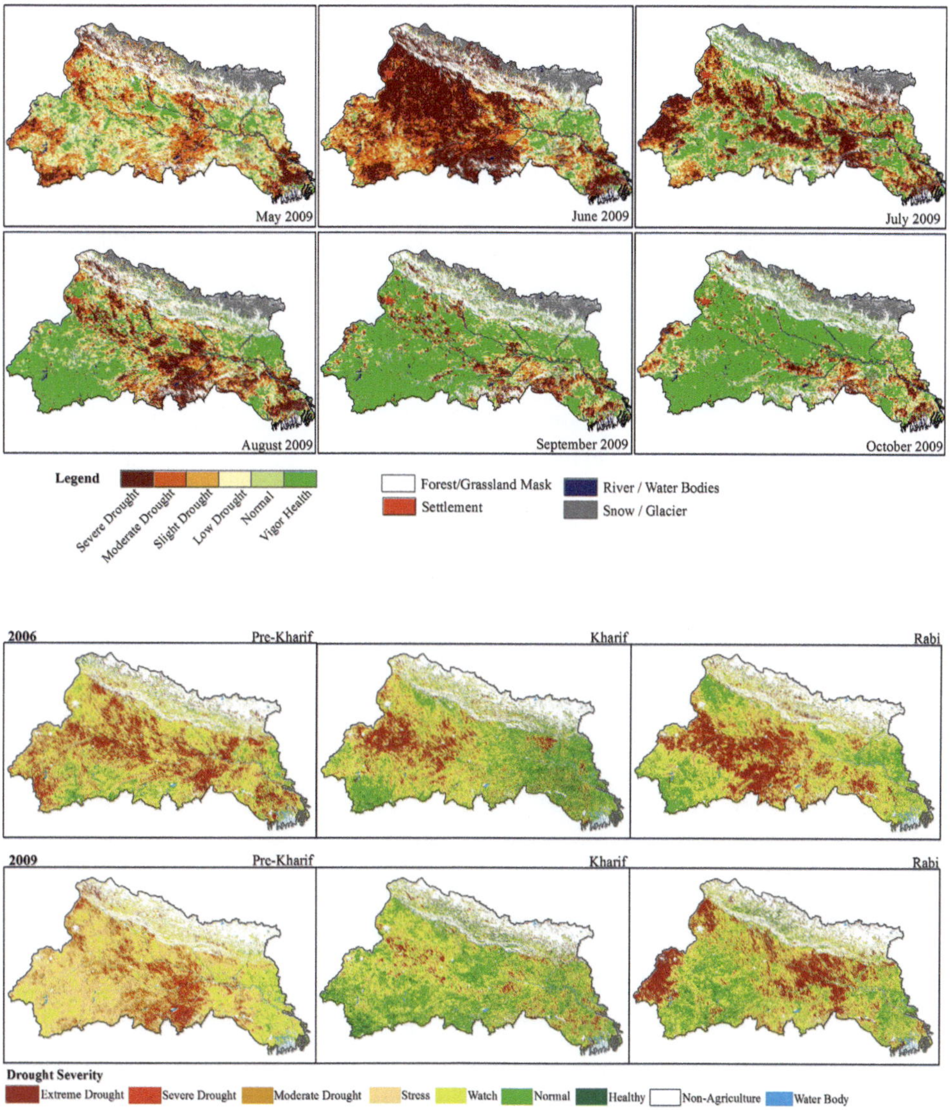

Figure 5.11 Integrated Drought Severity Index (IDSI) for Ganges Basin. (Top) Seasonal variation of drought severity for 2009 extreme event; (bottom) drought extent for different cropping seasons for two major drought events of 2006 and 2009

Source: IWMI

Developing processes that assure that drought data is shared with decision makers in a timely manner will demonstrate its value. This may also allow rapid drought relief to farmers, through compensation in advance of the next growing season. Countries in the region lack the capacity to aggregate and synthesize drought impact data and translate that into useable information for decision makers. Without a strong connection to decision makers, collected or reported data cannot be used to its full potential.

Table 5.2 Summary of major drought events in the Ganges Basin with areas affected, km² by crop season (pre-*Kharif* 16 March–15 July; *Kharif* 16 July–15 Oct.; *Rabi* 16 Oct.–15 March) for the period between 2001 and 2014

Year	*Season*	*Extremely Severe*	*Moderately Severe*	*Severe*
2001	Pre-*Kharif*	273,203	120,798	101,198
	Kharif	222,640	72,692	68,197
	Late *Rabi*	239,421	72,301	67,981
2002	Pre-*Kharif*	165,350	63,496	67,763
	Kharif	218,654	49,421	54,769
	Late *Rabi*	214,361	61,710	66,434
2003	Pre-*Kharif*	285,052	65,278	71,020
	Kharif	167,520	27,199	32,176
	Late *Rabi*	223,884	46,202	51,174
2006	Pre-*Kharif*	215,870	123,267	114,820
	Kharif	81,810	52,456	60,515
	Late *Rabi*	131,874	72,716	71,830
2007	Pre-*Kharif*	59,515	28,233	40,377
	Kharif	23,496	19,446	24,379
	Late *Rabi*	99,244	45,236	53,190
2009	Pre-*Kharif*	293,681	102,076	102,692
	Kharif	68,413	32,452	41,897
	Late *Rabi*	120,647	45,598	51,590
2010	Pre-*Kharif*	403,488	143,653	79,484
	Kharif	149,287	28,413	35,708
	Late *Rabi*	67,595	24,316	31,561
2012	Pre-*Kharif*	209,794	89,350	101,936
	Kharif	188,093	56,950	65,161
	Late *Rabi*	117,826	33,708	42,567

Population exposure and impact on agriculture

Analysis of the combined population exposure to floods (Figure 5.13) and droughts (Figure 5.14) in the Ganges Basin between 2000 and 2014 suggests that approximately 178 million people were severely affected, of which floods contributed to 93 million affected and droughts to 85 million (Table 5.4). India has the most exposure, with floods affecting 75 million people and drought around 84 million.

To assess the impact on agriculture, freely available SPOT-VEGETATION Land Use and Land Cover (LULC) data were used. It is estimated that an area of approximately 1.7 million km² is regularly affected by floods and droughts in the Ganges Basin. Floods in India affect 78,000 km² and droughts 1.6 million km². Uttar Pradesh and Madhya Pradesh are particularly affected. In Bangladesh, flood impact on agriculture is estimated to be 21,895 km² while drought affects only 9,713 km², while the least is Nepal with more areas affected by droughts than floods.

Table 5.4 shows that natural climate variability such as ENSO continues to occur as climate change progresses. This means that droughts and floods due to El Niño and La Niña still occur due to natural climate variability, but some of the impacts may intensify due to climate change. Large areas of the basin routinely suffer from both floods and droughts (Figures 5.13 and 5.14).

Table 5.3 Summary of drought classes in Bangladesh by crop season (million ha), 2006 drought

Drought class	*Rabi season*	*Pre-Kharif season*	*Kharif season*
Very severe	0.45	0.40	0.34
Severe	1.71	1.15	0.74
Moderate	2.95	4.76	3.17
Slight	4.21	4.09	2.9
No drought	3.17	2.09	0.68

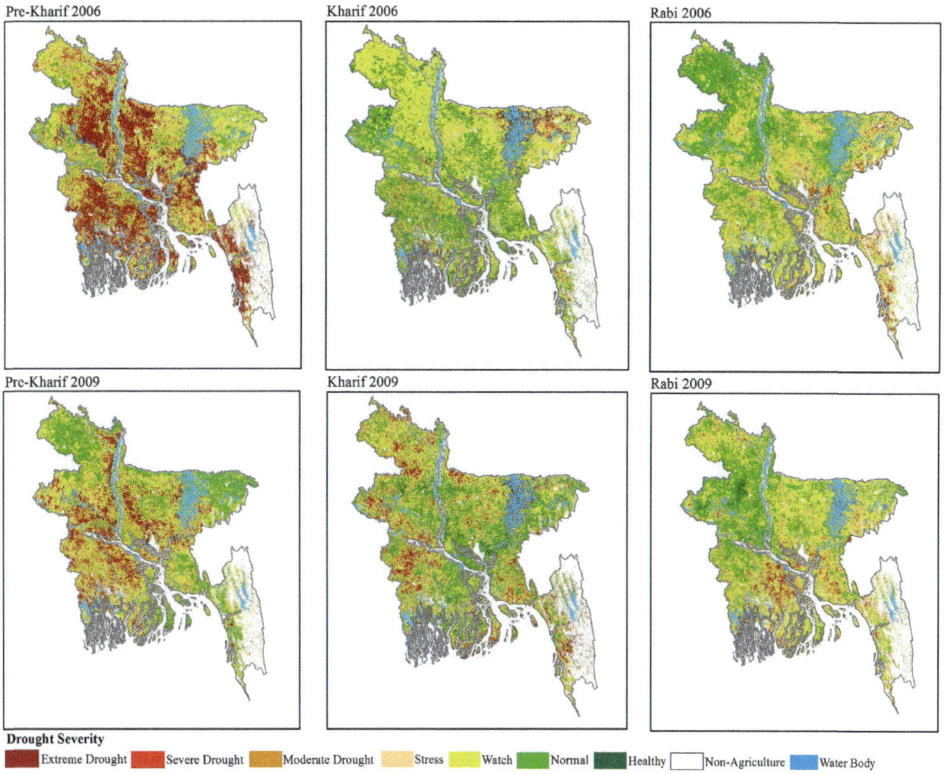

Figure 5.12 Pre-*Kharif*, *Kharif* and *Rabi* drought-prone areas of Bangladesh
Source: IWMI

Conclusions

The assessment of floods and droughts presented in this chapter highlights the complexity of the Ganges Basin and clearly provides up-to-date information useful to managing risks. Significant opportunities remain, however, to improve data acquisition, public data access, forecasting techniques using integrated ground and satellite systems, modelling and communication. In addition to technological innovations, robust institutional arrangements both within and between basin countries will be needed to diminish annual loss of lives and livelihoods caused by natural hazards.

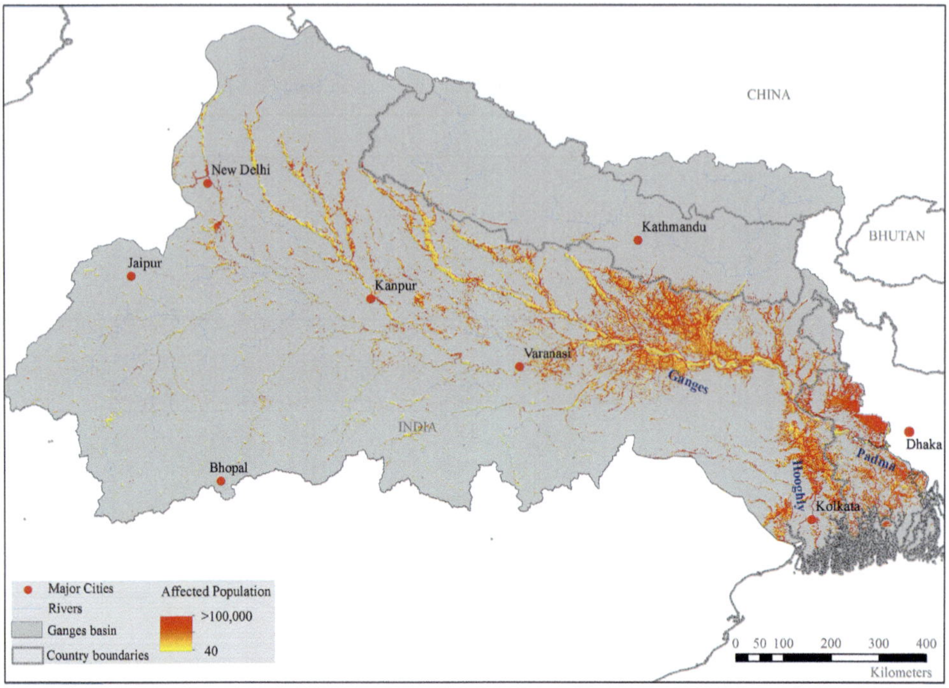

Figure 5.13 Flood-affected population in the Ganges Basin. The analysis was performed by overlaying flood extent (2001–2013) derived from remote sensing data and 2014 population count obtained from LandScan Global Population Data

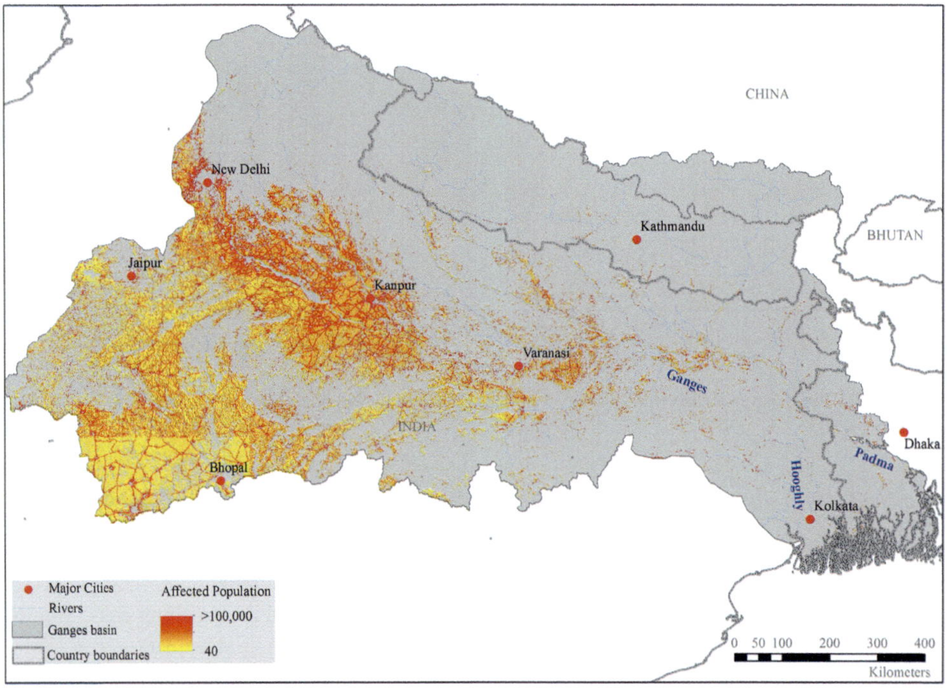

Figure 5.14 Drought-affected population in the Ganges Basin. The analysis was performed by overlaying drought extent (2001–2013) derived from remote sensing data and 2014 population count obtained from LandScan Global Population Data

Table 5.4 Summary of population exposure and impact on agriculture related to floods and droughts in the Ganges Basin

Countries covering basin area	*Population exposed*		*Agricultural losses km²*	
	Floods	*Droughts*	*Floods*	*Droughts*
Nepal	1,601,940	280,450	3,752	6,946
India	74,987,781	84,523,363	78,614	1,649,905
Bangladesh	16,470,496	384,116	21,895	9,713
Sub-total	93,060,217	85,187,929	104,261	1,666,564
Total	178,248,146		1,770,825	

Regional cooperation between Nepal, India and Bangladesh could take many forms, from a network of national institutions with an agreed information-sharing protocol, to a dedicated multilateral institution that would gather, analyse and disseminate crucial hydro-meteorological and climate data. A real-time regional hydro-meteorological information system would provide the scientific information needed by planners to sustainably manage and develop the basin. It is envisioned that such a system would eventually move into the public domain to assist farmers to enhance productivity and food security, allow disaster risk-management professionals to safeguard lives and assets, and allow climate researchers to understand, predict and adapt to the changing and challenging climate in the Ganges Basin.

Globally, there has been a phenomenal increase in access to open source satellite data, value-added information, open source models, high-end software and hardware systems to support visualization of high resolution satellite images. This has brought a paradigm shift in extending open information services to support different areas of decision making. Work in the Ganges Basin should develop such monitoring systems and continue to promote regional cooperation in data sharing.

Acknowledgements

The first author acknowledges funding provided by the CGIAR Research Program on Climate Change, Agriculture and Food Security (CCAFS); Water, Land and Ecosystems (WLE) and the International Water Management Institute (IWMI). The contributions of Niranga Alahacoon and Peejush Pani of IWMI in preparation of maps and statistics used in this chapter are also gratefully acknowledged.

Notes

1 www.droughtmanagement.info/idmp-activities/south_asia
2 www.droughtmanagement.info/wp-content/uploads/2015/07/SADMS-TechnicalDevelopment-Phase1.pdf

References

AghaKouchak, A. and Nakhjiri, N. (2012). A near real-time satellite based global drought climate record. *Environ. Res. Lett.*.

Amarnath, G. (2013). An algorithm for rapid flood inundation mapping from optical data using a reflectance differencing technique. *J Flood Risk Management* 7: 239–50.

Amarnath, G. (2014). Earth observation data: Monitoring floods and drought. *Geospatial Today* 6(June): 15–19.
Amarnath, G., Mohamed, A., Pramod, A., Smakhtin, V. (2012). 'Detecting spatio-temporal changes in the extent of seasonal and annual flooding in South Asia using multi-resolution satellite data'. In Civco D.L., Ehlers M., Habib S., Maltese A., Messinger D., Michel U., Nikolakopoulos K.G. and Schulz K. (eds). *Earth Resources and Environmental Remote Sensing/GIS Applications III: Proceedings of the International Society for Optics and Photonics*, 8538(1–6); Amsterdam: International Society for Optics and Photonics (SPIE).
Amarnath, G. and Rajah, A. (2015). An evaluation of flood inundation mapping from MODIS and ALOS satellites for Pakistan. *Geomatics, Natural Hazards and Risk* 1–12.
Ashouri, H., Hsu, K.-L., Sorooshian, S., Braithwaite, D.K., Knapp, K.R., Cecil, L.D., Nelson, B.R. and Prat, O.P. (2015). PERSIANN-CDR: Daily precipitation climate data record from multisatellite observations for hydrological and climate studies. *Bulletin of the American Meteorological Society* 96(1): 69–83.
Bollasina, M.A., Ming, Y. and Ramaswamy, V. (2011). Anthropogenic aerosols and the weakening of the South Asian summer monsoon. *Science* 334(6055): 502–5.
Bradshaw, C.J.A, Sodhi, N.S., Peh, K.S.H. and Brook, B.W. (2007). Global evidence that deforestation amplifies flood risk and severity in the developing world. *Global Change Biology* 13: 2379–2395.
Brakenridge, G.R., Nghiem, S.V., Anderson, E. and Chien, S. (2005). Space-based measurement of river runoff. *EOS, Transactions of the American Geophysical Union* 86(19): 185–8.
Dahm, C.N., Baker, M.A., Moore, D.I. and Thibault, J.R. (2003). Coupled biogeochemical and hydrological responses of streams and rivers to drought. *Freshwater Biol.* 48: 1219–1231.
Dasgupta, S., Huq, M., Khan, Z.H., Masud, M.S., Ahmed, M. Mukherjee, N. and Pandey, K. (2011). Climate proofing infrastructure in Bangladesh: The incremental cost of limiting future flood damage. *The Journal of Environment & Development* 20(2): 167.
DFO. Global Active Archive of Large Flood Events (2013). Available online at: www.dartmouth.edu/~floods/ (accessed 24 January 2013).
Dilley, M., Chen, R.S., Deichmann, U., Lerner-Lam, A.L., Arnold, M., Agwe, J., Buys, P., Kjekstad, O., Lyon, B. and Yetman, G. (2005). *Natural Disaster Hotspots: A Global Risk Analysis*. Washington, DC: International Bank for Reconstruction and Development/The World Bank and Columbia University.
EM-DAT. (2015). *The OFDA/CRED International Disaster Database*. Brussels, Belgium: Centre for Research on the Epidemiology of Disasters (CRED), Université Catholique de Louvain. Available online at: www.emdat.be (accessed July 31 2015).
Gerlak A.K., Lautze, J. and M. Giordano. (2013). *Greater Exchange, Greater Ambiguity: Water Resources Data and Information Exchange in Transboundary Water Treaties*. GWF Discussion Paper 1307, Global Water Forum, Canberra, Australia.
Hao Z. and AghaKouchak, A. (2013). Multivariate standardized drought index: A parametric multi-index model. *Adv. Water Resour.* 57: 12–18.
Hirpa, F.A, Hopson, T., De Groeve, T., Brakenridge, G.R., Gebremichael, M. and Restrepo, P.J. (2012). Upstream satellite remote sensing for river discharge prediction: Application to major rivers in South Asia. *Remote Sen. of the Environment*.
Hossain, M.M. (2015). *Basin-Wide Approach to Flood Risk Management in Bangladesh*. E-proceedings of the 36th IAHR World Congress, 28 June–3 July 2015, The Hague, The Netherlands.
Huffman, G.J., Bolvin, D.T., Nelkin, E.J., Wolff, D.B., Adler, R.F., Gu, G., Hong, Y., Bowman, K.P. and Stocker, E.F. (2007). The TRMM Multi-satellite Precipitation Analysis (TMPA): Quasi-global, multiyear, combined sensor precipitation estimates at fine scales. *J. Hydrometeorol.* 8: 38–55.
IPCC (Intergovernmental Panel on Climate Change) (2007). *Climate Change 2007: Synthesis Report*. Contribution of Working Groups I, II and III to the Fourth Assessment Report of the Intergovernmental Panel on Climate Change, R.K. Pachauri, and A. Reisinger (eds). Geneva, Switzerland: IPCC.
Islam, A.K., Bala, S.K. and Haque, A. (2010). Hydrologic characteristics of floods in Ganges-Brahmaputra-Meghna (GBM) Delta. *Natural Hazards* 54(3): 797–811.
Islam, M.M. and Sado, K. (2000). Development of flood hazard maps of Bangladesh using NOAA-AVHRR images with GIS. *Hydrol Sci J* 45: 337–55.
Jain V. and Sinha, R. (2003). Geomorphological manifestations of the flood hazard: A remote sensing based approach. *Geocarto International* 18(4): 51–60.
Jain, S.K., Saraf, A.K., Goswami, A. and Ahmad, T. (2006). Flood inundation mapping using NOAA AVHRR data. *Water Resour Manag.* 20: 949–59.
Jeuland, M., Harshadeep, N., Escurra, J., Blackmore, D. and Sadoff, C. (2013). Implications of climate change for water resources development in the Ganges basin. *Water Policy* 15: 26.

Jonkman, S.N. (2005). Global perspectives on loss of human life caused by floods. *Natural Hazards* 34(2): 151–75.

Joyce, R.J. and Xie, P. (2011). Kalman Filter-Based CMORPH. *Journal of Hydrometeorology* 12(6): 1547–1563.

Khan, H., Dawe, P., Khan, A. and Puestow, T. (2011). *Innovative Approaches to Monitoring for Transboundary Water Governance.* International Conference on Environment Science and Engineering IPCBEE, vol.8. IACSIT Press, Singapore.

Khan, S.I., Hong, Y., Wang, J., Yilmaz, K.K., Gourley, J.J., Adler, R.F., Brakenridge, G.R., Policelli, F., Habib, S. and Irwin, D. (2011). Satellite remote sensing and hydrological modeling for flood inundation mapping in Lake Victoria Basin: Implications for hydrologic prediction in ungauged basins. *IEEE Transactions on Geoscience and Remote Sensing* 49: 85–95.

Manjusree, P., Prasanna Kumar, L., Bhatt, C.M., Rao, G.S. and Bhanumurthy, V. (2012). Optimization of threshold ranges for rapid flood inundation mapping by evaluating backscatter profiles of high incidence angle SAR images. *International Journal of Disaster Risk Science* 3(2): 113–22.

Mishra, V. and Cherkauer, K.A. (2010). Retrospective droughts in the crop growing season: Implications to corn and soybean yield in the midwestern United States. *Agric. For. Meteor.* 150: 1030–1045.

Mishra, V., Smoliak, B.V., Lettenmaier, D.P. and Wallace, J.M. (2012). A prominent pattern of year-to-year variability in Indian summer monsoon rainfall. *Proc. Natl. Acad. Sci. U.S.A.* 109(19): 7213–7217.

Moench M. and Dixit A. (2007). *Adaptive Capacity and Livelihood Resilience: Adaptive Strategies for Responding to Floods and Droughts in South Asia.* Kathmandu, Nepal: Institute for Social and Environmental Transition.

Mu, Q., Zhao, M., Kimball, J.S., McDowell, N.G. and Running, S.W. (2013). A remotely sensed global terrestrial drought severity index. *Bull.Amer. Meteor. Soc.* 94: 83–98.

Nishat B. and Shams S. (2013). Towards improved data and information exchange in the Ganges basins. *Issue Brief. The Asia Foundation.* September 2013: 1–8.

Nishat B. and S.M. Rahman (2009). Water resources modeling of the Ganges-Brahmaputra-Meghna River Basins using satellite remote sensing data. *Journal of American Water Resources Association.* 45(6): 1313–1327.

Parthasarathy B., Rupa Kumar K. and Munot A.A. (1991). Evidence of secular variations of Indian summer monsoon rainfall – circulation relationships. *Journal of Climate* 11: 165–76.

Pervez, M.S., Henebry, G.M. (2015). Spatial and seasonal responses of precipitation in the Ganges and Brahmaputra river basins to ENSO and Indian Ocean dipole modes: Implications for flooding and drought. *Natural Hazards and Earth System Science* 15(1): 147–62.

Rodell, M. and Coauthors (2004). The Global Land Data Assimilation System. *Bull. Amer. Meteor. Soc.* 85: 381–94.

Rojas, O., Vrieling A. and Rembold F. (2011). Assessing drought probability for agricultural areas in Africa with coarse resolution remote sensing imagery. *Remote Sensing of Environment* 115(2): 343–52.

Sheffield, J. and Wood, E.F. (2008). Projected changes in drought occurrence under future global warming from multi-model, multi-scenario, IPCC AR4 simulations. *Climate Dyn.*,31: 79– 105.

Shrestha, M.S. (2008). Impacts of floods in South Asia. *Journal of South Asia Disaster Studies* 1(1): 85–106.

Shrestha, M., Artan, G.A., Bajracharya, S.R., Gautam, D.K. and Tokar, S.A. (2011). Bias-adjusted satellite-based rainfall estimates for predicting floods: Narayani Basin. *Journal of Flood Risk Management* 4(4): 360–73.

Shrestha, M.S., Grabs, W.E., Khadgi, V.R. (2015). Establishment of a regional flood information system in the Hindu Kush Himalayas: Challenges and opportunities. *Int. J. Water Resour. D.* 31(2): 238–52.

Shukla, S., Steinemann, A.C. and Lettenmaier, D.P. (2011). Drought monitoring for Washington State: Indicators and applications. *J Hydrometeorol* 12(1): 66–83.

Smakhtin, V.U., Thenkabail, P. and Gamage, N. (2006). 'Developing on-line near-real time drought monitoring system for southwest Asia'. In Savin, I. and Negre, T. (eds). *Agro-meteorological monitoring in Russia and Central Asian Countries.* European Commission, Directorate General Joint Research Center. Institute for the Protection and Security of the Citizen (IPSC). Istra, Italy, pp. 104–18.

Smakhtin, V.U., Pavelic, P., Amarnath, G., McCartney, M. and Campbell, B. (2014). 'Managing water variability: Floods and droughts'. In van der Bliek, J., McCornick, P., Clarke, J. (eds). *On Target for People and Planet: Setting and Achieving Water-Related Sustainable Development Goals.* Colombo, Sri Lanka: International Water Management Institute (IWMI), pp. 34–7.

Smith, L.C. (1997). Satellite remote sensing of river inundation area, stage, and discharge: A review. *Hydrological Processes* 11(10): 1427–1439.

South Asian Association for Regional Cooperation-Disaster Management Centre (SAARC-DMC) (2010). *SAARC workshop on Drought Risk Management in South Asia.* 8–9 August 2010, Kabul, Afghanistan.

Stimson Center (2010). *Fresh Water Futures: Imagining Responses to Demand Growth, Climate Change, and the Politics of Water Resource Management by 2040*. The Stimson Center.

Verhaar, P.M., Biron, P.M., Ferguson, R.I. and Hoey, T.B. (2011). Implications of climate change in the twenty-first century for simulated magnitude and frequency of bed-material transport in tributaries of the Saint-Lawrence River. *Hydrological Processes* 25(10): 1558–1573.

Vicente-Serrano, S.M., Beguería, S., López-Moreno, J.J. (2010). A multiscalar drought index sensitive to global warming: The standardized precipitation evapotranspiration index. *J Clim* 23(7): 1696–1718.

World Bank (2013). *Ganges Strategic Basin Assessment: A Discussion of Regional Risks and Opportunities*. Washington, DC: The World Bank.

6

Agriculture and water use

Implications for sustainable intensification in the Ganges Basin

Bharat Sharma, Alok K. Sikka, Ram Pratap Sah and Xueliang Cai

Key messages

- The actual primary productivity of agriculture, forestry, fisheries and livestock is only low-to-medium in the Ganges Basin as integrated development and utilization of its natural resources have not yet been practised by the riparian countries.
- Agriculture is highly varied in the upper catchments, Gangetic Plains, Central Highlands, and the delta regions of the basin.
- Crop and water productivity of most of the crops (rice, wheat, maize, millets, pulses and oilseeds) is low through most of the basin and improving the productivity of available farmland is the only option for improving rural economies.

- Farming systems need to be reconfigured basin-wide for sustainable intensification of ecosystem-based crop production.
- Rejuvenation of farmer-managed irrigation systems, development of multiple-use systems and revival of mountain springs is important for intensification and high-value diversification of agriculture in the Himalayan catchments.
- Large-scale adoption of resource conservation technologies and system of rice intensification can improve productivity in the Gangetic Plains. Investing in improvement of rural power supply can be the most effective stimulus plan for the agrarian economy in the poor eastern regions.
- Strategies for transforming the water-stressed rainfed agriculture in the Central Highlands may include decentralized rainwater harvesting and development of low-cost irrigation infrastructure to meet the critical water needs.
- Agriculture in the delta region may be improved through rice-based system intensification and diversification, innovative homestead production systems and intensification of coastal aquaculture.

Introduction

Availability of abundant water resources, fertile soil and suitable climate in the Ganges Basin have supported a large agriculture-based civilization and one of the most densely populated regions of the world. Ironically, the Ganges Basin is also among the poorest regions in the world despite its rich natural endowments of land, water, people and culture. The basin covers a large area of 1.01 million km^2 shared by India, Nepal, China and Bangladesh. The main physical sub-divisions of the Ganges Basin are: the northern high mountains comprising the Himalayan ranges and the foothills; the vast Gangetic fertile plains between the Himalayas and the Deccan plateau; the Central Highlands lying to the south of the Gangetic Plains; and the delta. The basin receives high annual average rainfall of more than 1,150 mm spatially averaged across the basin, and 2,000 mm or more in the upper northern Himalayan catchments. In absolute terms, the total amount of water in the Ganges Basin is enough to meet the social, economic and environmental needs of the riparian countries. However, there are large variations in the temporal and spatial distributions of water resources. Water is plentiful during the monsoon period and flooding generally occurs, particularly in the downstream reaches in India and Bangladesh. In contrast, during the non-monsoon summer and winter dry period, parts of the basin become water stressed and flows may be inadequate to meet irrigation demand. The basin has few dams and a series of barrages across the main stem and tributaries, but these have not prevented flooding or water shortages. Additionally, long stretches of the river have become highly polluted due to unregulated disposal of domestic sewage and industrial effluents, and contaminated by arsenic. The integrated development and utilization of the basin's huge natural resources have not yet been practised by the riparian countries due to differences in perceptions and lack of mechanisms for sharing the potential benefits (Sharma *et al.*, 2010a).

Agriculture in the Ganges Basin: an overview

More than 60 per cent of the people in the basin depend on agriculture. The actual primary productivity of agriculture, forestry, fisheries and livestock is only low-to-medium. As conditions are extremely heterogeneous, it is necessary to assess potential and actual productivity separately in the upper Himalayan catchments, Gangetic Plains, Central Highlands and the delta (Table 6.1). While the challenges in these sub-regions vary, large yield gaps (two- to four-fold) between the current yields and achievable potential are common across the basin (Sharma *et al.*, 2008).

Table 6.1 Potential and actual productivity in different parts of the Ganges Basin

Region	*Agriculture*		*Forestry*		*Fisheries*		*Livestock*	
	Potential	*Productivity*	*Potential*	*Productivity*	*Potential*	*Productivity*	*Potential*	*Productivity*
Upper Catchments	Low-to-medium	Low	High	Low-to-medium	High	Low	High	Low
Gangetic Plains	Medium-to-high	Medium	Low	Low	Medium-to-low	Low	High	Medium
Central Highlands	Medium	Low	Medium-to-high	Medium-to-low	Low	Low	High	Low
Gangetic Delta	Medium-to-high	Low	Low	Low	High	Medium	Low	Low

Source: Sharma, Amarasinghe and Sikka (2008)

Upper catchments

This region covers the Uttarakhand state of India and about 77 per cent of the area of Nepal in the form of hills and high mountains. The agricultural potential of the upper catchments, and associated energy and water resources, have not yet been sufficiently developed. Most of the hill and mountain systems in the basin are still practising low productivity rainfed subsistence farming. The green revolution, which brought change in the northern plains, was not able to penetrate the hills and mountains because high input technologies were not feasible in the uplands. The green revolution also mainly focused on rice, wheat and maize and overlooked many of the existing cropping systems in the hills, which could have benefited from research leading to increases in efficiency and productivity. The important reasons for slow growth and low yields are use of old varieties and seeds, limited use of fertilizers, water and other inputs, and lack of access to competitive markets and services (Sah, 2013). As a result, the region continues to have a food deficit and be a net importer of food grains to meet its demands (NRB, 2014). On the other hand, the hills and mountain systems, which include forests, grasslands, wetlands and high-value horticulture, are also providing ecosystem services to low-land areas through freshwater that is captured, stored and purified in mountain regions. Mountain regions are hotspots of biodiversity; and from a societal point of view, mountains are of global significance as key destinations for tourist, spiritual and recreation activities. Therefore, the challenge is to identify areas where traditional subsistence systems can be transformed into market-oriented agricultural ones through diversification with high-value crops like vegetables, flowers, spices and fruits, productive livestock and fisheries and agro-forestry systems – to make uplands eventually more productive, profitable, resilient and attractive to out-migrating youths. At the same time the challenge is to manage these upland systems without drastically impacting the ecological services that they provide. As the rains are large and distributed and the region remains cool during a large part of the year, irrigation water needs are small but critical to manage dry spells during the rainy season and meet the water needs of high value agriculture and livestock during the non-rainy season. Traditional water management in upper catchments is at a different scale through farmer-managed small diversion and lift irrigation schemes, use of natural springs and runoff storage in lined ponds, tanks and cisterns which play a pivotal role in improving productivity of mountain and hill agricultural systems.

Gangetic Plains

Spread largely over the states of Uttar Pradesh, Bihar and West Bengal in India, and the terai (Madhesh) region in Nepal and upper districts of Bangladesh, the Gangetic Plains is one of the most fertile and populous regions of the world – mean population density is more than 10 times the global average. With a major share of rice in the rainy season and wheat in the winter season, maize, sugarcane, pulses and oilseeds, and fruits and vegetables are also vastly cultivated in the plains. Livestock, dairy and riverine and freshwater fisheries form an integral part of the agricultural system and are more popular among the smallholder and marginal farmers and landless labourers. In the small western part of the Gangetic Plains the combined rice–wheat productivity is high and estimated above 8 t/ha/year, and potato and sugarcane productivity is also high (Cai *et al.*, 2010). Productivity of winter maize is also very high in the irrigated districts of Bihar and parts of Bangladesh. However, in large parts of the eastern Gangetic Plains, in spite of the rich alluvial soils, abundant surface and groundwater resources, favourable climate and large agricultural workforce, agricultural productivity remains low and as such the northern Gangetic Plains may be described as a '*low-productivity–high potential*' region (Sharma *et al.*, 2008). In large pockets in eastern Uttar Pradesh and Bihar, 55 per cent of people live below the poverty line and are vulnerable to annual floods and intermittent droughts. Other factors of low productivity include inadequate drainage, unfavourable land tenure and smallholdings, poor electrification, and inadequate infrastructure and institutional arrangements including marketing. The irrigated parts of the plains now face second generation problems with sustaining productivity, largely due to degradation of land and surface waters. Yet, the massive groundwater resources remain underused, mainly due to non-availability of pumps and the high cost of pumping with diesel-operated and rented pumping units with little or no access to affordable electric or solar power (Kishore *et al.*, 2014). Mounting salinity, arsenic contamination, waterlogging as well as an imbalance in nutrients and loss of bio-diversity pose additional threats for improving agricultural productivity.

Central Highlands

The Yamuna, Tons, Sone and Punpun tributaries of the Ganges flow through the semi-arid Central Highlands and cover parts of the states of Uttar Pradesh, Rajasthan, Delhi, Madhya Pradesh, Chattisgarh and Jharkhand in India. These parts receive insufficient monsoon rains, temperatures remain high and the annual deviation between potential evaporation and rainfall varies from (-) 561 mm to as high as (-) 967 mm (Table 6.2). Shallow, rocky, sandy and undulating soils have low water-holding capacity and poor aquifer characteristics. As such the groundwater resources are deep, discontinuous and limited and also afflicted with salinity, alkalinity and fluoride. Rainfed minor millets, pulses, oilseeds, wheat and barley are the main crops and have low productivity. Vast areas are also covered under degraded grasslands, pastures, scrub, open forest, ravines and wastelands. Water insecurity is a constant threat for crops, large herds of unproductive livestock and the poor and tribal population dependent on these resources. Only about 15 per cent of the net sown area is irrigated by different sources and as such value of agricultural output per hectare is only about one-third of the national average (Phansalkar and Verma, 2005). A major drawback of Ganges' basin irrigation and agriculture strategy has been the neglect of the catchment areas and the people living there. These Central Highlands offer the unique opportunity for enhancing rural livelihoods through water-centric investments. Facing severe water insecurity in this region, the communities are increasingly obliged to evolve a lifestyle of rainfed farming followed by distress migration. While large public irrigation systems have left these regions by the wayside and groundwater resources are scarce and costly, there is a big opportunity in imaginative small-scale interventions for improved water control, which can produce a dramatic impact on the agricultural productivity and dependability of the rural livelihood systems (Phansalkar and Verma, 2004, 2005).

Delta

The enormous delta of the Ganges, further joined by the Brahmaputra River in Bangladesh and surrounding areas of India, is the lifeblood of the largest populations on earth. The coastal region in the delta of the River Ganges in India and Bangladesh is known as the 'Sunderbans'. Although the delta has been extensively impacted by human activity for hundreds of years, the combination of river channel mobility and the persistence of traditional agricultural and settlement practices has retarded modern river control practices. The Ganges Delta is under increasing environmental pressure today in response to the needs of the rapidly growing and modernizing population. During the 1960s a major geophysical intervention took place in the coastal areas of Bangladesh and to a much lesser extent in the Indian part (Choudhury *et al.*, 2004). The embankments were built on the riversides to serve a dual purpose: first to protect the coastal communities from daily tidal surges, flooding and natural calamities, and, second, to increase agricultural productivity in the coastal regions. Over the years the polder infrastructure, with its embankments, canals and sluice gates, became a key source of water for multiple uses like irrigation, aquaculture, domestic use and drainage (Das *et al.*, 2015). In the Indian part, spread across 19 blocks of the two southernmost districts of West Bengal – North 24 Parganas and South 24 Parganas – only flood control levies have been built and it is one of the most disadvantaged and poverty-stricken areas in India. Farmers are mostly very resource-poor, marginal and smallholder with highly fragmented land holdings (Burman *et al.*, 2015). The productivity of the land is very low due to soil and water salinity.

Agriculture in the Sundarbans of India is mainly mono-cropped with *kharif* rice (*aman*) grown in the monsoon season. Scarcity of freshwater for irrigation during the post-monsoon period is the main constraint for growing *boro* rice and other crops. Despite huge investments in the polders of the coastal zone over the past 50 years, crop productivity remains much lower than in the rest of Bangladesh. Most farmers in the polders cannot adopt high-yielding varieties of *aman* rice due to the risk of submergence after transplanting. The local *aman* varieties mature late, delaying establishment of *rabi* (non-rice) crops, which, as a consequence, are often damaged and destroyed by pre-monsoon rains or cyclones prior to harvest (Mondal *et al.*, 2015). Innovative water management is one of the most imperative issues for the Ganges Delta. The importance of inland fisheries for the livelihood and food security of the poor and landless is widely acknowledged in the Ganges Delta. Fish from vast inland waters are vital to millions of poor people but landings and species diversity are believed to be declining due to siltation and change-over to agriculture, increasing fishing pressure and an acute shortage of dry season wetland habitat (Hughes *et al.*, 1994; Ahmed *et al.*, 2002).

Basin water balance and agricultural water use

The mean annual input by precipitation to the Ganges Basin totals about 1,170,000 mcm (1,158 mm). After meeting the needs of all the water uses, storage changes and losses in the basin, net runoff from the basin is about 429,000 mcm (429 mm) or about 37 per cent of the total precipitation, which is significant compared to some other major river basins (e.g. 10 per cent in Indus – Eastham *et al.*, 2010). Annual potential evaporation is the next largest component and varies, between years, around 1,560 mm. Rainfall and potential evapotranspiration vary spatially, from the drier and hotter west to the wetter and cooler Himalayan regions (Table 6.2). The large difference between annual precipitation and evaporation in the middle and lower Ganges sub-basins (Figure 6.1) indicates the need for irrigation – for higher agricultural productivity and intensification.

All the agriculturally important sub-basins, except Paksey in Bangladesh, have higher annual potential evaporation values than the annual precipitation, with the largest deviation of 970 mm in Chambal, India. In the basin as a whole, and especially in the catchments with large unmet potential evaporation, rainfed agriculture is the most extensive land use, covering 52 per cent of the basin. Its water use is also correspondingly higher, an annual average of about 372,000 mcm, or 32 per cent of the total water used. Irrigated agriculture covers 25 per cent of the basin, with 17 per cent of the total area irrigated from surface water resources and 8 per cent from groundwater. The estimated mean annual water use by irrigated agriculture is about 210,000 mcm, or 18 per cent of total water use. The majority of the irrigated water use is from crops irrigated from surface water resources (70 per cent), with the remaining 30 per cent from groundwater irrigated crops (Figure 6.1). Grasslands cover 14 per cent of the basin and consume about 100,000 mcm (8 per cent) of the water used. The remaining 5 per cent of the water

Table 6.2 Variations in mean annual precipitation and potential evaporation

Sub-basin	*Precipitation (mm)*	*Evaporation (mm)*	*Deviation (±, mm)*
Devghat	2,586	984	+1,602
Kampughat	2,520	1,055	+1,465
Chisapani	1,849	921	+928
Everest	1,162	784	+378
Paksey	1,860	1,528	+332
Farakka	1,410	1,595	–185
Lower Ganges	1,328	1,649	–321
Ghagra	1,228	1,592	–364
Ganges Source	939	1,437	–498
Upper Yamuna	892	1,453	–561
Sone	1,115	1,700	–585
Middle Ganges	965	1,760	–795
Yamuna	940	1,769	–829
Chambal	793	1,763	–970

Source: Adapted from Eastham *et al.*, (2010)

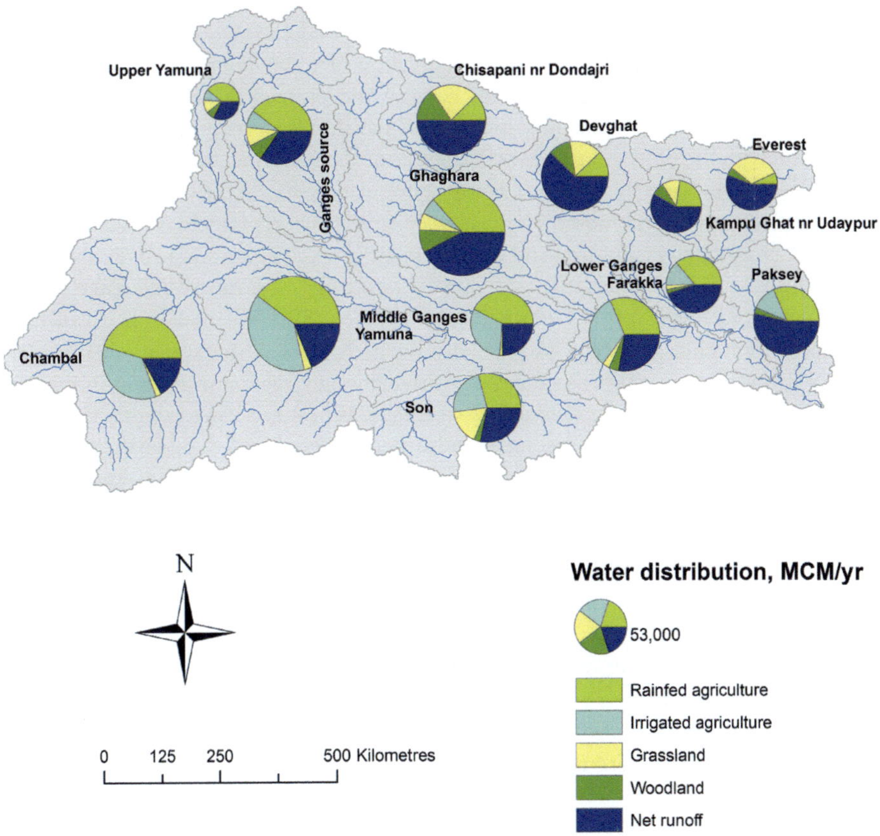

Figure 6.1 Agriculture and related water uses in the catchments of the Ganges Basin

Source: Published with permission from Eastham *et al.*, 2010 and CSIRO-Land and Water (about 57,000 mcm) is used by woodlands, forests, bare ground and urban areas etc. Rainfed or irrigated agriculture water use is the most important component of vegetation water use in the catchments in lower altitudes. Irrigated water use is most important in the Chambal, Sone and Lower Ganges catchments, as apposed to rainfed agriculture in the Upper Yamuna, Ganges Source, Middle Ganges, Farakka and Paksey catchments (Eastham *et al.*, 2010).

Agriculture and water productivity

There is very little prospect of expanding food production by bringing in more land under cultivation. Yet, sufficient opportunities do exist to augment the productivity per hectare of cultivated land and productivity per unit of applied inputs like water and fertilizers. In their seminal treatise, Hayami and Ruttan (1985) espoused 'internal' as opposed to 'external' land augmentation as a way of overcoming severe constraints of arable land in the densely populated basins. However, the technological innovation must shift from just land augmentation to environment saving by considering environment as a factor of production (Alauddinn and Sharma, 2013).

The crop types that can be grown in a basin are also determined by rainfall. Regions with low rainfall in the Yamuna and Chambal sub-basins grow crops like millets, pulses and oilseeds that require relatively less water, and crop mixtures that protect farmers against the uncertainty of rain. Rainfed cropping in regions with relatively high rainfall in the eastern part of the basin is also limited to the rainy season, but they grow water-intensive crops like paddy, maize and sugarcane. A distinguishing feature of this region is that, besides increasing the area under wheat and other *rabi* crops, it has led to extensive cultivation of water-intensive paddy in the *Kharif* season. Since the 1970s the area under paddy has grown exponentially and intensive production of irrigated rice, wheat and maize in some pockets has shown good promise, but in vast areas land and water productivity remain low. Mapping of the land and water productivity in the whole of the Ganges Basin by integrating crop census data with remote sensing and GIS helped in assessing productivity and food security at basin scale. The methodology included mapping of dominant crops using census data and existing maps of land use and land cover, interpolation of the calculated yields to pixel level (250 m * 250 m) using normalized difference vegetation index (NDVI) and employing a Surface Energy Balance (SEBAL) model for calculating the crop evapotranspiration and water productivity for each pixel (Sharma *et al.*, 2010a; Cai and Sharma, 2010; Cai *et al.*, 2010). Rice is the overwhelmingly predominant crop in the Ganges Basin cultivated from the upper catchments to the delta. Wheat to a lesser extent is the dominant crop in the *Rabi* season. Sugarcane is mostly cultivated in Uttar Pradesh and parts of Bihar state in India. Other crops like maize, potatoes, pulses, oilseeds, millets, fruits and vegetables scatter out in fragmentations.

Agricultural productivity

Agricultural productivity of most crops is low throughout most of the basin. The average rice yields are 2.53, 3.54 and 2.75 t/ha in the India, Nepal and Bangladesh parts of the basin (Figure 6.2). However, large variations exist in different parts of the basin. The 'bright spots' adjacent to the Indus Basin in India have an average yield of 6 t/ha. The 'hot spots' of low yield rice are found in the Indian states of Madhya Pradesh and Bihar, and the Dhaka division of Bangladesh. Yet, a pixel-level analysis indicates that the very low performance of Bihar also has a bright spot in a 37-km-diameter circle centred south-west of Bhojpur with an average yield of 4 t/ha. The trend of wheat yield variation is similar to that of rice. Western Uttar Pradesh and small pockets in the Madhya Pradesh and Rajasthan states of India are the bright spots with average

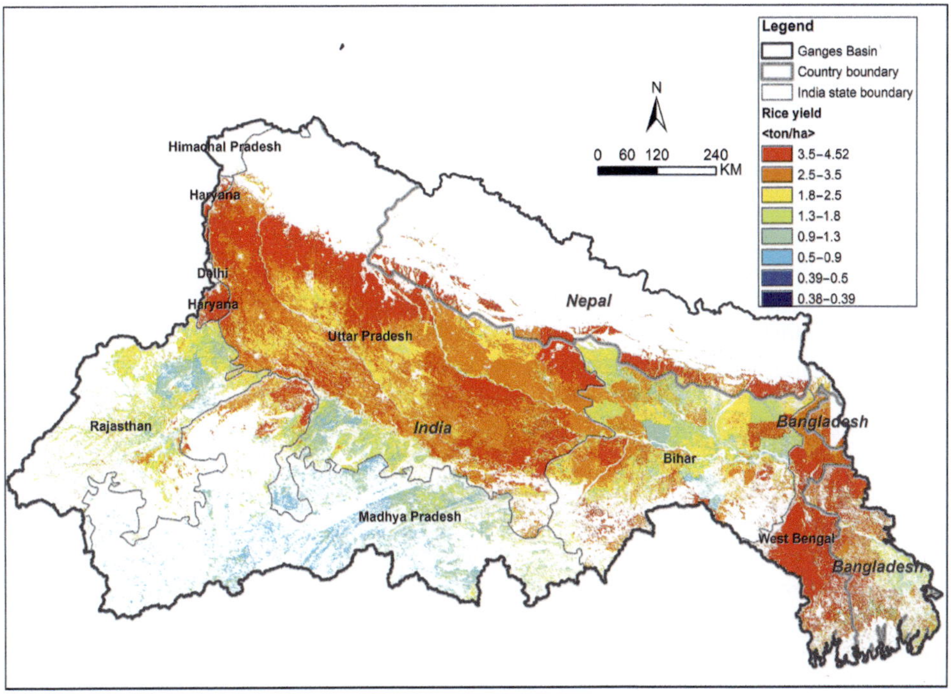

Figure 6.2 Variation of rice productivity in the Ganges Basin

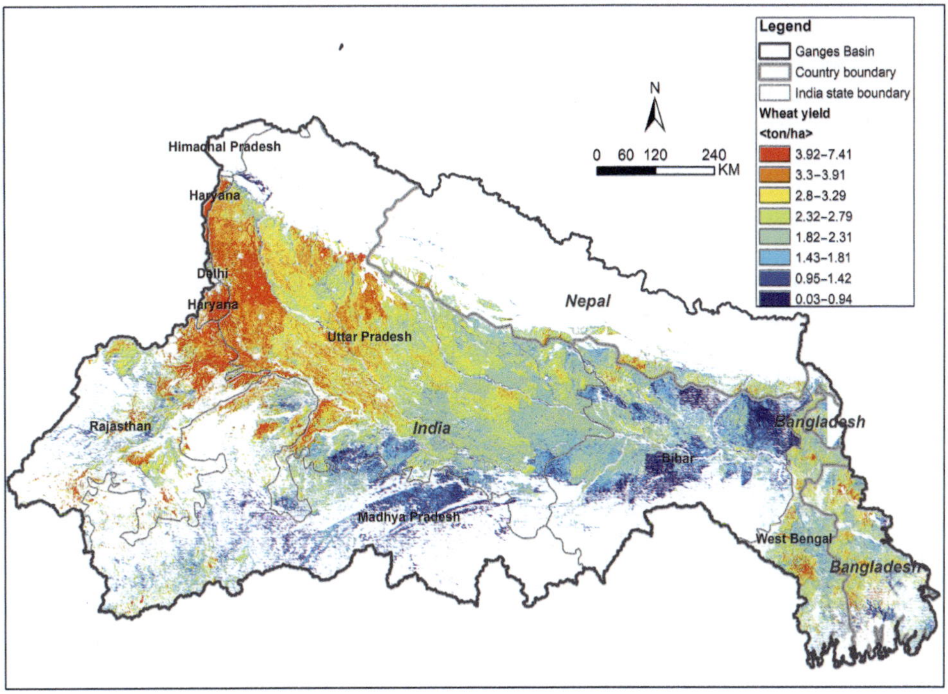

Figure 6.3 Variation of wheat productivity in the Ganges Basin

yield above 4 t/ha, and hot spots with yield as low as 0.70 to 1.58 t/ha appear in the Indian state of Bihar, parts of West Bengal and Bangladesh (Figure 6.3). The upper catchments in both India, Nepal and the Central Highlands in southern parts of the basin have low cropping intensity and agricultural productivity of cereals and other crops, largely due to water stress during non-rainy seasons.

The productivity of rice, maize and wheat in Nepal is 3.2, 2.3 and 2.4 t/ha, respectively with terai regions having about 10 per cent higher cereal productivity than the national average. Several rice farmers in Bihar and Bangladesh, having access to irrigation, have switched to growing maize during the non-rainy winter season. High-yielding hybrids help to drastically reduce diesel-pump operation hours and thus double the economic returns. This trend is particularly fast emerging in Bangladesh where, between 2000 and 2013, maize production increased from just 10,000 tonnes to 2.2 million tonnes and harvested area from 5,000 ha to 320,000 ha (FAO, 2015). Co-cultivation of fish in the paddy fields is also a popular practice in the high rainfall areas of West Bengal and delta areas in Bangladesh.

Water productivity

Assessment of water productivity variation in a basin is useful for developing an effective water management strategy. Water productivity of the rice and wheat crop for the entire Ganges Basin is depicted in Figures 6.4 and 6.5 (Cai *et al.*, 2010; Cai and Sharma, 2010). Water productivity variation of rice closely follows the pattern of yield variation with minimum and maximum values 0.20 to 2.04 kg/m^3. However, as much as 19 per cent of the total rice areas have water productivity less than 0.5 kg/m^3, which occurs mainly in the Indian states of Madhya Pradesh and Bihar and in the Dhaka division of Bangladesh. The high water productivity values in some southern regions of the basin do not necessarily suggest a satisfactory performance as these regions have low crop yield with a low water consumption (deficit irrigation) leading to high values of water productivity. Rather, it provides interesting clues to examine possible water stress which could be potentially reduced by supplying supplementary irrigation during the dry spell. Similar anomaly is also noticed for the water productivity of wheat crop. Due to extremely low evapotranspiration in the Indian states of Rajasthan and Madhya Pradesh, water productivity in these areas shows higher values despite low yield. These states still cultivate traditional low-yielding wheat varieties which incidentally have high cooking quality and fetch a premium price in the market. The growing season in these states is also shorter due to a shorter winter period and early maturity of the crop. The Indian state of Bihar has the largest areas with lowest water productivity, which means that there is significant scope for improvement in this area. The downstream areas of the Ganges Basin show relatively high water productivity despite high variability in yield. Large variations in water consumption and water productivity of rice and wheat could be explained by a number of factors, for example, land fertility, water supply, crop management and climatic conditions. Large areas in Nepal terai, Bihar and West Bengal have very shallow groundwater depth. Overall rice water productivity is relatively low – 0.300 to 0.459 kg/m^3 in Bangladesh – and this offers a significant scope for improvement. Technological diffusion was the key factor explaining inter-district water productivity differences in Bangladesh. Spread of high-yielding variety is an important determinant of rice (and wheat) water productivity but spread of this technology depends critically on the access to irrigation, more specifically groundwater resources. Widespread diffusion of high-yielding varieties during *Kharif* season is yet to occur. Technological breakthroughs, though available now, are yet to take firmer roots to circumvent these uncertainties (Alauddin and Sharma, 2013). This could be attributed to poor local crop and water management practices, especially low fertilizer use, use of traditional

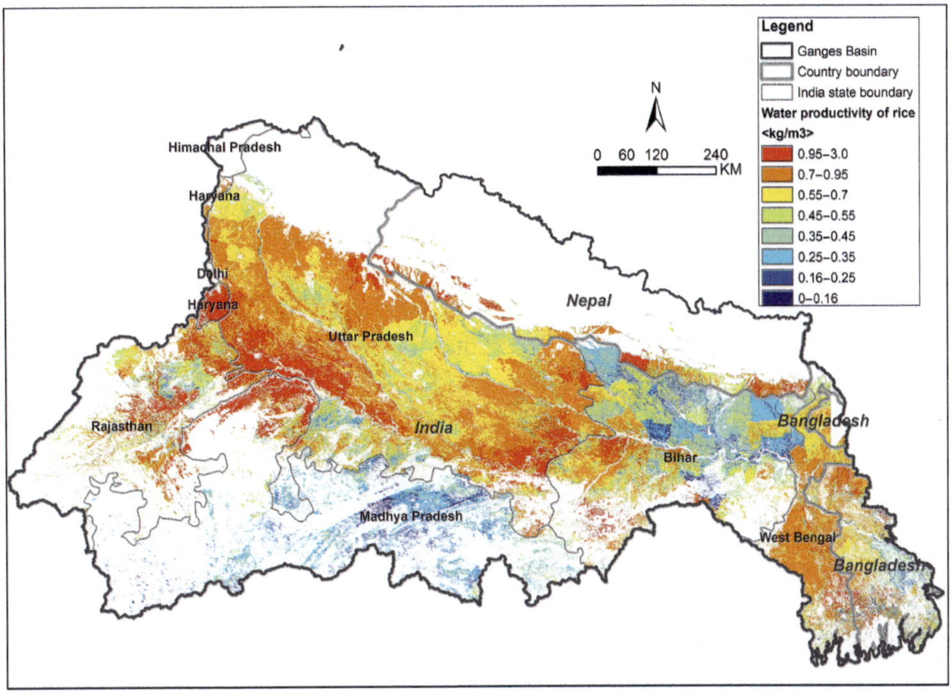

Figures 6.4 Variation in water productivity of rice in the Ganges Basin

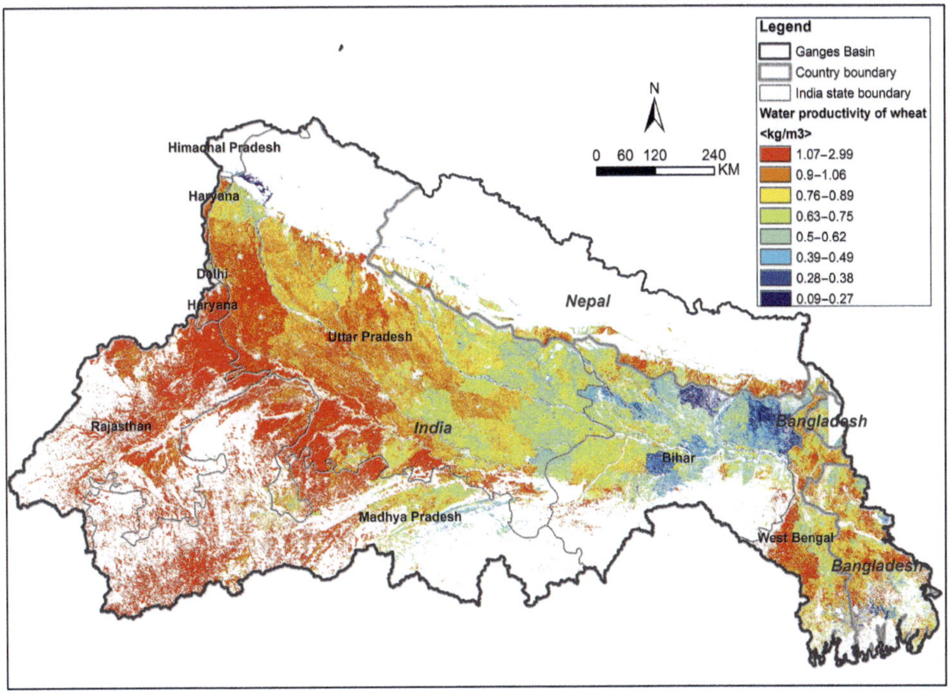

Figures 6.5 Variation in water productivity of wheat in the Ganges Basin

crop varieties, crop disease and pests, and poor extension services. Rainfall may occur at any time. Hence, higher rainfall areas have more water to evaporate but could still suffer from water stress during the critical crop growth period (especially the terminal grain filling stage), which drastically affects the final grain yield accumulation.

The main constraint in the Ganges Basin, however, is not water availability or climate, but the water infrastructures and energy policies, and irrigation and crop management practices (Biswas, 2008; Sharma *et al.*, 2010a). Well-developed irrigation and drainage systems, together with matching management practices, can help to supply water during the dry spell, and drain the excess water from rainfall and shallow groundwater. Other land and crop interventions such as resource conservation technologies and land levelling, insects and diseases control, fertilizer and improved variety, and farm credit and markets are also important factors to be considered along with water management.

Prospects for sustainable intensification of agriculture

The preceding analysis shows that low and stagnating yields from tiny parcels of land, under-developed and inefficient water infrastructure and farm energy, and insufficient technological diffusion threaten food security and the livelihoods of the roughly 635 million population of the Ganges Basin. This challenge is even more severe when seen in the context of increasing rainfall variability due to climate change, over-exploitation as well as under-utilization of groundwater resources due to inefficient energy policies and deteriorating water quality in the main stem and certain aquifers and tributaries of the Ganga River. Increasing the productivity of available farmland is the only option for improving rural economies and farmers' livelihoods, and saving the population at risk of hunger and food insecurity. Farming systems need to be reconfigured basin-wide for sustainable intensification through adoption of important components of conservation agriculture, soil health, high-yielding varieties, integrated pest management and above all efficient water management. But much more also needs to be done in improving governance, policies and institutions, and transboundary cooperation if agriculture is to deliver the full benefits of ecosystem-based crop production.

Upper catchments

Agriculture in the upper catchments is characterized by low, middle and high hills with little or no agriculture in the snow-covered mountains (> 2,300 m amsl) (Joshi *et al.*, 2012). Rugged terrain, remoteness and inaccessibility, and poor infrastructure and services keep the agriculture at subsistence level. Analysis of production constraints have duly reflected that the investments in agribusiness and production in Nepal and the Uttarakhand hills is constrained by timely availability of quality production inputs and competitive markets. In spite of the large availability of freshwater resources in the region, its access by scattered and smallholder poor farmers is challenging and is the main bottleneck for agriculture diversification and intensification. Only about 15 per cent of the cultivated farms are irrigated in the hill districts of Uttarakhand and 27 per cent in the low and mid hills of Nepal. Impacts of small but critical irrigation are phenomenal as cereal yields improve by more than 55 per cent as compared to rainfed crops (Sah, 2013). However, value addition and intensification of agriculture will happen through provisioning of necessary conditions to support the growing interest and opportunities for cultivation of high-value off-season vegetables, horticulture and floriculture, spices and medicinal plants, tea and coffee in the low and mid hills and temperate fruit orchards in the high hills and mountains. Water-regulation and irrigation of these productive plots is the first step to move up the ladder, and bright spots of such models have shown high returns of between US$3,000 and US$5,000 per hectare with benefit–cost ratio of above 3.5 (Sharma *et al.*, 2010c). Hilly areas also provide the niché production environments for the organic products which have witnessed an increasing demand in urban markets. The following small water solutions, along with the synchrony of back-end and front-end linkages, can potentially create big changes in the way agriculture is practised in the upper catchments of the Ganges Basin:

Rejuvenation of Farmer-Managed Irrigation Systems (FMIS): About 22 per cent of the total irrigation systems in the Nepal hills and a similar percentage in the hill districts of Uttarakhand are small FMIS. Important virtues of these systems are accumulated experiences, knowledge and skills on ecology, hydrology and layout, construction, operation and maintenance of the system achieving high water allocation equities and a strong communication system (Shivakoti and Shrestha, 2008). Most of these traditional systems now need to be refined and supported to achieve higher efficiencies.

Multiple Use Water Systems (MUS): In the upper catchments, only 5 per cent of the existing water resources are used for economic activities. MUS can provide reliable water supplies for household needs and more productive agricultural activities in the hilly regions of India and Nepal (Sharma *et al.*, 2010c). It is a participatory, integrated and poverty-reduction-focused approach that takes a community's diverse water needs as the starting point for providing services (Mikhail and Yoder, 2008). The designs for MUS are based on: groundwater/lake water lifting and distribution; rainwater collection and distribution; spring water distributed by gravity system; and stream/river water supply after treatment. Most MUS are designed to cover 10 to 40 households. In some cases, up to 80 households have been provided service from MUS. An evaluation of the 81 schemes installed in the low and mid hills of Nepal showed that they more than met the key criterion with the added benefit of low initial investment costs (approximately US$200 per household) and short cost-recovery periods. With MUS, households can earn an additional annual income of about US$190 through the sale of surplus produce, which means that the system has a payback period of only one year (Pant and Samad, 2011).

Revival of water springs in the low and mid hills: As streams and rivulets are in deep reliefs and difficult to access, water for life and livelihoods in the upper hamlets in the mid hills comes from natural springs. Most of the local drinking water needs are met by the springs and the surplus water is used for small-scale farming. Through appropriate technology the springs can be developed to meet the larger needs of the community and successful examples exist in the Kumaon and Garhwal regions of Uttarakhand where low-density polyethythene-lined tanks and small gravity-based drip irrigation systems have been integrated with the natural springs. In the mid hills of Nepal, spring density has been estimated at 5–7 springs/km^2. Unfortunately, the springs that once used to flow profusely are drying up at a fast rate due to unmindful developmental activities and loss of recharge areas. Revival of these springs through suitable technical and social engineering (Santiago *et al.*, 2007; ICIMOD, 2015) holds great promise for meeting the critical water needs and transforming the hill economy. The developed springs can also use solar-powered pumps for lifting of water from tanks and operation of micro-irrigation systems.

Gangetic Plains

The north-western plains of the basin are characterized by a largely intensified crop production, with high external inputs, high productivity and high market integration. In contrast, the eastern plains show poor crop yields, costly and scarce irrigation, high dependence on rainfall, and small and fragmented farm-holdings. Together with lack of institutional finance and extension services, poverty and uncertainties of rural livelihoods are much higher. Persistent poverty in the land corridor connecting Nepal terai, eastern India and Bangladesh is accentuated by inequalities based on class, caste, ethnicity and gender.

Increasing production and strong economic growth in agriculture – with small-scale farmers (that dominate agriculture in the region) at the centre of attention – are urgently needed to achieve food security and poverty reduction. The preservation and enhancement of the natural production potential of agriculture are essential to stabilize achieved high yields in favourable areas in the north-west, to realize more of the existing yield potentials in the east, and to increase the resilience of farming systems across the plains. Possible avenues for sustainable agricultural intensification in the Gangetic Plains may include the following.

Resource conservation technologies: Conservation agriculture aims to overcome today's intersecting challenges: raising crop productivity and ensuring food and nutrition security for all, while reducing agriculture's demands on natural resources, its negative impacts on the environment and its major contribution to climate change (FAO, 2011). Technologies include zero-tillage, laser-assisted land levelling, retention of surface residues, permanent bed planting, dry seeding of rice, surface seeding of wheat (Ladha *et al.*, 2009) and permanent raised beds, which produce more rice, wheat and income in less time, using less water, fertilizers and energy.

A major constraint to improved wheat productivity in the Gangetic Plains is its late sowing. Rice transplanting starts in July and often continues until late August, because of the uncertainty of rains, high cost of pumping groundwater and labour shortages. These delays result in a late rice harvest and sowing of the following wheat crop. The planting dates of both rice and wheat can be brought forward by new methods of crop establishment. Zero tillage and direct seeding of wheat economize on time and also on water and fuel, saving an estimated US$50 to US$70 per hectare (Gautam, 2008). On the eastern plains and terai region, where drainage is poor, pre-soaked wheat seeds are either broadcast or drum-seeded on saturated soils without tillage. For rice, substitution of long-duration rice cultivars with short duration ones, and direct wet or dry seeding, eliminate the need for transplanting. Aerobic rice and 'alternate wet and dry irrigation' of rice reduces water consumption by 30 to 50 per cent. Laser-assisted land levelling achieved by a tractor-mounted laser beam projector, operated by private contractors, allows precise levelling of fields at prices smallholders can afford. Recent studies in north-west India found that the technology is far more efficient than traditional levelling, reducing water applications by up to 40 per cent, improving the efficiency of fertilizer, and boosting rice and wheat yields by around 8 per cent. It is also scale-neutral, being equally profitable on all farm sizes (Gupta and Sayre, 2007; IRRI, 2009; Aryal *et al.*, 2014). To date, zero-tillage has been adopted mainly for the wheat component of the rice–wheat system. Applied to rice, it would lead to bigger reductions in the use of irrigation water, which are urgently needed (Erenstein, 2009). Numerous trials of zero-till dry-seeded rice have also shown that puddling is not essential for high yields (ACIAR, 2008).

System of Rice Intensification (SRI): Increasing numbers of farmers are adopting new crop, soil and water management practices which, together known as SRI, produce more rice and income using less water, less fertilizer and even less seed. Strikingly different to the traditional cultivation practices, 7- to 15-day-old seedlings are transplanted singly, often in grid patterns with wide spacing of 25 cm × 25 cm per plant. To promote moist, but aerated soil conditions, alternate wet and dry irrigation is followed with dry periods of three to six days. Weeding is done at regular intervals, and compost, farmyard manure and green manure are preferred to mineral fertilizer. Once the plants flower, the field is kept under a thin layer of water until 20 days before the harvest (Berkhout *et al.*, 2015; Uphoff, 2008). The system was found to improve grain yields by more than 42 per cent and water use was reduced from 25 to 47 per cent below that for the flooded system (Thakur *et al.*, 2009; Barah, 2009). SRI rice also made much more efficient use of mineral fertilizer and required less seed. In Nepal terai, seeding rates were reduced to 5 to 10 kg per ha, about five to 10 times less than in the traditional system (Dhital, 2011). SRI's high productivity is attributed to the fact that it capitalized on the existing potential of the rice genome that had been inhibited by sub-optimal management (Africare *et al.*, 2010). Farmers' per hectare net income had increased by an average of US$110, owing to a 40 per cent reduction in production costs (Dzung *et al.*, 2011).

The governments of India, Nepal and Bangladesh are providing good support for adoption and demonstration of SRI. An estimated 600,000 Indian farmers grow their rice using all or most of the recommended SRI practices. In Morang district, Nepal, farmers reported that SRI had often doubled their yields. In addition, their rice was maturing up to four weeks earlier, which saved water, reduced the risk of crop losses and made land available for the following crop. By growing more with less water, SRI will allow farmers to continue to cultivate rice in large rainfed areas, and in major irrigated areas where water supply is insufficient to meet the demand.

Agriculture and rural power for poverty reduction in the eastern Gangetic Plains: Rapid growth in agriculture is essential for poverty reduction when 90 per cent of people in eastern Uttar Pradesh, Bihar and Nepal terai still live in villages and 81 per cent of the workforce reports agriculture as the main source of income. Some innovative policies in the region need serious consideration for large-scale outscaling (Kishore *et al.*, 2014). Bihar became the first state in India to constitute an 'Agriculture Cabinet', with ministers from 17 departments, to improve coordination in implementation of agricultural projects. At the grassroots level, agriculture extension workers called *Bhumi Mitras* (Farmer Friends) have been hired in all village Panchayats of the state to bring new technology and practices to the farmers. The state government of Bihar is offering generous subsidies on farm implements, quality seeds and diesel for irrigation pumps, and has simplified the process for accessing these subsidies. The West Bengal government provided a good model of providing one-time capital subsidy to small and marginal farmers, along with metered electricity connection (Mukherji *et al.*, 2012). This will make irrigation much cheaper than the current option of diesel pumps. In the dry Barind tract of Bangladesh, technology-enabled cooperative tube wells have been successful. Each member farmer of the cooperative tube well can demand the groundwater as per his farm needs and pay through a pre-loaded smart water card (Zaman and Rushton, 1998).

Investing in improving rural power supply can be the most effective stimulus plan for the stagnant agrarian economy. The region needs intensification of affordable irrigation which will take place only if the variable cost of irrigation comes down and farmers do not economize on the critical needs of water.

Central Highlands

The Central Highlands, dotted with small hills and rolling topography, degraded forests, pastures and wastelands, variable rainfall and mixed farming practices, and inhabited by a largely tribal population, offer a unique opportunity for enhancing livelihoods through investments in land, livestock and water management. As irrigation covers less than 15 per cent of the cultivated areas, most parts remain rainfed. Regions with low rainfall grow crops that require relatively less water, and crop mixtures that protect farmers against the precarious uncertainty of rain. Rainfed paddy and maize cropping in the high rainfall eastern part of the highlands is also limited to the rainy season. The mean value of agricultural output per hectare in the Central Highlands is only about one-third of the national average for India. Strategies for transforming agriculture in the Central Highlands may include the following:

Decentralized water harvesting: Uncertainty of monsoon rains significantly reduces the yield of paddy and other rainy-season crops and forces farmers and livestock to migrate. Rainwater harvesting offers supplementary irrigation as well as protection against the variability. Most past efforts in rainwater harvesting have been initiated by the government on the community lands, and the outcomes show mixed results with unsuccessful projects outnumbering the successful ones (Sharma *et al.*, 2010b). Individual control over available water, by contrast, can enable a farmer to better plan agricultural operations, productively use the water resources, and maintain the structures for long-term use (Takeshima *et al.*, 2010; Pangare and Karmakar, 2003). The decentralized water harvesting approach emphasizes construction of structures on farmers' own land with their own resources. Creation of a large number of inexpensive decentralized water harvesting structures such as the '5 per cent Farm Pond', in which about 5 per cent of the farm at the most upstream spot is dug for capturing rainfall, will enable farmers to provide protective irrigation during critical periods of moisture stress. In the Dewas district in Madhya Pradesh, more than 6,000 farmers allocated 6 to 10 per cent of their lands for the ponds. The ponds are unlined but tractors are used to compact the soil in order to minimize seepage (Malik *et al.*, 2014). These have led to significant improvements in the availability of irrigation water and revival of the agricultural economy. In fact, it works as a land-augmenting intervention, enabling users to raise crops in both summer and winter, thereby increasing overall cropping intensity and productivity (Rijsberman, 2004).

Low-cost irrigation infrastructure: Lifting devices such as low-cost diesel pumps need to be encouraged and supported. The region is abundant in small tributaries that can be harvested through small-scale lift irrigation schemes and several of these have been successful in intensification of the production systems. At places where wells exist, low-cost micro-irrigation systems may be introduced to support women-managed nutrition gardens on homestead lands. The benefits can be further improved through treatment of the uplands for soil and water conservation measures and growing improved grasses and fodder trees for livestock, host plants for silkworms/lac and fruit trees. A series of such interventions undertaken together across the local topography can have a significant impact on rural livelihoods.

Coastal zone and delta

The Ganges coastal zone is characterized by a vast network of river systems, an ever-dynamic estuary, and is the drainage path of the mighty Ganges. The limited land and abundant water resources in the coastal zone of the Ganges Delta are vital for crop production, fisheries, ecosystem sustenance and livelihoods. The salinity front moves seasonally as well as spatially in the coastal area. Recent analysis indicates that the quantity of freshwater in the coastal zone is likely to decrease due to the combined effect of external drivers and inadequate transboundary flows that will increase river salinity during the dry season (Khan *et al.*, 2015). In the late 1960s, 139 polders were built in the coastal zone of Bangladesh, of which 54 are located in the Ganges coastal zone, for intensification of agriculture and livelihoods of coastal communities. On the Indian side, only critical sites were protected with levies and embankments.

Agriculture in the Sunderbans of India is mono-cropped with *kharif* rice in the monsoon season. Scarcity of fresh surface water for irrigation during the post-monsoon period is the main constraint on growing *boro* rice and other crops. As such the productivity of the land is very low and the region is one of the most disadvantaged and poverty-stricken areas in India (Burman *et al.*, 2013). In contrast, *boro* rice accounts for 57 per cent of Bangladesh's total annual food grain production of 33 million tonnes. It is produced under irrigated conditions using mainly groundwater (80 per cent). Many small farmers buy groundwater from neighbours to irrigate *boro* rice. Several innovative and sustainable agricultural systems have been designed for intensification of agriculture in the coastal areas:

Rice-based system intensification and diversification: Most agricultural land in the coastal zone grows a single rice crop (*aman*) during the rainy season, mainly using the local varieties with low yield potential. The first step towards intensification was to develop and introduce high-yielding *aman* varieties and this major biotechnological breakthrough has now been made through development of submergence tolerance varieties like Swarna-sub 1 and salinity tolerance varieties like BR11-Saltol which yield about 20 per cent more than the local cultivars. With a good choice of variety and drainage management, even the brackish water shrimp *ghers* (small bunded fields) after the shrimp harvest can produce about 4 t/ha of *aman* rice, improving livelihoods in the more saline areas (Rahman *et al.*, 2015). The next step was to develop a two rice crop *aus-aman* (summer-rainy season) system for increasing the productivity of moderately saline regions of the coastal zone. Combining a modern short duration *aus* variety sown in mid-April followed by a high-yielding *aman* variety can have system yields approaching 9 t/ha/year, much higher than the 2.5 t/ha/year of rice commonly obtained by farmers who grow a single rice crop (*aman*) using a local variety. Recent studies further show that in areas of year-round fresh water availability and using high-yielding short duration varieties, triple rice (*aus-aman-boro*) in a year is a possibility. Successful implementation will require community coordination, both for polder water management and for cropping system synchronization. Total in-field system duration ranged from 277 to 312 days with yield levels in excess of 15 t/ha/year (Saha *et al.*, 2015). For diversification, the third irrigated rice crop in the dry season can be replaced with high-value maize, oilseed and vegetable crops. Initial estimates show that more than 100,000 ha of low-intensity cropland could potentially be brought into intensified production using surface water irrigation.

Homestead production systems: Small homestead food production systems (HFS) are an important source of income, food and nutrition for millions of poor farmers in the Ganges coastal zone. HFS is mainly a need-oriented, self-provisioning, integrated, multi-species, economically sustainable and environmentally safe farming system around the house where the soil is enriched by homestead biological formulations (Bhattacharya *et al.*, 2013). This intensive micro-production system on a land area of 400–1,300 m^2 will typically have an aquaculture homestead pond (33 per cent of area), vegetable and fruit production, and poultry and small livestock with a very efficient integration and recycling of resources among different components of the system. HFS through its year-round intensive and diversified production from a very small piece of land has the potential to improve the household food security of the poor irrespective of their land-holding size (Dadabhau and Kisan, 2013). A survey of 2,000 HFS in West Bengal and Bangladesh showed an average annual income of about US$240. The value of inputs (fertilizer, seed, feed) used in HFS is much higher in Bangladesh than in West Bengal. Investing in this system can also help to mitigate the impact of climate variability. Despite being an important element of the coastal landscapes and playing a pivotal role in the bioresource-based rural economy, intensification of HFS is constrained by lack of skills, technology extension support and rural credit (McDermott et al., 2013) and needs to be mainstreamed in policies and programmes by the respective governments.

Intensification of coastal aquaculture: Coastal aquaculture to grow shrimp and fish has been a way of life in the Ganges Delta. Favourable ecological and climatic conditions, availability of wild shrimp post larvae, cheap labour, unused land and plenty of salt water contribute to the low cost of shrimp cultivation and an opportunity to integrate this region with the emerging international market. Shrimp farming has now become one of the key economic activities in coastal districts of Bangladesh. The favourable domestic policy environment and emerging global market helped in the commercialization of shrimp culture, but it has also helped rural households including small and marginal farmers and landless poor who benefited from a wide range of livelihood opportunities (Ahmed *et al.*, 2002). However, massive development of shrimp also had negative influences on ecological functioning, created conflict among fresh water agriculture and brackish water shrimp farmers and concerns have been raised about its long-term sustainability (Chowdhury *et al.*, 2006). In recent years, disease outbreak has been recognized as the biggest obstacle to the development of shrimp aquaculture in Bangladesh, causing massive economic loss.

Productivity of the coastal aquaculture can be substantially improved with the new technology of suitable species combination and moving from traditional practices of year-round mixed aquaculture to cyclic monoculture or polyculture (Wahab *et al.*, 2012). Saline water represents an asset to the coastal region that needs to be more efficiently used by both aquaculture and aquatic agricultural farming systems.

Conclusions

A unique feature of the Ganges Basin is the strong longitudinal and seasonal variation in both precipitation and potential evaporation. As irrigated agriculture covers only 25 per cent of the basin, crop and water productivity is low through most of the basin and is impacted to due both floods and water scarcity. Though several other factors are important, yet it is 'water' that remains the principal driver or the main set of brakes for development of agriculture in this vast – and the world's most populous – basin. Strategies for sustainable intensification of agriculture in the Ganges Basin need to be regionally differentiated for the upper catchments in the Himalayas, the northern Gangetic Plains, the Central Highlands and the fragile Gangetic Delta. Rejuvenation of farmer-managed irrigation systems, multiple use systems and revival of water springs in the mid and lower hill catchments; large-scale adoption of resource conservation technologies and systems of rice intensification, along with investments in improving rural power supply in the Gangetic Plains; decentralized rain water harvesting and development of low-cost irrigation infrastructure to meet the critical needs of water in the Central Highlands; and rice-based system intensification and diversification through flood and salt tolerant varieties, innovative homestead production systems for the functionally landless farmers and intensification of coastal aquaculture for the poor farmers in the Gangetic Delta – all are important strategies for sustainable intensification of agriculture and improvement of rural livelihoods in the Ganges Basin.

References

ACIAR. 2008. *Permanent Beds and Rice-Residue Management for Rice-Wheat Systems in the Indo-Gangetic Plain.* Proceedings of a workshop, Ludhiana, India, 7–9 September 2006. E. Humphreys and C.H. Roth (eds). Canberra, Australian Council of International Agricultural Research.

Africare, Oxfam America, WWF-ICRISAT Project. 2010. *More Rice for People, More Water for the Planet.* WWF-ICRISAT Project, Hyderabad, India.

Ahmed, A., Mallick, D.L., Liaquat, A.M. and Rahman, A. 2002. *Policy Research for Sustainable Shrimp Farming in Asia: Literature Review on Bangladesh Shrimp.* Dhaka: Bangladesh Centre for Advanced Studies.

Alauddin, M. and Sharma, B.R. 2013. Inter-district rice water productivity difference in Bangladesh: An empirical exploration and implications. *Ecological Economics*, 93: 210–18.

Aryal, J., Bhatia, M., Jat, M.L. and Sidhu, H.S. 2014. Impacts of laser land leveling in rice-wheat rotations of the North-western Indo-Gangetic Plains of India. Paper presented at the World Congress of Environmental and Resource Economists, 28 June–2 July 2014, Istanbul, Turkey.

Barah, B. 2009. Economic and ecological benefits of System of Rice Intensification (SRI) in Tamil Nadu. *Agricultural Economics Research Review* 22(July–December): 209–14.

Berkhout, E., Glover, D. and Kuyvenhoven, A. 2015. On-farm impact of the System of Rice Intensification (SRI): Evidence and knowledge gap. *Agricultural Systems* 132: 157–66.

Bhattacharya, A., Sundaray, J.K., Deo, A.D. and Ponniah, A.G. 2013. *Resource recycling in homestead farming: An innovation towards the inclusive development in Sunderban of West Bengal.* Proceedings National Seminar on Innovations in Science and Technology for Inclusive Development. Indian Science Congress Association, Bhubaneswar. 23–24 November, 2013, Bhubaneswar, India.

Biswas, A.K. (2008) 'Management of Ganga Brahmaputra Meghna system ways forward'. In O. Varis, A.K. Biswas and C. Tortajada (eds). *Management of Transboundary Rivers and Lakes Water Resources Development and Management*, Dordrecht: Springer.

Burman, D., Bandyopadhyay, B.K., Mandal, S., Mandal, U.K., Mahanta, K.K., Sarangi, S.K., Maji, B., Rout, S., Bal., A.R. Gupta, S.K. and Sharma, D.K. 2013. *Land Shaping – A Unique Techonology for Improving Productivity of Coastal Land.* Karnal, India: Technical Bulletin, Central Soil Salinity Research Institute.

Burman, D., Mahanta, K.K., Sarangi, S.K., Mandal, S., Maji, B., Mandal, U.K., Bandopadhyay, B.K. and Sharma, D.K. 2015. 'Effect of groundwater use on groundwater salinity, piezometric level and boro rice yield in the Sundarbans of West Bengal'. In E. Humphreys, T.P. Tuong, M.C. Buisson, I. Puskinskis and M. Philips (eds). *Revitalizing the Ganges Coastal Zone: Turning science into policies and practices.* Colombo: CGIAR Challenge Program on Water and Food (CPWF).

Cai, X.L. and Sharma, B.R. 2010. Integrating remote sensing, census and weather data for an assessment of rice yield, water consumption and water productivity in the Indo-Gangetic river basin. *Agricultural Water Management* 97(2): 309–16.

Cai, X., Sharma, B.R., Matin, M.A., Sharma, D. and Gunasinghe, S. 2010. *An assessment of crop water productivity in the Indus and Ganges river basins: Current status and scope for improvement.* Colombo, Sri Lanka: International Water Management Institute (IWMI Research Report 140).

Choudhury, N.Y., Paul, A. and Paul, B.K. 2004. Impact of coastal embankment on the flash floods in Bangladesh: A case study. *Applied Geography* 24: 241–58.

Chowdury, M.A., Shivakoti, G.P. and Salequzzaman, M. 2006. A conceptual framework for the sustainability assessment procedures of the shrimp aquaculture industry in coastal Bangladesh. *Int. J. Agric. Resourc. Governan. Ecol.* 5, 162–84.

Dadabhau, A.S. and Kisan, W.S. 2013. Sustainable rural livelihood security through integrated farming systems – A review. *Agri. Reviews* 34: 207–15.

Das, A., Buisson, M.C. and Mukherji, A. 2015. 'Predicting success in community driven water infrastructure maintenance: Evidence from publich goods games in coastal Bangladesh'. In E. Humphreys, T.P. Tuong, M.C. Buisson, I. Puskinskis and M. Philips (eds). *Revitalizing the Ganges Coastal Zone: Turning science into policies and practices.* Colombo: CGIAR Challenge Program on Water and Food (CPWF).

Dhital, K. 2011. *Study on System of Rice Intensification in transplanted and direct-seeded versions compared with standard farmer practice in Chitwan, Nepal.* Thesis submitted to the Tribhuvan University Institute of Agriculture and Animal Science, Rampur, Chitwan, Nepal.

Dzung, N.T., Minh, N.L., Yen, N.T.H., Linh, P.G., Ngoc, H.V., Nguyen, N.X., Quang, L.C. and Dzung, P.M. 2011. *Simple and Effective-SRI and Agriculture Innovation.* System of Rice Intensification website. Available at: http://sri.ciifad.cornell.edu/countries/vietnam/VN_SRI_booklet_Eng2012.pdf.

Eastham, J., Kirby, M., Mainuddin, M. and Thomas, M. 2010. Simple water-use accounting of the Ganges Basin. Water-use accounts in CPWF basins. CPWF Working Paper-BFP 05, Colombo, Sri Lanka: CGIAR Challenge Program on Water and Food.

Erenstein O. 2009. 'Reality on the ground: Integrating germplasm, crop management, and policy for wheat farming system development in the Indo-Gangetic Plains'. In J. Dixon, H. Braun, P. Kosina and J. Crouch (eds). *Wheat Facts and Futures.* Mexico: D.F., CIMMYT.

FAO. 2011. *Save and Grow. A policymaker's guide to the sustainable intensification of smallholder crop production.* Rome: Food and Agriculture Organisation.

FAO. 2015. FAOSTAT Statistical Database: Production. Available at: http://faostat3.fao.org/browse/Q/*/E.

Gautam, P. 2008. 'Emerging issues and strategies in the rice-wheat cropping system in the Indo-Gangetic Plains'. In Y. Singh, V. Singh, B. Chauhan, A. Orr, A. Mortimer, D. Johnson and B. Hardy (eds). *Direct Seeding of Rice and Weed Management in the Irrigated Rice-Wheat Cropping System of the Indo-Gangetic Plains.* Los Baños (Philippines), International Rice Research Institute, and Pantnagar, India, Directorate of Experiment Station, G.B. Pant University of Agriculture and Technology.

Gupta, R. and Sayre, K. 2007. Conservation agriculture in South Asia. Paper presented at the International Workshop on Increasing Wheat Yield Potential, CIMMYT, Obregon, Mexico, 20–24 March 2006. *Journal of Agricultural Science* 145: 207–14.

Hayami, Y. and Ruttan, V.W. 1985. *Agricultural Development: An International Perspective.* Baltimore, MD: John Hopkins University Press.

Hughes, R., Adnan, S. and Dala-Clayton, B. 1994. *Floodplains of Flood Plains.* London: International Institute for Environment and Development and Research and Advisory Services.

ICIMOD. 2015. *Reviving the Drying Springs: Reinforcing social development and economic growth in the midhills of Nepal.* Issue Brief. Kathmandu: International Centre for Integrated Mountain Development.

IRRI. 2009. *Revitalizing the Rice-Wheat Cropping Systems of the Indo-Gangetic Plains: Adaptation and Adoption of Resource-Conserving Technologies in India, Bangladesh, and Nepal.* Final report submitted to the United States Agency for International Development. Los Baños (Philippines), International Rice Research Institute.

Joshi, K.D., Conroy, C. and Witcombe, J. 2012. *Agriculture, Seed and Innovation in Nepal: Industry and Policy Issues for the Future.* Washington, DC: IFPRI/CSISA Study.

Khan, Z.H., Kamal, F.A., Khan, N.A.A., Khan, S.A., Rahman, M.M., Khan, M.S.A., Islam, A.K.M.S. and Sharma, B.R. 2015. 'External drivers of change, scenarios and future projections of the surface water resources in the coastal zone of Bangladesh'. In E. Humphreys, T.P. Tuong, M.C. Buisson, I. Puskinskis and M. Philips (eds). *Revitalizing the Ganges Coastal Zone: Turning science into policies and practices.* Colombo: CGIAR Challenge Program on Water and Food (CPWF).

Kishore, A., Sharma, B.R. and Joshi, P.K. 2014. *Putting Agriculture on the Takeoff Trajectory: Nurturing the Seeds of Growth in Bihar.* New Delhi: International Food Policy Research Institute – International Water Management Institute.

Ladha, J., Yadvinder-Singh, Erenstein O. and Hardy, B. (eds). 2009. *Integrated Crop and Resource Management in the Rice-Wheat System of South Asia.* Los Baños (Philippines): International Rice Research Institute.

McDermott, J., Ait-Aissa, M., Morel, J. and Rapando, N. 2013. Agriculture and household nutrition security – development practice and research needs. *Food Sci.* 5: 667–78.

Malik, R.P.S., Giordano. M. and Sharma, V. 2014. Examining farm-level perceptions, costs, and benefits of small water harvesting structures in Dewas, Madhya Pradesh. *Agric. Water Manage.* 131: 204–11.

Mikhail, M. and Yoder, R. 2008. *Multiple Use Water Services Implementation in Nepal and India: Experiences and lessons for scale-up.* Lakewood, USA: International Development Enterprises. Available at: www.ide.org/news/MUS_book.php.

Mondal, M.K., Saha, N.K., Ritu, S.P., Paul, P.L.C., Sharifullah, A.K.M, Humphreys, E., Tuong, T.P. and Rashid, M.A. 2015. 'Optimum sowing window for boro cultivation in the coastal zone of Bangladesh'. In E. Humphreys, T.P. Tuong, M.C. Buisson, I. Puskinskis and M. Philips (eds). *Revitalizing the Ganges Coastal Zone: Turning science into policies and practices.* Colombo: CGIAR Challenge Program on Water and Food (CPWF), p. 600.

Mukherji, A., Shah, T. and Banerjee, P. 2012. Kick-starting a second Green Revolution in Bengal, Commentary. *Economic and Political Weekly* 47(18): 27–30.

NRB. 2014. *Economic Outlook 2014.* Kathmandu: Nepal Rastra Bank.

Pangare, V. and Karmakar, D. 2003. *Impact on Livelihoods: PRADAN's collaboration study of the 5% technology, Purulia, West Bengal, India.* Poverty-Focussed Smallholder Water Management: An IWMI Research Project supported by DfID. Final report document 3 of 9. International Water Management Institute, Colombo, Sri Lanka. Available at: http://publications.iwmi.org/pdf/H043992.pdf.

Pant, D. and Samad, M. 2010. Synthesis of IMWI Work in Nepal. *Working Paper 138I. Colombo: International Water Management Institute.*

Phansalkar, S. and Verma, S. 2004. Improved water control as strategy for enhancing tribal livelihoods. *Economic and Political Weekly – review of Agriculture*, XXXIX: 3469–3476.

Phansalkar, S. and Verma, S. 2005. *Mainstreaming the Margins: Water-centric Livelihood Strategies for Revitalising Tribal Agriculture in Central India.* New Delhi: Angus and Grapher.

Rahman, M.A., Sarker, M.R.A., Sharma, N., Mondal, M.K., Islam, M.R., Gregorio, G.B., Humphreys, E. and Tuong, T.P. 2015. 'Challenges and opportunities for *aman* rice cultivation in *ghers* used for brackish water shrimp production'. In E. Humphreys, T.P. Tuong, M.C. Buisson, I. Pukinskis, M. Phillips (eds). *Revitalizing the Ganges Coastal Zone: Turning Science into Policy and Practices Conference Proceedings*. Colombo, Sri Lanka: CGIAR Challenge Program on Water and Food (CPWF), pp. 333–41.

Rijsberman, F. 2004. 'Sanitation and access to clean water'. In B. Lomborg (ed.). *Global Crises, Global Solutions*. Oxford: Oxford University Press.

Sah, R.P. 2013. *Role of Seeds in Transforming Agriculture in Nepal*. Washington, DC: International Food Policy Research Institute (IFPRI/ReSAAKS).

Saha, N.K., Mondal, M.K., Humphreys, E., Bhattacharya, J., Rashid, M.H., Paul, P.C. and Ritu, P.C. 2015. 'Triple rice in a year: is it a feasible option for the low salinity areas of coastal zone of Bangladesh?' In E. Humphreys, T.P. Tuong, M.C. Buisson, I. Pukinskis and M. Phillips (eds). *Revitalizing the Ganges Coastal Zone: Turning Science into Policy and Practices Conference Proceedings*. Colombo, Sri Lanka: CGIAR Challenge Program on Water and Food (CPWF), pp. 421–35.

Santiago, V., Strohmann, C. and Khaka, E. 2007. *Sourcebook of Alternative Technologies for Freshwater Augmentation in Some Countries in Asia*. Nairobi: United Nations Environment Program.

Sharma, B.R., Amarasinghe, U.A. and Sikka, A. 2008. Indo-Gangetic River Basins: Summary situation analysis. CPWF Basin Focal Projects Background Papers, IWMI, Colombo, Sri Lanka Available at: http://cpwfbfp.pbworks.com/f/IGB_situation_analysis.PDF. Accessed on: 31 August 2015.

Sharma, B.R., Rao, K.V., Vittal, K.P.R., Ramakrishna, Y.S. and Amarasinghe, U. 2010a. Estimating the potential of rainfed agriculture in India: Prospects for water productivity improvements. *Agricultural Water Management* 97(1): 23–30.

Sharma, B.R., Riaz, M.V., Pant, D., Adhikary, D.L., Bhatt, B.P. and Rahman, H. 2010b. *Water Poverty in the North-Eastern Hill Region (India): Potential Alleviation Through Multiple-Use Water Systems – Cross Learnings from Nepal Hills*. New Delhi, India: International Water Management Institute (IWMI-NAIP Report 1).

Sharma, B.R., Amarasinghe, U.A., Cai, X., de Condappa, D., Shah, T., Mukherji, A., Bharati, L., Ambili, G., Qureshi, A.S, Pant, D., Xenarios, S., Singh, R. and Smakhtin, V. 2010c. The Indus and the Ganges: River basins under extreme pressure. *Water International* 35(5): 493–521.

Shivakoti, G.P. and Shrestha, S.G. 2008. Farmer Managed Irrigation Systems in Nepal: Balancing water uses and environment and conservation for sustaining livelihoods. *Himalayan Jurnal of Sociology and Anthropology* (2008): 64–74.

Takeshima, H., Salau, S., Adeoti, A. and Okoli, S. 2010. *Economics of Farmers' Demand for Private Irrigation in Nigeria*. Policy Note No. 23. Nigeria: International Food Policy Research Institute.

Thakur, A., Uphoff, N. and Antony, E. 2009. An assessment of physiological effects of System of Rice Intensification (SRI) practices compared with recommended rice cultivation practices in India. *Expl Agric.* 46(1): 77–98.

Uphoff, N. 2008. Farmer Innovations Improving the System of Rice Intensification (SRI). Available at: www.future-agricultures.org/farmerfirst/files/T1a_Uphoff.pdf.

Vaidyanathan, A. 2013. *Water Resources of India*. New Delhi: Oxford University Press.

Wahab, M.A., Nahid, S.A., Ahmed, N., Haque, M.M. and Karim, M. 2012. Current status and prospects of farming the giant river prawn (Macrobrachium rosenbergii) in Bangladesh. *Aquaculture Research* 43: 970–83.

Zaman, M.A.U. and Rushton, K.R. 1998. 'Barind: Predicted a failure, proved a success'. In H. Wheater and C. Kirby (eds). *Hydrology in a Changing Environment* – Volume II. Sussex, UK: John Wiley.

7

Ganges hydropower: status and challenges

Dwarika N. Dhungel, Santa Bahadur Pun, Sonali Mittra and M. Monirul Qader Mirza

Key messages

- The Ganges riparian countries are experiencing energy shortages, which hamper socio-economic development. However, there is significant hydropower (HP) potential in Nepal and India. Further development of hydropower could create various co-benefits for all riparian countries. Countries must consider the differing development priorities of their neighbours in the broader context of the water-energy-food nexus and not hydropower alone.
- Despite large, estimated, technically feasible hydropower potential in Nepal (of about 42,000 MW) and large markets in India and Bangladesh, Nepal has not been able to develop its

hydropower potential due to a number of reasons: these range from limited financial resources and lack of national consensus on key water resources development, to the complex socio-political climate in the region. This adversely affects transboundary collaboration in the hydropower sector – collaboration which is imperative to hydropower development in Nepal. There are hopes of increasing the export of hydropower to India by 2020–21 when the Arun III and Upper Karnali plants are expected to be commissioned. It is also hoped that the 2014 Indo-Nepal Power Trade Agreement may help to open up the market in India.

- Over 576 million Indians – almost 50 per cent of India's population – live in the Ganges Basin. Food security of this population is the cornerstone of basin development. The Ganga Basin represents only around 25 per cent of the total Indian hydropower potential, with its major part concentrated in Uttarakhand state. Much attention is given to the Ganga hydropower development plans due to a significant power deficit, especially in northern India. Hydropower capacity additions to the national and basin energy balance, however, have been limited in recent decades. This is due to land acquisition problems, naturally occurring obstructions, administrative lapses, inadequate information and financial issues. Basin transboundary disputes over water and energy sharing are centred around unreliable, absent or unshared data, and inequitable sharing of benefits from hydropower projects.
- In Bangladesh, hydropower constitutes only about 4 per cent of the total installed generation capacity and overall, the potential for further development is limited. However, the country is actively engaged in transboundary dialogues related to hydropower development, with a particular emphasis on how such development can increase Ganges lean period flow, and contribute to the alleviation of the recurrent problem of flooding.

Nepal

Hydropower potential and markets

Nepal has a theoretical potential of 83,000 MW of hydropower (Water Resources Strategy 2002), but a review of this number is long overdue. According to Sharma (1997), the technically feasible hydropower potential in Nepal is about 42,000 MW, with Karnali, SaptaKoshi and Gandaki Basins holding around 24,000, 10,800, and 5,270 MW respectively.

While 787 MW is the total installed capacity of Nepal in 2014, the system peak load registered that year was around 1,200 MW. The latter is projected to increase over four times by 2030 (NEA, 2013). Domestic development in Nepal's hydropower sector can be attributed to the 22 MW Chilime plant, a public–private hydropower venture that was commissioned in the early 2000s. It kick-started confidence among domestic financial institutions and associated construction and consulting agencies in Nepal's hydropower sector. The government's quest for hydropower development from the private sector has resulted in a large flurry of projects, mostly of small (under 10 MW) capacity. As of 2014, 39 projects with a total of over 250 MW capacity have been commissioned. Another 62 projects ranging from under 1 MW to 456 MW, totalling 1,206 MW, are in various stages of construction. Overall, by the fiscal year 2020–21 (the year of commissioning of the Arun III and Upper Karnali projects, both of 900 MW capacity), 118 hydropower projects, totalling 3,859 MW, are expected to come onstream. While the energy market in Nepal is still small, it is huge in next-door India and Bangladesh. With 'non-discriminatory access' promised in the 2014 Indo-Nepal Power Trade Agreement, foreign developers and investors are expected to open up Nepal's hydropower development.

The 2015 hydropower capacity of 631 MW makes around 67 per cent of the total available power of 940 MW (inclusive of thermal and India import) in Nepal. This hydropower capacity if predicted to increase by over 120 per cent to 1,385 MW in 2018. It is also predicted that due to HP development, the deficit, even at increased peak demand of 1,740 MW, will drop more than 10 times by 2018. Yet, capacity shortage will continue to prevail during the dry season this calling for more storage projects of suitable sizes (NEA, 2013).

In India, of the total 223,000 MW installed capacity, 130,000 MW (58 per cent) is provided by coal/lignite-fired thermal stations. India does have substantial hydropower potential, mostly in the northern Himalayan ranges. A series of medium-sized hydro plants are under construction in the higher reaches of the Ganges. Plans are underway to develop larger ones on the Brahmaputra in the north-eastern states. Renewable energy sources, chiefly wind power along the coastal ranges, at 27,000 MW (12 per cent), claim the third position. India also has ambitious plans for nuclear power, which is planned to increase from the current 4,800 MW to 63,000 MW by 2032.

By the end of 2017 India is expected to have a peak load of 208,000 MW, requiring annual energy production of 1,355,000 GWh. It is forecast that by the end of 2022, while the all-India peak load will rise to 295,000 MW, annual energy requirements will increase to 1,905,000 GWh. Nepal's hydropower development will be influenced by developments in Indian states that border Nepal. These are the northern states (Uttar Pradesh, Rajasthan, Punjab, Haryana, Kashmir, Himanchal and Uttaranchal) and eastern (Bihar, Jharkhand, Orissa, West Bengal and Sikkim). The northern region has the highest projected peak load, at 87,000 MW, among the regions by 2022. Similarly, with vast industrial complexes, the northern region remains at the top in 2022 with projections of 594,000 GWhs. Nepal's hydropower potential is also concentrated in its western region, adjunct to India's northern region. India's eastern region, which Nepal will shortly be connected to through the Dhalkebar–Muzaffarpur link, has smaller peak load and annual energy than the northern region. Both Uttar Pradesh (UP) and Bihar states have annual deficits of about 15,000 GWh and 2,000 GWh respectively, thus creating an opportunity for energy export from Nepal. However, Nepal's 'cheap hydropower' is currently more expensive than prevailing market prices in India.

Policies, plans and strategies on hydropower development

Nepal's first 500 KW Pharping hydropower plant was commissioned in 1911. There were some private sector attempts in the 1940s and 1950s, but otherwise electricity/hydropower development remained in the public sector domain. With the Electricity Act (1992), the Electricity Regulations (1993) and the Electricity Tariff Fixation Rules (1994), the government opened the doors to the private sector. The government's 1992 Hydro Power Development Policy emphasized the importance of small isolated hydroelectric projects for remote hilly regions, and the extension of transmission and distribution networks from the national grid to supply electricity to the rest of the country. The policy encouraged both national and foreign investors, and the licensing system for prospective developers, along with the Build, Own, Operate and Transfer (BOOT) mechanism, came into existence. The Electricity Development Center (EDC) was established in July 1993 as a staff agency of the Water Resources Ministry, concerned with implementing the government's overall policies related to the water resources/electricity sector. In 2001, the EDC was upgraded to department level as the Department of Electricity Development. The major functions of the department are to ensure transparency of the regulatory framework and to accommodate, promote and facilitate the private sector's participation in the power sector. This is achieved by providing a 'one window' service and licensing to power projects (www.doed.gov.np/about_us.php). Recently, it has also been given the controversial job of implementing hydropower projects.

Hydropower development, with its associated massive structures and long, high-voltage transmission lines, has serious environmental implications. Nepal's Environment Protection Act (1997) and Environment Protection Rules (1997) make it compulsory for all hydropower developers to prepare an Initial Environmental Examination (IEE) and, when necessary, an Environmental Impact Assessment (EIA) as well.

The new Hydro Power Policy of 2001 emphasizes the need to undertake 'large multipurpose storage projects in such a way that downstream benefits would yield maximum benefits to the nation and encourage export of hydropower by taking into account bilateral and regional cooperation'. The stated objectives of the policy are:

> To generate electricity at low cost by utilizing the water resources available in the country, to extend reliable and qualitative electric service throughout . . . Nepal at a reasonable price, to tie-up electrification with the economic activities, to render support to the development of rural economy by extending the rural electrification and to develop hydropower as an exportable commodity.
>
> *(MoWR, GoN, 2001)*

The Water Resources Strategy Nepal (WRS) 2002, prepared with the support of the World Bank and Canadian International Development Agency (CIDA), followed the objectives and policies laid down by the Hydro Power Policy of 2001. The strategy sets targets in the electricity sector to be met within 5-, 15- and 25-year periods. Additionally, it outlines the measures to be taken during these periods, aiming by 2027 to provide 60 per cent of households with access to electricity and 22,000 MW developed by that time. The strategy also stresses the need for cooperation with neighbouring countries at bilateral and regional levels, and sets targets for exports in the three periods. For example, mid-term targets include the aims that by 2017, 400 MW is to be exported, riparian issues should be sorted out and Nepal should be receiving some benefits from the treaties signed with India. The National Water Plan Nepal (NWP) further set 13 hydropower-related targets to be achieved during the short, medium and long term. Notable ones are: (i) supply of electricity to 100 per cent of households (up from 60 per cent coverage target of WRS); (ii) achieve per capita electricity consumption of 400 KWh; and (iii) substantial amounts of electricity exported to earn national revenue by 2027. Other related objectives include monitoring existing treaties for the achievement of expected benefits, establishing a functioning mechanism for regional cooperation, and the export/exchange of a substantial amount of hydropower by 2027. The 13th Plan, running from 2013 to 2016, emphasizes the development of reservoir-type hydropower projects where a number of projects, with a total capacity of about 4,560 MW, will be undertaken.

Institutions and administration

There are four types of hydropower-related organizations in Nepal. They are: (i) the overall policy planning and coordinating bodies; (ii) the facilitating and regulatory bodies; (iii) the line ministries, departments and their field offices; and (iv) the autonomous and local governance bodies.

Policy planning and coordinating bodies include the National Development Council (NDC), the National Planning Commission (NPC), the National Water Resources Development Council (NWRDC) and the Water and Energy Commission (WEC). Of these, the NDC and NPC are the overall national planning bodies. The others are water resources and energy-related sectoral policy, planning and coordinating bodies. The NDC, chaired by the Prime Minister, is the country's highest policy and planning body. The Council members are drawn from political parties, civil society and professionals. The Council guides the NPC in preparing the national plan and annual programmes, and exercises supervisory and monitoring responsibilities in the implementation of the plans and programmes. The secretariat of the NPC, headed by a full-time secretary, provides professional and secretarial support to both the NDC and NPC.

The NWRDC is the policy-planning and inter-sectoral coordinating body in the water resources sector, including the development of hydropower. It was established in 1993 by executive order, under the chairmanship of the Prime Minister with a membership made up of various agencies and stakeholders. The council is mandated to: (i) create a conducive environment for national consensus on water resources development; (ii) formulate policy in order to maximize benefits to the nation; (iii) give direction to the government for promotion of better understanding at national and international level for water resources development; and (iv) identify the basis to establish coordination among the various sectors and institutions for the development of water resources (Pun, 2012). In other words, the Council is mandated to build consensus among different shades of opinion and provide coordination among the different water-related agencies. The secretarial service to the Council is provided by the secretariat of the Water and Energy Commission (WEC).

The other body that is responsible for coordination among the various water-related agencies of government, and providing water sector policy feedback to the government, is the Water and Energy Commission (WEC). Since the Water Resources Ministry was abolished in 2009, it is now chaired by the Minister for Energy and has the representation of various government as well as non-government agencies in its composition. The WEC is also mandated to look after bilateral and regional water resources-related matters, including hydropower. It has its own Water and Energy Commission Secretariat (WECS).

Facilitating and regulatory bodies include the Investment Board Nepal (IBN) and aim to 'create a framework for selecting and evaluating projects, provide incentives to encourage investments, and engage in investment promotion activities' (www.investmentboard.gov.np/page/about_us/investment_board_nepal.html). With regard to hydropower, the board is mandated to negotiate and finalize the investment required for hydropower projects of 500 MW capacity, or more, with prospective investors. The Environment Protection Council (EPC) was established to provide policy guidance and suggestions to the government with regard to environmental protection, and coordination among different agencies on the management and protection of the environment. Although it has not been effective for some time, the EPC still exists with its secretariat, at the Ministry of Science, Technology and Environment (MoSTE). Councils like the EPC and NWRDC have not been effective mainly due to prolonged political instability since 1996, a highly politicized bureaucracy and non-use of available professionals within and outside government.

The Industrial Promotion Board (IPB) within the Ministry of Industry (MoI) is another regulatory body with regards to energy in general and hydropower in particular, that deals specifically with approval of foreign investments. The Tariff Fixation Commission (ETFC) is tasked with fixing the electricity tariff and other charges by the electricity providers. Without its consent, electricity providers are not allowed to fix or make any changes in electricity tariffs. The Securities Board of Nepal (SEBON) advises the government on matters related to the development of the capital market (www.sebon.gov.np/). The Hydroelectricity Investment and Development Company Ltd (HIDC Ltd) is a public investment company with the primary purpose of helping the country in addressing unmet power demand through the mobilization of funds for investment in the generation, transmission and distribution of hydroelectricity.

Autonomous and local governance bodies include the Nepal Electricity Authority (NEA) – the only electricity related public utility in the country. The NEA is responsible for the generation, transmission and distribution of electricity, but because of the opening of electricity production to the private sector, the NEA is no longer the sole electricity-generating provider. The NEA is, however, the sole power-purchasing agency for electricity generated in the private sector. There are also country-wide, two-tier local governance bodies – the Village Development Committee (VDC), and District Development Committee (DDC). The VDC and municipalities are empowered to generate and distribute electricity. The DDCs are authorized regarding identification, planning, implementation, operation, distribution and maintenance of small rural hydroelectric projects, as well as other energy-related projects (Local Self Governance Act 1999)

Challenges

There are a number of domestic and external challenges regarding hydropower development in Nepal. Domestic challenges include, but are not limited to, the following.

Consistency in pursuance of approved policies, strategies and plans

While there are government-approved policies and plans for the hydropower sector, (Hydropower Policy 2001, Water Resources Strategy 2002 and National Water Plan 2005) based on extensive national consultations, successive governments were found to be interested in going their own way, instead of pursuing the approved policies and plans or having them reviewed. Related to this, there is a need for clear protocols on the relevant provisions of the laws concerned, and details of the procedures to be followed by investors in developing hydropower projects in Nepal (WECS, 2005).

Clarification on whether hydropower projects are for electricity only, or for other purposes (irrigation, flood control etc.) too

One school of thought is that hydropower projects should be developed purely as a power project, and should not be linked with other uses of water. The World Bank's Ganges Strategic Basin Assessment (World Bank, 2014) indicated that the Himalayas have large hydropower potential that can be considered as a source of 'domestic energy supplies as well as a source of export revenues for Nepal where potential supplies far outstrip potential demand'. The same study indicated that although many sites are attractive for the development of multipurpose water storage infrastructure,

> the steep terrain and deep gorges allow surprisingly little water to be stored behind even very tall dams. Developing the full range of structures under consideration . . . would provide additional active system storage equivalent to only about 18 per cent of annual average flow, which is not very significant on a basin wide scale.

The World Bank's study did not discuss India's river linking master plan which, without large storage projects in Nepal (Sapta Koshi, Budhi Gandaki, Karnali Chisapani, Pancheshwar etc.) would be impossible to realize. And Nepal should not also forego the accrued downstream benefits while developing such large multipurpose projects.

Complexity of inter-agency relationships and limited capacity of many such agencies

This normally leads to delays in starting hydropower projects. The decision-making process within several of the agencies involved is lengthy and cumbersome – and it sometimes takes years to get clearance. Such delays naturally result in implementation delays and increased project costs.

Demands of local communities

VDCs and DDCs have been empowered with substantial authority regarding natural resources available at the local level. They have the power to receive, and are receiving, royalties from hydropower projects. In addition, local communities have claimed shares in completed projects and those that are currently under construction. In the absence of firm governmental guidelines and policies, the complexities of these inter-relations have led to tensions during hydropower implementation.

External challenges

These are primarily related to the complex socio-political climate in the Ganges Basin and larger region. The first Indo-Nepal bilateral hydropower projects in Nepal – under the 1954 Koshi agreement – aimed primarily to alleviate the 'sorrow of Bihar' in India through flood control and irrigation, and had a 20 MW hydropower component as well. According to the agreement, Nepal was entitled to 50 per cent of the generated hydroelectric power on payment of mutually agreed tariff rates. Koshi power only became available in 1971 and even that was very irregular, affecting the industries concentrated in Nepal's eastern belt. Low power generation, undistributed compensation for lands taken by the project over 50 years ago and poor maintenance leading to the embankment breaching by Koshi in 2008, causing massive damages in Nepal and Bihar, are some of the major issues.

Another bilateral project which has not met expectations is the 1959 Gandak Agreement. The Gandak barrage was completed in 1969 but the 15 MW hydropower plant was only commissioned in 1979. The treaty stipulated that the power house, though located inside Nepal, would be handed over to Nepal only when Nepal developed a load of 6,000 KW from it. Until the assistance of the Asian Development Bank which commissioned the high-voltage transmission lines in 1981, Nepal could not evacuate such bulk power. This clause resulted in Nepal paying India for two years, for electricity generated within Nepal's own territory. Similar to Koshi, power generation from Gandak is poor (Ministry of Water and Power, 1975, 1996) as the Government of India had irrigation and flood control, not hydropower, as uppermost priorities when the two projects were formulated.

Both the Koshi and Gandak treaties became very controversial in Nepal. In 2004, Jagat Mehta, India's former Foreign Secretary, admitted that the Koshi and Gandak projects 'led to some submergence in the Terai (Nepal)' and that 'An alternative location could have greatly increased the benefits to Nepal'. Ramaswamy Iyer, India's former Water Resources Secretary, wrote that the Koshi and Gandak agreements 'coloured all subsequent dealings between India and Nepal. Suspicion and mistrust grew and became a massive impediment to good relations between the two countries' (Iyer 1999).

The 1996 Mahakali Treaty focused on the 6,480 MW Pancheshwar multipurpose project that India planned to augment the flows of the Yamuna River – a Ganges tributary – in order to water the parched farmlands of Rajasthan. The parties agreed that the Detailed Project Report (DPR) of Pancheshwar be finalized within six months 'from date of entry into force of the treaty' and the project would be 'completed within eight (8) years from the date of the agreement for its implementation subject to the provision of the DPR'. The treaty came into force in 1997 when the two countries exchanged the instruments of ratification. However, 17 years later, in 2014, India and Nepal were still directing their governments to 'set up the Pancheshwar Development Authority within 6 months and finalise the DPR of Pancheshwar Development Project and begin implementation of the Project within one year' (MoWR, 1996). Besides the determination of the price of energy to be exported to India, and equal sharing of waters of the Mahakali River, an equally important issue that does not figure much in the Mahakali Treaty discussion is the 'sharing of project costs according to the accrued benefits'. The two countries, despite having signed the treaty in 1996, have yet to agree on what exactly those accrued benefits are.

The way forward for Nepal's hydropower sector

Water, as the most valued natural resource, is the most talked about, as well as the most sensitive issue in Nepal. Whenever a new project is being negotiated, especially for export purposes, the issue generally generates a polarization of views. Doubts are raised whether Nepal would be able to secure the best deal or be deceived again. The political parties of Nepal take one position when in government, but when out of power, often take the opposite which sends negative signals to prospective investors. It is important to break this vicious cycle and develop a national consensus for the hydropower sector. The NWRDC, constituted in 1993, although ineffective for a long time, could play that important role in building political consensus in the form of a national vision for water-related matters, including hydropower, among political actors and professionals. The WEC and its secretariat should be restructured in such a way that it becomes an attractive professional arm of the NWRDC.

The hydropower sector also needs the enactment of new laws, entities and protocols to replace the old ones, including the National Electricity Regulatory Commission, New Electricity Act and One Window system. An integrated manual containing relevant provisions of the related laws and procedures in hydropower development is urgently needed, and in its absence there is a need for a simplification of procedures within the concerned organizations.

In the light of growing water scarcity and a burgeoning population in the Ganges Basin, water, food and energy security have become real concerns. Water-related matters need to be seen from a holistic point of view and not in a compartmentalized manner. Furthermore, Nepal is known for the availability of good hydroelectric sites which could also store valuable freshwater for irrigation, flood control, drinking water etc. They need to be treated as essential components of hydropower development in the country. From this perspective, the need to re-establish the Water Resources Ministry has become more important. The role of such a ministry would involve dealing with neighbouring countries and donors on water-related matters, looking after national projects in the water sector (including hydropower), and facilitating provincial and local governance bodies. The new Constitution of Nepal (2015) has made provision for three levels of government under the federal polity – central, provincial and local – each having power in the development of water-related projects falling under their functional jurisdictions. Mediating issues or disputes that might arise between provincial and local bodies is therefore an important function of the ministry.

The concept of distributing shares in hydropower projects for locals of the district, first pioneered during the 22 MW Chilime project, needs to receive further attention. This concept, with locals becoming stakeholders in the project, can generate much goodwill during both construction and operation phases.

In October 2014, Nepal and India signed the Electrical Power Trade, Cross-Border Transmission Inter-connection and Grid Connectivity Agreement (PTA), initially with a 25-year lifespan. The PTA aims to provide for cooperation between the two countries in the power sector 'including developing transmission interconnection, grid connectivity, power exchange and trading through the governmental, public and private entrepreneurs of the two countries on mutually acceptable terms' (GoN and GoI, 2014). The pertinent issue in relation to the implementation of the PTA is the legal and regulatory framework for effective operation of transmission systems and energy-trading market mechanisms between the two countries. India has its own energy regulator but Nepal is yet to pass a Bill on this. Both countries need to harmonize legal and regulatory frameworks dealing with cross-border trade. The PTA has also opened access for energy trade with countries other than India. It remains to be seen how this will work (Dhungel and Ghising, 2014).

The April 2015 earthquake (7.8 magnitude) caused loss of over 8,600 lives and the destruction of 800,000 houses in Nepal besides serious damage to hydropower plants both in operation and under construction. This earthquake is an eye-opener for hydropower development in Nepal's seismically active zones. According to the Ministry of Energy's Post Disaster Needs Assessment Report, while 115 MW of various operating hydropower plants have been 'severely damaged', 60 MW have been 'partially damaged'. The total cost of physical damage from the earthquake to the energy sector has been assessed at around US$180 million. About 1,000 MW of projects under construction have been partially damaged, thus pushing back their commissioning dates.

Despite a century-old hydropower history, despite a huge hydropower potential and despite a huge market across the border, Nepal has not been able to develop its hydropower potential. As hydropower development is capital-intensive, with a long gestation period, Nepal's hydropower can move ahead only when its hydra-headed politics stabilizes and comes to a minimum national consensus on the way forward. For this, institutions like the National Water Resources Development Council, now in a coma, must play the role of catalyst, whether hydropower is for domestic or transboundary uses.

India

Hydropower in the context of basin development

Hydropower in the Indian part of the Ganga Basin sits at the crossroads between developmental aspirations, environmental conservation and sharing of the river across political borders. Development of hydropower is, therefore, determined by a complex set of factors. The basin, traversing through mountainous ranges, fertile plains and ecologically rich delta regions, is densely populated with inadequate levels of socio-economic development and huge energy deficits. It has therefore been the focus for technological interventions to build irrigation capacity and generate electricity. Central to this is the requirement to meet the basic needs of the population in terms of energy access, food security, access to drinking water and the aspiration towards double digit economic growth.

A dominant belief of the Government of India has been that hydropower is an opportunity for economic growth and a necessity in providing access to its energy-starved population. This is visible in the Five Year Plans, the 50,000 MW Hydropower Development Initiative, the 2008 Hydropower Policy, and the National River Interlinking Plan which aims to construct more than 32 GW of installed capacity from hydropower. Scepticism remains on both the effective implementation and cumulative impact of these programmes.

An alternative group of specialists, belonging to civil societies and grassroots NGOs, are of the opinion that hydropower is risky for ecology, environment and communities living along the banks of the river (SANDRP, InternationalRivers.org, Thirdpole.net, Centre for Environment and Science etc.). They voice their concerns over development initiatives by the government and demand alternative plans. There are certain merits to their arguments, based on empirical evidence. There is a focus on questions as to how hydropower policies are formulated, what data sets feed into decision making, and whether marginalized voices are being addressed in policy making.

The third cluster of opinions is of those experts who attempt to find synergy between the two approaches, i.e. proposing wider political and social reforms to reformulate the problem and finding sustainable solutions for developing water resources (Sadoff and Grey, 2002; D'Souza and Mittra, 2013). They question the principles of water control, allocation, distribution and conservation. They also highlight the role of institutions that are required to deal with complex and interlinked issues in hydropower development. The task is daunting and requires serious attention if sustainable river development is to be achieved.

Hydropower potential

The total hydropower potential of the Indian part of the Ganga Basin is estimated to be 20,711 MW (Central Water Commission, 2015), which is much lower than the potential in the Indus (33,832 MW) or Brahmaputra Basin (66,035 MW), but higher than west-flowing, east-flowing and central Indian rivers. In comparison to the total economically exploitable hydroelectricity in India (84 GW) at 60 per cent load factor (CWC, 2015), the Ganga Basin represents only 24.6 per cent of the total potential. However, much attention is given to the Ganga hydropower plans. This may be due to two reasons. Firstly, India has a huge power deficit, especially in the north. It recorded a peaking power deficit of -4.7 per cent and a shortage of 3.6 per cent of the total energy demand in 2014–15 (CEA, 2015a). The rationale for the development of large hydropower plants in the basin is usually backed up by this argument. Secondly, development of hydropower in the Indus and Brahmaputra is comparatively more challenging due to political and technical difficulties. While 50 per cent of the Indus Basin is already exploited, more than 87 per cent remains unexploited in the Brahmaputra Basin. The Indus Basin hydropower development is, more often than not, impacted by the water-sharing agreement between India and Pakistan, and in the past has experienced multiple cases of arbitration on controversial hydropower plants. Brahmaputra, on the other hand, encounters challenges due to difficult terrain, inadequate connectivity and weak infrastructure. Therefore, it can be argued that the Ganga Basin offers a potentially more feasible alternative.

In the larger context of India's electricity profile, hydropower contributes 16 per cent to the total installed capacity in the country (CEA, 2015a). The potential to increase the share of hydropower in the electricity mix of the country remains large. Cognizant of this, and the fact that India will need to use all the potential energy sources to meet its energy demands and power deficits, the 12th Five Year Plan targets a hydropower capacity addition of 11 GW, which makes 12 per cent of the total capacity addition in the plan period, 2012–17 (Planning Commission, 2013). This target has a direct bearing on the Ganga Basin with multiple schemes being proposed on its tributaries.

The 2,500 km-long Ganga River crosses 11 states in India. The basin is diverse in geographical characteristics, hydrogeology and rainfall patterns, which are usually grouped under three main segments: the upper reaches which have mountainous terrain, the middle reaches that fall in the fertile plains and the delta. The distinction is useful in understanding the potential for hydropower development and the numerous challenges associated with it.

The upper reaches of the Ganga have maximum hydropower potential owing to the geology, turbulent flow and high velocities. There are six major hydropower projects in operation, another six under construction and 24 projects proposed to be developed on the Alaknanda River, a tributary of the Ganga. These projects range from 0.2 MW (Tharali dam) to 860 MW (Utiyasu dam). The size of the hydropower projects is comparatively larger on the Bagirathi and Bhilangana Rivers, another two tributaries of the Ganga. Three major hydropower projects exist with an installed capacity of between 99 MW and 1,000 MW. Six more projects, all with capacity of more than 300 MW, are under construction and nine more are proposed (CEA Consortium of 7 IITs, 2010). The middle and lower reaches of the Ganga Basin mostly utilize the waters for agricultural activities through an extensive network of canals, barrages and storage dams. Uttarakhand, a Ganga basin state, has the most hydropower potential. Only 26.3 per cent of the total capacity of the Indian part of the Ganga is currently operational, another 7.61 per cent is under construction and 66.1 per cent remains unexploited (CEA, 2015a). Over the past five years, capacity addition in the basin has been slow (CEA, 2010, 2013, 2015b), despite numerous policy initiatives.

Hydropower policies and programmes

Indian national laws and policies of the energy sector have a direct impact on Ganga hydropower development. The Constitution of India guarantees the state's ownership of water and electricity (produced within the state) as per Entry 17 of the VII Schedule. Inter-state rivers are under the jurisdiction of central government. In other words, hydropower development of more than 25 MW in the Ganga Basin states, like Uttarakhand, is contingent upon approval of central government.

The Government of India has always been cognizant of the hydropower potential of the country and has undertaken a number of wide and specific initiatives to develop hydropower. Continued decline in the share of hydropower in India's energy mix, liberalization of the economy and persistent peak power shortages necessitated reforms in the energy sector and led to a specific policy on hydropower development in 1998. The 1998 Hydropower Policy stated hydropower to be 'a renewable, economic, non-polluting, and environmentally benign resource' (Ministry of Power, 1998, p. 1). The objective of the policy was to accelerate the pace of hydropower development and ensure targeted capacity addition in the 9th Five Year Plan, to promote small and mini hydel projects, strengthen the role of public sector utilities and state electricity boards and increase private sector investment (Ministry of Power, 1998). The policy was successful in flagging key obstacles in the development of hydropower such as technical difficulties, land acquisition, inadequate resettlement and rehabilitation, lack of investment, administrative lapses and poor management. The effectiveness of the policy did not include achieving the targets in the 9th and the consequent five-year plans. Lack of adequate monitoring of the policies and programmes failed to produce renewed plans or to disseminate lessons learnt from previous failures.

In 2003, the Electricity Act improved the structure of the industry, allowing direct interaction between the generating companies and traders, and open access through state/central utilities – suited for hydropower peak power generation (Ministry of Power, 2003). The Act mentions: 'the proposed river-works will prejudice the prospects for the best ultimate development of the river or its tributaries for power generation consistent with requirements of drinking water, irrigation, navigation, flood-control, or other purposes . . .' (Ministry of Power, 2003, p. 11).

This is consistent with National Water Policy (1987, 2002 and 2012). Upon satisfaction that these requirements are met, the CEA, in consultation with the state and central government, will sanction hydropower projects. Besides the CEA, the Ministry of Water Resources and Central Water Commission examine technical and legal aspects of proposed hydropower projects. Geological aspects related to the plant are examined by the Geological Survey of India. In effect, the Electricity Act further clarified the principles, procedures and institutions for review, assessment and evaluation of all hydropower projects in the country, including Uttarakhand.

The National Electricity Policy of 2005 addressed issues pertaining to long-term financing, rehabilitation and resettlement (R&R), land acquisition and the participation of central and state agencies in hydropower development (Ministry of Power, 2005). The National Rehabilitation & Resettlement Policy came into effect in 2004 and was revised in 2007 to provide basic minimum requirements for project-affected people (Ministry of Rural Development, 2007). In 2008, the Mega Power Project Policy gave further thrust to hydropower development with features like exemption from sales tax, local levies and customs duty on imported equipment, and a 10-year tax holiday (Ministry of Power, 2009). In the same year, the Hydropower Policy was revised to attract private sector investment, earmark 1 per cent free power to the Local Area Development Fund and an additional 1 per cent from the state's share of 12 per cent free power from the project (Ministry of Power, 2008, pp. 35–6). Project-affected families were to get 100 units of electricity per month for a period of ten years from the date of commissioning of the project. Synchronization with electrification programmes such as the Rajiv Gandhi Rural Electrification Programme was also sought through the revised policy. The recent Land Acquisition Act 2014 has provided further transparency on issues of procurement of land for hydropower development and R&R policies (Ministry of Rural Development, 2014).

Another significant initiative which affected Uttarakhand's hydropower development was the launch of the 50,000 MW Hydroelectric Initiative by the then Prime Minister Shri Atal Bihari Vajpayi (CEA, 2003). The dire state of implementation of hydropower in the country, coupled with growing demand and power shortages, required an accelerated programme to deal with the existing challenges. The features of this initiative included additional budgetary support for ongoing and new hydro projects under Central Public Sector Undertakings, the basin-wide development of hydro potential and comprehensive ranking studies for 399 schemes. The initiative focused on the preparation of feasibility reports and detailed project reports (DPR) for 162 schemes totalling up to 50,000 MW installed capacity. Of these schemes, 39,000 MW are located in four states, namely, Uttarakhand, Himachal Pradesh, Jammu and Kashmir, and Arunachal Pradesh (CEA, 2003). As of August 2015, DPRs have been prepared for 40 schemes with an installed capacity of 19,376 MW, and 114 schemes (23,957 MW) are held up on account of non-allotment, clearances and local agitation (CEA, 2015c). Of those whose DRPs are ready, 11 schemes with 1,644 MW capacity are located in Uttarakhand on the Ganges (CEA, 2015c).

Despite the institutional and legal reforms related to hydropower development, several challenges persist, including land acquisition, R&R issues, environmental concerns, local agitation, clearance issues, law and order, long gestation periods and geological surprises (Ministry of Power, 2012). As per the mid-term review of the 11th Five Year Plan, 21,802 MW of total installed capacity missed the target, leading to the downwards revision of the target to 62,374 MW from 78,700 MW. These issues are not unique to any one energy sector but are fairly common in the thermal, hydro and nuclear power sectors. For the hydropower sector, the list of issues does not come as a surprise given that the larger discourse on sociological and environmental aspects of hydropower generation is usually ignored in planning for capacity additions. Despite the recognition of these issues, the 12th Five Year Plan has set a target of adding 10,897 MW capacity from hydropower plants between 2012 and 2017 (CEA, 2015a).

Challenges

Conventional planning and management of hydropower in the Ganga Basin has been a frustrating experience. Official targets for hydroelectric generation have been missed, environmental and ecological damage is becoming increasingly visible, and local agitation against dams is turning violent. A simplistic problem analysis would indicate faulty policies and programmes, and ineffective implementation. However, the issues run deeper – a flawed approach to problem analysis, faulty methods of data collection and interpretation, a lack of multidisciplinary and all-inclusive planning, variously contribute to the difficulties faced in hydropower development.

At the heart of the environmental quagmire is a troubling history of river management and development in the Ganga region. As a result of nineteenth-century-inspired supply-side hydrology interventions, reductionist engineering and comprehensive river control, the Ganga system has been drawn into a set of irreconcilable and competing demands (D'Souza and Mittra, 2013). At present, there are more than 47 major and medium irrigation projects (Ministry of Environment and Forests, 2009); 12 major storage and diversion projects (National Institute of Hydrology, 2010); and 564 small, medium and large hydropower projects operational in the Ganga Basin (Central Water Commission, 2015). Future targets for hydropower development are even more ambitious than other sectoral uses of water in the Ganga Basin. To meet the 12th Five Year Plan targets, Uttarakhand has 170 schemes of 12,475 MW of installed capacity under construction on different tributaries of the Ganga including the Alaknanda and the Bhagirathi (AHEC, IIT Roorkee, 2010).

The status of hydropower in the country is reflective of the challenges in the Ganga Basin. As is visible from Figure 7.1, growth in the installed capacity from hydropower was comparable to thermal power plants till 1990. By the end of the 10th plan, hydropower capacity addition was almost half of thermal and continued to decline. At present, hydropower additions are equivalent to renewable energy additions. There are multiple reasons for such a decline, but

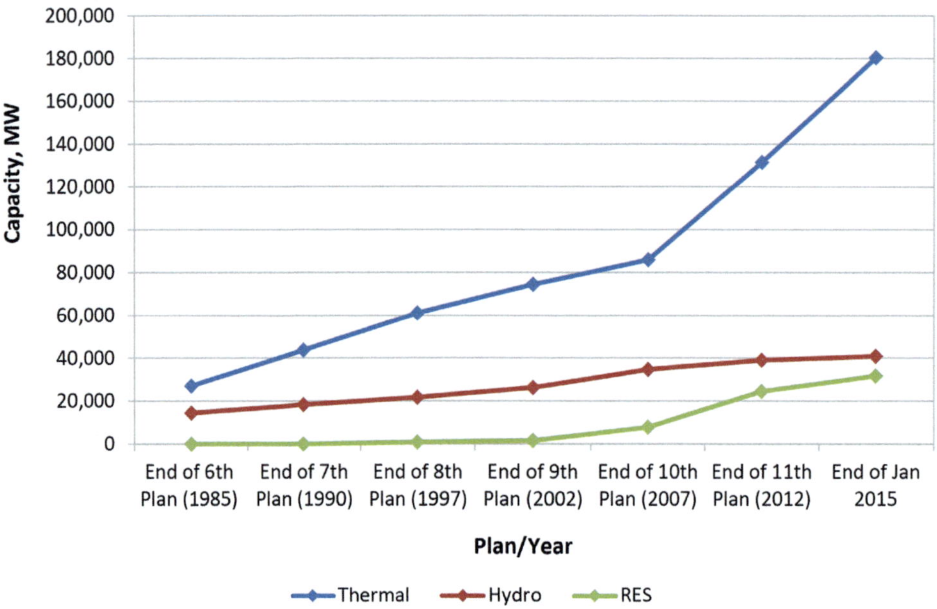

Figure 7.1 Growth in thermal and hydro capacity in India
Source: CEA (2015a)

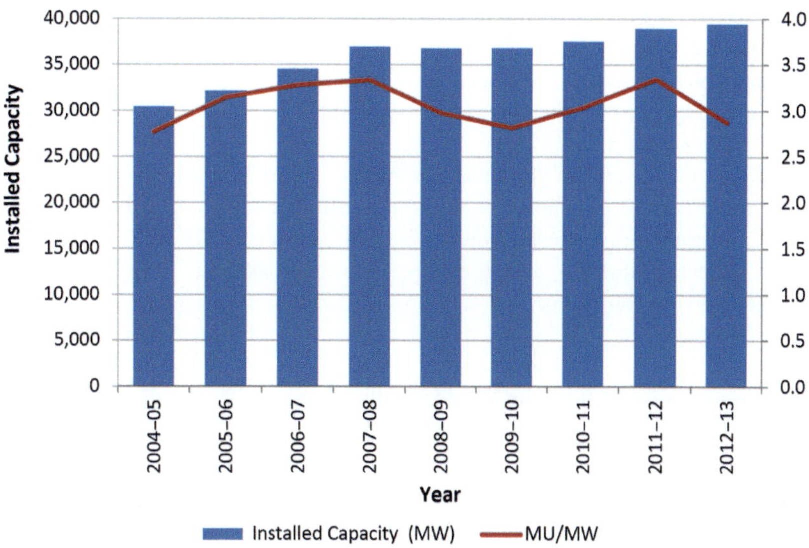

Figure 7.2 Calculated from *Report on Hydro Performance Review* (CEA, 2014)
Note: MU – million units of electricity; MW – megawatts

the primary one relates to land acquisition problems. Obstructions due to natural causes, administrative lapses, inadequate information and financial issues contribute further to the delay in clearance processes (Chaturvedi *et al.*, 2014). Sixty per cent of capacity slipped due to poor geology, weather conditions and accessibility, and 330 MW slipped the target due to protests against large dams (KPMG, 2010; Ministry of Power, 2012).

While there may be some sense in increasing the installed capacity each year to meet demand, the focus on performance, efficiency and maintenance of these projects, and adaptive strategies for changing climatic conditions, remain poorly addressed in policy design and planning. Combined installed capacity of hydropower projects in operation (74 schemes) is 2,894 MW (Ministry of Power, 2012). This cumulative installed capacity does not imply impressive generation capacity or efficiency. According to a study by the South Asian Network of Dams, Rivers and People (SANDRP) in 2011, hydropower plants are recording a decline in their generation efficiency. In 2011, generation (million units) per MW of installed capacity witnessed a drop of 22.5 per cent on average from the highest per MW generation achieved by different plants (CEA, 2014).

The story is no different for the years that follow, nor unique to the Ganga Basin. Figure 7.2 shows the generation vs. installed capacity ratio for all the hydropower plants in operation on different river streams in India. 'Technological fixes' suggested to improve the generation efficiency are a tall order and will require substantial investment in research and development.

While generation efficiency might be fixed through improved technical solutions, studies on cumulative impacts of multiple hydropower projects on the same river streams are defective. The Government of India commissioned Alternate Hydro Energy Centre and IIT Roorkee (AHEC, IIT R) to study the cumulative impacts of cascading hydel projects on the Alaknanda and Bhagirathi Rivers (AHEC, IIT Roorkee, 2010). The study was severely criticized by NGOs for being biased towards hydropower development, and technically incorrect in places (SANDRP, 2013). The study ignored cumulative impacts on hydrological flows, sedimentation, aquatic and terrestrial flora, greenhouse gas emissions, mining and other critical aspects. The Expert Appraisal Committee on River Valley Projects (MOEF) was also critical of the AHEC-IITR report (SANDRP, 2013). The Supreme Court of India expressed concerns over cumulative impacts of cascading dams on the Ganga river system:

> Prima facie, we are of the view that the AHEC report has not made any in-depth study on the cumulative impact of all project components like construction of dam, tunnels, blasting, power-house, muck disposal, mining, deforestation etc. by the various projects in question and its consequences on Alaknanda as well as Bhagirathi river basins so also on Ganga.
>
> *(Supreme Court of India, 2013)*

Unlike the report by AHEC, the World Wildlife Institute (WWI) reported severe impacts on aquatic and terrestrial ecosystems, the physical environment and alterations due to hydel projects (Rajvanshi, Arora *et al.*, 2012). All but one of 24 projects were scrapped by the Supreme Court, based on the final report by the MOEF expert committee. Kotlibhl 1A (3 × 35 MW) was the only hydropower project approved on condition of mitigating adverse environmental impacts.

In the aftermath of the 2013 flash floods in Uttarakhand, serious concerns were raised over the impact of hydropower in the upper Ganga Basin. The Supreme Court ordered MOEF to conduct detailed analysis of the cause–effect of flash floods with respect to existing and under-construction hydropower (Supreme Court of India, 2012). The expert committee drew similar conclusions to the WWI report and re-emphasized the damaging impact of 24 (earlier) proposed projects on forest land, biodiversity, national parks, sanctuaries and the environment (MOEF, 2014).

Another major point of contention on hydropower has been large-scale displacement of river communities and others dependent on the river resource for livelihood. Displaced people usually belong to lower income groups and have limited capacity to withstand the trauma of eviction (Agarwal, 2013). As a result, they are forced to live in poverty with few survival skills in areas unfamiliar to them. Despite revisions (mentioned in the section above), the R&R policies are still inadequate. There is a rising tide of inter-state disputes and state versus society conflicts surrounding most of the hydropower projects.

The social context of hydropower projects is marked deeply by issues of resource ownership, entitlement, culture and traditions around rivers and much more. On one hand, where political economy of hydropower deals with allocation, consumption and the role of the state in these processes, a sociological perspective draws upon the broader principles of sustainable development. As mentioned earlier, hydropower development in the Ganga Basin is more to do with meeting India's growing energy demands than with the energy demands of the local population. While the Ganga will be exploited to meet national targets, the benefits might not be redistributed in the basin states. It is important to mention that the electricity produced by these hydropower plants is fed into the main central grid and thereafter distributed to different states. Therefore, the direct beneficiaries of the hydropower plants are not the river communities, the displaced/project-affected people or the villages/cities in the region. Despite this, some of the direct benefits from the projects come in the form of 100 units of electricity to the displaced population, 12 per cent of the power produced to the state, jobs at the construction sites and improved access through roads, among other smaller benefits. It is also understood that by way of strengthening the central grid, improved connectivity and rural household electrification schemes, the benefits of development of hydroelectricity (and other thermal power plants) will ultimately have a trickle-down effect and reach energy-deprived communities in India, including the Ganga Basin states. However, the social and cultural significance of the Ganga River has proved greater than what was imagined during policy formulation, in the energy-economic growth calculus of hydropower development.

The Alaknanda and Bhagirathi basin areas, also known as 'Dev Bhoomi', are home to many religious sites and are a popular destination for pilgrims and tourists (AHEC, IIT Roorkee, 2010). The existence value of the Ganga is high due to its religious connotations across different religions in India. Religious leaders have long protested against the damage caused to the pristine beauty of the river valleys. The 600 MW Loharinag Pala project was scrapped after public protests and a 'fast-unto-death' by Professor G.D. Agarwal, an eminent scientist (www.thehindubusinessline.com/todays-paper/tp-corporate/article1042757.ece). Work on the 480 MW Pala Maneri and 350 MW Bharav Ghatti projects was stalled and the Vishnu Sagar project of 440 MW installed capacity was scrapped in 2010 by the National Ganga River Basin Authority (NGRBA) (Bhaduri, 2012).

With 330 MW dropped in the 11th Plan due to protests from the local population (Planning Commission, 2013), it is now seen as a serious matter hampering growth in the sector. Inadequate rehabilitation and resettlement policies for project-affected people and the exclusion of locals in project planning and execution stirred these protests. Until early 2010, opposition to hydropower projects was comparatively peaceful, but 2012 saw violent protests especially in north-east India, which could well be the reality for the Ganga Basin in the coming years, if strict reforms on R&R are not undertaken. Due attention to the large area of land required to be submerged for building dams and reservoirs, causing not only environmental degradation but displacement of population, is yet to be given in India's strategic planning for hydropower.

Besides local agitation, the major reasons for slippages in capacity addition are delays in project planning and initial activities, and gaps in survey and investigations (CAG, 2012). According to the same source, there is also lack of transparency in awarding the contracts. The CAG (2012) report was criticized however for not producing stronger recommendations. Environmental groups indicated that it did not address critical issues like the ecological and social costs of developing hydropower projects or the flawed process of environmental impact assessments. The India case study report by the World Commission of Dams (WCD, 2000) listed similar findings and recommendations. The report highlighted the cost over-runs of hydropower projects, which have important consequences for public and private budgeting. The WCD and Comptroller Auditor General reports highlighted issues that could have helped improve the policy and practice in the hydropower sector. There has been limited follow-up action, and Ganga Basin hydropower now stands at the crossroads of more complex and dynamic issues of social, environmental, economic and political processes.

Further to the issues of domestic hydropower development, the Ganga Basin is also confronted with challenges to transboundary resource management. For instance, one of the major areas of concern over cooperative hydropower development between India and Nepal is the relevance of political limitedness and dominance over mainstream economics and sociology. Nepal is still struggling with its political crisis. The high political risk has discouraged Indian private investment in the hydropower sector. With China proactively entering Nepal's market through 'generous' bilateral investment, Indian experts have further shown concerns over its impact on the already fragile and sensitive agreements signed between India and Nepal.

The geo-strategic importance of Nepal (Ministry of External Affairs, 2012) to India's energy demands during dry seasons, peak hours and shared river systems requires India to encourage hydro-potential development in Nepal (Maxwell, 2012). Equally, India provides a vast market for Nepal's electricity exports, and an integrated basin approach to internal water management issues would require support and assistance from India. Therefore, both countries need to remove the obstacles in joint hydro-potential development and achieve a balance between inclusive interests and sustainable regional development.

Indo-Bangladesh relations over the shared water issues are finally showing signs of a rational and realistic approach to conflict resolution. In 2012, the first Indo-Bangladesh joint sub-group meeting finalized essentials to begin joint studies on the controversial Tipaimukh dam project. The then Prime Minister Manmohan Singh's assurance to Sheikh Hasina 'to ink the proposed Teesta-Water-Sharing deal as soon as feasible', sent positive signals across the border (MEA, 2013). Indo-Bangladesh water disputes are not limited to the sharing of Ganga or Teesta waters. The ongoing controversy regarding the construction of Tipaimukh dam on the embankment of the Barak River in Manipur remains unresolved (IPCS, 2012). Bangladesh experts claim that the dam would lead to possible drying and devastation of the wetland in the downstream areas, affecting fisheries and agriculture. Bangladesh apprehensions over the dam can be argued to be genuine 'lower riparian anxiety'. It gains critical importance due to the other pending disputes, especially Teesta water sharing. The recent visit of Prime Minister Narendra Modi to Bangladesh in 2015 could not resolve the issues regarding Farakka barrage, Tipaimukh dam or the Teesta River. Fortunately, both sides recognize that more rigorous and reliable study needs to be undertaken to build trust and mutually share the benefits of the project. The success of the joint sub-group commission meeting reflects growing cooperation between India and Bangladesh (*Daily Star,* 2012). With finalized terms of reference for the joint study and assurance by India to provide supplementary data, there is an optimism in Bangladesh towards the possibility of more concrete cooperative arrangements. Furthermore, India has asked Bangladesh to invest in the project to gain benefit from the electricity produced. The result of the study may or may not favour the dam, but this step towards building trust and lowering riparian sensitivity is a positive sign for the region's geo-political dynamics and sustainable management of the Ganga Basin.

Way forward for India's hydropower sector

The Ganga Basin offers immense potential for hydropower generation, irrigation, economic development and cross-border cooperation. Unplanned exploitation, unilateral decisions on development and a uni-disciplinary approach to river management has led to severe consequences. The 2013 flash floods in Uttarakhand, intense public outcry against dams on the Alaknanda and Bhagirathi Rivers, under-development in the basin states and non-cooperation over water among riparian nations stand as plausible indications that the current policies, programmes and institutions for hydropower development have largely failed. Clearly, the approach to hydropower development in terms of technical assessment, social transformation and economic feasibility needs to be redefined to achieve sustainable development and build resilience against climate change.

Increasing capacity addition through hydropower in the Ganga Basin can no longer be rationalized based on the energy poverty narrative. Multi-dimensional factors such as social viability, inclusive cost–benefit analysis and ecological integrity need to be considered for existing and proposed projects. Substitutability factors and alternative energy resources must be taken into account, especially for projects which are found to be socially or economically unviable.

The Ganga Basin, like other Himalayan rivers, lacks real-time data and assessment studies. The hydro-geology is complex and the science is still indeterminate. Decisions on Ganga, therefore, will have to be more sensitive, presumptive, precautionary and judicious. The challenge lies in redefining the notion of river management so as to accommodate dissent and marginalized voices, conserve river ecology and pursue economic development at the same time. Given that changing perceptions is a lengthy process, a combination of short- and medium-term solutions will be required to reverse the impacts of the already existing dams and the ones under construction. On the technical side, focus on improving the efficiency of already existing hydropower plants is a short-term solution.

Hydropower plants are known to have huge operating benefits to meet peak electricity demand. However, declining performance efficiency of the power plants is counter-productive to such an advantage. Extensive engineering solutions may be required to improve designs and deal with problems imposed by natural geological restraints in the basin. In this sense, upgrading existing plants gains critical importance. More investment needs to be diverted towards the research and development of hydropower, especially in the ecologically sensitive Alaknanda and Bhagirathi Basins. Appropriate incentives must be given to producers for multipurpose, innovative and efficient designs. In the planning stage, possible structural interventions should be analysed and evaluated for mitigating the impact of climate change on the Himalayan landscape.

Other technical solutions involve improving managerial processes, such as effective monitoring, and research and development on hydropower designs. The major shortcoming of appraisals in the sector is the lack of baseline data and impact monitoring during construction and operation of the projects. The last CAG report on the hydropower sector was considered to be inadequate in that it overlooked systemic, institutional, social and environmental issues. Monitoring mechanisms have to be made more comprehensive in order to provide information on the characteristics and functions of environmental and social variables in the sector. Processes and practices of audit need to be more stringent and regular. Periodic reviews will act as a feedback mechanism, allowing corrective measures to be taken in a timely manner. Third-party auditing may improve the transparency and reliability of the reports. Most importantly, social impact assessment of hydropower projects (especially multipurpose projects) should be made compulsory in order to establish baseline conditions and to assess changes in parameters over the period of project operation.

As seen in the case of the Alaknanda and Bhagirathi protests, social inclusion is crucial in effective project development. It would not only create a sense of ownership towards the project but would also aid in producing real-time data, reducing mistrust and in turn would contribute to informed decision making by planners. Calculations should be made for the loss of livelihood and opportunities for both the displaced and downstream communities, along with ensuring policies are flexible enough to maintain or improve living standards throughout the resettlement process. Policies should be able to adhere to the International Covenant on Economic, Social and Cultural Rights. Customary laws around water should be acknowledged and traditional management of the watershed should be promoted.

International transboundary disputes over water sharing are burdened with political overtones. Most of the issues in all international basins are rooted in trust deficit, unreliable data, inequitable sharing and prior appropriation of use. This limits institutional cooperation and puts riparian nations in an adversarial position. Data sharing is a prerequisite for cooperation among riparian nations. Trust building can be initiated through transparent data sharing, joint data collection and research programmes. Cooperative arrangements over joint hydropower projects that are mutually beneficial and environmentally benign should be proposed and promoted on transboundary rivers. Enabling interventions need to be made at the river community level to re-model water management policies in contested river basins. A joint responsibility to protect and preserve the integrity of the basin should be undertaken by all riparian countries.

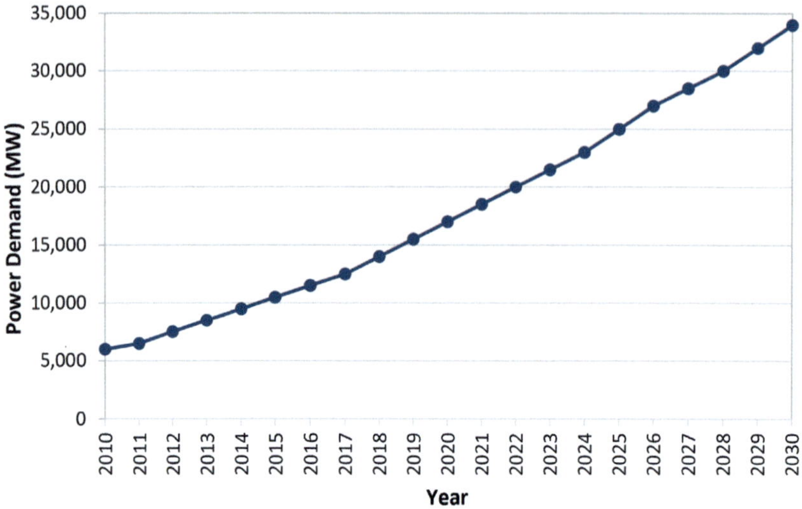

Figure 7.3 Projection of electricity demand in Bangladesh until 2030
Source: Bangladesh Power Development Board

Bangladesh

Bangladesh has recently been elevated by the World Bank to the level of a low-middle income (GNI $1,046 to $4,125) country (World Bank, 1 July 2015). This achievement occurred due to the country's 6–7 per cent sustained economic growth over the last decade and a half. Although the key socio-economic indicators look promising, per capita electrical energy use in Bangladesh is one of the lowest (169 in the global ranking) in the world. Electricity consumption and economic growth are strongly interlinked. Electricity consumption contributes to economic growth as well as investment (Masuduzzaman, 2012).

More than half of the population of Bangladesh was without electricity in 2011. The Government of Bangladesh has targets to bring electricity coverage to the entire population by 2021. This is certainly an ambitious plan as the country's electricity demand is projected to increase threefold from 2010 levels by 2020 (Figure 7.3).

Energy security

Bangladesh's future electricity security challenge lies with its energy mix. Presently more than 80 per cent of power plants are run on natural gas. The power sector alone consumes 60 per cent of produced gas. Driven by increased demand, it is projected that the existing gas reserve would be exhausted by 2023 unless new gas fields are discovered. If reserves are exhausted, Bangladesh is likely to confront a major setback in power generation, leading to setbacks in industrial and agriculture production, resulting in an inability to sustain economic growth. Energy mix diversification with local resources is also limited. The country has very limited hydropower potential due to lack of sites. Presently hydropower constitutes only about 4 per cent of the total installed generation capacity. Hydropower potential for further development is also very limited. The Bangladesh Power Development Board (BPDB) is considering increasing the production of the Karnafuli hydropower project from 230 MW to 330 MW. Two additional sites have been selected for hydropower plants at the Sangu (140 MW) and Matamuhuri

(75 MW) Rivers. The country has only some 1,207 Kw small hydropower potential (Ullah *et al.*, 2012). Overall, hydropower development potential is very limited compared to growing demand. The share of coal in power generation is only 2.6 per cent (GoB, 2011). A 1,320 MW coal-fired power station is planned at Rampal Upazilla near the ecologically sensitive Sundarbans mangrove forest in the south-west region of the country. The project faces criticism from environmental and civil society groups for its potential to harm the Sundarbans. Diesel-powered stations are expensive. Bangladesh is also planning to construct two 2,400 MW capacity nuclear power plants. They are scheduled to be commissioned in 2021 and 2025. Experts have recommended that due to a lack of technical and regulatory expertise, Bangladesh should not adopt nuclear power generation until the required capacity is created (Siddiky, 2015). It is evident that Bangladesh does not have enough local fuel resources for power generation. Therefore, in the long-run, Bangladesh's best option is the import of cheaper hydropower from Nepal. Apart from ensuring affordable energy security, the development of hydropower in Nepal will generate many co-benefits for both India and Bangladesh.

Augmentation of the Ganges flow

Hydropower development in Nepal will contribute to augmentation of the Ganges River flow in the dry season. As early as 1974, both Bangladesh and India recognized that during periods of minimum flow in the Ganga there might not be enough water to meet the needs of the Calcutta Port or Bangladesh, so Ganga flow in the lean months would have to be augmented (Abbas, 1984). It was agreed that the problem should be approached so that the interests of both countries are fairly met. It was, accordingly, decided that the best means of such augmentation should be researched by the Indo-Bangladesh Joint Rivers Commission (JRC). The Commission would make suitable recommendations to meet the requirements of both countries.

Since 1974, the JRC has been studying two proposals. The *Indian proposal* was to divert water from the Brahmaputra through a 320 km-long canal through India and Bangladesh and finally linking with the Ganges River upstream of Farakka. Under this plan, effectively 54.5 billion m^3 of water storage would be created by building dams at Dihang (5,710 MW), Subansiri (1,789 MW) in the Brahmaputra Basin and Tipaimukh in the Meghna/Barak (390 MW) Basin. According to this plan, diversion of water through the link canal would increase the Ganges flow at Farakka by 2,830 m^3/sec (Indo-Bangladesh Task Force on Flood Management Report, 1990).

Bangladesh hoped to solve the augmentation problem within the Ganges Basin as a by-product of hydropower development. Bangladesh proposed the construction of seven dams in Nepal and argued that with rising dam heights, flow could be substantially augmented. Compared with the average natural flow of the Ganges at Farakka of 1,445 m^3/sec from January to May, the mean flow could be increased by 1,656 m^3/sec and 5,339 m^3/sec under the unraised and raised conditions, respectively. Therefore, raising the height of four dams (Chisapani, Trisulganga, Seti and Sapta Koshi) would create an additional flow of 3,683 m^3/sec (Indo-Bangladesh Task Force on Flood Management Report, 1990). The augmentation would facilitate the construction of the proposed Ganges barrage in Bangladesh. This barrage has the potential to generate 113 MW of hydropower as a co-benefit (GoB, 2014).

Building hydropower dams in Nepal has many merits. First, from an engineering point of view, the construction of dams in Nepal would be far less challenging than India's inter-basin water transfer proposal. Second, Nepal could sell the generated energy to the India and Bangladesh markets. Third, water from the dams would be directly released to the tributaries.

There would be no requirement for cross-drainage structures. Additionally, Nepal could get access to the sea through a connection between the Gandak and Koshi Rivers in Nepal, the Mahananda in West Bengal and the Karatoya in Bangladesh. As well as increasing dry season flow in the latter two rivers, this canal would provide Nepal with a navigable link to the sea (Crow, 1995).

Hydrologic modelling under a recent World Bank study (Sadoff *et al.*, 2013) has found a strong potential of augmentation from the proposed water reservoirs in Nepal. The study concluded that in the driest months, the flow volume in the Ganges could be doubled even while storing a small portion of monsoon flows in the reservoirs. It supported the findings of Mirza (1998) regarding the link between low flow in the Ganges and Gorai Rivers and high salinity in the south-west region of Bangladesh. Mirza (1998) argued that salinity intrusion could only be checked by increasing dry season flows. Sadoff *et al.* (2013) concluded that in addition to contributing to agricultural productivity, low flow augmentation could better manage saline intrusion, sustain the Sundarbans ecosystem and maintain navigation services.

Alleviating the flood problem

From upstream Nepal to downstream Bangladesh, the entire basin is prone to floods. However, the main source of floods in Bangladesh is cross-border runoff from Nepal and India. Bangladesh acts as a drainage conduit for the entire basin, as the country has to flush out more than 80 per cent of the total runoff generated outside of its borders. As flooding in Bangladesh is a regional issue, the Government of Bangladesh approached the co-basin countries to examine the problem after the devastating floods of 1987 and 1988. However, the appeal for regional effort that would involve all riparian countries did not succeed and instead bilateral task forces were set up.

The Bangladesh–Nepal joint flood study team was set up in October 1988. The team examined both structural and non-structural options for flood mitigation. Storage reservoirs for flood waters were extensively investigated and found to be likely to contribute to flood moderation. The possibility of flood reduction was explored by the Indo-Bangladesh Task Force on Flood Management (1990). The Bangladesh–Nepal Joint Study Report recommended 'creation of reservoirs at upstream reaches for optimal and multiple use of water resources of the region which inter-alia could also achieve flood peak attenuation at lower reaches' (Bangladesh-Nepal Joint Study Team, 1989). In addition to the structural measures, the team recommended immediate action programmes on the development of flood forecasting and warning systems, the initiation of catchment management, an afforestation programme aimed at soil conservation and the coordination of development activities to reduce drainage congestion.

A World Bank project (Sadoff *et al.*, 2013) has recently expressed scepticism regarding the flood mitigation potential of the reservoirs in Nepal. The project team examined storage potentials of 23 sites in Nepal including Koshi, Chisapani and Pancheswar. The report found that these sites would generate an additional active storage equivalent to only 18 per cent of the entire basin's annual average flow. The total estimated active storage in the basin in Bangladesh, India and Nepal would be somewhere in the range of 130–145 billion m^3. This is only 26–29 per cent of the 500 billion m^3 average annual flow of the Ganges River (Sadoff *et al.*, 2013). However, it is argued in this chapter that it is not logical to compare the active storage in terms of annual average flow, as the objective is the reduction of flood flow in the monsoon (July–September). Previous studies estimated water-holding capacities of the reservoirs in terms of the monsoon in Nepal and India (e.g. Bangladesh-Nepal Joint Study, 1989; Indo-Bangladesh Joint Study). For example, the water-holding capacity of the Sapta Koshi, Gandak, Karnali, Mahakali and Southern river basins was estimated to be 45, 55, 124, 43 and 93 per cent, respectively (Bangladesh-Nepal Joint Study Team, 1989).

References

Abbas, A.T. (1984). *The Ganges Water Dispute*. Dhaka: University Press Limited.

Agarwal, R. (2013). Hydropower projects in Uttarakhand displacing people and destroying lives. *Economic and Political Weekly* XLVIII(29).

AHEC, IIT Roorkee (2010). *Assessment of Cumulative Impact of Hydropower Projects in Alaknanda-Bhagirathi Basins*. New Delhi: Ministry of Environment and Forests, Government of India.

Bangladesh-Nepal Joint Study Team (1989). *Report on Flood Mitigation Measures and Multipurpose Use of Water Resources*. Government of the People's Republic of Bangladesh and His Majesty's Government of Nepal.

Bhaduri, A. (2012). *Twenty-four hydro-electric power projects on Ganga in Uttarakhand need to be assessed to prevent environmental damage, says a report by Wildlife Institute of India*. India Water Portal. June.

Central Electricity Authority (2003). *50,000 MW Hydro Electric Initiative*. CEA, New Delhi: Ministry of Power, Government of India.

Central Electricity Authority (2010). *Status of Hydro-electric Potential Development – Basinwise. Monthly Hydro Reports – August*. CEA, New Delhi: Ministry of Power, Government of India.

Central Electricity Authority (2013). *Status of Hydro-electric Potential Development – Basinwise. Monthly Hydro Reports – August*. CEA, New Delhi: Ministry of Power, Government of India.

Central Electricity Authority (2014). *Review of Performance of Hydropower Stations 2013–14*. New Delhi: Ministry of Power, Government of India.

Central Electricity Authority (2015a). *Executive Summary of the Power Sector*. CEA, New Delhi: Ministry of Power, Government of India.

Central Electricity Authority (2015b). *Status of Hydro-electric Potential Development – Basinwise. Monthly Hydro Reports – August*. CEA, New Delhi: Ministry of Power, Government of India.

Central Electricity Authority (2015c). *Status of 50,000 MW Hydro Electric Initiative*. CEA, New Delhi: Government of India.

Central Electricity Authority and Ministry of Power (2013). *Growth of Electricity Sector in India from 1947–2013*, CEA, MoP, New Delhi, July 2013.

Central Electricity Authority Consortium of 7 IITs (2010). *River Ganga at a Glance: Identification of Issues and Priority Actions for Restoration*. CEA. Available at: www.moef.nic.in/sites/default/files/ngrba/001_GEN_DAT_01.pdf. Last accessed: 10 July 2015.

Central Water Commission (2015). *Water and Related Statistics*. CEA, New Delhi: Water Resources Information System Directorate, Information System Organisation, Government of India. p.153.

Chaturvedi, V., Gupta, V., Choudhary, N., Mittra, S., Ghosh, A. and Sugam, R. (2014). *State of Environmental Clearances in India: Procedures, Timelines and Delays Across Sectors and States*.

Comptroller and Auditor General of India (2012). Performance Audit of Capacity Expansion in Hydro Power Sector by CPSEs, Ministry of Power. *Report no. 10 of 2011–12*. Available at: http://saiindia.gov.in/content/report-no-10-2011-12-%E2%80%93-performance-audit-capacity-expansion-hydro-power-sector-cpses. Last accessed: 15 March 2015.

Crow, B. (1995). *Sharing the Ganges: The Politics and Technology of River Development*. Dhaka: University Press Limited.

The Daily Star (2012). Tipaimukh Dam: Dhaka, Delhi finalise joint study terms, *The Daily Star*, 29 August 2012. Available at: www.thedailystar.net/newDesign/news-details.php?nid=247472. Last accessed: 3 January 2013.

Dhungel D.N. and Ghising K.M.S. (2014). *Nepal and India Energy Cooperation – Prospects, a paper in a Dialogue on A South Asian Cooperative Architecture: The Indo-Nepal Dialogue (I)*. Jointly Organized by the Delhi Policy Group and Nepal Institute for International and Strategic Studies held at India Habitat Center in New Delhi between 21–23 December 2014 with the support of B.P. Koirala India Nepal Foundation, pp. 9–11.

D'Souza, R. and Mittra, S. (2013). Fishing in troubled waters and the cusec-megawatt river. *The Seminar* (online edn 652).

GoB/GoI (1990). *Report: Indo-Bangladesh Task Force on Flood Management*. Dhaka and New Delhi: Government of the People's Republic of Bangladesh (GoB) and Government of India (GoI).

GoN and GoI (2014). *Agreement between the Government (GoN) of Nepal and The Government of the Republic of India (GoI) on Electric Power Trade, Cross Border Transmission Interconnection and Grid Connectivity*. Available at: www.moen.gov.np/pdf_files/PTA-English-21-Oct-2014.pdf. Last accessed: 21 May 2016.

Government of the People's Republic of Bangladesh (2011). *Power and Energy Sector Road Map: An Update*. Finance Division, Ministry of Finance. GoB, Dhaka, p. 66.

Institute for Defence Analysis and Studies (IPCS) (2012). *Bangladesh Country Review* 8(4) New Delhi: IPCS.
Iyer, R. (1999). Conflict-resolution: Three river treaties, *Economic and Political Weekly*, 34(24), 12–18 June 1999. Mumbai. *Special Article and South Asian Journal*, 4(3), 65–78.
Jha, R. (2010). Total-run-of-river type hydropower potential of Nepal. *Hydro Nepal: Journal of Water, Energy and Environment* 7.
KPMG (2010). *Power Sector in India*. White paper on Implementation Challenges and Opportunities. Switzerland: KPMG International.
Masuduzzaman, M. (2012). Electricity consumption and economic growth in Bangladesh: Co-integration and causality analysis. *Global Journal of Management and Business Research* 12(11): 1–11.
Maxwell, D.M. (2012). Exchanging power: Prospects of Nepal-India cooperation for hydropower development. *Keck Journal of Foreign Affairs*. Claremont, 28 October.
Ministry of Environment and Forests (2009). *Status Paper on River Ganga, National River Conservation Directorate*. New Delhi: MOEF, Government of India.
Ministry of Environment and Forests (2014). *Expert Committee Report on Assessment of Environmental Degradation and Impact of Hydroelectric projects during the June 2013 Disaster in Uttarakhand*. New Delhi: MOEF, Government of India.
Ministry of External Affairs (2013). *India–Nepal Relations*. Available at: http://mea.gov.in/Portal/Foreign Relation/Nepal-February-2012.pdf. Last accessed: 15 July 2015.
Ministry of Power (1998). *Policy on Hydro Power Development*. New Delhi: Government of India, pp. 1–19.
Ministry of Power (2003). *The Electricity Act*. New Delhi: Government of India, pp. 1–11.
Ministry of Power (2005). *National Electricity Policy*. New Delhi: Government of India, pp. 1–3.
Ministry of Power (2008). *Policy on Hydro Power Development*. New Delhi: Government of India, pp. 28, 29, 35, 36.
Ministry of Power (2009). *Mega Power Project Policy*. New Delhi: Government of India, pp. 1–2.
Ministry of Rural Development (2007). *National Rehabilitation & Resettlement Policy*. Land Reforms Division, Department of Land Resources. New Delhi: Government of India, pp. 1–26.
Ministry of Rural Development (2014). *Land Acquisition Act 2014*. Department of Land Resources. New Delhi: Government of India, pp. 1–8.
Ministry of Water and Power (1975). *Agreement between His Majesty's Government of Nepal and the Government of India on the Gandak Irrigation and Power Project*, MoWP, Government of Nepal, Singhdurbar, Kathmandu.
Ministry of Water and Power (1975, 1996). *Revised Agreement between His Majesty's Government of Nepal and the Government of India on the Kosi Project*, MoWP, Government of Nepal, Singhdurbar, Kathmandu.
Ministry of Water Resources (1987). *National Water Policy*. New Delhi: Government of India.
Ministry of Water Resources (1996). *Treaty between His Majesty's Government of Nepal and The Government of India Concerning the Integrated Development of the Mahakali River including Sarada Barrage, Tanakpur Barrage and Pancheswor Project*, MoWR – The Hydro Power Development Policy, 1996, Government of Nepal, Singhdurbar, Kathmandu.
Ministry of Water Resources (2001). *Hydro Power Development Policy 2001*, MoWR, Government of Nepal, Singhdurbar, Kathmandu.
Ministry of Water Resources (2002). *National Water Policy*. New Delhi: Government of India.
Ministry of Water Resources (2012). *National Water Policy*. New Delhi: Government of India.
Mirza, M.M.Q. (1998). Diversion of the Ganges Water at Farakka and its effects on salinity in Bangladesh. *Environmental Management*, 22(5): 711–22.
National Institute of Hydrology (2010). Available at: www.nih.ernet.in/rbis/india_information/irrigation% 20demands.htm. Last accessed: 15 August 2015.
Nepal Electricity Authority (2013). *Corporate Development Plan 2013*, Kathmandu: NEA.
Planning Commission (2013). *Twelfth Five Year Plan (2012–2017)*. Volume II. New Delhi: SAGE Publications India Pvt Ltd. Government of India, p. 146.
Pun, S.B. (2012). 'Nepal's Water Resources in 2030'. In S.S. Raj, B.R. Upreti and K. Pyakuryal (eds). *Nepal 2030: A Vision for Peaceful and Prosperous Nation*. Kathmandu University, Kathmandu: South Asia Regional Office of the Swiss National Centre of Competence in Research (NCCR North–South) and Department of Development Studies, p. 54.
Rajvanshi, A., Arora, R. *et al.* 2012. *Assessment of Cumulative Impacts of Hydroelectric Projects on Aquatic and Terrestrial Biodiversity in Alaknanda and Bhagirathi Basins, Uttarakhand*. Wildlife Institute of India, Technical Report. p. 203.
Sadoff, C.W. and Grey, D. (2002). Beyond the river: The benefits of cooperation on international rivers. *Water Policy* 4(5): 389–403.

Sadoff, C., Harshadeep, N.R., Wu, X., O'Donnel, A., Jeuland, M., Lee, S. and Whittington, D. (2013). Ten fundamental questions for water resources development in The Ganges: Myths and realities. *Water Policy* 15: 147–64.

Sharma C.K. (1997). *Treaties on Water Resources in Nepal*. Kathmandu: Mass Printing.

Siddiky, I.A. (2015). The Rooppur Nuclear Power Plant: Is Bangladesh really ready for nuclear power? *J World Energy Law Business*, Jan12: jwu040.

South Asia Network on Dams, Rivers & People (SANDRP) (2013). *Comment on IMG (B.K. Chaturvedi) Committee Report on Upper Ganga Hydro and the River*. SANDRP. Available at: http://sandrp.in/IMG_report_on_Ganga_has_Pro_Hydro_Bias_June2013.pdf. Last accessed: 31 August 2015.

Supreme Court of India (2012). *Alaknanda Hydro Power Co. Ltd. Vs. Anuj Joshi & ORS*. Available at: http://judis.nic.in/supremecourt/imgs1.aspx?filename=40641. Last accessed: 25 July 2015.

Supreme Court of India (2013). *Alaknanda Hydro Power Co. Ltd. Vs. Anuj Joshi & ORS*. Available at: http://supremecourtofindia.nic.in/scr/2013/2013_v11_piii.pdf. Last accessed: 10 July 2015.

Ullah, M.H., Hoque, T. and Hasib, M.M. (2012). Current Status of Renewable Energy Sector in Bangladesh and a Proposed Grid Connected Hybrid Renewable Energy System. *International Journal of Advanced Renewable Energy Research*, 1(11): 618–27.

Water and Energy Commission Secretariat (2002). *Water Resources Strategy Nepal 2002*. WECS, Government of Nepal, Singhdurbar, Kathmandu.

Water and Energy Commission Secretariat (2005). *National Water Plan 2005*. WECS, Government of Nepal, Singhdurbar, Kathmandu.

World Bank (2014). IBRD-IDA (SAWI/South Asia Water Initiative). *Ganges Strategic Basin Assessment: A Discussion of Regional Opportunities and Risks*, Report No: 67668 SAS. Washington, DC: The World Bank, p. xvi.

World Commission on Dams (2000). *Dams and Development: A New Framework for Decision-making: The Report of the World Commission on Dams*. London: Earthscan.

8

The water–energy–food nexus in the Ganges Basin

Challenges and opportunities

Christopher A. Scott, Arica B. Crootof, Bhuwan Thapa and Rashmi Kiran Shrestha

Key messages

- Food and energy security in the Ganges Basin are inextricably linked to water resources, particularly flood and drought cycles, and pollution, as well as socio-economic inequalities in access to water.
- Three types of nexus examples are considered in this chapter: the groundwater–electricity nexus in the Ganges Plains, the hydropower–irrigation nexus in the headwaters, and the wastewater–energy nexus in urban centres.

- Groundwater and therefore energy (particularly electricity) used to pump it are crucial determinants of food security. Policy measures that ensure reliable rural electricity supply, even with rationing, are essential.
- Expanding hydropower development in the basin often has negative social and environmental impacts; however, tradeoffs between hydropower and irrigated food production can be addressed when user groups for electricity and irrigation coordinate management actions.
- Water pollution and food security of downstream peri-urban irrigation can be exacerbated when intermittent electricity supply causes the breakdown of urban wastewater treatment systems. Waste stabilization ponds and aerated lagoons, which are lower in cost and energy intensity than oxidation-ditch systems, may be preferred technologies in the absence of continuous electricity supply.

Introduction

The inter-relations among water, energy and food are intrinsic to human well-being and planetary sustainability. The interaction among these three resources is referred to as the water–energy–food (WEF) nexus, a conceptual approach that is increasingly understood to influence economic development, social policy and environmental planning. The WEF nexus is more than just an integrative resource-management tool; it provides opportunities to internalize waste and reduce inefficiency while setting benchmarks for societal and environmental goals. The nexus provides understanding of the conceptual linkages between water, energy and food as resources, especially in the pursuit of the twin goals of livelihood security and environmental quality. Attainment of these goals faces difficult though not insurmountable challenges in the South Asian context. Global environmental change with extremes in hydroclimatic conditions (flood–drought cycles), inequitable access to resources and markets, as well as urbanization and human migration across the region pose significant challenges to livelihood security. And in the process of meeting livelihood needs for food and energy, particularly, water is a unifying thread. Thus, the river basin becomes the crucial domain for management and policy; this is evidenced by the important contributions of various chapters in this book.

In the South Asian context, and particularly the Ganges River Basin, numerous past studies have addressed nexus inter-relations, although in rather piecemeal fashion. The primary focus has been on the electricity–groundwater irrigation nexus, beginning with Shah *et al.* (2003). Emphasis in early studies was placed on improving viability of state-run electricity companies, which were being bankrupted as a result of low tariff collection. Additional studies, however, compared diesel and electricity supply for groundwater pumping, farmer access to irrigation, water markets and associated local-level adaptations (Mukherji, 2007; Kumar *et al.*, 2011). As interest in the nexus as an analytical tool with basin-wide management and policy implications increased, Scott and Sharma (2009) assessed implications of nexus dynamics for groundwater resources. More recent focus on the nexus in South Asia, and the Ganges Basin in particular, has emphasized food security (Kumar *et al.*, 2014) and transboundary challenges (Rasul, 2014). There is emerging interest in solar pumping potential to decouple irrigation from grid power (Shah and Kishore, 2012).

Yet all of these studies focus on electricity for water. Little attention has been paid to the water-for-electricity loop of the nexus, e.g. hydropower or thermoelectric cooling. Additionally, the crucial electricity requirements for urban water supply, sanitation and wastewater management have not yet been adequately considered. In this context, our study aims to synthesize findings of the groundwater–electricity nexus research in the Ganges Basin, while expanding

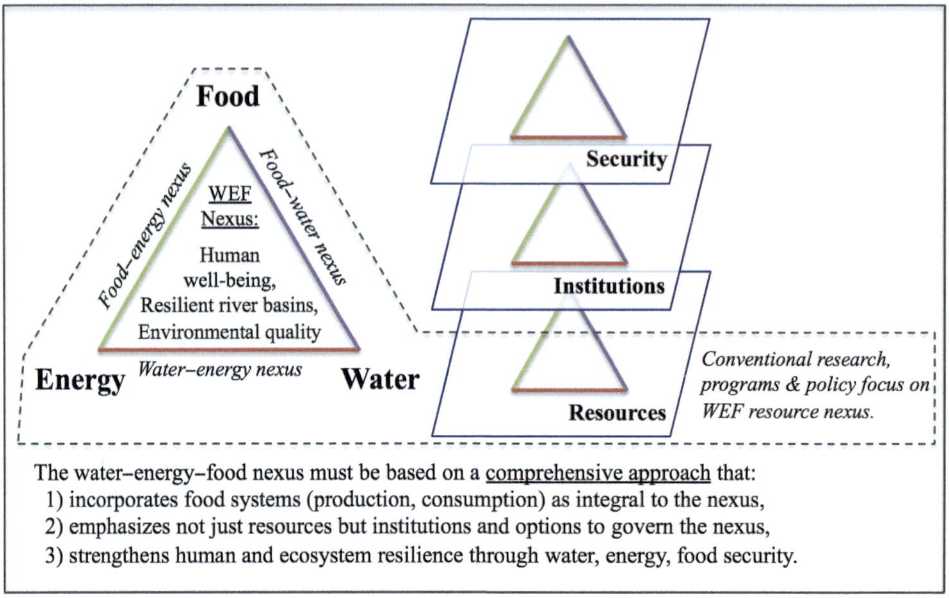

Figure 8.1 The water–energy–food nexus: resources, institutions, security

nexus understandings by documenting two of these additional interactions, i.e. the hydropower–irrigation nexus and the wastewater–electricity nexus. As shown in Figure 8.1, we consider the nexus from the perspective of (a) interlinked resources (Hellegers *et al.*, 2008), (b) institutions for their management and governance (Kurian and Ardakanian, 2015; Scott *et al.*, 2011), as well as (c) the implications for human security (Beck and Walker, 2013; Asthana and Shukla, 2014).

Challenges and opportunities: resources, institutions and security

Livelihoods in the Ganges Basin are predominantly sustained through agriculture, which is highly reliant on water and energy resources. Water, energy and food are inter-related and inter-dependent. Agriculture both demands water and is influenced by water. Similar to global trends, water for irrigation in the Ganges is the greatest user of water resources; approximately 100 billion cubic metres (BCM) of water is used annually for irrigation (Rasul, 2014). Agriculture also demands energy inputs. Energy resources are needed to develop water resources, to grow crops, to transport water and to distribute food for consumption. Developing (i.e. wells or canals), transporting (i.e. pumps) and treating water resources for irrigation and human use are energy-intensive. Similarly, energy is needed to grow and distribute food for consumption. Therefore, we focus on water, energy and food resources as the core components in our nexus to promote well-being in the Ganges Basin.

While a nexus approach is a tool to better understand and study the inter-related and inter-dependent factors between water, energy and food resources, integration alone is not sufficient to address the security challenges of these three resources. Governance (including formal and informal institutions) is needed to address the equitable distribution and sustainable use of water, energy and food across scales and sectors. This is particularly true for transboundary systems, which, although they are beyond the scope of the present analysis, add further complexity and challenges to governance. Examining the role of both formal and informal institutions is necessary to understand the interplay of WEF resources as well as the obstacles facing WEF security and the opportunities to promote it.

By expanding the water–energy–food nexus to incorporate not only resources but also institutions and security, this nexus approach can strengthen human and ecological resilience. Resilience, as defined by Walker and Salt (2012), is the ability of the system to respond and adapt to disturbances while maintaining the same structure and identity. Using a nexus approach to identify complex linkages, system governance and related security challenges can offer valuable insights to support water, energy and food policies that build livelihood security to local and global changes. The following three case studies from the Ganges Basin illustrate how a nexus approach can be utilized to promote human well-being through the sustainable and equitable development of water, energy and food resources.

To demonstrate nexus linkages in practice on the ground, we explore how water, energy and food influence each other throughout the Ganges Basin. We begin by examining the groundwater–electricity nexus in the Ganges Plains, India, followed by a case study highlighting the hydropower–irrigation nexus in the mountainous headwaters of the Kali-Gandaki tributary of the Ganges in Nepal. Finally, we present a case study of wastewater management in the urban centre of Kathmandu, Nepal.

Water–energy–food nexus case examples

Groundwater–electricity nexus in the Ganges Plains

The challenges surrounding groundwater resources are highly spatially variable across the Ganges Plains. For example, groundwater depletion is widespread in the Upper Plain while water quality issues are prevalent in the Middle and Lower Plain. Moving downstream, there is generally more surface water available, while groundwater quality deteriorates. Portions of the Middle Plain and much of the Lower Plain have toxic concentrations of naturally occurring arsenic in shallow (<150 m) alluvial aquifers (Ahmed *et al.*, 2004; Ahamed *et al.*, 2006; Chakraborti *et al.*, 2003; Saha and Sahu, 2015). While groundwater quality poses a significant human and environmental threat, the water quality issues associated with groundwater are beyond the scope of this section. This section focuses on reviewing the groundwater–energy nexus in the upper Ganges Plains in northern India where power subsidies and groundwater depletion are tangled in a vicious cycle.

Global changes are reshaping the water, energy and food landscape in the Ganges Plains. This dynamic region, which covers approximately 250,000 km^2, is experiencing significant population growth, urbanization and industrialization. For example, changes in land use as well as cropping patterns and irrigation and drainage networks have significantly changed surface water flows. Additionally, climate change is increasing the variability of surface water flows, contributing to a growing reliance on groundwater resources to sustain food production, particularly in times of drought (Shah, 2009). These changes, coupled with energy policies, are placing greater stress on water resources. As a result, groundwater resources in the Upper Plain are being exploited faster than they are being replenished (Gleeson *et al.*, 2012). Because extracting groundwater requires energy, the groundwater–electricity nexus provides a lens to examine water, energy and food resources and to better understand the role of governance for water, energy and food security in the Ganges Plains.

Declining groundwater levels are having rippling effects both locally and globally. Locally, water availability is a growing concern for farmers, particularly small-scale farmers who are disproportionally affected by the financial investment required for new pumps or deeper wells

(Ahmed, 2013). In addition to affecting agriculture production, declining groundwater levels are also increasing energy demand and thus, greenhouse gas (GHG) emissions. While both diesel and electric pumps are prevalent throughout the Ganges Plains, once water levels fall below 15 metres it becomes more efficient to transition from diesel to electric pumps. However, this can also have a significant impact on energy consumption and GHG emissions. For example, Shah (2009) found that although less than 15 per cent of the groundwater pumps in India were electric, they accounted for over two-thirds of the carbon emissions from groundwater pumping. Solar pumps offer an attractive alternative to fossil fuel-based energy. However, as Bassi (2015) discusses, substituting solar pumps without changing the current energy policy context will do little to mitigate falling water tables.

Energy policies in the Ganges Plains have and continue to be a significant driver of groundwater development. Governmental policies beginning in the 1930s initiated the use of groundwater for irrigation; these were continued as part of the Green Revolution from the 1960s to the present. Power subsidies helped spur a rapid rise in the number of electric groundwater pumps. Between 1980 and 2010, over 15 million wells with electric pumps were installed in India. As groundwater levels decline, the need for power subsidies increases as more infrastructure and energy is now needed to access the deepening groundwater (Shah *et al.*, 2012). While substituting solar pumps would reduce greenhouse gas emissions, the power subsidies – which are required for solar to be competitive with conventional electric and diesel pumps – would not address local water availability or equity issues (Bassi, 2015). Revising energy policies to address energy and water supply and demand will be critical to support the long-term viability of groundwater resources in the region.

Integrating water and energy management and policies is considered the most effective option to promote more sustainable use and equitable access to groundwater resources (Scott and Sharma, 2009; Aggarwal *et al.*, 2004). Aggarwal *et al.* (2004) encourage incorporating policies specific to irrigation, land use and migration to address the complexity of the socio-ecological systems associated with water, energy and food. The importance of technical feasibility, economic viability and equity of access is further stressed by Bassi (2015). Although water quality is not assessed in this section, it is another integral component to consider. The water and electricity case study below illustrates the connections between energy and water treatment for the supply and treatment of human and environmental health, which is a pressing concern in a basin that has naturally high concentrations of arsenic, fluoride and manganese, and is experiencing groundwater pollution from nutrients and pesticides (Saha and Alam, 2014).

The interdependences between groundwater and electricity support the need for a nexus approach in the Ganges Plains. Although this review focuses on the groundwater–electricity nexus in relation to food production, the nexus as a conceptual tool has broader application. The nexus is equally important to consider for municipal water as well as industrial and mining use. Formal and informal governance systems that account for the interplay between groundwater and energy can support sustainable and equitable water, energy and food policies in the Ganges Plains for a more water-secure future.

Hydropower–irrigation nexus in the Ganges headwaters

Hydropower is rapidly expanding throughout the Himalayas (Ahlers *et al.*, 2015) and the Ganges headwaters, in particular. The abundance of water resources and high relief of the Himalayas offer enormous hydropower potential in a region where energy demand largely exceeds energy supply. Therefore, this technology is being promoted for its low GHG emissions and its ability to provide rural electrification at a competitive cost. However, even small hydropower development can affect water availability for irrigation. Run-of-the-river hydropower designs reduce water availability in streams by diverting water from the river for power generation. Water is first diverted at a headgate or weir, then transported through a pipe or tunnel for a specified distance (varies, based on project) for power generation, and is finally returned to the river at a downstream location. So although there are benefits from these hydropower projects, there are also tradeoffs. Here we examine the tradeoffs and benefits of utilizing water for energy and food production in the Ganges headwaters, using the run-of-the-river Andhikhola hydropower project (AHP) in western Nepal as an example.

AHP is a multipurpose hydroelectric project designed to minimize tradeoffs by providing electricity to local residents, selling surplus electricity to the national grid and supplying water for new irrigation. AHP was designed to ensure that the local communities were the beneficiaries of this hydropower plant. To ensure that the local communities benefited from the project, the financial donors insisted that the hydropower plant only be connected to the central grid if local communities could access the electricity first and then sell surplus energy to the grid. Furthermore, they delineated the impacted area and developed land- and water-sharing institutions, which are discussed in more detail below.

AHP is located in the lower Galyang village development committee (VDC) of Syangja district. This run-of-the-river project extends from upstream of Galyang village to just below the Andhikhola–Kali-Gandaki confluence. Although AHP has undergone recent renovations

Figure 8.2 Andhikhola hydropower-irrigation multipurpose project (diversion weir)
Photo: Rashmi Kiran Shrestha

(increasing its capacity from 5.1 to 9.4 megawatts [MW]), it still comes under Nepal's 'small' hydropower classification, which is defined as 1–10 MW. AHP was built by Butwal Power Company (BPC) under the aegis of United Mission to Nepal (UMN). The construction began in 1982 and power generation came into commercial operation in July 1991. The 60 m wide and 6 m tall concrete diversion weir (Figure 8.2) draws water from the Andhikhola River just north of Galyang village. The diverted water then flows down a 1.3 km tunnel to a 237 m vertical shaft that fuels the electric power station. The diverted water is then returned downstream and discharged into the Kali-Gandaki River.

After two decades of operation, the turbines in the power station needed to be replaced. At this time, BPC decided to rehabilitate and upgrade the existing hydropower project from 5.1 MW to 9.4 MW. AHP was closed for two and a half years for construction. The construction included expanding the underground powerhouse by 8 metres, enlarging the tail-race tunnel and adding an additional 250 metres of penstock pipe. Furthermore, a rubber dam, which is a new technology in Nepal, was implemented at the headgate. The improved AHP now has three intake gates, two desilting basins, a headrace tunnel, a penstock pipe, a powerhouse and a tailrace tunnel. It became fully operational on 7 April 2015. Financing for this upgrade came from several sources. The International Finance Corporation (IFC) invested Nepali Rupees (Rs.) 500 million (US$4.9 million), while Mega Bank and BPC each invested Rs. 450 million (US$4.4 million).[1]

The upgraded AHP is projected to produce 70.73 MWh yearly.[2] The additional energy production will be supplied to the grid, which BPC believes will help decrease load-shedding hours in Nepal. According to the distribution head of BPC in the project office, this project serves 31,000 consumers in the district of Syangja and the neighbouring district of Palpa. Out of 31,000 consumers, 418 are small enterprises. Similar to the far-reaching rural electrification from AHP, the Andhikhola irrigation system (AIS; Figure 8.3) also has far-reaching impacts.

The main irrigation canal that supplies water for the AIS was built with the financial and technical support of UMN during the construction of AHP. UMN managed this canal until 1995 when its management was handed over to the newly formed, community-run Andhikhola Multipurpose Water Users Association (AKWUA).[3] Initially, the total command area was 282 hectares and irrigation water was supplied to villages and households that were both directly and indirectly affected by AHP. For example, Lalyang ward, where the tailrace tunnel was built, was directly impacted by the construction of AHP and received and continues to receive water from AIS. Similarly, five other nearby wards, which were not in the path of AHP construction, also receive water from AIS.

Today, AKWUA is comprised of 13 members (nine males and four females) representing different VDCs and wards. This organization oversees the management of AIS, which supplies water to 850 households and irrigates 330 hectares of land year-round. This irrigation system serves six wards of two VDCs from Syangja district and two wards of Hungi VDC from Palpa district. Irrigation pipelines run across the suspension bridge over the Kali-Gandaki River to irrigate VDCs in Palpa district.

Water allocations between AHP and AIS were negotiated between BPC and AKWUA and a formal agreement was signed in 2004. The agreed flow of water in the irrigation canal varies seasonally between the monsoon and dry season. From June to October the agreed rate of flow is 684 litres per second (lps) and from November to May the agreed rate of flow is 300 lps. During the month of April, BPC supplies an additional 50 lps of water for maize planting.

UMN, in cooperation with local farmers, researched the amount of land each person would require for year-round food security. Based on their study, it was concluded that 0.036 hectares of land per person was needed to grow a year's worth of food. Households that had more land

Figure 8.3 Andhikhola irrigation canal
Photo: Rashmi Kiran Shrestha

than this minimal requirement were encouraged to sell 10 per cent of their 'excess' land to AKWUA at a minimum rate. This plan allowed AKWUA to acquire 11.2 hectares of land. A total of 71 households purchased land at a nominal price, which could then be paid for in instalments. This kind of practice, if implemented elsewhere, would allow community members to acquire enough land to meet their basic food needs. Although there have been political discussions to acquire barren lands and distribute these lands to poor farmers, no formal decisions regarding these barren lands have yet been made.

In 1995, when the agricultural command area was established, 25,000 shares were issued to equitably distribute water. Each person received four shares, with each share equalling 0.0275 lps. There were two ways of earning additional shares. Local people could contribute labour in canal construction or pay for shares. For example, a person could gain one share by contributing five working days in canal construction, could send five workers for one day of canal construction, or buy a share for Rs. 33 (US$0.32). Later in 2005, it was determined that four shares per person was no longer enough to meet irrigation demand, especially now that the area had increased from 282 to 330 hectares. Beginning in 2005, eight shares per person were distributed and additional shares could still be obtained.

AKWUA is financially supported by BPC as well as from user fees. At present, farmers pay Rs. 20 (US$0.20) per share per month to AKWUA for water. Of the initial 25,000 shares, AKWUA has sold 20,500. BPC also supports AKWUA's administrative costs. Prior to the upgrade of AHP, BPC paid AKWUA Rs. 250,000 (US$2,456 as per 2015 rate) per year. Now with the completion of the AHP upgrade from 5.1 to 9.5 MW, BPC is providing Rs. 400,000 (US$3,930) per year as administration support to AKWUA.

Despite this financial support, AIS is in need of extensive maintenance and renovation. The irrigation canal infrastructure is 20 years old. The Government of Nepal, with support from the World Bank, is implementing the Irrigation and Water Resources Management Project (IWRMP) to improve irrigated agriculture productivity and management of selected irrigation schemes and enhance institutional capacity for integrated water resource management. The Andhikhola Irrigation System was chosen to receive funding from this project and received Rs. 50 million (US$491,000) for construction, maintenance and repair of the canal system. Furthermore, AKWUA received Rs. 9.7 million (US$95,300) from the Western Region Irrigation Development Division 2 for the maintenance of old canals.

Focus group discussions with men and women from two wards in Lalyang and Jimuwa in Jagat Devi and Tulsi Bhanjyang VDCs reveal that AHP and associated AIS have proved to be a gift to these VDCs. Before the construction of AHP and AIS, farmers could hardly harvest one crop. Now farmers grow three major crops per year. This was a result of AIS, which increased the availability of water for irrigation and brought once barren lands into cultivation. Furthermore, women said they had to walk four to five hours to buy rice in Galyang Bazaar. Since the development of AHP and AIS, women can now sell surplus rice, vegetables and other crops in Galyang Bazaar. The access road, which was built for AHP, also makes it easier for farmers to bring their agricultural products to sell in the main market. In the present context of increasing rice imports from India, and the roughly 25 per cent of agricultural lands which did not receive enough 2015 monsoon rainfall for rice plantation, the hydropower–irrigation nexus principally demands more attention for food and energy security. Ensuring irrigation facilities in the project-affected areas has been one of the major demands of the public. These demands are likely to be increasingly important for developers to address as efforts to upscale hydropower development in Nepal and the central Himalayas continues. For example, in the Karnali 900 MW project that is being developed by an Indian company for energy exports to India, the main demand of downstream communities is the unconditional release of water for irrigation.

Although AHP, and associated AIS, have brought many benefits to the region, including increased food security, other challenges have emerged. Focus group discussions with women in the villages indicate that women carry out most of the agricultural activities. This change is a result of out-migration. Approximately 75 per cent of the males from these VDCs have gone to Gulf countries as migrant workers and women are disproportionally left to attend to agricultural activities.

This hydropower–irrigation nexus case study indicates how a nexus approach can help illuminate the benefits and tradeoffs among water, energy and food. Despite many communities experiencing negative impacts from small hydropower implementation, the AHP case study highlights how small run-of-the-river hydropower can support food security as it has done in the Syangja and Palpa districts. As one local development officer stated, well-designed, multi-purpose run-of-the-river projects that focus on local benefits, like AHP, could be the best option for water, energy and food security. The Government of Nepal recently declared that the country will be able to meet its energy demands by 2017, reducing the country's load shedding to zero. However, the recent April 2015 earthquake has had a tremendous impact on the hydropower sector, extensively damaging infrastructure and resulting in the delay or diversion of investments in energy. Despite these setbacks, many hydropower projects remain under construction and continue being planned. Additionally, Nepal has an increasing trade deficit with India, the biggest share of which meets fuel and food needs. To balance trade with India, Nepal needs to produce more energy and food; the hydropower–irrigation nexus can play a vital role. However, this calls for more in-depth research and policy support to ensure institutional arrangements that maximize the local benefits of hydropower while minimizing negative tradeoffs.

Wastewater-energy nexus in urban centres

Urban wastewater is a growing management challenge. The biophysical study of the nexus focuses on the energy intensity of wastewater treatment plants and the energy recovery during the treatment process. It aims to minimize losses and maximize the plants' efficiency; address externalities across sectors like agricultural production downstream and GHG reduction from energy recovery; and promote interactions among sectors including water, wastewater and energy that are generally managed in isolation (Liu *et al.*, 2015; Lal, 2015; Scott *et al.*, 2015).

The wastewater–energy nexus is an emerging challenge in South Asia. Centralized wastewater treatment, which is energy-intensive due to pumping and conveyance in addition to treatment, is gradually expanding in the urban areas in the region. In the Ganges Basin in India, Nepal and Bangladesh, more than 53 centralized wastewater treatment plants are currently in operation. The number of plants is increasing in the basin as the sewerage coverage expands in many cities, including 75 major cities with the population exceeding 50,000 (NRCD and MEF, 2009; Muzzini and Aparicio, 2013). While these systems require continuous supply of energy, the region is suffering from persistent power shortages. The annual electricity deficits are: 24,077 GWh (2.1 per cent) in 2015 in India, 982 GWh (20.3 per cent) in 2011 in Nepal, and 292 MW (4.5 per cent) in 2011 in Bangladesh (CEA, 2015; HIDCL, 2015; MOF, 2011).

One of the burgeoning urban centres in the highland region of the Ganges Basin is the Kathmandu Valley. Comprising the five major cities – Kathmandu, Lalitpur, Bhaktapur, Kirtipur and Madhyapur Thimi (Figure 8.4) – the valley that is growing at the annual rate of 5 per cent faces challenges in wastewater management that are similar across many developing countries

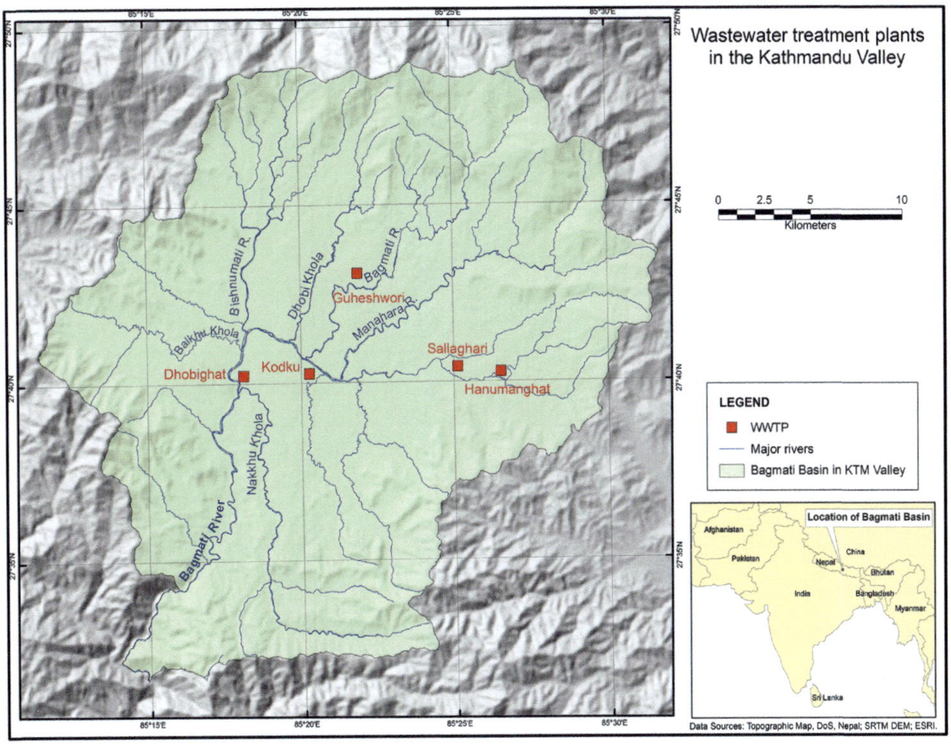

Figure 8.4 Wastewater treatment plants in the Kathmandu Valley

Table 8.1 Major WWTP in the Kathmandu Valley

Name	*Year*	*Type*	*Treatment Capacity (MLD)*	*Land Area (ha)*	*Energy Intensity*	*Remarks*
Guheshwori	2002	Oxidation ditch	16.4	5	High	Partially functional
Sallaghari	1985	Aerated lagoon	1	3.3	Minimal	Partially functional; aeration system removed
Kodku	1982	Waste stabilization pond	1.1	6.5	Minimal	Partially functional
Dhobighat	1982	Waste stabilization pond	15.4	19.4	Minimal	Not operational since 1982
Hanumanghat	1975	Aerated lagoon	0.5	0.7	Minimal	Partially functional; aeration system removed

Source: ADB (2009), Jha and Bajracharya (2014), Pradhan (2012) in the region (Muzzini and Aparicio, 2013). These include inadequate management, insufficient infrastructure capacity, ambiguous institutional responsibilities, poor financial recovery mechanisms and weak human resources capacity (ADB, 2009). While food is not specifically addressed in this nexus case study, it is important to note the connection between wastewater and food production. As described in Rutkowski *et al.* (2007) wastewater is used for irrigation throughout the Kathmandu Valley, as well as other regions in South Asia, which can have serious implications for public health.

Recognizing the emerging needs for centralized wastewater management services in the Kathmandu Valley, the Government of Nepal, with support from international donor agencies, developed four wastewater treatment plants in the 1980s (Table 8.1). Unfortunately, all four systems are non- or partially functional due to poor operational management, improper regulation of inflow and outflow, and institutional negligence (ADB, 2009; Regmi, 2013). In 2002 the Government of Nepal built Guheshwori Wastewater Treatment Plant (WWTP) in order to improve the water quality of the Bagmati River at the religious and culturally significant Pashupatinath area (GON and NTNC, 2009). It is the only plant that is partially operational in the Kathmandu Valley. The plant is operated and maintained by the High Power Committee for Integrated Development of Bagmati Civilization (HPCIDBC).

From the resource perspective, the wastewater–energy nexus has two broad elements: (i) reducing the energy intensity of WWTP through efficiency improvements or retrofits and (ii) energy recovery from wastewater treatment plants. These two elements are examined here in Guheshwori WWTP, and the broader implications of the nexus in the Ganges Basin are discussed.

Energy is crucial for the proper functioning of Guheshwori WWTP. The plant is an extended aeration system using an oxidation ditch. All three treatment units – primary, biological treatment and secondary treatment units – are highly mechanical systems and require energy. The two oxidation ditches are the most energy-intensive unit, consuming about 60 horsepower (HP). The system requires about 14,171 KWh/day (ADB, 2009). In addition, the performance of the oxidation ditches is very sensitive to power supply. Breakage of power for even a short duration can create anoxic conditions and kill the microorganisms in the aeration unit. The performance of the plant had been severely altered by extended power outages of up to 80 hours a week during the summer season due to the national power deficit (Development Bureau, 2014). Fortunately, the plant has now received a 24-hour power supply line 13 years after its establishment.

The energy intensity of the Guheshwori WWTP can be improved by different mechanisms. Energy audits can help identify opportunities for energy efficiency. Changes in operation and maintenance activities and system retrofits can help improve energy efficiency. For example, altering the design from Activated Sludge Unit to Sequencing Batch Reactor (SBR) can improve energy efficiency from 2.3 unit/kg BOD to 0.77 unit/kg BOD (Regmi, 2013). Installing a catchment to hold high silt load can also improve the efficiency of the plant and reduce energy demand (ADB, 2009).

Options of energy recovery from wastewater treatment plants are installation of biogas digestion and bio-solids incineration. In the case of Guheshwori WWTP, there are proposals for energy recovery from sludge digestion or gasification (ADB, 2012). Lohani *et al.* (2013) argue that upflow anaerobic sludge blanket (UASB) and anaerobic baffle reactor (ABR) can recover the energy from wastewater.

Besides the energy efficiency and recovery mechanisms, the urban wastewater WEF nexus is driven by social, political, institutional, economic and engineering factors: namely the land availability, power shortages, financial sustainability and institutional capacity. A major constraint in installing the wastewater treatment in the Kathmandu Valley is the lack of land. Within two decades, the price of land in and around urban areas has sky-rocketed, making it too expensive for the government to purchase large plots that are required for the plant installation. This has forced the government to consider centralized treatment processes that require less land and perform system retrofits in the existing treatment plants, all of which are generally more energy-intensive (ADB, 2009).

The effluent load heavily influences the choice of treatment system. The Kathmandu Valley has a population density of about 98 people per hectare (Muzzini and Aparicio, 2013). The average wastewater generated from the highly dense population is about 150 million litres a day (MLD) (ADB, 2012). Only the centralized and highly mechanical plants can handle such a large volume of wastewater in a limited area. Though the Decentralized Wastewater Treatment System (DEWATS) is a less energy-intensive system, it can only support a population of 100–200 households with inflow ranging from 20 to 50 m^3/day and hence is suitable only in peri-urban areas (ENPHO, 2010).

Institutions to manage wastewater are still in a nascent stage in Nepal. Guheshwori WWTP is the only wastewater treatment plant that is operated by an inter-governmental agency, i.e. HPCIDBC. The plant receives its operation and maintenance fund from land sales tax within the valley. The new treatment plants will be operated by Kathmandu Upatyaka Khanepani Limited (KUKL), which is the public–private entity that has a legal mandate to operate water and wastewater systems in the Kathmandu Valley. Unfortunately, this institution is currently suffering from net financial loss. In such a situation centralized treatment plants can add a huge financial burden to the institution.

In summary, the wastewater–energy nexus is an emerging challenge in South Asia because of rapid urbanization that leads to an increase in energy demand for wastewater management, with implications for downstream water quality and urban and peri-urban irrigation. At the resource level, the nexus can be strengthened by reducing energy intensity and conducting energy recovery from the treatment plant. The nexus in the region is further challenged by acute power deficit, land availability for new treatment plants, and financial and institutional challenges. Addressing the wastewater–energy nexus can enhance food, energy, water and environmental securities.

Conclusions

The Ganges River Basin, which is the mostly heavily populated large river basin in the world, highlights the critical links among water resources and food and energy security for hundreds of millions of people. Seeking to contribute to the basin-wide approach of this book, we explicitly consider water as a unifying thread in the nexus. As illustrated in the sections above on the groundwater–electricity nexus, the hydropower–irrigation nexus and the wastewater–energy nexus, there is an urgent need to support more integrative management to promote water, energy and food security. However, this is a difficult task. With global change rapidly transforming the basin, and climate-induced variability in flow regimes (flood and drought cycles), environmental degradation (particularly from pollution), and socio-economic inequalities (across local and transboundary scales), numerous serious obstacles remain. There is an urgent need to better understand specific interactions among water, energy and food as well as the institutions that govern them in order to enhance human well-being, economic development and environmental quality. The nexus approach provides an entry point to examine and address these complex and compounding challenges.

Considering resources, institutions and security (in human, environmental and state security terms), this chapter explores how the water–energy–food nexus can be leveraged to assist with integrating natural resources management, maximizing benefits and minimizing tradeoffs, and implementing energy conservation and recovery to enhance human well-being, economic development and environmental quality. In the Ganges Plains fractured governance systems have led to overlapping or absent policy measures that fail to effectively address groundwater depletion. Because groundwater depletion and energy policies are interlinked in a vicious cycle, integrative water, energy and land management will be crucial to find solutions that can sustain productive agriculture. In the Ganges headwaters, where dam development has brought criticism due to the social and environmental impacts, the Andhikhola hydropower project provides an example of how a nexus approach can identify and enhance social and environmental benefits while minimizing negative tradeoffs. Finally, the wastewater–energy nexus is an emerging challenge in South Asia because of rapid urbanization that leads to an increase in energy demand for wastewater management and because wastewater is routinely used downstream for irrigation of food crops. As described in the wastewater–energy section, wastewater treatment plants can significantly reduce energy needs, while also recovering energy from the treatment process. Opportunities to reduce energy intensity in centralized wastewater treatment plants in the Ganges Basin can be achieved by performing energy audits, system retrofits and changes in operation and maintenance activities. Energy recovery from the plant can also be utilized through biogas digestion and bio-solids incineration, which will reduce the net energy demand of the treatment plant. As wastewater treatment expands to meet growing demand, it will be imperative that these centralized plants reduce their energy demand through more efficient operations as well as develop energy-recovery opportunities. Ideally, this emphasis on energy efficiency and local production could be a model for other industries that could help mitigate future energy demands in the Ganges Basin.

By examining multiple dimensions of the WEF nexus in the Ganges Basin, in this chapter we have aimed to explore resource linkages – the multiple ways in which water, energy and food are bound together. In the process, however, we have uncovered a significant degree of policy disarticulation across sectors, e.g. how water management does not adequately account for energy inputs or power generation, or alternately how energy or food production imperatives make unreasonable assumptions on the availability and timing of water resource availability. In future research, we aim to address nexus transboundary challenges through documentation of resources disputes as well as cooperation. Here, we have attempted to identify WEF nexus gaps, especially in electricity-based groundwater irrigation, hydropower as renewable energy with food production tradeoffs, and urban wastewater management based on reliable power supply that has implications for the sustainability of downstream agriculture and food production.

Acknowledgements

This research is supported by the CGIAR Research Program on Water, Land and Ecosystems, a global research programme that promotes a new approach to sustainable agricultural intensification in which a healthy functioning ecosystem is seen as a prerequisite to agricultural development, resilience of food systems and human well-being. Additional support was provided by Lloyd's Register Foundation, a charitable foundation helping to protect life and property by supporting engineering-related education, public engagement and the application of research. The authors are also grateful to ICIMOD for providing core funds which were contributed by the governments of Afghanistan, Australia, Austria, Bangladesh, Bhutan, China, India, Myanmar, Nepal, Norway, Pakistan, Switzerland and the United Kingdom. The authors would also like to thank the volume's editors and Dr Prachanda Pradhan for their thoughtful insights. The views and interpretations in this publication are those of the authors and are not necessarily attributable to the organizations they are affiliated to.

Notes

1 Interview with the senior officer at BPC.
2 BPC press release, 6 April 2015. For preparing this case study, field visits were done from 22 to 28 March 2015 in the project sites and affected village development committees. Interviews were conducted with staff of the BPC head office and project office, district government officers, community members, and women's groups in project-benefited as well as affected areas.
3 Interview with the chairman of AKWUA.

References

ADB (Asian Development Bank) (2009). *Kathmandu Valley Water Supply and Wastewater System Improvement Conceptual Waster Master Plan TA 4893-NEP (Draft)*. Report published by GHD and ICON for Asian Development Bank: Manila.

ADB (Asian Development Bank) (2012). *Proposed Loan and Grant and Administration of Loan and Grant Nepal: Kathmandu Valley Wastewater Management Project*. Project Number: 43524. Asian Development Bank: Manila.

Aggarwal, P.K., Joshi, P.K., Ingram, J.S.I. and Gupta, R.K. (2004). Adapting food systems of the Indo-Gangetic plains to global environmental change: Key information needs to improve policy formulation. *Environmental Science and Policy*, 7: 487–98.

Ahamed, S., Sengupta, M.K., Mukherjee, A., Hossain, M.A., Das, B., Nayak, B. & Chakraborti, D. (2006). Arsenic groundwater contamination and its health effects in the state of Uttar Pradesh (UP) in upper and middle Ganga plain, India: A severe danger. *Science of the Total Environment*, 370(2): 310–22.

Ahlers, R., Budds, J., Joshi, D., Merme, V. & Zwarteveen, M. 2015. Framing hydropower as green energy: Assessing drivers, risks and tensions in the Eastern Himalayas. *Earth System Dynamics*, 6: 195–204.

Ahmed, I., Al-Othman, A.A. & Umar, R. (2013). Is shrinking groundwater resources leading to socioeconomic and environmental degradation in Central Ganga Plain, India? *Arabian Journal of Geosciences*, 7(10): 4377–4385.

Ahmed, K.M., Bhattacharya, P., Hasan, M.A., Akhter, S.H., Alam, S.M., Bhuyian, M.H., Khan, A.A. & Sracek, O. (2004). Arsenic enrichment in groundwater of the alluvial aquifers in Bangladesh: An overview. *Applied Geochemistry*, 19(2): 181–200.

Asthana, V. & Shukla, A.C. (2014). *Water Security in India: Hope, Despair, and the Challenges of Human Development*. Bloomsbury Publishing.

Bassi, N. (2015). Irrigation and energy nexus. *Economic & Political Weekly*, 50(10): 63.

Beck, M.B. & Walker, R.V. (2013). On water security, sustainability, and the water–energy–food-climate nexus. *Frontiers in Environmental Science and Engineering*, 7(5): 626–39.

CEA (Central Electricity Authority) (2015). *Load Generation Balance Report 2015–16*. Central Electricity Authority, Government of India: New Delhi.

Chakraborti, D., Mukherjee, S.C., Pati, S., Sengupta, M.K., Rahman, M. M., Chowdhury, U.K. & Basu, G.K. (2003). Arsenic groundwater contamination in Middle Ganga Plain, Bihar, India: a future danger? *Environmental Health Perspectives*, 111(9): 1194.

Development Bureau (2014, Dec 23). Melamchi's tunnel vision. Ekantipur. Available at: www.ekantipur.com/2014/12/23/development/melamchis-tunnel-vision/399398.html.

ENPHO (Environment and Public Health Organisation) (2010). *Factsheet on Decentralized Wastewater Treatment Plant*. Kathmandu: Environment and Public Health Organisation.

Gleeson, T., Wada, Y., Bierkens, M.F. & van Beek, L.P. (2012). Water balance of global aquifers revealed by groundwater footprint. *Nature*, 488(7410): 197–200.

GON & NTNC. (2009). *Bagmati Action Plan (2009–2014)*. Government of Nepal & National Trust for Nature Conservation. Kathmandu, Nepal.

Green, H., Poh, S.C. & Richards, A. (2003). Wastewater Treatment in Kathmandu, Nepal (unpublished Master of Engineering dissertation). Massachusetts: Massachusetts Institute of Technology.

Hellegers, P., Zilberman, D., Steduto, P. & McCornick, P. (2008). Interactions between water, energy, food and environment: Evolving perspectives and policy issues. *Water Policy*, 10(1): 1–10.

HIDCL (Hydroelectricity Investment & Development Company Limited) (2015). *Nepal Hydropower Overview*. HIDCL website, Kathmandu, Nepal. Available at: www.hidcl.org.np/nepal-hydropower.php.

Jha, A.K. & Bajracharya, T.R. (2014). *Wastewater Treatment Technologies in Nepal*. Proceedings of IOE Graduate Conference, 2014. Kathmandu: Institute of Engineering, Tribhuvan University.

Kumar, M.D., Scott, C.A. & Singh, O.P. (2011). Inducing the shift from flat-rate or free agricultural power to metered supply: Implications for groundwater depletion and power sector viability in India. *Journal of Hydrology*, 409: 382–94.

Kumar, M.D., Bassi, N., Narayanamoorthy, A. & Sivamohan, M.V.K. (eds) (2014). *The Water, Energy and Food Security Nexus*, London: Earthscan.

Kurian, M. & Ardakanian, R. (eds). (2015). *Governing the Nexus: Water, Soil and Waste Resources Considering Global Change*, Berlin: Springer.

Lal, R. (2015). 'The Nexus Approach to Managing Water, Soil and Waste under Changing Climate and Growing Demands on Natural Resources'. In M. Kurian and R. Ardakanian (eds). *Governing the Nexus: Water, Soil and Waste Resources Considering Global Change*, Berlin: Springer, pp. 39–60.

Liu, Y., Yu, H.Q., Ng, W.J. & Stuckey, D.C. (2015). Wastewater-Energy Nexus. *Chemosphere*, 140: 1.

Lohani, S.P., Chhetri, A., Adhikari, J. & Bakke, R. (2013). Sustainable biogas production potential from urban wastewater in Nepal. *International Journal of Environmental Science and Development*, 4: 5.

MOF (Ministry of Finance) (2011). *Power and Energy Sector Road Map: An Update*. Ministry of Finance, Government of the People's Republic of Bangladesh: Dhaka. Available at: www.mof.gov.bd/en/budget/11_12/power/power_energy_en.pdf.

Mukherji, A. (2007). The energy-irrigation nexus and its impact on groundwater markets in eastern Indo-Gangetic basin: Evidence from West Bengal, India. *Energy Policy*, 35(12): 6413–6430.

Muzzini, E. & Aparicio, G. (2013). *Urban Growth and Spatial Transition in Nepal: An Initial Assessment*. Washington, DC: World Bank.

NRCD (National River Conservation Directorate) and MEF (Ministry of Environment and Forests). (2009). *Status paper on River Ganga – State of Environment and Water Quality*. New Delhi: National River Conservation Directorate and Ministry of Environment and Forests.

Pradhan, P. (2012). 'Results of unplanned programmes: Drinking water and sanitation in Bhaktapur, Nepal'. In A. Prakash, V.S. Saravanan, J. Chourey (eds). *Interlacing Water and Human Health: Case studies from South Asia*, New Delhi: Sage Publications, pp. 381–403.

Rasul, G. (2014). Food, water, and energy security in South Asia: A nexus perspective from the Hindu Kush Himalayan region. *Environmental Science and Policy*, 39: 35–48.

Regmi, S. (2013). Wastewater treatment in Kathmandu Management, Treatment and Alternative (Unpublished Bachelor's Thesis). Mikkeli University of Applied Science, Mikkeli.

Rutkowski, T., Raschid-Sally, L. & Buechler, S. (2007). Wastewater irrigation in the developing world – Two case studies from the Kathmandu Valley in Nepal. *Agricultural Water Management*, 88(1): 83–91.

Saha, D. & Alam, F. (2014). Groundwater vulnerability assessment using DRASTIC and Pesticide DRASTIC models in intense agriculture area of the Gangetic plains, India. *Environmental Monitoring and Assessment*, 186(12): 8741–8763.

Saha, D. & Sahu, S. (2015). A decade of investigations on groundwater arsenic contamination in Middle Ganga Plain, India. *Environmental Geochemistry and Health*, 1–23.

Scott, C.A. & Sharma, B. (2009). Energy supply and the expansion of groundwater irrigation in the Indus-Ganges Basin. *International Journal of River Basin Management*, 7(2): 119–24.

Scott, C.A., Kurian, M. & Wescoat, Jr., J.L. (2015). 'The water-energy-food nexus: Adaptive capacity to complex global challenges'. In M. Kurian and R. Ardakanian (eds). *Governing the Nexus: Water, Soil and Waste Resources Considering Global Change*, Berlin: Springer, pp. 15–38.

Scott, C.A., Pierce, S.A., Pasqualetti, M.J., Jones, A.L., Montz, B.E. & Hoover, J.H. (2011). Policy and institutional dimensions of the water–energy nexus. *Energy Policy*, 39: 6622–6630.

Shah, T. (2009). *Taming the Anarchy: Groundwater Governance in South Asia*. Washington, DC: RFF.

Shah, T. & Kishore, A. 2012. Solar-powered pump irrigation and India's groundwater economy: A preliminary discussion of opportunities and threats. IWMI-Tata Water Policy Research Highlight, 26(7).

Shah, T., Giordano, M. & Mukherji, A. (2012). Economia políticía do nexo energia – águas subterrâneas na Índia: Exploração das questões e avaliação das oções políticas. *Hydrogeology Journal*, 20(5): 995–1006.

Shah, T., Scott, C.A., Kishore, A. & Sharma, A. (2003). *Energy-irrigation nexus in South Asia: Improving groundwater conservation and power sector viability*. IWMI Research Report No. 70. International Water Management Institute, Colombo, Sri Lanka.

Walker, B. & Salt, D. (2012). *Resilience Practice: Building Capacity to Absorb Disturbance and Maintain Function*. Washington, DC: Island Press.

9

Ganges Water Machine

One solution to basin water problems?

Upali A. Amarasinghe and Lal Mutuwatte

Key messages

- Groundwater is the key to water security in the Ganges River Basin (GRB). More than 75 per cent of the process consumptive water use, i.e. the evapotranspiration from irrigation, domestic and industrial sectors, in the GRB at present is from groundwater. The reliance on groundwater will increase further due to limited capacity to construct additional surface water storages in the GRB.
- Accelerated use of groundwater in the dry season in shallow groundwater zones will deplete aquifers and increase their capacity to store subsequent monsoonal rainfall. The pump-

deplete-recharge-pump cycle, often referred to as the Ganges Water Machine (GWM), can create sustainable groundwater supply for irrigation.

- Of the 21 sub-basins of the GRB, making the GWM work is not possible in four sub-basins – Banas, middle and upper Yamuna, and upstream of Ramganga confluence – due to over-allocation of surface water and groundwater resources. In these basins, implementing the GWM may decrease water availability for other uses and for the environment, and may lead to unsustainable groundwater use.
- There is, however, a potential for partial or full 'operation' of the GWM in other sub-basins of the GRB. It can provide groundwater to increase irrigated area in either the *Rabi* (November to March) and hot weather (April to June) seasons, or both. The estimated monsoon surface runoff is adequate to recharge the sub-surface storage.
- Realization of the full potential of the GWM, however, depends on other factors: the ability to recharge the sub-surface storage by managed aquifer recharge (MAR) during the monsoon period, availability of energy and land, and people's willingness to increase cropping intensity beyond the present level.

Introduction

Recurring floods and droughts associated with climate variability are common in the land and waterscape of the Ganges River Basin (GRB). Climate change may exacerbate these hazards in the future (Hosterman *et al.*, 2012; Immerzeel *et al.*, 2010; Sharmila *et al.*, 2015). The GRB will encounter population and economic growth, which will further increase demand for water (Amarasinghe *et al.*, 2007), and lead to deterioration of the basin conditions (NMCG, 2014). Finding sustainable water management solutions for the GRB is imperative (Sharma *et al.*, 2010).

The Ganges Water Machine (GWM), a term and concept coined 40 years ago by Revelle and Lakshminarayana (1975), proposed to buffer the variability and magnitude of floods during the monsoon period and the droughts in the non-monsoon period while increasing underground storage in the GRB. The concept has never been implemented at scale, but the idea has never died either. With increasing water shortages in the basin, which were foreseen even 40 years ago, the concept may need to be revisited, and prospects and limits of its practical implementation examined in detail in the current physical and socio-economic context of the basin.

Monsoon rains between June and September generate more than 80 per cent of the GRB flow, and contribute to over 340 billion cubic metres (BCM) of mean annual surface runoff (Amarasinghe *et al.*, 2015). On the one hand, the runoff from the upstream combined with intense rainfall makes the downstream of the basin highly flood prone; flood damage to property and crops constitutes millions of dollars annually (CWC, 2013). Yet, more than 90 per cent of water demand, especially from irrigation, is in the non-monsoon period (Amarasinghe *et al.*, 2015). The low rainfall and high water demand create substantial pressure on the water resources in the non-monsoon period. This twin problem of floods and water scarcity is still the major stumbling block in GRB development.

The original concept of the GWM envisaged addressing this twin problem. It entailed first pumping water before the monsoon for depleting consumptively, in the irrigation and other sectors, to create an additional sub-surface storage (SSS) in the basin. Next, using carefully planned recharge structures, the GWM proposed recharging the SSS from the monsoon runoff. The cycle of pump-recharge-pump increases storage for run-off water, which mitigates the impacts of floods and droughts.

The basin has undergone substantial changes since the 1970s. Population and cropped area increased substantially to the present level of 650 million people and 50 million ha (Mha) respectively. Groundwater is already meeting 60 per cent of the irrigation consumptive water use at present (Amarasinghe *et al.*, 2015). Despite an increase in groundwater use since the 1970s, as envisaged in the GWM, there has been only limited effort to increase recharge beyond what occurs naturally. As a result, the groundwater table in some areas is falling at an alarming rate (CGWB, 2013); many rivers and tributaries in the basin are running dry for much of the year, and much of the GRB is under water stress (Sharma *et al.*, 2010).

In spite of the water stress and over-exploitation in some regions, demand for water in the basin will increase substantially in the next few decades. Amarasinghe *et al.*, (2007) project demand to double by 2050. Yet, potential to increase surface storages within the basin is very low. This is especially true in the Indian part of the Ganges, where the current surface storage capacity of 50 BCM is close to the potential surface storage capacity. The potential for surface storage is high in Nepal, but is unlikely to materialize with the present socio-economic and environmental constraints. Jeuland *et al.* (2013), based on hydro-economic analysis of the surface storage in the Ganges, document that even if much of the surface storage potential of Nepal is harnessed, there is still only a limited ability to control the peak flows and floods downstream. Bangladesh has low potential for surface storage.

Therefore, it is logical that more focus will be placed on the idea of storing more water underground to meet increasing demand. What could benefit the Ganges Basin is an integrated water resources development plan with an improved groundwater management component (Sadoff *et al.*, 2013). This was the main reason behind the interest of global and regional discourses for application of the GWM concept in the GRB (Smakhtin, 2013). The proposal of Chaturvedi and Srivastava (1979) to increase pumping along perennial and non-perennial tributaries of the Ganges, and in irrigation canals prior to the onset of the monsoon, further contributed to original GWM concept development.

However, over the past few decades, population expansion and economic growth have led to tremendous changes in land and water-use patterns, as well as to water depletion. Because of the substantial socio-economic, hydrological and environmental changes over the last 40 years, the development of a sustainable GWM, at scale, should necessarily satisfy the following three conditions.

- There should be sufficient demand for water for consumptive needs, and if met by an increase in groundwater pumping, it would create additional SSS.
- There should be an adequate monsoon surface runoff in excess of environmental flow requirements available for recharge.
- It should be feasible to recharge the SSS through both natural and artificial means.

This chapter examines these conditions for the 21 sub-basins of the GRB (Figure 9.1; Table 9.2) and, overall, aims to generate a renewed attention to the concept of the GWM that may help, in the long run, to alleviate water availability and variability issues emerging in the GRB.

Basin water accounting

The partial water accounting (Molden, 1997) that reflects the present level of water resources development in GRB sub-basins includes renewable water resources (surface and groundwater) and water depleted through evapotranspiration (ET). The ET includes the process consumptive water use (CWU) from irrigation, domestic and industrial withdrawals, and the non-process

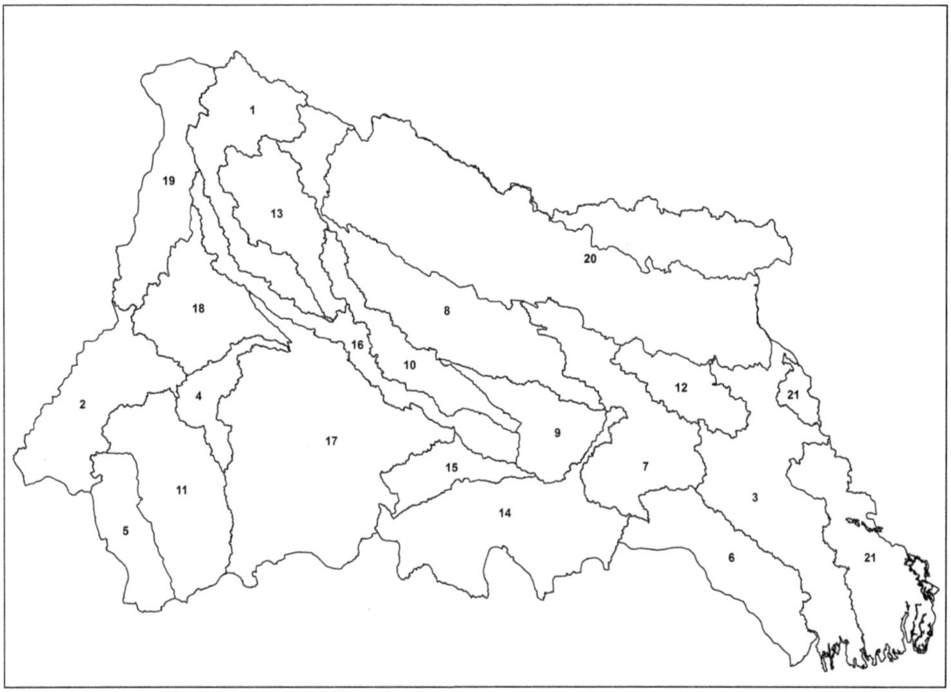

Figure 9.1 Sub-basins of the GRB used in this study

Note: The Indian part of the GRB has 19 major sub-basins, which are the major water development and management units considered by the Central Water Commission. Names of sub-basins are listed in Table 9.2

ET from reservoirs, homesteads and bare surface etc. The other parts of the water accounts are flows to sink (flows of which the quality has deteriorated for any further use), committed flows (environmental flows, inter-basin transfers) and uncommitted flows (flows to sea). Renewable water resources and ET are important parameters for assessing the potential for increasing ET under a GWM concept: when groundwater meets the incremental ET, it creates an additional SSS that surface runoff of the monsoon can subsequently recharge.

As per the official estimates (GoI, 1999; FAO, 2015), the total renewable water resources (TRWR) of the GRB is about 552 BCM. Sixty-eight per cent of TRWR comes from surface runoff, and the rest from groundwater recharged from rainfall. In 2008–2010, the two recent years for which the estimated CWU are available, the process CWU depleted 123 BCM or about 22 per cent of the TRWR (Table 9.1). Irrigation is the largest contributor to process CWU (about 93 per cent), for which the groundwater had a substantial contribution (about 78 per cent). The non-process ET consists of another 5 per cent of the TRWR. Thus, the total water depletion as process and non-process ET of the basin at present is only 27 per cent of the TRWR.

Importantly, the water accounts show that only 7 per cent (i.e. 25.5 / 375) of the surface water resources are depleted (process CWU) now (Table 9.1). However, this depletion could be a slight underestimate, because a part of the groundwater CWU from the reuse of return flows of surface water withdrawals are not accounted for here, and part of the non-process ET is also from surface withdrawals. Nevertheless, it is clear that a large part of the surface runoff, most of which is generated during the monsoon period, is uncommitted (either not used or not committed as environmental flows, navigation etc.) flows at present. Amarasinghe *et al.*

Table 9.1 Water accounts of GRB (in BCM per year)

Country	*Internally renewable water supply*		*Inflow*	*Total renewable water supply*	*Process CWU in 2008–2011*				*Non-process ET*	*Total ET*
					Irrigation		*Domestic & total industry*			
	SW	*GW*			*SW*	*GW*				
Nepal	198	20[a]	12[c]	210	2.6	0.9	0.04	3.6	–	3.6
India	143	172	210[d]	525	21.7	82.9	7.8	112.4	27.2	139.6
Bangladesh	22	5[b]	525[e]	552	1.2	4.7	1.2	7.1	–	8.3
GRB	375[f]	177	–	552	25.5	88.6	9.0	123.1	27.2	150.3

Sources: AQUASTAT database (FAO, 2015); GoI, 1999; process CWU and non-process ET are authors' estimates

Notes: a. all overlap with surface water; b. no overlap with surface water; c. inflow from China to Nepal; d. inflow from Nepal to India; e. inflow from India to Bangladesh

(2015) showed that mean annual Ganges river flow observed at the Hardinge Bridge in Bangladesh, just below the Indian border, since 2000 was 347 BCM.

Environmental flows (EF) – the water that needs to be left in a river for the maintenance of river ecology and social/spiritual requirements – shall be seen as committed water flows. However, there is no EF allocation policy at present in the GRB, or elsewhere in South Asia, although significant work in India, for example, is emerging (see Chapter 10, 'Environmental flows'). The EF for maintaining the river in a 'moderate' condition, were estimated to be about 152 BCM at the Indian-Bangladesh border (Smakhtin and Anputhas, 2006). Therefore, even if significant EF requirements are satisfied, the GRB (at the outlet from India to Bangladesh) still has around 200 BCM (i.e. 375–26–152) of surface runoff that can be depleted (withdrawn from the river) for other productive activities. However, the magnitude of surface flow, and the mechanisms to capture and beneficially use it, vary across the sub-basins.

There are no official estimates of surface water resources available/accessible for the sub-basins of the GRB, although the potential additional depletion through process CWU can be estimated (as will be shown later in this chapter as unmet demand). Several modelling efforts to simulate GRB hydrology with varied spatial resolution and accuracy are known (see Chapter 2, 'Surface water resources'). However, the detailed results for most of them are not available in the public domain. For illustrative purposes, this chapter uses the simulations of

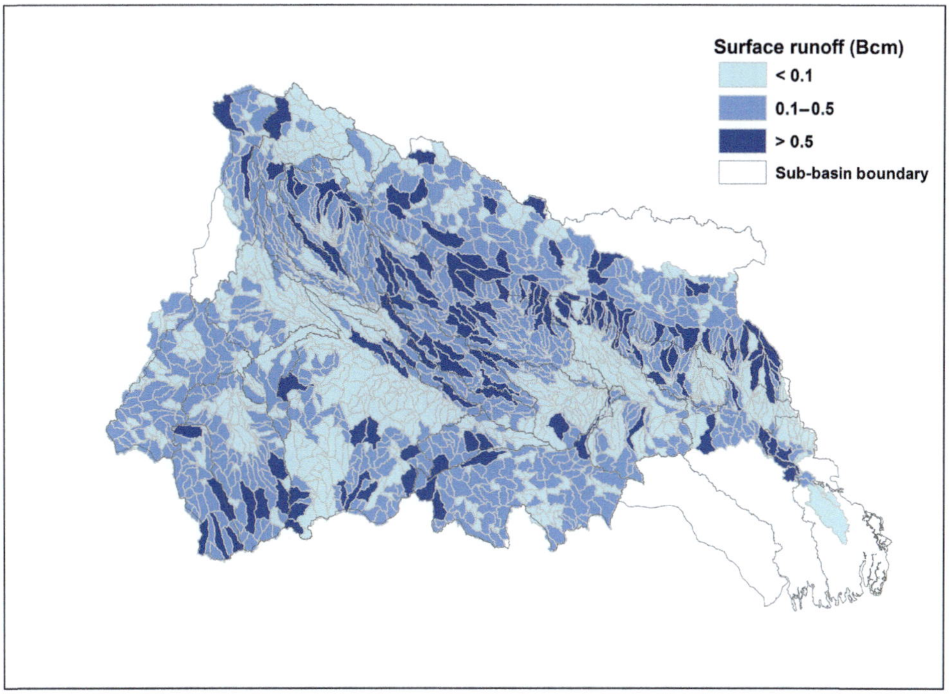

Figure 9.2 A map of simulated surface runoff for 1,684 sub-catchments

Source: Muthuwatta *et al.* (2015)

Note: The Ganges River Basin was delineated using 3,000 ha as the minimum area threshold that resulted in 1,684 catchments. The model was initially developed to study stream flow entering Bangladesh, and the overall GRB boundary was subsequently revisited. Therefore, the spatial domain of this Ganges SWAT model does not entirely cover the entire basin area. Also, due to limited observed flow data, the model was calibrated only at a few locations in Nepal, and at Farakka barrage. It is naturally possible to improve the accuracy of analyses described in this chapter when/if other simulations of GRB hydrology (considered or perceived as more accurate) are available

Table 9.2 Water accounts and potential for increasing irrigation CWU in GRB sub-basins

	Sub-basins	*Surface runoff (BCM)*			*Monsoon surface runoff, % of total*	*Ground water resources in 2008–2011 (BCM)*	*Total ET in 2008–2011 (BCM)*		*Surface water ET (2008–2011) – % of SR_{p75}*	*Ground water ET (2008–2011) – % of groundwater resources*	*Uncommitted water resources in 2008–2011 (BCM)*	
		Mean	*SD*	*SR_{p75}*			*SW*	*GW*			*SW*	*GW*
	C1	*C2*	*C3*	*C4*	*C5*	*C6*	*C7*	*C8*	*C9*	*C10*	*C13*	*C14*
1	Upstream of Ramganga confluence	10.0	5.0	5.5	81	5.2	0.8	5.6	14	107	4.7	–0.4
2	Banas	9.9	7.1	3.5	94	2.6	0.4	3.0	10	116	3.2	–0.4
3	Bhagirathi	–		–	–	21.7	1.8	2.7	–	12	Na	19.0
4	Chambal Lower	2.2	1.4	1.2	95	1.3	0.4	0.8	36	64	0.8	0.5
5	Chambal Upper	8.7	3.0	6.6	90	4.0	0.7	3.1	10	78	5.9	0.9
6	Damodar	–		–	–	9.7	1.0	1.1	–	12	Na	8.6
7	Gandak and others	16.0	6.6	11.8	86	13.0	1.4	3.4	12	26	10.4	9.6
8	Ghaghara	35.6	17.6	23.3	84	20.5	1.8	10.5	8	51	21.6	10.0
9	Between Ghaghara and Gomti confluences	4.7	2.1	3.3	88	7.7	2.2	2.9	68	37	1.1	4.8
10	Gomti	13.6	7.3	9.8	91	8.5	1.8	4.8	18	56	8.0	3.8
11	Kali Sindh	15.5	6.6	10.5	81	5.9	1.9	4.0	18	67	8.6	1.9
12	Kosi	9.4	4.0	6.8	73	6.3	0.5	1.8	7	28	6.3	4.5
13	Ramganga	15.6	7.8	10.1	83	7.8	1.0	7.8	10	99	9.1	0.0
14	Son	19.5	7.9	14.1	85	9.3	1.3	1.1	9	12	12.8	8.1
15	Tons	6.8	2.5	5.2	89	1.6	0.5	0.7	10	43	4.7	0.9
16	Upstream of Gomti confluence up to Muzaffanagar	9.4	4.8	5.7	88	9.7	2.2	6.8	38	71	3.5	2.8
17	Lower Yamuna	22.4	10.8	15.2	94	16.9	4.9	7.6	32	45	10.3	9.3
18	Middle Yamuna	4.8	3.7	2.1	79	5.4	1.3	6.3	60	116	0.9	–0.9
19	Upper Yamuna	7.2	3.9	4.5	83	8.5	3.7	8.9	82	105	0.8	–0.4
	Nepal	63.2	11.6	54.4	89	20.0	2.6	0.9	5	5	51.8	19.1
	Bangladesh	22.0	–	–		5.5	1.2	4.7	5	87	20.7	0.7

Source: Runoff estimates (C2–C5) are from Mutuwatte *et al.* (2015). Groundwater resources are district-wise estimates of CGWB (2013). ET estimates are from Amarasinghe *et al.* (2015)

Notes: C9=C7/C4, C10= C*/C6, C13=C4-C7, C14=C6-C8

surface runoff (excluding groundwater) provided by Mutuwatte *et al.* (2015) who applied a widely used SWAT model to the GRB, splitting it into 1,684 sub-catchments (Figure 9.2).

SWAT is a semi-distributed hydrological model developed by the Agricultural Research Service of the United States Department of Agriculture (USDA) over the last 30 years and is available in the public domain model (Arnold *et al.*, 1998; Gassman *et al.*, 2007). The model describes all basin hydrology processes and allows for detailed representation of spatial heterogeneity of basin soils, land use and topography.

The estimate of surface runoff of a large sub-basin is the sum of surface runoffs from the small catchments within the sub-basin. The estimated surface runoff of the sub-catchments is shown in Figure 9.2, and for the sub-basins are given in Table 9.2, which also summarizes the details of water accounts for these 21 sub-basins. The surface runoff generated during a monsoon period is the water available to recharge SSS. The mean surface runoff and the runoff that has 75 per cent exceedance probability (i.e. the flow that may be expected on average in three out of four years) are listed in columns C2 and C4 in Table 9.2, respectively.

The ratio of total ET (process and non-process) from surface water withdrawals to average surface runoff varies from as low as 7–9 per cent in the Kosi, Ghaghara and Son sub-basins to more than 82 per cent in the upper Yamuna sub-basin (Table 9.2, column C9). In fact, the ratio of ET from surface withdrawals to dependable surface runoff is below 10 per cent in 11 sub-basins, and is below 25 per cent in all sub-basins except for the middle and upper Yamuna and Ghaghara confluence to Gomti confluence sub-basins. Therefore, most sub-basins have sufficient surface runoff to meet an increase in process ET without tapping into EF.

However, many basins have substantially high level of groundwater use (Table 9.2, column C10). The ratio of groundwater ET to groundwater resources is at or above 100 per cent in five sub-basins (above Ramganga confluence, Banas, Ramganga, middle and upper Yamuna). These sub-basins have widespread or large pockets of over-exploited groundwater resources. Of these, the middle and upper Yamuna sub-basins have very high groundwater and surface water depletions. Any further increase of ET in these sub-basins would leave very little water for EF, and only aggravate the unsustainable water use. Therefore, the middle and upper Yamuna sub-basins have no potential to implement the GWM concept in practice.

The other three sub-basins – above Ramganga confluence, Banas and Ramganga – have high groundwater ET (Table 9.2, column C10), but low surface water ET (Table 9.2, column C9) with respect to the renewable water resources. These sub-basins can have aggressive (MAR) programmes, the still missing part of the GWM approach, to reduce over-exploitation and increase their groundwater storage capacity.

Many other basins can increase process groundwater ET to increase the SSS and capture part of the uncommitted surface runoff (columns C13 and C14 in Table 9.2). However, the need for additional water to meet an increase in ET (unmet irrigation demand) needs to be established in these basins.

Unmet irrigation demand

Irrigation is the largest contributor to total (process and non-process) ET (93 per cent in 2008–2011). Therefore, the potential for further increase in process ET depends on the potential unmet CWU demand in the irrigation sector. The demand for additional irrigation CWU depends on the potential increase in land-use intensity in irrigated areas, which is well below 100 per cent at present (Figure 9.3). In the GRB, the net sown area is 48 Mha, but the maximum cultivated area is only about 37 Mha (Figure 9.3A). The cropped areas in January, February, March and April are only 31, 25, 4 and 9 Mha, respectively. The net irrigated area of the GRB

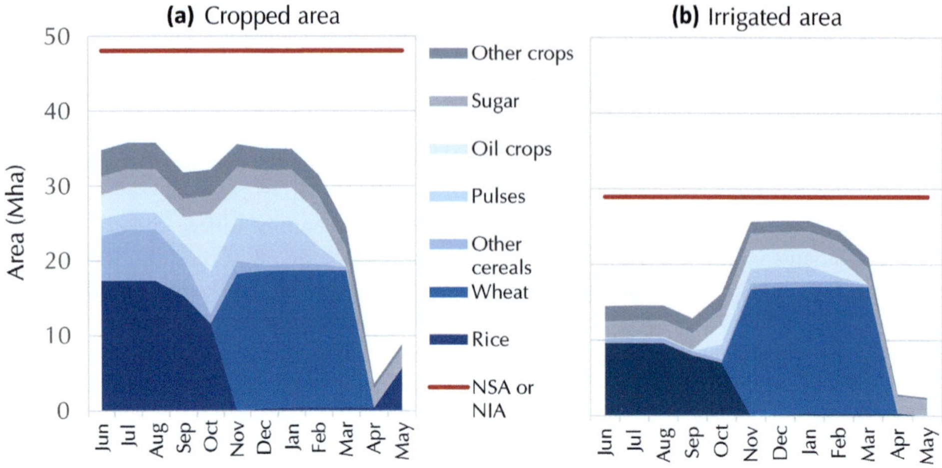

Figure 9.3 Monthly cropping and irrigation pattern in the Indian part of the GRB in 2008–2011
Source: Authors' estimates based on the district-level data on cropping and irrigation land pattern in GoI (2015)
Note: NSA is net sown area (i.e. cultivable area) and NIA is net irrigated area (i.e. irrigable area)

is 30 Mha, but the maximum irrigated area at present is about 26 Mha (Figure 9.3B); and the irrigated areas in January, February, March and April are only 24, 21, 3 and 2 Mha respectively. These statistics show that there is scope to expand irrigated lands in the *Rabi* (November to March) and hot-weather (April to June) seasons.

In the GRB, irrigation demand is low during the *Kharif* season (June to October), but high in the *Rabi* and hot-weather seasons (Figure 9.4). Between June and October, rainfall meets 70 per cent of the total monthly CWU. Between November and March, rainfall meets only 30 per cent of the total monthly CWU. Irrigation requirement is very high during April to May. Therefore, the most feasible strategy for increasing irrigation CWU is to increase the irrigated area in the *Rabi* and summer seasons. If groundwater meets the additional irrigation demand, it can deplete a substantial volume of groundwater as process CWU, and can create an additional SSS before the onset of monsoon.

Two possible scenarios to expand irrigation in the entire GRB for the *Rabi* (November–March) and hot-weather period (April–May) may be considered:

- **Scenario A**: Provide irrigation to the total irrigable area, i.e. increase irrigated area in the *Rabi* season from 26 Mha (current irrigated area in this season) to 30 Mha (irrigable area), and in the hot-weather season from 3 Mha (current irrigated area in this season) to 30 Mha (irrigable area), respectively.
- **Scenario B:** Provide irrigation to the total cropped area. At present, not all cropped area is equipped for irrigation, i.e. irrigable area (30 Mha) is less than the cropped area (35 Mha). Therefore, Scenario B is to increase irrigable area in order to increase irrigated area from 26 to 35 Mha in the *Rabi* season and from 3 to 35 Mha in the hot-weather season, respectively.

First, the scenarios assess the potential increase in irrigated area in the *Rabi* and hot-weather seasons, which under scenario A is the difference between net irrigated area and the maximum monthly irrigated area, and under scenario B is the difference between maximum irrigated area

Table 9.3 Potential increase in irrigated area

	Sub-basins	NIA (Mha)	Maximum monthly[1] irrigated area (Mha)			Maximum monthly[1] cropped area (Mha)			Increase in irrigated area[2] (Mha)				Irrigation CWU (mm)	
									Scenario A		Scenario B			
			Jun–Oct	Nov–Mar	Apr–May	Jun–Oct	Nov–Mar	Apr–May	Nov–Mar	Apr–May	Nov–Mar	Apr–May	Nov–Mar	Apr–May
	C1	C2	C3	C4	C5	C6	C7	C8	C9	C10	C11	C12	C13	C14
1	Upstream of Ramganga confluence	1.35	0.80	1.35	0.36	1.22	1.51	0.37	0.00	0.99	0.16	1.15	274	173
2	Banas	0.99	0.48	0.99	0.00	1.71	1.64	0.01	0.00	0.98	0.72	1.70	276	123
3	Bhagirathi	1.78	1.70	0.50	0.42	4.75	2.12	0.92	1.27	1.35	4.24	4.32	230	124
4	Chambal Lower	0.41	0.22	0.39	0.00	0.40	0.53	0.00	0.02	0.41	0.14	0.53	272	190
5	Chambal Upper	1.08	0.50	0.92	0.01	1.57	1.38	0.01	0.16	1.07	0.65	1.57	333	190
6	Damodar	0.96	0.96	0.10	0.10	2.89	0.96	0.20	0.86	0.86	2.79	2.79	231	201
7	Gandak and others	1.55	1.00	1.18	0.08	1.91	1.63	0.24	0.37	1.47	0.73	1.83	269	284
8	Ghaghara	3.01	1.76	2.95	0.49	3.35	3.50	0.68	0.06	2.52	0.55	3.01	288	197
9	Between Ghaghara and Gomti confluences	1.39	1.10	1.10	0.04	1.29	1.28	0.05	0.29	1.35	0.19	1.25	294	186
10	Gomti	1.48	1.03	1.36	0.16	1.21	1.52	0.19	0.11	1.32	0.15	1.36	303	173
11	Kali Sindh	1.96	1.04	1.50	0.01	2.71	2.21	0.01	0.46	1.95	1.21	2.70	307	127
12	Kosi	0.70	0.45	0.65	0.10	1.05	0.87	0.23	0.05	0.60	0.40	0.94	241	150
13	Ramganga	1.68	1.36	1.68	0.42	1.60	1.84	0.44	0.00	1.25	0.16	1.42	298	197
14	Son	0.74	0.43	0.51	0.02	2.69	1.35	0.07	0.23	0.72	2.19	2.68	314	165
15	Tons	0.32	0.14	0.28	0.00	0.59	0.65	0.00	0.03	0.32	0.37	0.65	311	184
16	Upstream of Gomti confluence to Muzaffanagar	1.95	1.15	1.95	0.23	1.55	2.17	0.24	0.00	1.72	0.21	1.93	278	171
17	Lower Yamuna	3.86	1.71	3.64	0.05	4.53	6.19	0.05	0.22	3.82	2.55	6.15	282	187
18	Middle Yamuna	2.14	1.04	2.14	0.06	1.44	2.46	0.06	0.00	2.08	0.32	2.40	249	164
19	Upper Yamuna	2.76	1.65	2.76	0.52	2.10	3.23	0.54	0.00	2.24	0.47	2.71	231	166
	Indian sub-basins	30.1	18.5	25.9	3.1	38.6	37.0	4.34	4.15	27.0	18.1	41.1	274	173

Source: C15–C27 are from Amarasinghe *et al.* (2015)

Note: 1 The months June to October, November to March and April to May are considered as *Kharif*, *Rabi* and hot-weather seasons respectively

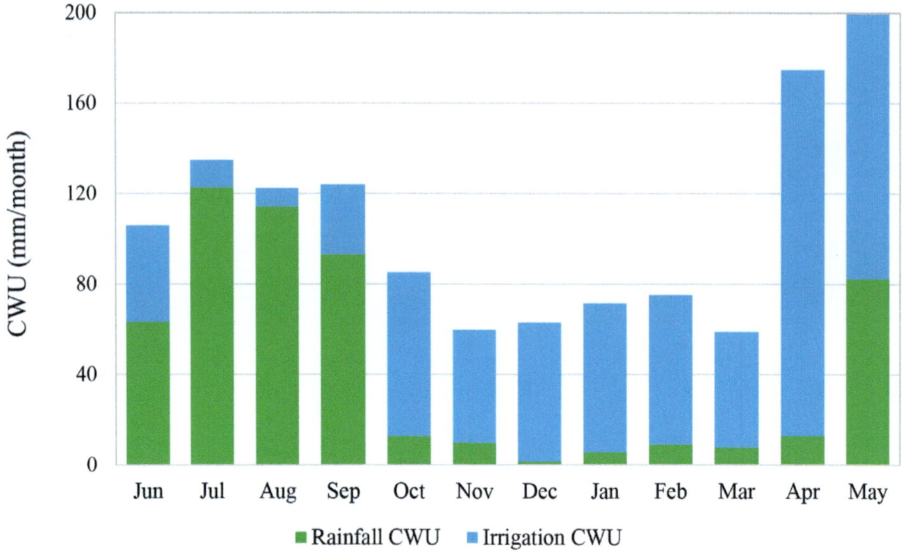

Figure 9.4 Irrigation and rainfall CWU in irrigated crop areas in the GRB
Source: Authors' estimates

and maximum cropped area. Next, the unmet irrigation demand or the potential increase in ET from irrigation is obtained by multiplying the potential increase in irrigated areas by the average irrigation CWU at different periods. The potential increases in irrigated area and irrigation CWU are shown in Tables 9.3 and 9.4 respectively.

The potential increases in irrigated area in the *Rabi* and hot-weather seasons under scenario A are 4.15 and 27 Mha, respectively, and under Scenario B, 18 and 41 Mha respectively (columns C9–C12 in Table 9.3). The corresponding unmet irrigation demand under the two scenarios is 59 and 118 BCM respectively (Table 9.4, columns C2–C5). However, the realization of this potential is difficult given the current water use and availability in different sub-basins.

For example, the middle and upper Yamuna sub-basins have the potential to increase irrigated area by 2.1–2.7 Mha and 2.2–3.2 Mha (Table 9.3, columns C9–C12). To realize this potential, these basins require 3.4–4.8 BCM and 3.7–5.6 BCM of irrigation CWU respectively (Table 9.4, columns C2–C5). However, these basins have already exhausted their annual groundwater resources (116 per cent and 106 per cent of groundwater depletion; Table 9.2, column C10) and a substantial part of their surface water resources (60 per cent and 82 per cent of surface water depletion; Table 9.4, column C9). Any further increase in irrigation ET would reduce the uncommitted surface runoff available for environment flows and only exacerbate the unsustainable water use.

On the other extreme are sub-basins with low water depletion at present, but with high potential to increase irrigated area and CWU in the future. For example, the Ghaghara and Bhagirathi sub-basins can potentially increase their irrigated area by 2.62–8.56 Mha (Table 9.3, columns C9–C12) under the two scenarios, and for this, it requires 4.10–15.13 BCM of irrigation CWU (Table 9.4, columns C2–C5). The sub-basin has about 19 BCM of uncommitted groundwater resources, which itself is sufficient to meet the unmet CWU demand. The Ghaghara basin has the potential to increase the irrigated area by 2.68–3.56 Mha, and the irrigation CWU by 5.11–7.49 BCM under the two scenarios. However, these are small portions of the uncommitted surface water and groundwater resources of the basin (33.8 and 7.1 BCM).

Table 9.4 Potential increase in irrigation CWU in *Rabi* (November to March) and hot-weather (April-May) seasons

	Sub-basin	*Potential increases in irrigation CWU[3] (BCM)*												*Potential for GWM*
		Total				*% of uncommitted groundwater resources*				*% of uncommitted total water resources*				
		Scenario A		*Scenario B*		*Scenario A*		*Scenario B*		*Scenario A*		*Scenario B*		
		Nov–Mar	*Apr–May*	*Nov–Mar*	*Apr–May*	*Nov–Mar*	*Nov–May*	*Nov– Mar*	*Nov–May*	*Nov–Mar*	*Nov–May*	*Nov–Mar*	*Nov–May*	
	C1	*C2*	*C3*	*C4*	*C5*	*C6*	*C7*	*C8*	*C9*	*C10*	*C11*	*C12*	*C13*	*C14*
1	Upstream of Ramganga confluence	0.00	1.71	0.45	1.99	0.0	–452	–118	–645	0	39	10	56	4
2	Banas	0.00	1.21	1.99	2.10	0.0	–285	–468	–961	0	44	73	150	4
3	Bhagirathi[1]	2.93	1.67	9.78	5.35	15.4	24	51	80	15	24	51	80	1
4	Chambal Lower	0.06	0.77	0.38	1.00	12.3	181	84	302	5	67	31	111	3
5	Chambal Upper	0.53	2.04	2.17	2.97	61.4	296	250	592	8	38	32	76	3
6	Damodar[1]	1.99	1.74	6.44	5.62	23.1	43	75	140	23	43	75	140	1
7	Gandak and others	0.99	4.17	1.97	5.20	10.3	54	21	75	5	26	10	36	1
8	Ghaghara	0.16	4.95	1.58	5.91	1.6	51	16	75	0	16	5	24	1
9	Between Ghaghara and Gomti confluences	0.85	2.51	0.56	2.33	17.7	70	12	60	14	57	10	49	2
10	Gomti	0.34	2.29	0.47	2.36	9.1	70	12	75	3	22	4	24	1
11	Kali Sindh	1.42	2.48	3.71	3.43	73.5	202	192	370	14	37	35	68	3
12	Kosi	0.13	0.90	0.97	1.42	2.9	23	21	53	1	10	9	22	1
13	Ramganga	0.00	2.46	0.49	2.79	0.0	5189	1031	6901	0	27	5	36	3
14	Son	0.73	1.19	6.87	4.41	8.9	24	84	139	3	9	33	54	2
15	Tons	0.10	0.58	1.15	1.19	11.9	78	130	265	2	12	21	42	2
16	Upstream of Gomti confluence up to Muzaffarnagar	0.00	2.93	0.60	3.30	0.0	104	21	138	0	46	9	61	3
17	Lower Yamuna	0.62	7.13	7.18	11.5	6.7	83	77	201	3	40	37	95	2
18	Middle Yamuna	0.00	3.41	0.79	3.93	0.0	–395	–92	–547	0	–33156	–7703	–45918	4
19	Upper Yamuna	0.00	3.72	1.08	4.50	0.0	–874	–254	–1311	0	1030	299	1545	4
20	Nepal													1
21	Bangladesh													2

Source: C28–C31 are from Amarasinghe *et al.* (2015); C32–C40 are authors' estimates

Note: 1 The total water resources of the Bhagirathi and Damodar sub-basins used in columns C10 to C13 are only groundwater resources

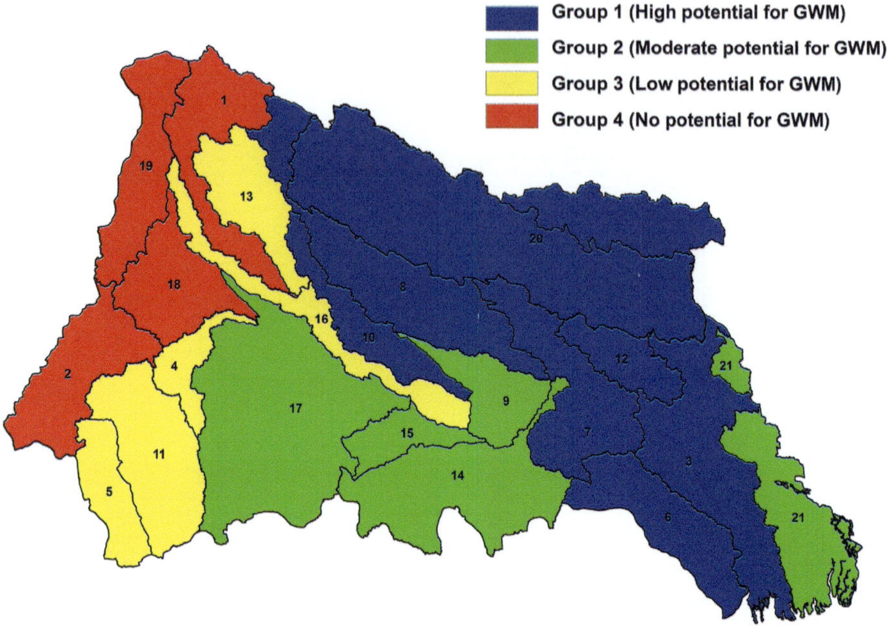

Figure 9.5 Estimated potential for application of the GWM concept in the GRB sub-basin

Based on the potential increase in irrigated area, and the uncommitted surface water and groundwater use, the sub-basins have been categorized into four groups (Figure 9.5, based on column C14 in Table 9.4).

Group 1 – High potential for GWM. This group has the highest potential for application of the GWM concept and includes six sub-basins, mainly in the eastern part of the GRB, including Gandak, Ghaghara, Gomti, Kosi, Bhagirathi and Damodar, and also Nepal. The cycle of pump-recharge-pump is possible without any detrimental effect on the environmental flows or unsustainable groundwater use. The present levels of groundwater development in these basins are low, and they have sufficient uncommitted groundwater resources to meet the increased irrigation CWU under both scenarios. In these basins, natural interactions between groundwater and surface water can recharge the SSS created by the depletion of groundwater resources.

Group 2 – Moderate potential for GWM. This group has the next highest potential for application of the GWM concept, and includes the lower Yamuna, Sone, Ghaghara, between Ghaghara and Gomti confluences, and Tons sub-basins. The uncommitted groundwater resources of these sub-basins are sufficient to meet the increased CWU under Scenario A, but not sufficient under Scenario B. However, the uncommitted surface water and groundwater resources in these basins can meet the increased irrigation CWU under both scenarios. The potential for increasing groundwater CWU depends on the ability of MAR programmes to capture the uncommitted monsoon surface runoff.

Group 3 – Low potential for GWM. This group includes five sub-basins: lower and upper Chambal, Kali Sindh, upstream of Gomti confluence, and Ramganga. The uncommitted groundwater resources of these sub-basins cannot meet the additional irrigation CWU of either scenario. However, the uncommitted groundwater and surface water are large enough to meet the irrigation CWU under Scenario A. As in Group 1, these sub-basins also require MAR to capture the uncommitted surface runoff.

Group 4 – No potential for GWM. Application of the GWM concept is not possible in this group, which contains the Banas, middle and upper Yamuna, and upstream of Ramganga confluence sub-basins. The current levels of water use in these sub-basins are very high and the uncommitted water resources are not sufficient to meet any additional irrigation CWU. If they do, it will be at the expense of the available water for other sectors (domestic, industrial and environmental flows), or it will aggravate the unsustainable groundwater use. Although the lower Chambal sub-basin is in this group, the total uncommitted water resources are, however, adequate to meet the increased CWU in the *Rabi* season. However, potential increase in irrigated area and CWU and the resultant SSS is very small.

It is noted that, with surface runoff, the total water availability of Bhagirathi and Damodar would be much higher than those available only from groundwater. However, if surface runoff were also included in total water resources, the total CWUs as percentages of total water resources in these two basins would be much lower than indicated in Table 9.4, columns C10 to C13. Therefore, these two basins are included in Group 1. Similarly, based on uncommitted surface and groundwater resources, and present CWU, the Nepal and Bangladesh riparian regions are placed in Groups 1 and 2 respectively. Although the above grouping shows the broad picture of the potential for application of the GWM concept, there is substantial spatial variation of groundwater development and recharge within sub-basins at present. Even the sub-basins in Groups 2 and 3 can have locations where natural interactions are sufficient to recharge the SSS created through groundwater depletion. The identification of these locations requires further analysis briefly discussed in the next section.

Groundwater depletion and recharge

The groundwater depletion and recharge in the GRB are monitored through a large network of bore wells (CGWB, 2013). Figure 9.6 shows the groundwater depth (below ground level) for a few sub-basins in the GRB, where reliable observations are available. It shows the groundwater depth before (end of May) and after (end of October) monsoon in 2000 and 2010.

Figures 9.6A and 9.6B demonstrate large fluctuations of groundwater depth by depletion and recharge in 2000. Groundwater recharge from rainfall and other sources (such as surface irrigation), especially in the eastern and north-eastern parts of the basin, were sufficient to recover from the large depletion before the monsoon. However, the situation has changed over time. By 2010, there are many pockets where depth to groundwater is still large after the monsoon (Figures 9.6C and 9.6D), indicating faster depletion than recharge.

For example, the depth to groundwater in many locations, such as at point A in Figure 9.6, has declined over time both before and after the monsoon (Figure 9.7A). In these locations, the natural recharge during the monsoon is sufficient to compensate for the depletion before the monsoon. Further groundwater use is not sustainable in these types of locations without MAR. However, at location B, there was no significant decline in the depth to groundwater between 2000 and 2010 (Figure 9.7B). At location B, the natural interactions between surface water and groundwater are sufficient to recharge the depleted aquifer. Locations similar to B have the potential for application of the GWM. In areas where there are declining trends of groundwater depth, MAR is necessary for a sustainable groundwater use, provided these areas have adequate uncommitted surface runoff.

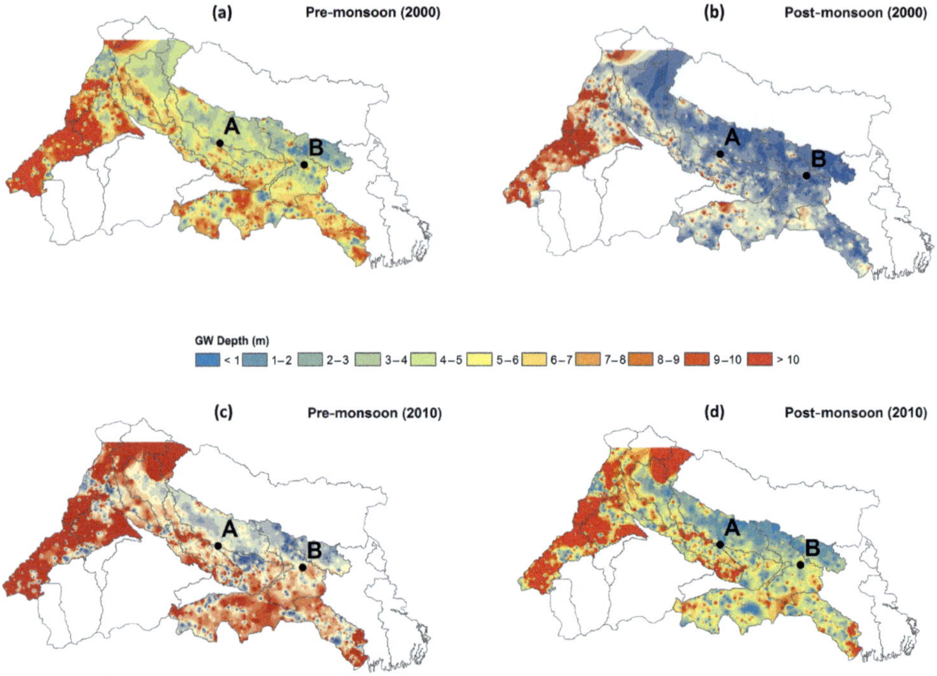

Figure 9.6 Groundwater depth before and after monsoon in parts of the GRB

Note: The depth to groundwater analysis used for the Indian riparian region is based only on the observations that are permissible to download from the WRIS (Water Resources Information System) open access database of the Ministry of Water Resources (www.india-wris.nrsc.gov.in/wris.html). The data for the Nepal and Bangladesh riparian regions are not available for this analysis

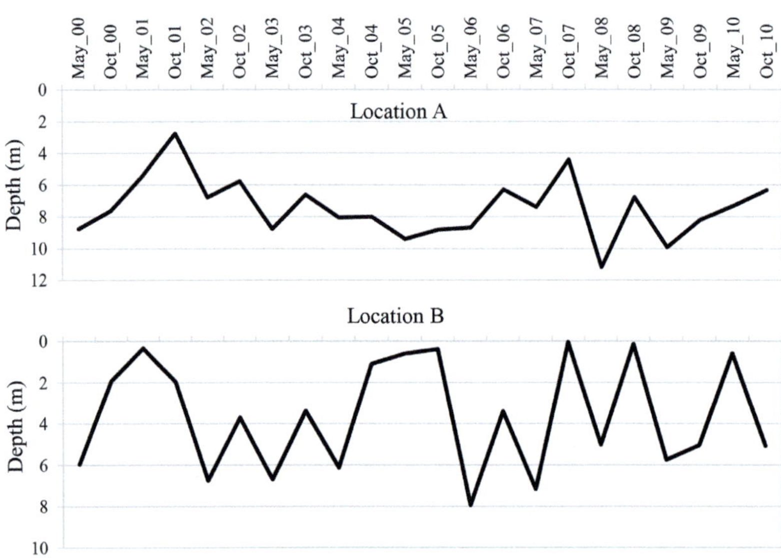

Figure 9.7 Trends of the depth to groundwater before and after monsoon at two locations in the GRB

Note: The pre- and post-monsoon observations are taken at the end of May and end of October respectively

Discussion and conclusions

Groundwater is the main source of water supply for process CWU in the Ganges. The reliance on groundwater will further increase due to increasing demand for water during non-monsoon seasons, and limited potential to develop surface storage further. Although it is now 40 years since the notion of the GWM was first put forwards, there has hardly been a structured progress in its implementation. There are parts of the basin where groundwater has been developed, resulting in steady decline in groundwater level, which could be arrested if recharge had been enhanced beyond natural rates. On this backdrop, application of the GWM may be the way forward for meeting additional water needs and mitigating the impacts of floods and droughts in the basin.

Implementation of the GWM is precluded by a lack of knowledge regarding some of its key drivers. From a purely biophysical point of view, spatial variation in unmet demand for irrigation, excess runoff generated during monsoon, and the capacity of the aquifers to store water (SSS) are not adequately known. This chapter only assessed the first two elements of unmet water demand and excess runoff of sub-basins, due to limited data availability, and difficulty in modelling surface water and groundwater interactions for a large basin such as the Ganges. The detailed surface water and groundwater modelling are being conducted now for the Ramganga sub-basin.

This preliminary and macro-scale analysis shows that there is substantial unmet demand for increasing the irrigated area. However, the water development to date in four sub-basins – Banas, middle and upper Yamuna, and upstream of Ramganga confluence – was so intense that there is no more water available for further process depletion. If additional development depletes more water, it will be at the expense of CWU in some other sectors or environments.

However, the analysis also shows that the application of the GWM concept is possible partially in some sub-basins (lower and upper Chambal, Kali Sind, upstream of Gomti, Ghaghara confluence to Gomti confluence, Tons, lower Yamuna, and Bangladesh), and fully in others (Bhagirathi, Damodar, Gandak, Ghaghara, Gomti, Kosi, Son, and Nepal). Due to the patterns of irrigation requirement, the additional irrigation in the *Rabi* and hot-weather seasons has the highest potential for depleting groundwater resources and creating SSS. The potential unmet demand under the two scenarios considered in this chapter range from 59 to 119 BCM. However, due to supply constraints in Group 1 basins, the available water resources can meet only about 45 to 84 BCM. Potential in the sub-basins can be characterized as:

- The sub-basins Bhagirathi, Damodar, Gandak, Ghaghara, and Kosi can use only their naturally recharged groundwater resources to meet the unmet irrigation demand for fully utilizing the current net irrigated area.
- The application of the GWM and MAR can provide irrigation to utilize all net irrigated area in all basins except upstream of Ramganga confluence, Banas, Ghaghara confluence to Gomti confluence, upper and middle Yamuna.
- The application of the GWM can provide irrigation to utilize all net sown area in Gandak, Ghaghara, Ghaghara confluence to Gomti confluence, Gomti, Kosi, Ramganga and Tons.

The ability to realize this potential depends on many other conditions, which include:

- Properties of the soil, and the 'crop holidays' (a period of time when the cultivation of a particular crop does not take place) required for the soil in between intensive cropping in the *Rabi* and *Kharif* seasons.
- People's willingness to increase cropping and irrigation intensities from about 135 per cent now to 300 per cent. The questions of why people do not increase cropping intensity even

in locations with adequate groundwater resources, and whether any socio-economic and institutional factors constrain them from doing so, need to be examined.

- Most importantly, the access to energy for additional pumping, and access to land for MAR interventions. Population pressure in the GRB is already high and will only increase in the future. Therefore, any displacement of people or submergence of land due to MAR, or, for that matter, for any other infrastructure development, would be major issues of contention for implementation of the GWM, because many of the people who would be affected by such development activities are the stallholders and marginalized population in the GRB.

Whether the application of the GWM concept can realize the SSS potential requires further hydrogeological and economic analysis. This includes agronomic feasibility, energy demand and supply, especially on how to reduce the dependency on electricity for pumping and feasibility of using alternative energy sources such as solar, and analysis of the social and economic costs, benefits and trade-offs of various surface and sub-surface storage plans.

The benefits of developing only a small portion of such potential storage are likely to be enormous. Besides irrigation benefits, it can buffer rainfall variability and reduce extreme flooding, especially in downstream regions. SSS can increase river flow during the low-flow months, either through base flow or through re-allocation of canal irrigation. Importantly, it can mitigate the negative effects of floods and water scarcity in the same year, which often affects the most vulnerable people in society – women and children, the poor and other disadvantaged social groups. While it may not be a panacea for water availability and scarcity issues throughout the basin, the GWM approach can surely help alleviate these issues in parts of the Ganges, and needs to receive more attention now than it has in the past.

References

Amarasinghe, U.A., Shah, T., Turral, H. and Anand, B.K. 2007. *India's water future to 2025–2050: Business-as-usual scenario and deviations.* IWMI Research Report 123. Colombo, Sri Lanka: International Water Management Institute (IWMI).

Amarasinghe, U.A., Mutuwatte, L., Surinaidu, L., Anand, S. and Jain, S.K. 2015. Reviving the Ganges Water Machine: Why? *Hydrol. Earth Syst. Sci. Discuss.*, 12: 8727–8759. Available at: www.hydrol-earth-syst-sci-discuss.net/12/8727/2015/.

Arnold, J.G., Srinivasan, R., Muttiah, R.S. and Williams, J.R. 1998. Large area hydrologic modeling and assessment part 1: Model development. *Journal of the American Water Resources Association*, 34(1): 73–89.

CGWB (Central Groundwater Board). 2013. *Dynamic groundwater resources of India (as 31 March 2011).* Available at: www.cgwb.gov.in/documents/Dynamic-GW-Resources-2011.pdf.

Chaturvedi, M.C. and Srivastava, V.K. 1979. Induced groundwater recharge in the Ganges basin. *Water Resources Research*, 15(5): 1156–1166.

CWC (Central Water Commission). 2013. *Water and Related Statistics.* Available at: www.cwc.nic.in/main/downloads/Water%20and%20Related%20Statistics-2013.pdf.

FAO (Food and Agriculture Organization of the United Nations). 2015. *Ganges-Brahmaputra-Meghna River Basin.* Available at: www.fao.org/nr/water/aquastat/basins/gbm/gbm-CP_eng.pdf.

Gassman, P.W., Reyes, M.R., Green, C.H. and Arnold, J.G. 2007. The soil water assessment tool: Historical development, applications, and future research directions. *Transactions of the ASABE*, 50(4): 1211–1250.

GoI (Government of India). 1999. *Integrated water resources development. A plan for action.* Report of the National Commission for Integrated Water Resources Development Volume I. New Delhi: Ministry of Water Resources, GoI.

GoI (Government of India). 2015. *Agriculture at a Glance.* Directorate of Economics and Statistics website. Available at: http://eands.dacnet.nic.in/.

Hosterman, H. R., McCornick, P.G., Kistin, E.J., Sharma, B. and Bharati, L. 2012. Freshwater, climate change and adaptation in the Ganges River Basin. *Water Policy*, 14(1): 67–79.

Immerzeel, W.W., Van Beek, L.P. and Bierkens, M.F. 2010. Climate change will affect the Asian water towers. *Science*, 328(5984): 1382–1385.
Jeuland, M., Rao, H., Escurra, J., Blackmore, D. and Sadoff, C. 2013. Implications for climate change for water resources development in the Ganges basin. *Water policy*, 15(S1): 26–50.
Molden, D. 1997. *Accounting for Water Use and Productivity*. Colombo, Sri Lanka: International Irrigation Management Institute (IIMI).
Muthuwatta, L., Amarasinghe, U.A., Sood, A. and Lagudu, S. 2015. Reviving the 'Ganges Water Machine': Where and how much? *Hydrol. Earth Syst. Sci. Discuss.*, 12: 9741–9763.
NMCG (National Mission for Clean Ganga). 2014. *Aims and Objectives*. Available at: https://nmcg.nic.in/AboutGangaManthan.aspx.
Revelle, R. and Lakshminarayana, V. 1975. The Ganges water machine. *Science*, *188*: 611–16.
Sadoff, C., Nagaraja, R.H., Blackmore, D., Xun, W., O'Donnell, A., Jeuland, M., and Whittington, D. 2013. Ten fundamental questions for water resources development in the Ganges: Myths and realities. *Water policy*, 15(S1): 147–64.
Sharma, B., Amarasinghe, U., Xueliang, C., de Condappa, D., Shah, T., Mukherji, A. and Smakhtin, V. 2010. The Indus and the Ganges: River basins under extreme pressure. *Water International*, 35(5): 493–521.
Sharmila, S., Joseph, S., Sahai, A.K., Abhilash, S. and Chattopadhyay, R., 2015. Future projection of Indian summer monsoon variability under climate change scenario: An assessment from CMIP5 climate models. *Global and Planetary Change*, 124: 62–78.
Smakhtin V. 2013. The 'Ganges Water Machine': A sexy title or the concept of global relevance? Available at: https://wle.cgiar.org/thrive/2013/05/21/%E2%80%9C gangeswater-machine%E2%80%9D-sexy-title-or-concept-global-relevance.
Smakhtin, V. and Anputhas, M. 2006. *An assessment of environmental flow requirements of Indian River basins*. Research Report 107. Colombo, Sri Lanka: International Water Management Institute (IWMI).

PART II

Environment

10

Environmental flows

Keeping the basin rivers alive

Vladimir Smakhtin and Luna Bharati

Key messages

- Environmental flows (EF) practice is still in its infancy in Ganga riparian countries, but is emerging. There has been a particularly significant increase in interest in EF in India, where several environmental flow assessments of various complexity have been accomplished in the last decade, notably for the upper Ganga, and where the concept of environmental flow allocations has been taken seriously by the Ministry of Water Resources and Ganga Rejuvenation. In Nepal and Bangladesh, EF is beginning to enter the discussions on river development.

- The upstream basin in India and many Ganga tributaries originating in Nepal still have underdeveloped water resources. While this is changing, there is still a chance to set up proactive environmental flow measures and policies – before such rivers are seriously degraded. This should not contradict development, as environmental flows may and should be seen as an integral part of sustainable basin planning.
- A characteristic feature of the Ganga is the high importance of religious/spiritual water needs – in addition to the requirement of the aquatic and riparian ecosystems, or tourism. Spiritual water requirements, which form part of environmental flows, may satisfy most of the other instream requirements in many parts of the Ganga Basin.
- Environmental flows cannot be properly estimated, and their presence in basin rivers cannot be ensured, without access to and sharing of existing hydro-meteorological and hydraulic data – within and between riparian countries. Lack of such access and sharing at present also affects all other aspects of water resources management in the Ganga Basin.
- Environmental flow assessments are important, but they are only one part of the overall environmental flow management programme for a river basin. In the Ganga, there is a need and potential to go from basin-wide assessment to analysis of practical options that ensure that environmental flows are physically present in a river, including in real time, that maintenance of environmental flows is economically viable, and that regulatory mechanisms are firmly in place to ensure environmental flows.

Introduction

Many countries face a number of water-related challenges, including an increasing demand for water on the one hand, and the need to allocate a share of water to maintain the functioning of freshwater-dependent ecosystems in a river basin on the other. This share is normally referred to as environmental flows (EF).

EF is a scientifically established, ecologically acceptable flow regime that is designed to maintain a river in an agreed or predetermined state. EF is normally defined in terms of discharges of particular magnitude, frequency and timing necessary to ensure a healthy river system from environmental, economic and social perspectives. All components of the hydrological regime have a certain ecological significance, and maintaining elements of natural hydrology in rapidly developing river basins is essential for overall river health and functions (Poff *et al.*, 1997). High flows of different frequency are important for channel maintenance, bird breeding, wetland flooding and maintenance of riparian vegetation. Moderate flows are critical for cycling of organic matter from river banks and for fish migration, while low flows of different magnitudes are important for algae control, water quality maintenance and the use of the river by local people. Worldwide, EF is seen progressively more and more as a need – in cases when rivers are subject to flow regulation and competition from multiple water users. EF therefore is always a compromise between water for 'development' and water for nature.

EF may indeed be perceived as environmental water demand, similar to water demand of other users – irrigation, industry, domestic – but is more complex to estimate. EF is estimated through a process known as environmental flow assessment (EFA). EFAs often consider not just water demand for the maintenance of aquatic and riverine ecology, but also livelihood requirements which are dependent on the ecosystem services provided by the river and not explicitly included in water infrastructure development plans. These environmental water uses and requirements are defined on the basis of observed direct linkages between changes in ecosystem character and the delivery of important ecological services to people (e.g. fish as food, high-quality drinking water, riparian trees for house construction and recreation, religious/cultural practices etc.)

EFA methods differ significantly in the level of accuracy and input information required, and range from (i) quick, low-confidence, desktop type to (ii) comprehensive multidisciplinary assessments (http://waterdata.iwmi.org/applications/efm/efm_home.php; Acreman and Dunbar, 2004). The first are using primarily hydrological indices or analysis of hydrological time series. They are more suitable for initial, reconnaissance-level assessments, and for river basins where the pressure on water resources is only starting to grow. While such methods provide estimates of low confidence, they help to set feasible limits for future water resource exploitation that can be fine-tuned later. The second type – comprehensive EFAs – are required for detailed studies, particularly in stressed water systems. They adopt a whole-ecosystem view in assessing EF, whereby ecologically and/or socially acceptable flow regime is defined by a multidisciplinary panel of experts, and/or by hydro-ecological models. These methods, and multidisciplinary holistic EFA in particular, normally require substantial field work, time and resources to complete for a single river basin. Ideally, both types of EFA approaches should be present in national environmental water policies. Such a two-tier framework can help establish EF over a range of management objectives, spatial scales and available resources.

Deciding on EF is not only a scientific question, although the science of EF is an important component of EFA and is constantly developing. Ensuring that EF are actually maintained in a river is a technical, social and political issue. Estimating and maintaining an EF regime should involve many different actors. Researchers and experts can provide technical advice, whereas communities and water decision makers need to understand and agree on the EF objectives. In an ideal world, EF would be a dynamic process written into legislation, provided by water managers, and monitored by a government body or river basin agency with assistance from research organizations (Hirji and Davies, 2009).

Environmental flows in the Ganga riparian countries

The environmental status of the streams and rivers in the Ganga Basin depends, to a large extent, on their location. In upstream parts of Ganga in India, and its tributaries in Nepal and China, there are still rivers that are free flowing and largely unmodified, with some abstraction for irrigation – mainly in more densely populated areas. There are, however, major plans to develop hydropower infrastructure as well as multipurpose storage projects, which provide water for irrigation and domestic purposes. In the plains, apart from deteriorating water quality, mainly due to industrial, domestic and agricultural effluents, flow regulation through dams and barrages has already changed the river flow regime. Large canal diversions reduced river flows and their assimilative capacity during the dry season. Future increases in domestic and industrial water demand in the basin will further deteriorate this situation in the whole basin, especially in the urban centres and densely populated areas. The health of the Ganga is impacted equally by both quantitative flow changes and pollution. EF concepts and practices ensure that important *quantitative* aspects of flow regime are maintained, in the first place.

Ganga riparian countries still lack technical and institutional capacity to establish appropriate EF allocation practices. The known EFA methods are either not tailor-made for the specific local conditions, or not tested in such conditions. Understanding of the links between flow and ecology, or other components of EF such as cultural flows, is relatively weak, and data that supports such analysis is lacking. There is still a widespread incorrect (although slowly changing) perception of environmental water demand being the least important and hence 'minimum'.

There are, however, some distinct differences between Ganga riparian countries in terms of perceptions of EF, levels of scientific understanding and attention to this concept from the implementing agencies.

Nepal

Rivers in Nepal are important sources of domestic and irrigation water, mechanical and hydroelectric power, aquatic resources (such as fish and sand), as well as centres of biodiversity. Nepal also boasts an extraordinary natural beauty, which stimulates international tourism. Maintaining a healthy aquatic environment is therefore important from this perspective as well. Recreational activities such as rafting are also dependent on flows and are important aspects of tourism. The need to minimize adverse environmental and social impacts of projects such as hydropower and irrigation development is high on the international agenda, which has direct implications for Nepal.

In Nepal, environmental protection was not required before 1992. The Water Resources Act of 1992 then mandated that:

> A person or a corporate body, who desires to conduct a survey or to utilize water resources, shall be required to submit an application to the prescribed officer or authority along with the economic, technical and environmental study report and with other prescribed particulars.

The introduction of the Environment Protection Act in 1997 added the need for carrying out the Initial Environmental Examination and Environmental Impact Assessment (EIA) for projects based on their installed capacity. In 2001, the Hydropower Development Policy was introduced which then specifically mentioned the amount of minimum flow that must be released. It states that 'Downstream release shall be maintained, either 10 per cent of minimum mean monthly discharge or the quantum identified in the EIA study whichever is higher' and 'Implementation of EIA recommendations shall be emphasized' (Rijal and Alfredsen, 2015). However, in some smaller streams with hydropower plants, during peak hours, even 10 per cent flow is not maintained. As in many low-income countries, Nepal has a relatively high rural population living a predominantly subsistence livelihood. Consequently, programmes for establishing EF should consider the interaction of rural communities with the river flow and ecosystem. A characteristic feature of the country from an EF perspective is that EF thresholds and practices can still be established prior to major water resources development – a rare possibility to actually ensure sustainable and environmentally responsible water resources development.

EFAs in the country have been limited. Bhattarai *et al.* (2002) attempted to suggest an EF to minimize possible ecological damage in Indrawati Basin, although the specifics of the method used were not detailed. Smakhtin *et al.* (2006) examined the applicability of several desktop hydrology-based EFAs in the East Rapti River that drains a 3,500 km^2 catchment, and flows through the Royal Chitwan National Park – one of the major tourist attractions and biodiversity refuges of Nepal. Some of these techniques were modified, following the discussion of their limitations. It was stated that hydrology-based methods of EFA represent a necessary first step in planning for environmental allocations in Nepal rivers, and it was shown that use can be made of complementary features of existing EFA techniques to arrive at justified estimates of EF, even in the conditions of limited basin-specific eco-hydrological knowledge.

Recent agreements between the governments of Nepal and India on hydropower development may catalyse water resources development in Nepal, and may bring EF into the spotlight as well (GoN and GoI, 2014). It is critical therefore that a sufficient knowledge and evidence base supported by adequate assessment tools is made available to inform decision making around specific projects and locations. Some organizations such as the IFC have also started to support the inclusion of EF in the design of water projects (ESSA Technologies Ltd, 2014). The environmental and tourism lobbies have also joined forces to confront the powerful energy and irrigation ministries for inclusion of EF in future projects. So far, however, the Environmental Impact Assessments do not make EFA compulsory. Due to the lack of EF information for particular rivers as well as a general and still dominant lack of understanding of EF among various stakeholders, a constructive national dialogue that pays due attention to EF has been restricted.

India

India has a large network of perennial rivers. Some of them have become highly degraded and polluted (Gopal, 2000), and Ganga is one such example. There is, however, a growing recognition of the need to improve and maintain a healthy aquatic environment in the country, and in the Ganga in particular. The National Environment Policy (2006) of India (www.moef.gov.in/sites/default/files/introduction-nep2006e.pdf) in its section on 'Freshwater Resources', calls for promotion of:

> integrated approaches to management of river basins by the concerned river authorities, considering upstream and downstream inflows and withdrawals by season, interface between land and water, pollution loads and natural regeneration capacities, to ensure maintenance of adequate flows, in particular for maintenance of in-stream ecological values, and adherence to water quality standards throughout their course in all seasons.

This essentially accepts EF as an inherent attribute of water resources planning and management. In addition, National Water Policy (2012; http://wrmin.nic.in/writereaddata/NationalWaterPolicy/NWP2012Eng6495132651.pdf) stipulates that ecological needs of rivers should be determined through scientific study, duly accommodating development needs. Yet this policy still uses primarily the terminology of 'minimum flow' when it comes to EF.

The Ganga Basin upstream of Haridwar has relatively limited water development to date. The major projects include Tehri multipurpose reservoir on the Bhagirathi tributary, and consumptive water diversion through the Upper and Eastern Ganga Canal at Haridwar. The construction of the Tehri dam is believed to have augmented average flow in non-monsoon months downstream, and attenuated flood peaks. This part of the basin, due to its ecological and spiritual importance, attracts most attention to date in EF assessments.

In 2003–2004, the Water Quality Assessment Authority of India (jointly comprised of the Ministry of Water Resources [MoWR] and Ministry of Environment and Forests [MoEF]) set up, for the first time, a working group to advise on 'minimum flow in rivers' and to suggest 'studies required for it' (B. Gopal, National Institute of Ecology, pers. comm.) A working group recommended that, for Himalayan rivers, including parts of Ganga, 'minimum flows shall be not less than 2.5 per cent of 75 per cent dependable annual flow, and that one flushing flow during monsoon with a peak not less than 250 per cent of 75 per cent dependable annual flow is provided. Smakhtin and Anputhas (2006) estimated, using a desktop analysis and limited publicly available monthly time step flow data for major Indian rivers, that EF at the outlet of the entire Ganga vary from around 20 per cent to 68 per cent of mean annual runoff (MAR) – for various (acceptable from management perspective) river conditions. An MoWR

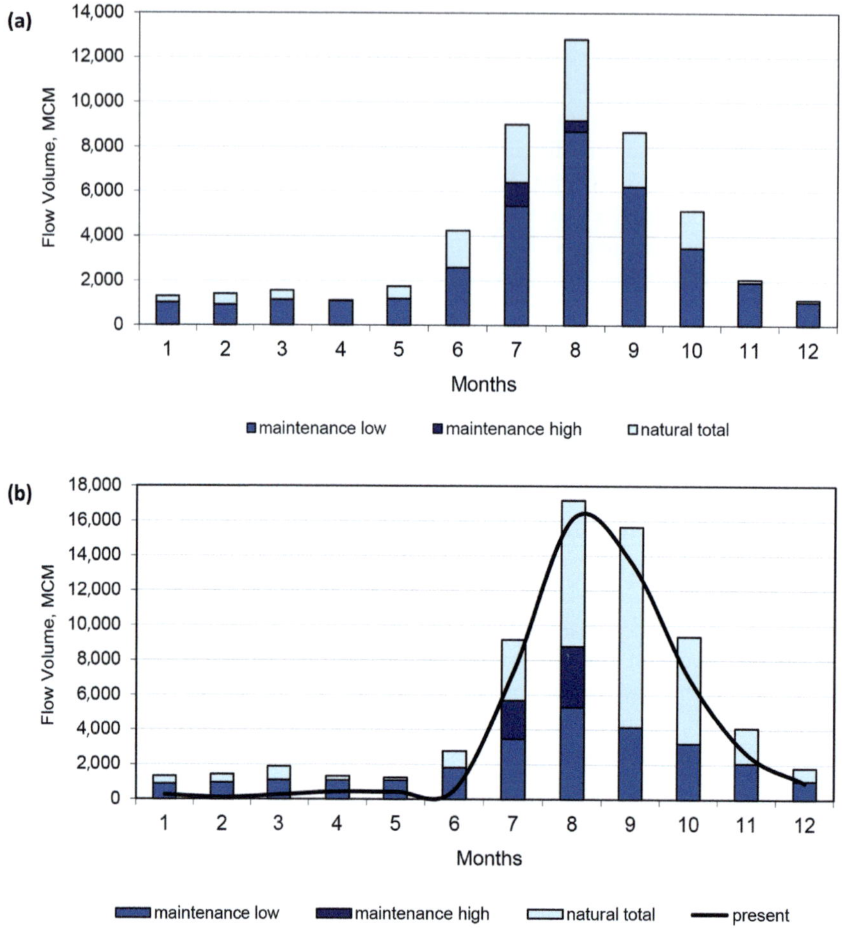

Figure 10.1 EFA recommendations for Kaudiyala (top) and Bithoor (bottom) EF sites

Source: WWF-India, 2012

Notes: 'Maintenance' is the methodology term used to denote normal years. 'Low' and 'high' are the low-flow and high flow requirements estimated separately. The total estimated EF for Kaudiyala and Bithoor sites were 72 and 47 per cent of mean annual natural runoff at the sites respectively

sub-committee on 'Minimum Environmental/Ecological Flows in Rivers' in 2010–11 suggested that an allocation of as little as 10 to 20 per cent of MAR may be reserved for environmental needs of Himalayan rivers, and that provision of flushing flows (that provide cues to fish and other species to start spawning/regeneration) may not be necessary unless major storage dams exist with substantial high flow-release facilities.

More recently, MoEF has formed a consortium of Indian Institutes of Technology (IIT) for the preparation of a Ganga River Basin Management Plan under the aegis of the National Ganga River Basin Authority. The consortium's role was to quantify reach-wise ecological parameters required for more detailed assessment of EF in the Ganga. Based on 10 years of flow data representing non-regulated conditions, the consortium suggested, for parts of the upper Ganga Basin, that EF can vary from 40 to 75 per cent of natural flows during the lean period (January and February), and from 29 to 36 per cent of natural flows during the wet period

(Tare, 2013). For other reaches of the river – up to Kanpur – the consortium referred to the results of the first ever comprehensive EFA in the Ganga, which was conducted by the WWF-India, IWMI and UNESCO-IHE (Kaushal and Babu, 2013; WWF-India, 2012). In this study, EF were assessed at Kaudiyala, Kachla Ghat and Bithoor sites representing Gangotri to Rishikesh, Narora to Farrukhabad and Kannauj to Kanpur reaches respectively. The hydrology of the study basin was simulated using a widely used SWAT model (Bharati *et al.*, 2011a, 2011b; Sapkota *et al.*, 2013). Recommended EF (Figure 10.1) varied from 45 to 72 per cent of MAR in a normal year, and from 14 to 44 per cent of MAR – in a drought year (WWF-India, 2012).

During this assessment, it was shown, also for the first time, that cultural/spiritual water requirements in such rivers as the Ganga – of high importance for pilgrims and religious bodies – can be very significant (Lokgariwar *et al.*, 2013; Kaushal and Babu, 2013). River flows naturally support direct livelihood uses such as fishing, ferrying, sandbank farming, rafting and others. But rivers also support the quality, the 'way of life' and the existence of riparian communities. Many riparian communities have developed customs, rituals and philosophies that are synchronic with and reflective of the natural rhythms of the river. The ability to practise such rituals (e.g. bathing and religious festivals) depends on availability of certain flows in certain times of the year. Also, folklore is attached to varying levels of flow, thus the appearance of the culturally important and iconic rivers should match their description in folklore and mythology (see also Foreword). During the comprehensive EFA in the upper Ganga, local riparian communities were interviewed and asked to provide information on historical flows, desirable flow levels in various seasons, and the magnitude of periodic events – in the context of the

Figure 10.2 Estimated desired flow levels for spiritual water requirements in the Ganga at Kachla Ghat site – for two contrasting months during 'maintenance' (normal) years – located on the photo of the site showing the left bank

Source: Lokgariwar *et al.* (2013)

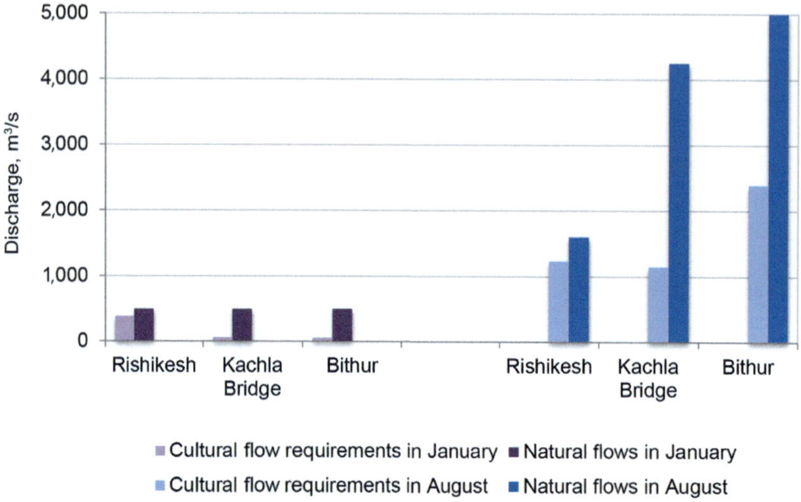

Figure 10.3 Total cultural flow requirements vs naturalized flows at selected sites in the upper Ganga Basin during normal years

Source: Lokgariwar *et al.* (2013)

flows' ability to help people satisfactorily carry out rituals (Figure 10.2), maintaining aesthetic qualities they expected of a revered river, and carrying out other ecosystem functions. The results of the surveys were then interpreted in the context of a naturalized hydrological regime (Figure 10.3). These cultural/spiritual water requirements were found to be (on average across the study area), 38 per cent of the natural flow in the lean flow season and 47 per cent of natural flow during monsoons. The interesting outcome was that the flows favoured by the participants in the cultural survey generally matched very closely with those required for the maintenance of the biota and natural processes, and that such spiritual requirements in parts of the Ganges may be the greatest – catering for other EF components as well.

Several other EFAs were carried out in the Alaknanda-Bhagirathi Basin of the upper Ganges upstream of Devprayag. The study by IIT-Roorkee aimed to determine EF at various hydropower projects and religious sites (AHEC, 2011). It was based, however, on hydraulic mean depth and flow velocity required by freshwater biota only, and recommended EF of 5–25 per cent of MAR. However, the EF during lean season were almost equal to natural flows, leaving hardly any water for other uses. A study by the Wildlife Institute of India (WII), for the same region, recommended, based on estimated Environmental Management Class of rivers and ecological needs of trout, EF in the range of 20 to 30 per cent of seasonal runoff – for three seasons (dry, interim and high-flow seasons) (Rajvanshi *et al.*, 2012). The Inter-Ministerial Group, constituted by the GoI (IMG, 2013), generally agreed with the WII recommendations but suggested modification, e.g. provision of up to 50 per cent of daily flow in winter in modified rivers – to meet societal needs. IMG also recommended making EF mandatory anywhere in the Ganga.

Overall, the above EFAs varied in methods and underlying data, levels of transparency and, naturally, the results. However, most of these EFAs did not seem to take into account spiritual flow needs that have been clearly shown to be dominant at least in the major part of the upper Ganga (Lokgariwar *et al.*, 2013) but instead focus on the needs of some target fish species (e.g. https://sandrp.wordpress.com/2015/04/15/mowr-report-on-assessment-of-e-flows-is-welcome-needs-urgent-implementation/). Hence some such recommendations tend to be on the low side. An important and overall positive observation, however, is that such EFAs are happening and that they are happening at the high institutional level with all major policy structures of the country actively involved. Arriving at some agreed EF is still a challenge primarily due to lack of quantitative data on relationships between river flows and ecological characteristics in the Ganga, and limited access to required observed hydrological and hydraulic data – a typical limitation for implementing EF in many parts of the world, but particularly acute in India (at least in terms of access to observed data).

One perception of EF that currently exists seems to be that allocation of water for environmental needs should not be decided in isolation, and that assessment of requirements from other competing demands such as domestic, industrial, irrigation, hydropower etc. in the basin should also be taken into account. This is true, as long as EF are estimated independently of pressures from other uses, serve to ensure the agreed environmental objective (condition of the river), and are not adjusted, iteratively, for the sake of requirements of other uses.

More details on recent dramatic developments in the EF arena in the Ganga and the whole of India may be found in Gopal (2013), and Jain and Kumar (2014). The most recent development of 2015 has been the explicit instruction by the Minister of Water Resources of India to consider EF in all future water resources projects on Ganga or any of its tributaries (http://indianexpress.com/article/india/india-news-india/govt-takes-in-principle-decision-to-ban-all-construction-on-ganga/). While this may still be contested at the Supreme Court, and it remains to be seen to what extent this instruction will be maintained consistently in the future, the very fact that such events happen at the highest level in India may have a major positive impact on sustainable, environmentally sensitive water resources development in the basin.

Bangladesh

Bangladesh is at the 'receiving end' of two major Asian rivers, the Ganges and the Brahmaputra. Apart from problems of frequent flooding and water pollution associated with shrimp hatcheries and oil spills from marine traffic (Khondker, 1995), a major area of concern is the status of the Sundarbans – the largest deltaic mangrove forest area in south-east Asia. Other related concerns include sea water intrusion – a threat to the Sunderbans as well as to coastal agriculture. This intrusion is also partially a result of progressively reducing river flows from the upstream basin. Information on EF research or policy in Bangladesh is rather limited. Bari and Marchand (2006), and Meijer (2007) reported on several EFAs with necessary socio-economic and other adaptations to local conditions. The study sites, defined through a dialogue with a number of Bangladesh institutions, were conducted on the Surma-Kushiyara, Teesta and Gorai Rivers, and employed a range of EFA methods – from hydrological ones to complex physical habitat modelling combined with field visits, discussion with local fishermen, some on-site measurement of depth and observation of velocity, and expert consultations. Research focused on the EF required to stabilize the bifurcation points of the Surma and Kushiyara Rivers, to maintain fisheries downstream of existing barrages on the Teesta River, and to ensure an adequate freshwater supply to the Sundarbans (Gorai). Meijer (2007) also developed a conceptual model for relationships between human well-being and EF and applied it in the Surma-Kushiyara river system in Bangladesh. Bari and Marchand (2006) emphasized that all EF applications at that time were preliminary, largely due to the approximate nature of the fish habitat-related input data, and that they should not be seen as EF recommendations for rivers studied.

More recent studies apply either a range of hydrological EF methods (Hossain and Hossain, 2011) or more comprehensive methods (Akter and Ali, 2012) to other rivers in Bangladesh.

The National Water Management Plan (WARPO, 2004) aims to ensure freshwater inflow to the Sundarbans to maintain productivity of mangroves and moderate salinity.

Conclusions and way forward

The work on EF in most Ganga riparian countries is at the early stages at present and primarily focuses on EFAs only. While this is an inevitable and, naturally, important stage, it is necessary to go significantly further and establish scientific and practical principles to actually ensure that EF in rivers throughout the Ganga Basin are physically maintained. Recent efforts of the GoI to reinforce EF in the process of water infrastructure development in the Ganga Basin are strong steps in this direction. There are many issues, of course, that need to be addressed on the way to firmly embed EF practice in basin management. Some specific questions that will need to be answered include:

- What are the scientifically justified EF in different parts of the Basin? This warrants a *basin-wide EFA* that has not yet been carried out. While the enormity of the challenge is clear, EF can still be quantified for a limited number of feasible management objectives/scenarios that cover, amongst others, interstate boundaries and stretches of particular concern from the river health perspective. It is also necessary and possible to apply a two-tier EFA – combining desktop EFAs (such as global or Ganga EF calculators: www.iwmi.cgiar.org/resources/models-and-software/environmental-flow-calculators/) with comprehensive EFAs. The first type is applicable for preliminary estimates, the second for more accurate and 'defendable in court' estimates.
- What practical options are available to physically meet the above, scientifically determined EF throughout the basin, especially during the dry season – when competition for water is particularly high – and in river reaches that are already stressed? This calls for assessment of technical and policy *options for actual provision of EF*. This should consider such approaches as canal water allocation, use of less water-intensive crops, improving irrigation systems performance through improved efficiency and water-saving technologies, EF releases from reservoirs, various forms of accelerated surface–sub-surface water interactions and others (Revelle and Lakshminarayana, 1975; Bharati *et al.*, 2011; Sapkota *et al.*, 2013).
- What are the *economics of EF, and trade-offs* associated with EF maintenance in a river? Which levels of EF are economically feasible to maintain, particularly if re-allocation from other sectors is considered? This would entail a comprehensive cost–benefit and trade-off analysis based on financial, environmental and socio-economic parameters, to determine which interventions are suitable in which parts of the basin. Preliminary ideas to this effect were explored in some studies (e.g. Amarasinghe *et al.*, 2012 in the upper Ganga). Assuming that EF in the upper Ganga equals 42 per cent of the MAR, and considering that current runoff during the low-flow months falls below this EF, the authors explored the economics of canal water re-allocation to meet the required EF. They estimated that the potential loss of irrigation benefits may be under 2 per cent of the gross value of crop production in the basin. This loss of benefits can be overcome with an increase in irrigation efficiency or marginal productivity. The message here, preliminary as it may be, is that there is a way to reconcile agricultural uses and EF. This idea is worth exploring in other parts of the basin.
- *How can EFs be operationalised in real time* – i.e. their actual presence in or release to the river ensured in any particular month of the *current year* under the likelihood of the anticipated

rainfall and monsoon progression and recession? This would require quantifying what EF will be needed in a river reach during the coming lean season, instead of just knowing long-term EF values. EF should be defined for the coming year to reserve a definite quantity of flow in the river and, subsequently, to estimate the balance which can be withdrawn from it. The approach would incorporate both an element of the flow forecast and the need to reserve more water for an aquatic ecosystem in wet years as opposed to drier ones.

- *To what extent can EF really be used as a pollution-control measure*? This is an important aspect, particularly in the context of the recent plans of some riparian countries (e.g. India) to clean the Ganga. In essence, EF do not address water quality issues and deal with quantity only. EF can only help to maintain or improve conditions of a river if the other non-flow-related aspects are also looked after (O'Keefe and LeQesne, 2009). It is further absolutely imperative to abandon the idea that dilution is a solution. EF can be combined with pollution concentrations to determine the required flow that could mitigate pollution in the river at environmentally critical periods and in the critical stretches – mostly to just clearly confirm that so much water will not be available. The environmentally critical periods generally coincide with low flow months, and environmentally critical locations are where assimilative capacity of the river is fast decreasing (e.g. Kanpur on the Ganga main stem and Moradabad on the Ramganga tributary). This should contribute to the overall policy shift to treating pollution at the source.
- What *policies and regulatory mechanisms* are needed in the basin to reinforce and ensure that EF do 'happen'? This analysis should include a transboundary dimension. This will need to be done through review and analysis of existing perceptions, policies and institutions (including those that determine infrastructure operations), working with stakeholders in riparian countries and various states/provinces of countries.
- Linked to the above, and to overall basin development, is the issue of *data sharing*. Proper EFA and practical implementation of EF are not possible without access to observed, daily hydrological data, river cross-sections' data, etc. If the countries of the basin do not establish proper mechanisms for such data access and sharing – no matter how limited such available data in the basin are – EF in the basin will remain largely an academic exercise. India has the major role to play in this regard if it makes its hydrology data archives freely accessible.

The set of questions above, if answered convincingly and if implemented, may significantly contribute to practical solutions of Ganga river health. So far, the EF–related effort in the Ganga Basin has been rather ad-hoc, assessment–only focused, and geographically limited to a few sub-basins (e.g. upper Ganga). The above is therefore proposing a huge quantum leap in EF theory and practice that should benefit all countries in the basin.

References

Acreman, M.C. and Dunbar, M.J. (2004). Defining environmental river flow requirements – A review, *Hydrology and Earth System Sciences* 8: 861–76.

AHEC (Alternate Hydro Energy Center) (2011). *Assessment of Cumulative Impacts of Hydroelectric Projects on aquatic and terrestrial Biodiversity in Alaknanda and Bhagirathi Basins*. Uttrakhand, India.

Akter, A. and Ali, M.H. (2012). Environmental flow requirements in the Halda River, Bangladesh. *Hydrol. Sci. J.* 57: 326–43.

Amarasinghe, U., Smakhtin, V., Bharati, L. and Ravinder, R.S.M. (2012). Reallocating water from canal irrigation for environmental flows: Benefits forgone in the Upper Ganga Basin in India. *Environment, Development and Sustainability* 15(2): 385–405.

Bharati, L., Lacombe, G., Gurung, P., Jayakody, P., Hoanh, C.T. and Smakhtin, V. (2011). *The impacts of water infrastructure and climate change on the hydrology of the Upper Ganges River Basin.* IWMI Research Report 142. Colombo, Sri Lanka: International Water Management Institute.

Bharati, L., Smakhtin, V., Jayakody, P., Kaushal, N. and P. Gurung (2011). 'Use of a distributed catchment model to assess hydrologic modifications in the Upper Ganges Basin'. In *River Basin Management VI*, UK: WIT Press, pp. 177–86.

Bhattarai, M., Pant, D., Mishra, V.S., Devkota, H., Pun, S., Kayastha, R.N. and Molden, D. (2002). *Integrated development and management of water resources for productive and equitable use in the Indrawati River Basin, Nepal.* Working Paper 41. International Water Management Institute, Colombo, Sri Lanka.

Bari, M.F. and Marchand, M. (2006). *Introducing environmental flow assessment in Bangladesh.* BUET-Delft UT linkage project, Phase iii. Dhaka–Delft.

ESSA Technologies Ltd. (2014). Environmental Flows Assessment Upper Trishuli-1 Hydropower Project, Nepal. Project Report Nepal Water and Energy Development Report and International Finance Corporation (IFC). Available at: http://ifcextapps.ifc.org/IFCExt/spiwebsite1.nsf/0/c0c5f97854bf6e4885257de90056f4b4/$FILE/UT-1%20Supplemental%20ESIA_Appendix%20E_2014_Environmental%20Flows%20Assessment.pdf. Last accessed: 21 May 2016.

GoN and GoI (2014). *Agreement between the Government (GoN) of Nepal and The Government of the Republic of India (GoI) on Electric Power Trade, Cross Border Transmission Interconnection and Grid Connectivity.* Available at: www.moen.gov.np/pdf_files/PTA-English-21-Oct-2014.pdf. Last accessed: 21 May 2016.

Gopal, B. (2000). 'River conservation in the Indian subcontinent'. In Boon, P.J., Davies, B.R. and Petts, G.E. (eds). *Global Perspectives On River Conservation: Science, policy and practice.* Chichester, UK: John Wiley & Sons, pp. 233–61.

Gopal, B. (2013). 'Environmental flows: The South Asian experience'. In Gopal, B. (ed.). *Environmental Flows: An introduction for water resources managers.* New Delhi: National Institute of Ecology, pp. 183–97.

Hirji, R. and Davies, R. (2009). *Environmental Flows in Water Resources Policies, Plans and Projects: Findings and recommendations.* Washington, DC: The World Bank.

Hossain, M.M. and Hossain, M.J. (2011). Instream flow requirements of Dudhkumar River in Bangladesh. *Asian Transactions on Engineering* 1: 42–49.

IMG (2013) *Report of Inter-Ministerial Group on issues relating to River Ganga.* GoI, New Delhi.

Jain, S.K. and Kumar, P. (2014) Environmental flows in India: Towards sustainable water management. *Hydrol. Sci. J.* 59: 751–69.

Kaushal, N. and Babu, S. (2013) 'Environmental Flows assessment in India: The Ganga experience'. In Gopal, B. (ed.). *Environmental Flows – An introduction for water resources managers.* New Delhi: National Institute of Ecology, pp. 207–26.

Khondker, M. (1995) 'Limnological research in Bangladesh'. In Gopal, B. and Wetzel, R.G. (eds). *Limnology in Developing Countries. Vol. 1.* New Delhi: International Association for Theoretical and Applied Limnology (SIL). International Scientific Publications, pp. 105–20.

Lokgariwar, C., Chopra, R., Smakhtin, V., Bharati, L. and O'Keeffe, J. (2013). Including cultural water requirements in environment flow assessment: An example from the Upper Ganges River, India. *Water International*, 39: 81–96.

Meijer, K.S. (2007) *Human Well-Being Values of Environmental Flows: Enhancing Social Equity in Integrated Water Resources Management.* Netherlands: IOS Press.

O'Keefe, J. and LeQesne, T. (2009) *Keeping Rivers Alive. A Primer on Environmental Flows and their Assessment.* Godalming, UK: WWWF-UK.

Poff, N., Allan, J.D., Bain, M.B., Karr, J.R., Prestegaard, K.L., Richter, B.D., Sparks, R.E. and Stomberg, J.C. (1997). The natural flow regime: A paradigm for river conservation and restoration. *Bioscience*, 47: 769–84.

Rajvanshi, A., Arora, R., Mathur, V.B., Sivakumar, K., Sathyakumar, S. Rawat, G.S., Johnson, J.A.., Ramesh, K. Dimri, N.K. and Maletha, A. (2012) *Assessment of Cumulative Impacts of Hydroelectric Projects on aquatic and terrestrial Biodiversity in Alaknanda and Bhagirathi Basins, Uttrakhand.* Technical Report, Wildlife Institute of India, Dehradun.

Revelle, R. and Lalshminarayana, V. (1975). The Ganges Water Machine. *Science*, 188: 611–16.

Rijal, N. and Alfredsen, K. (2015). Environmental Flows in Nepal – An Evaluation of Current Practices and an Analysis of the Upper Trishuli-I Hydroelectric Project. *Hydro Nepal*, 17: 9–17.

Sapkota, P., Bharati, L., Gurung, P., Kaushal, N. and Smakhtin, V. (2013). Environmentally sustainable management of water demands under changing climate conditions in the Upper Ganges Basin, India. *Hydrol. Process.*, 27: 2197–2208.

Smakhtin, V. and Anputhas, M. (2006). *An assessment of environmental flow requirements of Indian River basins*. Research Report 107. Colombo, Sri Lanka: International Water Management Institute.

Smakhtin, V.U., Shilpakar, R. L. and Hughes, D.A. (2006). Hydrology-based assessment of environmental flows: An example from Nepal. *Hydrol. Sci. J.*, 51: 207–22.

Tare, V. (2013). *The Ganga River Basin Environment Management Plan – Interim Report*. Reports of the Consortium of Seven IITs. Available at: https://nmcg.nic.in/writereaddata/fileupload/25_GRBMPInterim_Rep.pdf. Last accessed: 21 May 2016.

WARPO (Water Resources Planning Organisation) (2004). *National Water Management Plan*. 5 vols. Ministry of Water Resources, Government of Bangladesh, Dhaka.

WWF-India. 2012. *Assessment of Environmental Flows for Upper Ganga Basin*. New Delhi, India: World Wide Fund for Nature (WWF)-India.

11

Ecosystem services and conservation assessment of freshwater biodiversity

Asghar Nawab, Ravindra Kumar Sinha, Paul M. Thompson and Subodh Sharma

Key messages

- Ganga is the holiest river in the sub-continent, and the freshwater ecosystems of its basin provide a variety of ecosystem services. Fisheries are a big industry, and although the total riverine catches declined in the late 1980s, there has been some resurgence since 1996. Agriculture benefits from glacier-fed rivers, a system of irrigation canals, channels running from hand-dug wells, gravity canals, and electric and diesel-powered lifting devices. The basin hydropower potential is enormous, and estimates range from 51,700 to 128,700 MW, of which about 40 per cent is in India and the rest is in Nepal. Sundarbans mangrove forests form an important store of carbon – ranging between 160 and 360 t/ha depending on the vegetation type. An estimated value of ecosystem services flow to humans from 2050 to 2100 may be around US$63 trillion.

- The Ganga River Basin sustains a diverse biological spectrum – from micro-organisms to higher vertebrates, valuable as indicator species and key to maintaining river health. To maintain viable populations, critical biodiversity hotspots need to be identified and legally protected from anthropogenic pressures based on irreplaceability, species assemblage and connectivity considerations.
- Several species form the classic examples of endemism of the riverine ecosystems. They have particular attributes (e.g. narrow tolerances of some habitat/environmental parameter) that may enable them to signal changes in ecosystems and therefore may act as indicator species. Their occurrence is often associated with clean water and undisturbed habitat. They often play a vital role in balancing the freshwater ecosystems as a top carnivorous species, therefore significantly influencing the overall spatio-temporal dynamics of the eco-region over a long period of time.

Introduction

The River Ganga, with a total catchment area of 1,086,000 km^2 covers parts of four riparian countries (India, Nepal, China and Bangladesh), with the major part of the basin in India (http://gfcc.bih.nic.in/Basin.htm). Its mean annual flow is estimated to be close to 500 billion cubic metres, which is equivalent to some 25 per cent of India's total water resources. Ganga's seaward journey covers over 2,525 km goes from the Himalaya to the Bay of Bengal. Its flood plain is one of the largest wetland systems in southern Asia, including man-made reservoirs and a large number and variety of small wetlands spread across the basin, in the mountainous Himalayan region, the Himalayan terai region (in Nepal and India), the Ganga Plains, and the coastal deltaic region (in India and Bangladesh), which together support a large and diverse ecological system in different geophysical settings (Prasad *et al.*, 2002). The river breaks into a number of interlaced channels in the plains, forming meanders, oxbow lakes and swamps in its northern tributaries. The river carries an annual load of 1.46 million tons of sediment (Natarajan, 1989), second only to the Yellow River of China (Lisitzen, 1972). Silt and suspended solids are essential for creating diversified micro/macro habitats to sustain rich biodiversity. The basin freshwater ecosystems provide a variety of ecosystem services.

Ecosystem services

Provisioning services

Shrestha (1978), while considering the cold-water upland areas of the Ganga Basin in Nepal and northern India, gives an estimate of some 1.5 tonnes km^{-1} for average annual fish production, which is comparable to lowland African rivers of the same order. Fisheries in Nepal, although not a big industry, have been increasing at a rapid pace in recent times. National production of fishes in Nepal between 2000 and 2001 was recorded at 33,270 metric tons (Gurung, 2003). In the middle reaches of the Ganga Basin, the most important single group recorded are the major carps (Catla, Mrigal, Rui and Calbasu). In 1957 estimated total annual catches from

Figure 11.1 Fishing boats at the bank of the Ganges-Jamuna confluence, Bangladesh. October 2011
Photo: Paul M. Thompson

the Yamuna and the stretch from Allahabad to Farakka in the Ganga recorded 770 tonnes and 275 tonnes respectively, which translates into a yield of 0.75 tonnes km^{-1} and 0.77 tonnes km^{-1} for the two stretches (Natarajan, 1989). In addition to the fisheries of the main channel, there are also those of the northern tributaries, in particular the Rivers Gandak, Buhri-Gandak and Sapta-Kosi. All of these rivers meander through the alluvial plains in northern Bihar and, during the monsoons, often produce large floodwater areas known as the North Bihar Wetlands. Extensive fishing takes place in these floodplains, although not well documented (Ahmed and Singh, 1991). In 1965–66 the ox-bow lakes of the Burhi-Gandak sub-basin alone covered an area of 36,000 ha which provided a fishery of 2,900 tonnes yr^{-1} (Natarajan, 1989). Downstream the river spills out into the delta, which is characterized by a large floodplain and interaction with the estuarine zone. The floodplain is most extensive in Bangladesh (Figures 11.1 and 11.2), and production from the rivers and their associated floodplains varied from 460,000 tonnes in 1983–84 to 561,824 tonnes in 1996–97 (DOF, 1991). Total riverine catches have been declining from 90,000 tonnes in 1983–84 to 84,463 tonnes in 1988–89 (DOF, 1991), although there has been something of a resurgence to around 60,000 tonnes since 1996. Of this riverine total in Bangladesh, the Ganga (Padma, as it is called in Bangladesh) contributes around 4–5 per cent (Payne *et al.*, 2004).

The cultivated area of the Ganga Valley in Uttar Pradesh and Bihar benefits from a system of irrigation canals that has increased the production of cash crops – sugarcane, cotton and oilseeds. Large areas in Uttar Pradesh and Bihar are also irrigated by channels running from hand-dug wells. Irrigation in Bangladesh is based on both gravity canals and electrically powered lifting devices (www.britannica.com/place/Ganges-River). In Nepal, agriculture has long been based

Figure 11.2 A typical floodplain fish catch, north-eastern Bangladesh, February 2015
Photo: Paul M. Thompson

on subsistence farming, and is fully dependent on the water from the glacier-fed rivers (www.nationsencyclopedia.com).

The hydroelectric potential of the Ganga and its tributaries is enormous. Estimates range from 51,700 to 128,700 MW, of which about 40 per cent is in India and the rest is in Nepal (Shah, 2001; Rehman, 2004). This potential has been exploited in India, including hydroelectric developments on headwater tributaries of the Ganga in Uttarakhand along the Bhagirathi and Alaknanda Rivers, on the upper Yamuna River and its tributaries in Himachal Pradesh, and further downstream, along the Chambal (a right-bank tributary of the Yamuna) and Rihand Rivers. Rehman (2004) records that only a tiny fraction of Nepal's hydroelectric generating capacity has been exploited.

Regulating services

The Sundarbans are an important store of carbon – ranging between 160 and 360 t/ha depending on the vegetation type (Rahman *et al.*, 2015). Auerbach *et al.* (2014) quantify another large-scale environmental impact: polders to exclude floods and salinity have resulted in declining land levels compared with the dynamic equilibrium of the open Sundarbans mangrove forests. Polders enclosed by embankments in the 1960s have lost 1.0–1.5 m of elevation, whereas the land level in the neighbouring Sundarban mangrove forest has accreted and kept pace with increasing high tide levels. Although damage to trees occurs in storms, the mangrove belt along the coast, including the Sundarbans and planted mangroves in coastal chars, provides shelter during cyclones – helping to absorb storm surges thereby protecting settlements and embankments. The construction of the Farakka barrage has caused several disruptions to the ecosystem structure and functions further downstream, thereby disrupting the ecosystem services. There is some indicative evidence of how excessive sedimentation in the Farakka has led to stream-flow depletion in the natural course of the lower Ganga sub-basin, and has caused loss in the ecosystem services (Ghosh, 2015). On the other hand, the Kolkata Port, which this barrage was supposed to resuscitate, has nominal business – probably much less than the ecosystem service losses. While there are a few attempts to place a value on selected ecosystem services of the Sundarbans, most of them have been at the level of individual ecosystem services. Ghosh *et al.* (forthcoming) report that if the ecosystem services remain as of today, the current value of the flow of ecosystem services (considering carbon sequestration, fishery production, storm protection, tourism, fuelwood, honey, prawn larva collection and crab) flowing to the human community from 2050 to 2100 will be around US$63 trillion.

Cultural services

Ganga is the holiest river in the sub-continent – pilgrims travel from distant places to bathe in its water at least once in their lifetime, on certain occasions to invoke forgiveness of their sins, hence purifying their souls. In India in particular, it is not only its largest river but is an integral part of the country's mythology, geography and history (Misra, 2014). The river is worshipped by Hindus and personified as a goddess. The ancient scriptures mention that the water of the Ganga carries the blessings of Lord Vishnu's feet; hence Mother Ganga is also known as *Vishnupadi*, which means 'Emanating from the Lotus feet of Supreme Lord Sri Vishnu'. It is also believed that if the ashes of a deceased are immersed in the River Ganga, one's soul will be sent to paradise and will acquire salvation, 'Nirvana' or release from the mundane cycle of life, i.e. death and rebirth (Prasad, 2010).

Freshwater biodiversity: species composition, distribution and conservation status

According to the estimates of the Zoological Survey of India (ZSI), about 16,500 species i.e. >20 per cent of the total fauna (78,000 species) – in India are aquatic and the majority of these are freshwater (Gopal, 2013). The first comprehensive information on the aquatic biodiversity and the various ecological aspects of the river has been generated during the Ganga Action Plan (Phase I); the record of 268 fishes forms the first ever scientific documentation of the fauna of the river (Hamilton, 1822).

Diverse taxonomic groups of Zooplankton representing Protozoa, Rotifera, Cladocera, Copepoda and Ostracoda are reported from the upper Ganga (Khanna *et al.*, 2012). Nautiyal *et al.* (2014) record Zooplankton richness to increase from mountain zone (Bhagirathi-Ganga from Gangotri and Badrinath to Haridwar) to plain zone (downstream of Haridwar till Varanasi). Sinha *et al.* (1998) have recorded 11 taxa of Zooplankton in the stretch between Kanpur and Allahabad. In middle Ganga, Rotifers, Cladocerans and Copepods have been recorded to dominate the Zooplankton population throughout the year. Altogether 100 taxa of Zooplankton (62 Rotifers, 25 Cladocerans and 13 Copepods) are reported from the River Ganga and its tributaries in Bihar in different seasons during 2000–2007.

The Ganga River supports a rich and diverse benthic macro-invertebrate fauna representing Protozoa to Mollusca (Table 11.1). In the upper Ganga, benthic macro-invertebrate fauna in mountain zone (Bhagirathi-Ganga from Gangotri and Badrinath to Haridwar) and plain zone

Table 11.1 Species inventory of macro-invertebrates reported from Bhutan, Bangladesh, Nepal, northern India and northern Pakistan

Macro-invertebrates Species Composition			
Insecta	*Total*	*Non-insecta*	*Total*
Coleoptera	2,420	Bivalvia	63
Diptera	2,144	Gastropoda	115
Ephemeroptera	295	Hirudinea	63
Heteroptera	109	Nematoda	2
Megaloptera	1		
Odonata	282		
Plecoptera	182		
Trichoptera	1,159		
Others	1,265		

Source: Sharma (2008) (database managed as HKHdipsoftware)

(downstream of Haridwar to Varanasi) belong to the Phyla: Annelida, Arthropoda and Mollusc. Earlier studies by Singh *et al.* (1994) and Joshi (2005) reported 20 Families from eight Orders from Devprayag to Rishikesh, and 10 Families from Devprayag to Kanpur respectively. Nautiyal *et al.* (2014) record that Annelida is represented by one taxon each of Oligochaeta, Polychaeta and Hirudinea; Arthropoda consist of 10 Orders, 52 Families of Insecta; while 1 taxon each of Crustacea and Arachnida; Mollusca are found in the plain zone only and consist of Gastropoda (five Families) and Pelecypoda (two Families). Nesemann *et al.* (2007) documented benthic invertebrates species including polychaetes (6), oligochaetes (31), leeches (15), bivalves (36),

Table 11.2 Major groups of aquatic ecosystem fauna in the River Ganga and its major tributaries in Nepal (Nesemann *et al.*, 2007)

Major groups of aquatic ecosystem fauna in the River Ganga and its major tributaries		
1	Palaearctic and Holarctic Fauna	*Dendrodrilus rubidus, Tubifex, Bothrioneurum, Aulodrilus limnobius, Spirosperma nagarkotensis, Gammarus lacustris, Galba truncaluta, Pisidium casertanum, Helobdella stagnalis, Nais alpine, Nais elinguis, Lumbriculus variegatus etc.*
2	Endemic Himalayan Fauna	*Himalayanpotamon atkinsonianum, H. emphysetum, H. sunkoshiense, Erhaia sp., Tricula spp., Myxobdella nepalica, Haemadipsida zeylanica agilis, Haemadipsida zeylanica monivindicis, Perionyx fluviatilis, Pisidium (Ohdneripisidium) annandalei, Pisidium (Afropisidium) clarkeanum dhulikhelensis, Musculium goshaitanensis, Asiaticobdella punyamatensis, Lymnaea andersoniana simulans, Dero phewatalensis.*
3	Oriental Fauna	*Aulophorous indicus, A. hymanae, A. tonkinensis, A. flabelliger, Dero sawayai, Branchiodrilus semperi, B. hortensis, Allonais paraguyaensis, A. gwaliorensis, Gyraulus euphraticus, Indoplanorbis exustus, Camptoceras lineatum, Radix persica, Radix ovalis, Radiatula spp., Lamellidens spp., Neoniphargus indicus, Sartoriana spinigera, Barythephusa lugubris, Parathelphusa spec., Bothrioneurum iris, Perionyx excavates.*
4	Cosmopolitan Fauna	*Limnodrilus hoffmeisterai, Aulodrilus pluriseta, branchiura sowerbyi.*
5	Neozoa Fauna	*Physa (Haitia) mexicana, Physa acuta, Indoplanorbis exustus, Bellamya bengalensis, Eisenia fetida.*

gastropods (40), and crustaceans (10), odonates (52), heteropterans (35), diapterans (13), coleopterans (7), trichopteran (1), ephemeropterans (6) and plecopterans (2) from different rivers and rivulets in the Ganga Basin in Bihar and West Bengal. The Trans Himalayan zones of Nepal have Palearctic freshwater fauna distributed above 1,700–2,700 m above sea level. The knowledge of freshwater invertebrates is still very poor, with information only for a few popular groups like molluscs, and some terrestrial insects or Zooplankton crustaceans being attractive to scientific interest. Illustrated taxonomical keys have been compiled for Nepal; an updated version with description of new species was published by Nesemann *et al.* (2007) (Table 11.2). A study of macro-invertebrate diversity may help in understanding the position, role and value of the invertebrates in the present-day ecosystem, giving much explanation for their suitability as bio-indicators.

Aquatic mammalian fauna is represented by two species of river dolphins (Ganges River Dolphin [Figure 11.3] *Platanista gangetica* and Irrawaddy Dolphin *Orcaella brevirostris*) and three species of otters (Smooth-coated Otter *Lutra perspicillata*, Eurasian Otter *Lutra lutra*, and Small-clawed Otter *Aonyx cinereus*) (Figure 11.4). Ganges River Dolphin forms a classic example of endemism of the riverine ecosystem. The species has been declared as the National Aquatic Animal of India and accorded legal protection in the Wildlife (Protection) Act of India, 1972 as a Schedule I species.

In the 1980s, research conservation work under the Ganga Action Plan provided baseline information about the Ganga River Dolphin (hereafter referred to as Dolphin) distribution. Across the species distribution range, i.e. Ganga-Brahmaputra-Meghna and Karnaphuli river

Figure 11.3 Ganga River Dolphin near the edge of Bangladesh Sundarbans
Photo: Mansur/Wildlife Conservation Society, Bangladesh

systems of India, Nepal, Bhutan and Bangladesh between the foothills of the Himalayas and the tidal zones, Reeves *et al.* (2000) suggested 2,500 Dolphins, and out of this around 1,800 individuals were within the Indian territory. Smith *et al.* (2001) estimated 5,000–6,000 Dolphins, while Sinha and Kannan (2014) reported over 3,600 individuals (460 in Bangladesh, 50 in Nepal and 1,500 in the main stem of the Ganga and its tributaries in Bihar). In 2012, Behera *et al.* (2014) recorded 671 Dolphins in the 2,500 km river stretch surveyed across Uttar Pradesh state. In the middle Ganga, 808 Dolphins were recorded in December 2006 between Buxar and Maniharighat (500 km) in Bihar; and about 25 Dolphins recorded between Manihari ghat and Farakka. Sinha (1997) recorded 152 Dolphins in the Bhagirathi-Hooghly river system up to Kakdwip during 1994–1995 which declined to 125 in 2004–2006. Gopal Sharma (Pers. comm.) reported sighting 97 Dolphins in the Hooghly in 2008. Eighty-five Dolphins were recorded in the Kosi River in Bihar (Sinha and Sharma, 2003) and 257 Dolphins in the Gandak River (320 km) (Choudhary *et al.*, 2012). Heinen (1987) reported sighting a few Dolphins in the Kosi (Eastern Nepal) and Narayani (Western Nepal). WWF Nepal (2006) recorded about 20 Dolphins in the Karnali River. Smith *et al.* (2001, 2006) recorded 225 and over 200 Dolphins respectively from the Sundarbans area of Bangladesh. About 450 Irrawaddy Dolphins are also reported (Smith *et al.*, 2001) just in the Sundarbans, particularly around deeper scour holes around confluence points, part of the largest single population of about 5,800 in the Bangladesh delta coast.

In India the Smooth-coated Otter has a cosmopolitan distribution from the Himalayas southward and in most of their range they are sympatric with the Small-clawed Otter (Figure 11.4) and Eurasian Otter (Foster-Turley, 1992). The Smooth-coated Otter is accorded legal protection in the Wildlife (Protection) Act of India, 1972 as a Schedule II species. In the upper Ganga, their occurrence is reported from Dudhwa Tiger Reserve and Katerniaghat Wildlife Sanctuary (Hussain, 2002), Corbett Tiger Reserve (Nawab, 2007), Hastinapur Wildlife Sanctuary and Alaknanda–Ganga Basin (Khan *et al.*, 2014). In the middle Ganga, in recent times few Smooth-coated Otter have been sighted between Sultanganj and Bhagalpur; healthy populations are reported from Vikramshila Gangetic Dolphin Sanctuary in Bihar state (Choudhary *et al.*, 2012). In Bangladesh otters are used by specialized fishing households in an area north of the Sundarbans to help them catch fish; there Feeroz *et al.* (2011) recorded 176 otters of different age and sex classes belonging to 46 Families during 2003–2005. In the Sundarbans and in tidal rivers, Small-clawed Otter and the Smooth-coated Otter are reported but there are no reliable population estimates. Nepal also holds three species of Otters, representing 1.6 per cent of the mammals cited in the country. Their occurrence is reported from two lowland districts, Kailali and Kapilvastu, although information available on their population trends is inadequate.

Mugger Crocodile *Crocodylus palustris*, a Schedule I species under the Wildlife (Protection) Act of India, 1972, is found abundantly in slow-moving rivers, preferring deep pools, swamps, lakes and village ponds. In the upper Ganga, in Uttarakhand state, 129 individuals have been reported in River Ramganga and the reservoir area from Corbett Tiger Reserve (Nawab, 2007). Recent studies (Joshi, 2013) highlight this species distribution range extension in the upper catchment of the River Ganga near to Rajaji National Park and in parts of Haridwar and Lansdowne Forest Divisions. While in Uttar Pradesh state, this species is reported along a 180 km stretch of the River Ganga in Hastinapur Wildlife Sanctuary and the Ramsar site between Garhmukteshwar and Narora. In Bangladesh, a small population of Estuarine Crocodile *Crocodylus porosus* is reported in the Sundarbans; it is considered a nationally endangered species. In Nepal, Mugger Crocodile are common in and around Chitwan National Park and small populations are found in Bardia and Sukla Phanta, but they are largely absent from their historical range in other parts of Nepal Terai.

Figure 11.4 Small-clawed Otter, Sundarbans, Bangladesh, October 2009
Photo: M. Monirul H. Khan, Jahangirnagar University, Bangladesh

In its distribution range in the Indian sub-continent, Gharial *Gavialis gangeticus* was first recognized as an endangered species in the 1970s. However, populations rebounded in the 1980s and 1990s as a result of large-scale captive rearing and head-starting programmes, particularly in the protected areas of India and Nepal. The species is functionally nationally extinct in Bangladesh. Gharial is accorded legal protection as a Schedule I species under the Wildlife (Protection) Act of India, 1972 and listed as Critically Endangered under the IUCN Red List. In the upper Ganga, in Uttarakhand state, Gharial is the subject of a re-introduction programme in Corbett National Park; by January 1987, 27 young Gharial had been released in the park. Current estimates are 42 adults inclusive of 10 adult males and 59 individuals of smaller size classes in the River Ramganga and its tributaries flowing across the Protected Area (Chowfin and Leslie, 2013). This is the only Gharial population reported in Uttarakhand state. Rao (2001) reports that captive-reared Gharial were released in the River Ganga along Rishikesh and Kanpur in Uttar Pradesh state, but due to lack of protection these animals might have been killed because of fishing activities. Since 2009 WWF-India, in collaboration with the Uttar Pradesh Forest Department, has initiated a re-introduction programme wherein 571 captive-bred Gharial have been released. Other notable populations have been recorded along the 180 km stretch from Tasod to Chakar Nagar in the River Chambal. Numbers recorded over the years are 311 (in 2004; Sharma, 2004), 392 (in 2005; Sharma, 2005), 493 (in 2006; Sharma, 2006), 619 (in 2007; Sharma, 2007), 638 (in 2008; Anon, 2008) and 720 (in 2015; Anon, 2015) respectively. In 2014, 65 Gharial were re-introduced in Ghaghra (*The Pioneer*, 2014) and 25 Gharial are reported in the Girwa River within the Katerniaghat Wildlife Sanctuary (Hussain,

Figure 11.5 Mugger Crocodile basking, Rapti River, Chitwan National Park, Nepal, September 2006
Photo: M. Monirul H. Khan, Jahangirnagar University, Bangladesh

2009). In the middle Ganga, Gharial are occasionally sighted in the main stem of the Ganga in Bihar, between Bhagalpur and Sultanganj. Twenty-five Gharial (22 adults, one hatchling and two sub-adults) were sighted in the Gandak in December 2012, and three to four hatchlings and one adult in the Ganga at Patna in 2007. In Nepal, a single population was found in the Narayani and Rapti Rivers bordering Chitwan National Park (Figure 11.5). Several hundred captive-reared Gharial from Chitwan have been released in the Narayani and about 100 individuals have been released into the Kosi River in eastern Nepal. Most of the released population has moved downstream through the barrages at the Indo-Nepalese border which precludes their return (Shrestha, 1986).

Freshwater turtles (Figure 11.6) in the Ganga Basin are divided broadly into two categories: Hard-shell Turtles (Emydid turtles) and Soft-shell Turtles (Trionychid turtles); however, no exclusive studies on population count are available. In the upper Ganga, occurrence and inventory of 12 species along Rishikesh to Kanpur in Uttar Pradesh were first reported by Rao (2001), while Singh *et al.* (2009) have reported 13 species from the Ganga and its major tributaries (Chambal, Ghaghra, Gomti, Sarju, Yamuna) in Uttar Pradesh state. In the middle Ganga, ecological studies on freshwater turtles are lacking, although in the last ten years, on many occasions turtles have been confiscated in Bihar (Sinha *et al.*, 2014). Thirteen species of freshwater turtle are known from Bangladesh (Kabir *et al.*, 2009), though there is a lack of recent detailed information on their conservation status. Large rivers in lowland areas in Nepal support populations of Gangetic Softshell *Trionyx gangeticus*, Indian Flap-shell *Lissemys punctata* and Indian Saw-back Turtle *Kachuga tecta* which have been recorded in wildlife trade for eggs and flesh (Saharia, 1982).

Though the exact number of fish species in Ganga is still unknown, Talwar and Jhingran (1991) report 265 fish species, while Talwar (1991) gave a list of 382 fish species. Indian Major Carps, and large catfishes (*Aorichthys aor, A. seenghala, Wallago attu, Bagarius bagarius*), feather-backs (*Notopterus notopterus, N. chittala*) and murrels (*Channa marulius, C. punctatus*) are

Figure 11.6 Indian Roofed Turtle (*Pangshura tecta*), Keraniganj, Dhaka, November 2014
Photo: M. Monirul H. Khan, Jahangirnagar University, Bangladesh

of commercial value (Talwar and Jhingran, 1991). The fishery in the Haridwar to Sundarban stretch is rich and diverse, with cyprinids (176 species) and silurids forming the mainstay (Sinha and Khan, 2001). In earlier documented studies in the upper Ganga basin, Khanna and Badola (1994) recorded 30 fish species between Rishikesh and Haridwar; Sharma and Rajput (1986) recorded 26 fish species around Bijnor district; Rao (2001) and WWF-India Report (2004) report 82 fish species between Brijghat and Narora; Shukla and Asthana (1995) recorded 25 fish species around Kanpur; further downstream below Allahabad, in different studies, 13, 38 and 75 fish species respectively have been recorded (Nautiyal, 2005). In recent times Nautiyal *et al.* (2014) recorded 21 fish species between Devprayag and Rishikesh; 22 fish species at Narora; 27 fish species at Kachla Ghat; 31 fish species at Bithoor near Kanpur; and 88 fish species below Allahabad. Nautiyal *et al.* (2014) document that from the fisheries' perspective the ratio of Carp:Catfish:Others has changed in the last few decades (46:30:34 in the 1960s, 36:36:28 in the 1970s and 14:45:40 in the 1990s). Thus fisheries have shifted from carp- to catfish- to others-dominated. Similarly, Nepal fisheries also consist of carp, catfish, sheat-fish, feather-backs, eels and hill stream fish. The lowland terai is home to more than 100 species of fishes, with decreasing species richness as altitude increases (Jha, 2008). In the middle Ganga stretch at Allahabad, Patna and Bhagalpur, fish landings have declined sharply both qualitatively and quantitatively. The most affected group is the Indian Major Carp, yield of which at Allahabad decreased from 44.5 per cent in 1958–66 to 8.3 per cent in 1996–97, and *Tenualosa ilisha* (Hilsa) from 9.7 per cent to 4.2 per cent, while the yield of catfish increased slightly from 22.7 per cent to 24.1 per cent. The yield of miscellaneous fish (*Setipinna phasa*, *Chela* spp., *Mastacembelus* spp., *Puntius* spp., *Eutropichthys vacha*, *Clupisoma garua*, *Notopterus* spp., *Rita rita*, *Mystus vittatus*, *Ailia coila*,

Nandus nandus etc.) increased from 23.1 per cent to 63.4 per cent at the same centre and duration. Similarly, a decrease in major carp catches in the middle stretch of the Ganga was observed, from 26.62 kg/ha in 1958–61 to a low of 2.55 kg/ha in 1989–95 (Sinha, 2000). Other researchers also emphasized that stock of most of the economically important fishes, including the Indian Major Carps, in the Ganga system has dwindled in recent years (Hassan, 1999; Payne *et al.*, 2004). Bangladesh in general has diverse (about 265 species) and highly productive inland capture fisheries (Rahman, 2005; Siddiqui *et al.*, 2007) and over 450 species recorded in coastal and marine waters (Rahman *et al.*, 2009). Fish catches have declined substantially in the main Ganga-Padma River; for example, the riverine catch reported in national statistics in Faridpur (Padma) crashed from 12,500 t in 1983–84 to 1,500 t in 1988–89, and then gradually to 1,100 t in 2007–2008 (DOF, 2009). There are 204 species of fin fish and 26 species of shrimp reported from the Sundarbans, although detailed studies of fish catches/landings have recorded only 95 species of fish and shrimp over three years, dominated by shrimps, Hilsha and Poa (IPAC, 2013).

In the upper Ganga, Bashir *et al.* (2010) and Khan *et. al.* (2013) reported 55 and 95 aquatic birds species respectively. Shallow wetlands, associated agricultural land and open sandbanks support the largest breeding population of Sarus Crane *Grus antigone* and Indian Skimmer *Rynchops albicollis* (Khan *et al.*, 2013). Baral and Inskipp (2005) report an extensive avian diversity from the Nepal part of the Ganga Basin. In middle Ganga noteable records are those of Sinha (2013) which report 162 avian species in Bihar state, of which 97 are reported from Vikramshila Gangetic Dolphin Sanctuary. Along the Ganges-Padma River in Bangladesh, 159 bird species have been recorded (Bangladesh Bird Club and P. Thompson, unpublished data), and 254 bird species are reported from the Sundarbans in Bangladesh (Khan, 2005; Thompson *et al.*, 2014) including mangrove-dependent species. Several threatened species depend on wet riverine grasslands, including Swamp Francolin *Francolinus gularis* (vulnerable) at sites such as Kosi Tappu (Nepal) (Baral and Inskipp, 2005), and Bristled Grassbird *Chaetornis striatus* (vulnerable) at sites in all three countries. The river and its sandbanks have a distinctive avifauna that is increasingly threatened by declining dry season flows, human disturbance and conversion to crops. Several species dependent on this system for nesting have declined in recent decades, resulting in River Tern *Sterna aurantia* being categorized as near-threatened and Black-bellied Tern *Sterna acuticauda* as endangered, while one species – Pink-headed Duck *Rhodonessa caryophyllacea*, whose stronghold was the mid-lower Ganga wetlands – is already probably extinct (BirdLife International, 2015). The occurrence of the main species described above – throughout the Ganga Basin – is also shown in Figure 11.7.

Conservation threats: drivers of biodiversity change

Although the Indian government has embarked upon various restoration programmes in the Ganga River, the environmental conditions have apparently deteriorated and the species population depleted over much of its distribution range (Rao, 2001). Wildlife in general is becoming increasingly rare outside of protected areas. Much of the distributional records are either subjective or based on chance observations; therefore no concrete scientific database exists for monitoring population trends. Owing to increasing dependency on aquatic resources, in much of the distribution range aquatic species are often in direct conflict with fishermen who view them as vermin or competitors for fish and kill them. Indiscriminate fishing also reduces fish stocks which changes prey size and species composition (Nawab, 2007).

Pollution and changing land use practices with adverse impacts on water flow and quality have led to fragmentation of habitats, limiting species movement over much of their apparent

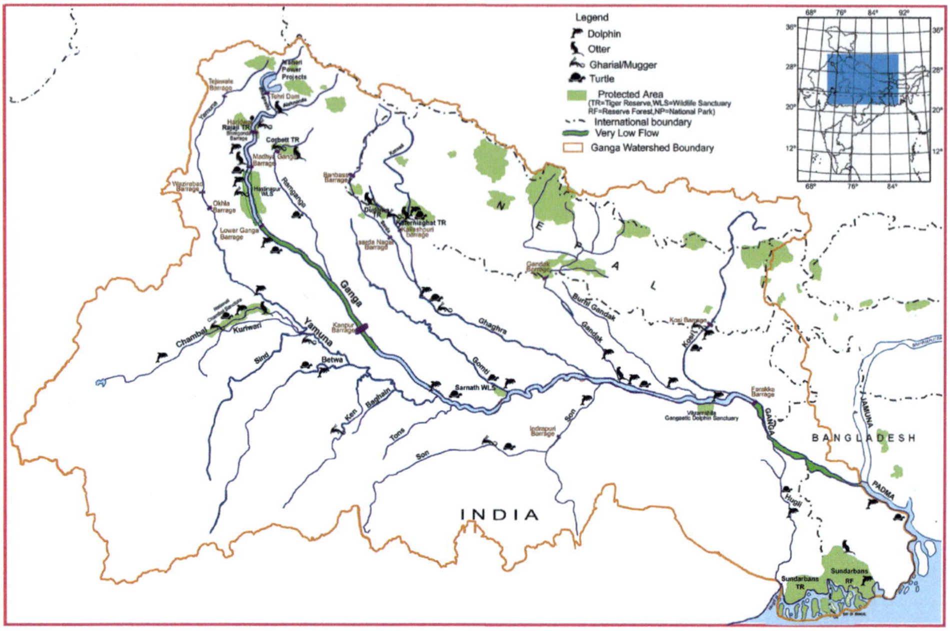

Figure 11.7 Occurrence of higher vertebrate fauna throughout the Ganga Basin

range, leading to poor dispersal and isolation of populations. In spite of general awareness of the trade in wildlife and its derivatives in India, there is little information on the extent and prevalence of illegal trade in lesser known species, like otters and turtles. Life history strategies of aquatic species are known to have evolved in response to natural flow variability (Bunn and Arthington, 2002). The impacts of river alterations manifest themselves in the imperilment of aquatic biota and in the reduced ability of rivers to provide valued ecosystem services upon which humans depend (Millennium Ecosystem Assessment, 2005). Quantifying environmental flows – the amount of water that a river must have to maintain its ecology in a certain condition – is important in this regard, but it is only developing in Ganga riparian countries (see Chapter 10, 'Environmental flows', for more detail).

Conservation implications: the way forward

Critical biodiversity hotspots need to be identified and legally protected from anthropogenic pressures, based on irreplaceability, species assemblage and connectivity considerations. Criteria for identifying irreplaceable and important riparian sites need to be based on identification of specific areas within the basin that require protection. As significant downstream impacts are due to changes in sediment or flow patterns, maintenance of longitudinal connectivity within the river basin should be considered critical to its ecosystem integrity.

Priority recommendations comprise:

- Degraded habitats should be restored under the ambit of a protected areas network through creation of community reserves in biodiversity hotspots. Critical habitats where species' populations still exist should be surveyed and aquatic species conservation management plans devised at landscape level.

- A holistic approach is needed to improve the design and management of water resources. This requires establishing inclusive governance and integrated approaches to basin management to adopt water conservation technologies that enhance water productivity.
- Local communities (like fishers and farmers) dependent on the river for their livelihoods are the important stakeholders. Large-scale awareness campaigns and capacity-building programmes should be conducted to encourage their participation, which can play a significant role in the long-term conservation of species and the habitat.
- Scarce conservation funds and conflict mitigation concerning aquatic species conservation need to be focused where they will most improve the survival odds of threatened species.

Acknowledgements

The lead author is grateful to Mr Ravi Singh (Secretary General and CEO, WWF-India) and Dr. Sejal Worah (Programme Director, WWF-India) for providing infrastructural support and encouragement. Mr Suresh Babu (Director-Rivers, Wetlands and Water Policy Programme, WWF-India) is thanked for his thoughtful suggestions and for reviewing the manuscript, which helped improve the content quality. We are grateful to Dr Nilanjan Ghosh for his inputs to the chapter. Help rendered by Mr Mohd. Shahnawaz Khan (Senior Project Officer-Rivers, Wetlands and Water Policy Programme, WWF-India) and Ms Devanshi Kasana (Senior Project Officer-Rivers, Wetlands and Water Policy Programme, WWF-India) in collating information is much appreciated. Dr G. Areendran (Director-IGCMC and IT, WWF-India) and Mr Krishna Raj (Senior Programme Coordinator-GIS and Remote Sensing, IGCMC and IT, WWF-India) are thanked for preparing the map. We sincerely thank Professor M. Monirul H. Khan of Jahangirnagar University, Bangladesh, for the provision of several high-quality photos for this chapter.

References

Ahmed S.H. and Singh A.K. 1991. River systems of Bihar: Scope prospects, potentialities and conservation of capture fisheries. *Fishing Chimes Patna*, (November): 51–6.

Anon. 2008. *Annual faunal census in the National Chambal Sanctuary, Madhya Pradesh*. Report submitted to Conservator of Forest, Gwalior by Prof. R.J. Rao, S.P. Sharma, S.C. Bhadoria and R.K. Sharma, pp. 1–20.

Anon. 2015. *Annual survey report on Gharial and Mugger in the National Chambal Sanctuary, Madhya Pradesh*. Report submitted to Conservator of Forest, Gwalior by V. Rahim, A. Pateria and R.K. Sharma, pp. 1–18.

Auerbach, L.W., Goodbred, S.L., Mondal, D.R., Wilson, C.A., Ahmed, K.R., Roy, K., Steckler, M.S., Small, C., Gilligan, J.M. and Ackerly, B.A. 2015. Flood risk of natural and embanked landscapes on the Ganges-Brahmaputra tidal delta plain. *Nature Climate Change*, 5(2): 153–7.

Baral, H. S. and Inskipp, C. 2005. *Important Bird Areas in Nepal: Key Sites for Conservation*. Kathmandu, Nepal and Cambridge, UK: Bird Conservation Nepal and BirdLife International.

Bashir, T., Khan, A., Gautam, P. and Behera, S.K. 2010. Abundance and prey availability assessment of Ganga River Dolphin Platanista gangetica in the Upper Ganga River, India. *Journal of Aquatic Mammals*, 36: 19–26.

Behera, S.K., Singh, H. and Sagar, V. 2014. 'Indicator species (Gharial and Dolphin) of riverine ecosystem: An exploratory of River Ganga'. In Sanghi, R. (ed.). *Our National River Ganga: Lifeline of Millions*, Switzerland: Springer International Publishing, pp. 121–40.

BirdLife International. 2015. IUCN Red List for birds. Downloaded from www.birdlife.org/globally-threatened-bird-forums/wp-content/uploads/2015/10/2015-Red-List-changes.xlsm on 15 December 2015.

Britannica. 2015. Ganges River. Downloaded from www.britannica.com/place/Ganges-River on 23 October 2015.

Bunn, S. E. and A. H. Arthington. 2002. Basic principles and ecological consequences of altered flow regimes for aquatic biodiversity. *Environmental Management*, 30: 492–507.

Choudhary, S., Dey, S., Dey, S., Sagar, V., Nair, G. and Kelkar, N. 2012. River dolphin distribution in regulated river system: Implications for dry season flow regimes in the Gangetic basin. *Aquatic Conservation: Marine and Freshwater Ecosystems*, 22(1): 11–25.

Chowfin, S.M. and Leslie, A.J. 2013. A preliminary investigation into nesting and nest predation of the critically endangered, gharial (*Gavialis gangeticus*) at Boksar in Corbett Tiger Reserve, Uttarakhand, India. *International Journal of Biodiversity and Conservation*, 5(2): 54–7.

DOF. 1991. *1998 Fish Catch Statistics of Bangladesh 1990–91*. Dhaka: Department of Fisheries, Government of Bangladesh.

DOF. 2009. *Fisheries Statistical Yearbook of Bangladesh 2007–2008*. Dhaka: Fisheries Resources Survey System, Department of Fisheries.

Feeroz, M.M., Begum, S. and Hasan, M. K. 2011. Fishing with Otters: A Traditional Conservation Practice in Bangladesh. Proceedings of 6th International Otter Colloquium, *IUCN Otter Spec. Group Bull.*, 28A: 14–21.

Foster-Turley, P. 1992. *Conservation ecology of sympatric Asian Otters Aonyxcinerea and Lutra perspicillata*. Ph.D. Thesis, University of Florida.

Gopal, B. 2013. 'Water and biodiversity: An overview'. In *Water and Biodiversity* (ed.), pp. 1–7. U.P. State Biodiversity Board.

Ghosh, N. 2015. Challenges to environmental security in the context of India-Bangladesh Trans-boundary water relations. *Decision*, 42(2): 211–28. Special Issue on Managing Critical Resources: Food, Energy and Water.

Ghosh, N., Danda, A., Bandyopadhyay, J. and S. Hazra. (forthcoming). *Planned Retreat and Ecosystem Restoration: Social, Economic, and Ecological Arguments of Climate Change Adaptation in coastal zones for Indian Sundarbans Delta*. PRI Working Paper No. 1, WWF-India.

Gurung, T.B. 2003. Fisheries and aquaculture activities in Nepal. *Aquaculture Asia*, 7(1): 14–19.

Hamilton, B. 1822. *An Account of Fishes found in the River Ganges and its Branches*. Edinburgh, UK: A. Constable and Co.

Hassan, S.S. 1999. *The current status of the fish stock of commercial importance in River Ganga in and around Patna*. Ph.D. Thesis (Patna University, Patna).

Heinen, J.T. 1987. *Interim Report for the KosiTappu wild buffalo/Human Interaction Project*. Unpublished. DNPWC and USAID, Kathmandu.

Hussain, S.A. 2002. 'Conservation status of otters in the Tarai and Lower Himalayas of Uttar Pradesh, India'. In Dulfer, R., Conroy, J., Nel, J. & Gutleb, A. (eds). *Otter Conservation – An Example for a Sustainable use of Wetlands. IUCN Otter Specialist Group Bulletin*, 19: 131–42. Trebon, Czech Republic.

Hussain, S.A. 2009. Basking site and water depth selection by gharial *Gavialis gangeticus* in National Chambal Sanctuary and its implication for river conservation. *Aquatic Conservation: Marine and Freshwater Ecosystems*, 18: 127–33.

IPAC. 2013. *Sundarbans fish catch monitoring study*. Integrated Protected Area Co-Management (IPAC) Project. Dhaka: International Resources Group.

Jha, S. 2008. Status and conservation of lowland Terai wetlands in Nepal. *Our Nature*, 6: 67–77.

Joshi, B.D. 2005. 'Studies on the eco-biology of tributaries of River Ganga between Devprayag and Rishikesh'. In Nautiyal, P., Bhatt, J.P., Gosain, O.P. and Dodriyal, A.K. (eds). *Biological Diversity in Freshwater Environments*. Transmedia, Srinagar Garhwal. pp. 19–35.

Joshi, R. 2013. Range extension of mugger crocodile *Crocodylus palustris* (Lesson, 1 831) in upper Ganges and tributaries, lesser Himalayan zone, north India. *J Biol Earth Sci*, 3(1): 100–9.

Khan, M.M.H. 2005. Species diversity, relative abundance and habitat use of the birds in the Sundarbans East Wildlife Sanctuary, Bangladesh. *Forktail*, 21: 79–86.

Khan, M.S., Aftab, Z. Syed, Nawab, A., Ilyas, O. and Khan, A. 2013. Composition and conservation status of avian species at Hastinapur Wildlife Sanctuary, Uttar Pradesh, India. *Journal of Threatened Taxa*, 5(12): 4714–4721.

Khan, M.S., Dimri, N.K., Nawab, A. *et al.* 2014. Habitat use pattern and conservation status of Smooth-coated Otters Lutrogale perspicillata in the Upper Ganges basin, India. *Anim. Biodivers. Conserv.*, 37(1): 69–76.

Khanna, D.R. and Badola, S. P. 1994. Foothill zone of River Ganga – habitat of adult Mahseer. In Nautiyal, P. (ed.). *Mahseer the Game Fish*. Jagdamba Prakashan, Dehradun. pp. 98–101.

Khanna, D.K., Bhutiani, R., Matta, G., Singh, V. and Bhadauriya, G. 2012. Study of planktonic diversity of river Ganga from Devprayag to Roorkee, Uttarakhand (India). *Environment Conservation Journal*, 13(1&2): 211–17.

Lisitzen, A.P. 1972. *Sedimentation in the World Oceans*. Society of Economic Paleontology Minerology. Spec. Publ. 17: 218.

Millennium Ecosystem Assessment 2005. Millennium Ecosystem Assessment. 2005. *Millennium Ecosystem Assessment Synthesis Report*. Washington, DC: Island Press.

Ministry of Water Resources. 2015. Ganga Basin. Downloaded from http://gfcc.bih.nic.in/Basin.htm on 23 October 2015.

Misra, M. 2014. 'What and where is Ganga?' In Sanghi, R. (ed.). *Our National River Ganga: Lifeline of Millions*. Switzerland: Springer International Publishing, pp. 375–7.

Natarajan, A.V. 1989. 'Environmental impact of Ganga Basin development on gene-pool and fisheries of the Ganga River system'. In Dodge, D.P. (ed.). *Proceedings of the International Large River Symposium*, Can. Spec. Publ. Fish. Aquat. Sci. 106: 545–60.

Nations Encyclopedia. 2015. Downloaded from www.nationsencyclopedia.com on 23 October 2015.

Nautiyal, P. 2005. *Spatial and microdistribution of benthic macroinvertebrate community in tropical mountain streams of Garhwal region*. Final Technical Report submitted to Department of Science and Technology, New Delhi (Ref. No. SP/SO/C-42/98) dated 22 July 1999.

Nautiyal, P., Verma, J. and Mishra, A.S. 2014. 'Distribution of major floral and faunal diversity in the mountain and upper Gangetic plains zone of the Ganga: Diatoms, macroinvertebrates and fish'. In Sanghi, R. (ed.). *Our National River Ganga: Lifeline of Millions*. Switzerland: Springer International Publishing, pp. 75–118.

Nawab, A. 2007. *Ecology of Otters in Corbett Tiger Reserve, Uttarakhand, India*. Ph.D. Thesis, Forest Research Institute (University), Dehradun, Uttarakhand, India, p. 174.

Nesemann, H., Sharma, S., Sharma, G., Khanal, S.N., Pradhan, B., Shah, D.N. and Tachamo, R.D. 2007. *Aquatic invertebrates of the Ganga river system [Mollusca, Annelida, Crustacea (in part)]* Vol. 1. H. Nesemann.

Payne A.I., Sinha, R., Singh, H.R. and Huq, S. 2004. 'A review of the Ganges basin its fish and fisheries'. In Welcome, R. and Petr, T. (eds). *Proceedings of The Second Symposium On the Management of Large Rivers for Fisheries*. Vol. I. FAO Regional Office for Asia and Pacific, Bankok, Thailand. RAP Publication 2004/16, pp. 229–51.

Prasad, P.M. 2010. *Livelihood Objectives for Maintaining Sustainable Flows in the River Ganga*. Unpublished starter document submitted for Upper Ganga Environmental flows assessment study under Living Ganga Programme, WWF-India. Department of Humanities and Social Sciences, Indian Institute of Technology Kanpur, India.

Prasad, S.N., Ramachandra, T.V., Ahalya, N., Sengupta, T., Kumar, A., Tiwari, A.K., Vijayan, V.S. and Vijayan, L. 2002. Conservation of wetlands of India: A Review. *Tropical Ecology*, 43(1): 173–86.

Rahman, A.K.A. 2005. *Freshwater Fish of Bangladesh*. Dhaka: Dhaka University.

Rahman, A.K.A., Kabir, S.M.H., Ahmad, M., Ahmed, A.T.A, Ahmed, Z.U., Begum, Z.N.T., Hassan, M.A. and Khondker, M. (eds) 2009. *Encyclopaedia of Flora and Fauna of Bangladesh*, 23. *Marine Fishes*. Dhaka: Asiatic Society of Bangladesh.

Rahman, M.M., Khan, M. N. I., Hoque, A.K.F. and Ahmed, I. 2015. Carbon stock in the Sundarbans mangrove forest: Spatial variations in vegetation types and salinity zones. *Wetlands Ecology and Management*, 23: 269–83.

Rao, R.J. 2001. Biological resources of the Ganga River, India. *Hydrobiologia*, 458: 159–68.

Reeves, R.R., Smith, B.D. and Kasuya, T. 2000. *Biology and Conservation of Freshwater Cetaceans in Asia*.

Rehman M.M. 2004. *Water versus power: Role of dams in geopolitics of Ganga Basin*. River Symposium, Australia, pp. 1–10.

Saharia, V.B. (ed.). 1982. *Wildlife in India*. Dehradun, Uttar Pradesh, India: Natraj Publication.

Shah, R.B. 2001. 'Ganges-Brahmaputra: The outlook for the twenty-first century'. In Biswas, A.K. and Uitto, J.I. (eds). *Sustainable Development of the Ganges-Brahmaputra-Meghna Basins*, pp. 17–45. Tokyo: United Nations University Press.

Sharma, M.K. and Rajput, D.B. 1986. Ichthyofauna of Bijnor District (Uttar Pradesh). *Journal of Bombay Natural History Society*, 83: 562–9.

Sharma, R.K. 2004. *Survey of aquatic animals in National Chambal Sanctuary, Morena, Madhya Pradesh*. Report submitted to Superintendent of National Chambal Sanctuary, Madhya Pradesh, pp. 1–30.

Sharma, R.K. 2005. *Survey of aquatic animals in National Chambal Sanctuary, Morena, Madhya Pradesh*. Report submitted to Superintendent of National Chambal Sanctuary, Madhya Pradesh, pp. 1–24.

Sharma, R.K. 2006. *Survey of aquatic animals in National Chambal Sanctuary, Morena, Madhya Pradesh*. Report submitted to Superintendent of National Chambal Sanctuary, Madhya Pradesh, pp. 1–26.

Sharma, R.K. 2007. *Annual faunal survey in National Chambal Sanctuary, Morena, Madhya Pradesh*. Report submitted to Superintendent of National Chambal Sanctuary, Madhya Pradesh, pp. 1–15.

Shrestha J. 1978. Fish fauna of Nepal. *J. Museum of Natural History, Tribuvan University, Kathmandu*, 2: 33–43.

Shrestha, T.K. 1986. Ecology of Gangetic Dolphin in the Karnali River. In Joshi, R.L. (ed.). *Nepal Himalaya.* Naini Taal, Uttar Pradesh, India: Himalaya Research Group.

Shukla, A.C. and Asthana, V. 1995. *Ganga – A water marvel.* New Delhi: APH Publishing.

Siddiqui, K.U., Islam, M.A., Kabir, S.M.H., Ahmad, M., Ahmed A.T.A, Rahman, A.K.A., Haque, E.U., Ahmed, Z.U., Begum, Z.N.T., Hassan, M.A., Khondker, M. and Rahman, M.M. (eds) 2007. *Encyclopaedia of Flora and Fauna of Bangladesh, 23. Freshwater Fishes.* Dhaka: Asiatic Society of Bangladesh.

Singh, H.R., Nautiyal, P., Dobriyal A.K., Pokhriyal, R.C, Negi, M., Baduni, V., Nautiyal, R., Agarwal, N.K. and Gautam, A. 1994. Water quality of River Ganga (Garwal Himalayas). *Acta Hydrobiol*, 36: 3–15.

Singh, S., Basu, D., Tripathi, A., Singh, L. and Singh R.P. 2009. *An evaluation of distribution, status and abundance of Freshwater Turtles in Uttar Pradesh.* Freshwater Turtles and Tortoises of India.

Sinha, R.K. 1997. Status and Conservation of Ganges river dolphin in Bhagirathi-Hooghly river systems in India. *Int. J. Ecol. and Env. Sc.*, 23: 343–55.

Sinha, M. 2000. Inland aquatic resources of India, Issues and Threats – A Fisheries Perspectives: 1–9. In Sinha, M., Jha, B.C., Khan, M.A. (eds). *Summer School on Environment Impact Assessment of Inland waters for sustainable fisheries management and conservation of Biodiversity.* Central Inland Capture Fisheries Research Institute, Barrackpore, 25 July–14 August 2000.

Sinha, M. and Khan, M. A. 2001. Impact of environmental aberrations on fisheries of the Ganga (Ganges) River. *Journal of Aquatic Ecosystem Health and Management*, 4(4): 493–504.

Sinha, R.K. and Sharma, G. 2003. Current status of Ganges dolphin, *Platanista gangetica* in River Son and Kosi in Bihar. *Journal of Bombay Natural History Society*, 100: 27–37.

Sinha, M., De, D.K. and Jha, B.C. 1998. *The Ganga Environment and Fisheries.* Barrackpore: CIFRI.

Sinha, R.K., Kedia, D.K. and Kumari, A. 2014. Overview of higher vertebrates in the Ganges-Brahmaputra-Meghna River Basin: Their status, threats and conservation, pp. 105–123. In *Rivers for Life – Proceedings of the International Symposium on River Biodiversity: GBM River System,* Ecosystem for Life, A Bangladesh–India initiative, IUCN International Union for Conservation of Nature.

Smith, B.D., Ahmed, B., Ali, M. E. and Braulik, G.T. 2001. Status of the Ganga River Dolphin or shushuk *Platanista gangetica* in Kaptai Lake and the southern rivers of Bangladesh. *Oryx*, 35: 61–72.

Smith, B.D., Braulik, G., Strindberg, S., Ahmed, B. and Mansur, R. 2006. Abundance of Irrawaddy dolphins (*Orcaella brevirostris*) and Ganges River dolphins (*Platanista gangetica gangetica*) estimated using concurrent counts made by independent teams in waterways of the Sundarbans mangrove forest in Bangladesh. *Marine Mammal Science*, 22: 527–47.

Talwar, P.K. 1991. Pisces. In *Faunal Resources of Ganga, Part – 1(General Introduction and Vertebrate Fauna).* 59–145. Zoological Survey of India, Calcutta.

Talwar, P.K. and Jhingran, A.G. 1991. *Inland Fisheries of India and Adjacent Countries*, 2 vols. Delhi: Oxford and IBH Publishing Co.

The Pioneer 2014. Forest Department Releases Kukrail Ghariyals in Ghaghra River. Wednesday, 5 March 2014, S.K. Pioneer News Service.

Thompson, P.M., Chowdhury, S.U., Haque, E.U., Khan, M.M.H. and Halder, R. 2014. Notable bird records from Bangladesh from July 2002 to July 2013. *Forktail*, 30: 50–65.

WWF-India Report. 2004. Information Sheet on Ramsar Wetlands (RIS). pp. 1–14. http://jncc.defra.gov.uk/pdf/RIS/UK61002.pdf.

WWF Nepal 2006. *Status, Distribution and Conservation Threats of Ganges River Dolphin in Karnali River, Nepal.* Nepal: WWF.

Britannica. 2015. Ganges River. Downloaded from www.britannica.com/place/Ganges-River on 23 October 2015.

Nations Encyclopedia. 2015. Downloaded from www.nationsencyclopedia.com on 23 October 2015.

12

Arsenic in the eastern Ganges Basin

Extent and impact on food chain and human health

Rajmohan Natarajan, Dipankar Chakraborti and Sanmugam Prathapar

Key messages

- Arsenic which migrated with fluvial sediments from the Himalayas has contaminated groundwater above drinking and irrigation standards in the eastern Gangetic Basin (EGB) along the main stem of the Ganges River. In general, shallow groundwater has more arsenic (As) presence than deeper groundwater.
- Food grown with contaminated water is also a route for As ingestion. Cultivation of rice, the main staple of the locals, under flooded conditions enhances arsenic prevalence and accumulation in the roots, which is subsequently transferred to the grains.

- An estimated 75 per cent of the EGB's population of more than 360 million live in the As-affected regions. Groundwater in 15 districts in Bihar, 14 districts in West Bengal, 61 districts in Bangladesh and 20 districts in Nepal Terai have arsenic levels above drinking standards set by the WHO.
- Arsenic causes cancer, cardiovascular, neurological, hematological, renal and respiratory problems. Arsenic toxicity to humans depends on its chemical and physical forms, the dosage, duration of exposure, dietary compositions of interacting elements, age, sex, the route by which it enters the body, nutritional status of the person, immunity level of the individual, etc.
- Arsenic contamination in Bihar and Nepal is not yet evaluated systematically. Information on arsenic accumulation in the food chain and human health issues is also sparse. It is critically important to establish systematic monitoring and analysis of As contamination in groundwater, soils and food across the EGB. A comprehensive long-term programme is needed to evaluate As contamination in the soil, water, foods (grains, vegetables, fruits, etc.), and its impact on human health.

Introduction

Arsenic is one of the most hazardous elements in the environment, causing serious health issues such as cancer in the skin, lungs, bladder, liver, kidneys etc, and dermal, cardiovascular, neurological, hematological, renal and respiratory problems. Arsenic contamination is widely reported worldwide (Rajmohan and Prathapar, 2014) and in the Ganges, particularly in the eastern Gangetic Basin (Figure 12.1).

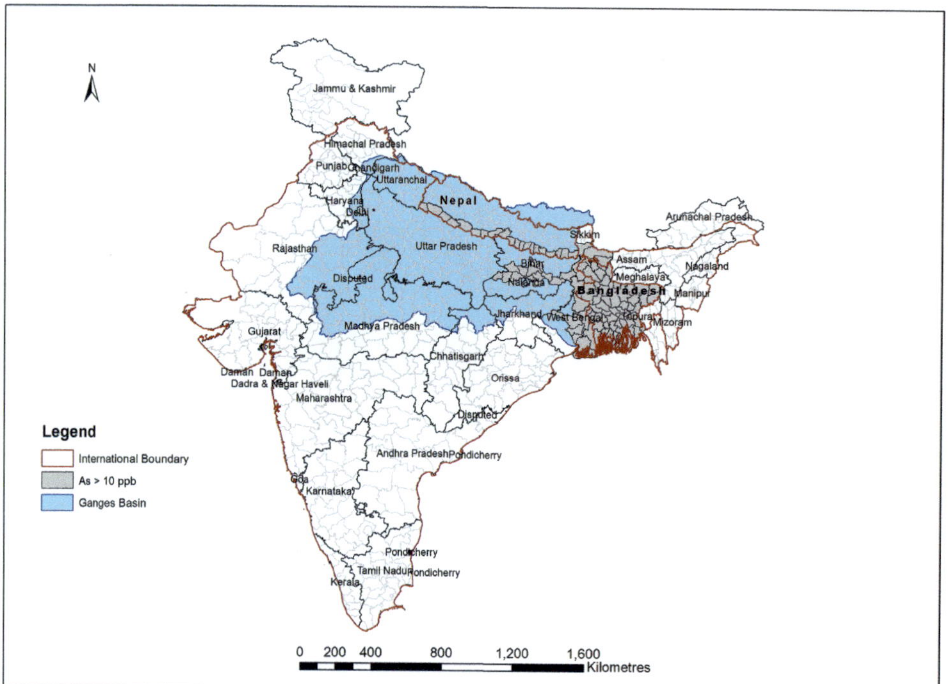

Figure 12.1 Arsenic-contaminated districts (As > 10 ppb) in the Ganges Basin

The total population in the EGB was around 360 million in 2011, 80 per cent of these living in rural areas depending on agriculture for their livelihoods. Groundwater is a major source of water for drinking, domestic use and irrigation, but its quality is questionable due to arsenic contamination. Several studies reported that As-contaminated drinking water is a major source of the above health issues in the region (Kunar *et al.*, 2009; Bhattacharya *et al.*, 2011). Other studies implied that food crops are also a major route for As exposure due to the usage of As-contaminated groundwater for irrigation (Singh and Ghosh, 2012; Mandal and Suzuki, 2002).

The groundwater wells with high As content are mostly located on younger deltaic depositions (Mallick and Rajagopal, 1996). It is widely accepted that As migrated with fluvial sediments from the Himalayas to the EGB (Gurung *et al.*, 2007). It is considered that arsenic, which is present in aquifer sediment with iron oxides, has been released by reductive dissolution via microbial respiration of organic carbon. Arsenic species and their behaviour, especially chronic toxicological effects, led the WHO, US-EPA and EU to lower the standard for drinking water from 50 μg/L to 10 μg/L in the groundwater (WHO, 2004; US Environmental Protection Agency, 2002; European Union, 1998).

This chapter provides a brief overview of arsenic sources, forms and toxicities, followed by a review of the extent of As contamination in groundwater and soil in the EGB. Although As is widely spread in the EGB, including parts of Uttar Pradesh and Jharkhand (Chakraborti *et al.*, 2009), the chapter primarily concentrates on Bihar, West Bengal, Nepal Terai and Bangladesh, where arsenicosis has become widespread. Finally, the chapter summarizes the impact of As on food and health, and concludes by identifying knowledge gaps to mitigate the As crisis in the EGB.

Arsenic sources, forms and toxicities

Sources

Arsenic may be of natural (geogenic) or anthropogenic origin. Geologically, it may originate from As minerals like arsenopyrite, orpiment, realgar, claudetite, arsenolite, pentoxide, scorodite and arsenopalledenite. Anthropogenic activities such as coal combustion, refinement of petroleum products, mining, smelting, ore processing and manufacturing of cement, pesticides, glass, leather, textiles, alloys and pigments contribute to As contamination (Smedley, 2005). Arsenic release and transport from these sources to the soil, water and air is explained by four major hypotheses: (1) oxidation of pyrite, (2) reductive dissolution of iron oxyhydroxides, (3) reduction and re-oxidation, and (4) competitive ion exchange. Arsenic generally occurs in four major oxidation states (+5, +3, 0 and −3) in the environments (Bhattacharya *et al.*, 2011; Ravenscroft *et al.*, 2009; Chakraborti *et al.*, 2001).

Arsenic forms in groundwater

In groundwater, As is identified in inorganic (arsenate ($H_2AsO_4^-$, pH < 7; HAO_4^{2-}, pH > 7) and arsenite (H_3AsO_3, pH < 9.2; $H_2AsO_3^-$)) and organic forms (methyl arsenic acid ($CH_3AsO[OH_2]$) and dimethyl arsenic acid ($(CH_3)_2AsO(OH)$) (Fazal *et al.* 2001). Arsenite is 60–80 times more toxic than arsenate due to a greater thermodynamic stability of the combination with the thiol (-SH) part of proteins (Fazal *et al.*, 2001). In the human body when arsenic (both As(III) and As(V)) in water is ingested, As(V) is reduced to As(III) by glutathione process so that –SH in enzyme, and proteins are blocked, resulting in arsenic toxicity. In other words, in the human body both As(III) and As(V) are equally toxic.

Arsenic forms in soils

Arsenic uptake by crops is regulated by soil arsenic content along with irrigation water. Globally, the baseline concentrations of As in uncontaminated soils are in the order of 5–10 mg/kg. The recommended upper limit of As in irrigation water is 100 µg/L (FAO, 1985). Arsenic is distributed in major soils and rocks and its concentration in rocks and in contaminated soils varies between 0.5 and 2.5 mg/kg and from 10 to 2,470 mg/kg, respectively (Hossain, 2006). As-contaminated irrigation water, mining, smelting, coal burning, wastes application, animal manures, arsenic-containing pesticides and herbicides enhance the As load in soils. Brammer and Ravenscroft (2009) calculated that if a soil irrigated with 1,000 mm water containing 100 µg/L As, the soil gets 1 kg of As/ha/year. Arsenic content in irrigation water does not reach the soil in irrigated land completely, as it is reduced during the flow in the irrigation channel (Roberts *et al.*, 2007). The variation in As in irrigation water during flow seems to be the adsorption of arsenic on Fe hydroxides, which is formed when the groundwater is exposed to aerobic condition (Brammer and Ravenscroft, 2009). Arsenic availability for plant uptake is regulated by several factors such as redox potential, pH, organic matter content, iron, manganese, phosphorus, calcium carbonate, plant species, soil microbes and the chemical form of arsenic and its content in the soil (Mahimairaja *et al.*, 2005).

Arsenic toxicities

Arsenic toxicity to humans varies with respect to several factors such as the chemical and physical form of the arsenic compound, dosage, duration of exposure, dietary compositions of interacting elements, age and sex of individuals, route by which it enters the body, nutritional status of the person, immunity level of the individual etc. Arsenic mostly affects the lungs and the skin, but also causes internal cancer in the liver, bladder, kidneys, lungs and skin (WHO, 2011). Based on toxicological study, Styblo *et al.* (2000) suggested that methylated trivalent arsenic is more toxic than inorganic arsenic. Earlier studies reported various ways of arsenic exposure, for example, inhalation, ingestion, and dermal contact (Mondal and Polya, 2008; Pal *et al.*, 2007; Rahman *et al.*, 2006b). Symptoms of acute toxicity caused by arsenic are vomiting, muscle cramps, colicky abdominal pain and tingling of the hands and feet. Long exposure to arsenic provokes arsenicosis, cardiovascular diseases, skin lesions etc. (Bissen and Frimmel, 2003). Groundwater in 15 districts in Bihar, 14 districts in West Bengal, 61 districts in Bangladesh and 20 districts in Nepal Terai have arsenic levels above drinking standards set by the WHO. This accounts for 75 per cent of the EGB's population living in the arsenic-affected regions.

Arsenic contamination in groundwater and soil in the EGB

Arsenic contamination in Bihar state, India

Arsenic in groundwater was first discovered in 2002 in two villages (Semariya-Ojhapatti and Doodhghat) in the Bhojpur district, which exceeded 50 µg/L (Chakraborti *et al.*, 2003). In Bihar, literature on As contamination of soil is not available at present. However, As contamination in groundwater has been studied extensively (Figure 12.2). Existing literature shows 15 districts located along the course of the Ganges are affected by arsenic contamination, and 10.4 million people (2001 census) are exposed to arsenic calamity. Maximum As concentration of 178 µg/L was found at depths between 20 and 35 m. Dug wells with shallow depth (< 8 m) and the aquifer below 80 m are free from As contamination. In Bihar, the As content in groundwater is largely varied with geological formations. The wells with high As content are located mostly at newer

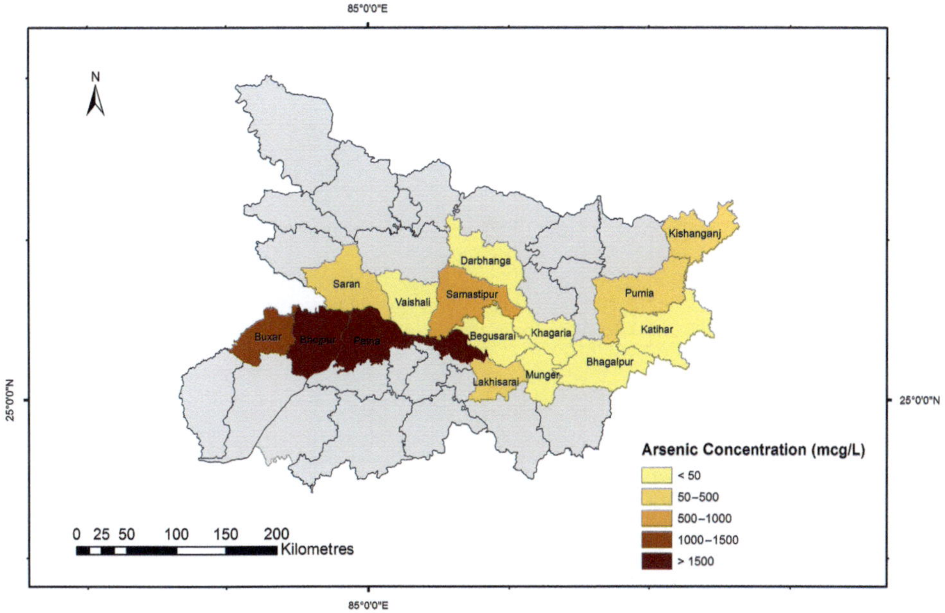

Figure 12.2 Arsenic concentration (μg/L) in groundwater in Bihar

alluvium (Bhojpur, Patna, Vaishali and Bhagalpur districts), along the course of the Ganges (Saha *et al.*, 2009; Saha *et al.*, 2010; Bhattacharya *et al.*, 2011).

Arsenic contamination in West Bengal state, India

Figure 12.3 shows the extent of arsenic-contaminated groundwater in West Bengal. Arsenic contamination in groundwater was first reported in West Bengal in 1978 (Dhar *et al.*, 1997). Bhattacharya *et al.* (2011) reported that the groundwater arsenic concentration is varied from <1 μg/L to 4,200 μg/L. The groundwater within 100 m depth, recharged within the last 50 years, is found to be arseniferous (Kunar *et al.*, 2009). Nine districts (Maldah, Murshidabad, Nadia, North-24-Parganas, South-24-Parganas, Bardhaman, Haora, Hugli and Kolkata) are severely affected and five districts (Koch Bihar, Jalpaiguri, Darjiling, North Dinajpur and South Dinajpur) are mildly affected. Groundwater arsenic in the other five districts (Bankura, Birbhum, Puruliya, Medinipur and West Medinipur) is low (<3 μg/L), which is within acceptable level. Total population in the nine severely affected districts and five mildly affected districts is 50.4 million and 11.4 million, respectively, in the 2001 census, and over 57.0 million and 13.2 million, respectively, in the 2011 census. According to the 2011 census, almost 77 per cent of the total population (91.35 million) of West Bengal is living in these 14 districts.

Bhattacharya *et al.* (2010a, 2010b, 2010c) reported that in the Nadia district, West Bengal, the total arsenic content in the soil is higher than the background value of arsenic in the soil, and the groundwater arsenic is strongly correlated with the arsenic in agricultural soil and plants. Norra *et al.* (2005) reported that As content in the top soil layers in paddy fields (38 mg/kg) irrigated with As-rich groundwater (500–800 μg/L) is five times higher than As content (7 mg/kg) in paddy soil irrigated with uncontaminated water. Furthermore, the arsenic in the paddy field decreased with depth (11 mg/kg at 1.0–1.1 m below ground level).

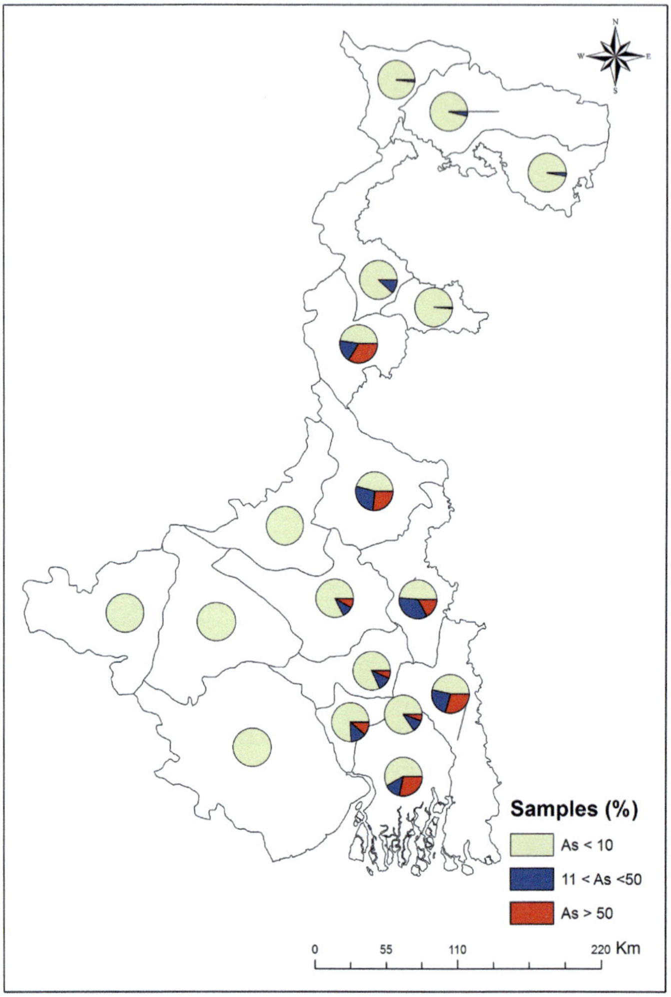

Figure 12.3 Arsenic concentration (μg/L) in groundwater in West Bengal

Arsenic contamination in Bangladesh

Figure 12.4 shows the extent of arsenic contamination in groundwater in Bangladeshi districts. Groundwater in 61 out of 65 districts in Bangaladesh is affected by arsenic contamination (BGS, 1999). Shallow tube wells with water table between 5 and 10 m below ground level are highly contaminated compared to deep aquifers (van Geen *et al.*, 2003). Duxbury and Zavala (2005) concluded that ten years of irrigation using As-contaminated water would add 5–10 mg of As/kg of soil. Dittmar *et al.* (2007) reported that arsenic concentrations remained unchanged at the start of two successive irrigation seasons, because arsenic accumulated during the first (*boro*) irrigation season was leached by floodwater during the following monsoon (*aman*) season. Yamazaki *et al.* (2003) reported that As concentration in the soil is dependent on the type of sediment. They reported that sandy sediments contained 3–7 mg/kg (median, 5 mg/kg), clay sediments contained 4–18 mg/kg (median, 9 mg/kg), while peat and peat-clay sediments

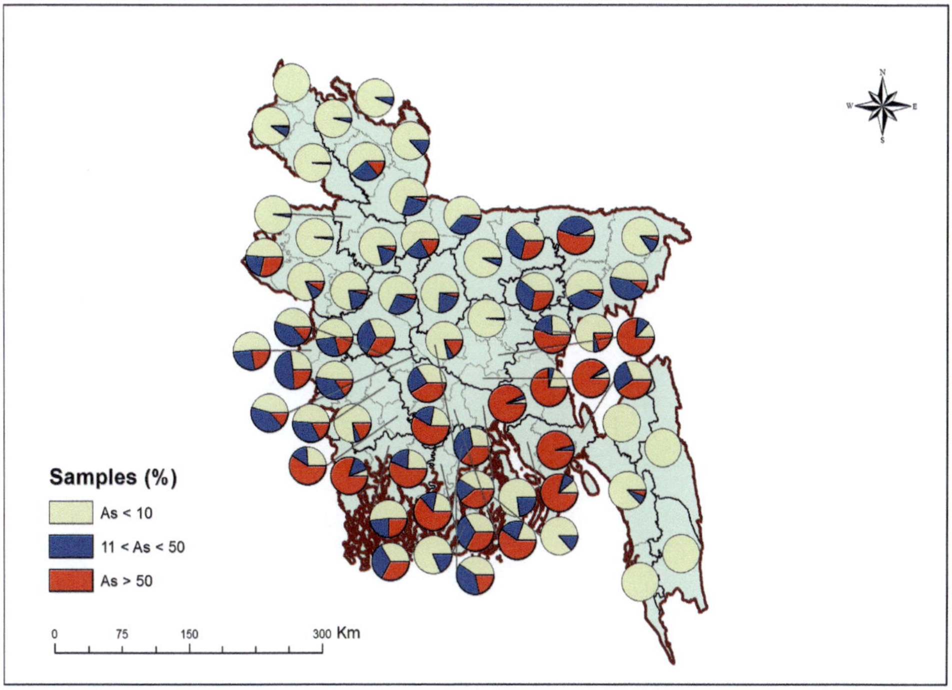

Figure 12.4 District-wise arsenic concentration (μg/L) in groundwater in Bangladesh

contained 20–111 mg/kg. Das *et al.* (2004) reported a positive correlation between As in groundwater from shallow tube wells and the soil. In summary, As in soil builds up with continuous irrigation of contaminated water, and the rate of build-up is higher in peaty-clay soils than in sandy soils.

Arsenic contamination in Nepal

Figure 12.5 shows the extent of arsenic concentration in groundwater in Nepali districts. In the Terai region of Nepal, 90 per cent of people depend on groundwater for their livelihoods. There are 20 districts and the groundwater in all these districts contains As>10 μg/L (Thakur *et al.*, 2011). As content in deeper tube wells was less than 10 μg/L (Smedley, 2005), while in tube wells at depth between 10 and 30 m it was higher than 10 μg/L (World Bank, 2005). Among the 20 districts, groundwater wells in the Rautahat, Nawalparasi, Kapilvastu and Banke districts are severely affected by As contamination. Approximately 7.4 per cent of tube wells exceeded As >50 μg/L; 16.3 per cent of tube wells contained 11–50 μg/L of As; 23.7 per cent of tube wells were above WHO guidelines (As >10 μg/L); and the majority of tube wells (76.3 per cent) were below WHO guidelines (Shrestha *et al.*, 2003b).

Gurung *et al.* (2005) reported an average arsenic concentration of 9 mg/kg in sediments, and stated that the distribution is not homogeneous. They observed that the arsenic enrichment in finer sediments (black clay, As = 31 mg/kg) is far greater than in coarse sediments (silt and fine sand, As = 3 mg/kg). Dahal *et al.* (2008b) concluded that continuous use of As-contaminated water for irrigation will add an average of 1.5 mg/kg/year of arsenic to agricultural soil. Furthermore, they insisted that the addition of arsenic may rise up to 10.19 mg/kg/year. Dahal *et al.* (2008a) reported that surface soils (<10 cm) had higher As than the deeper soils (10–20

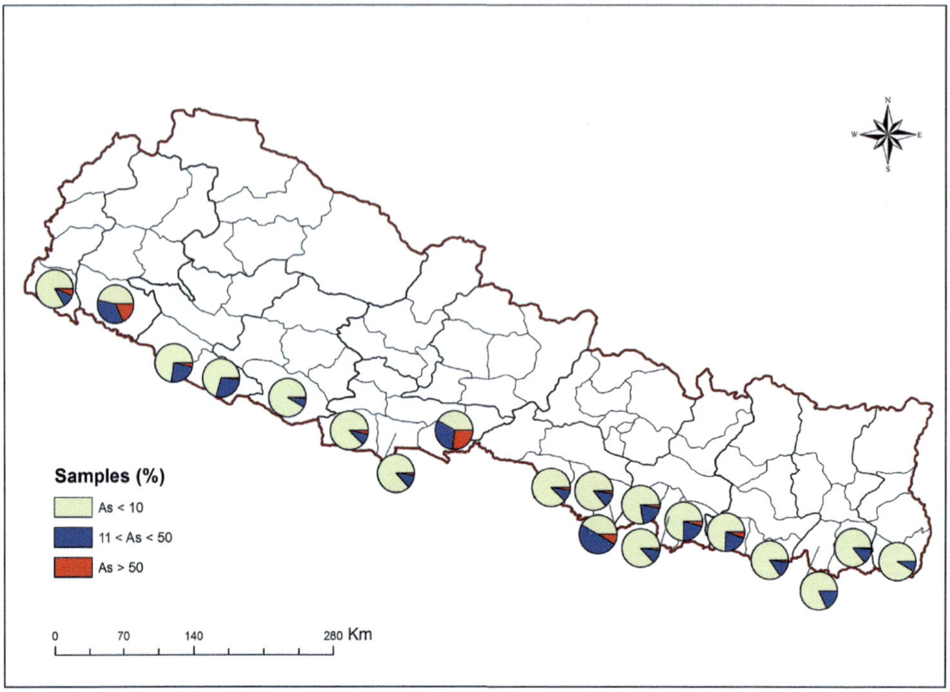

Figure 12.5 Arsenic concentration (μg/L) in groundwater in Nepal Terai districts

cm). In addition, they concluded that there is a strong correlation between irrigation water and soil arsenic contents. Observations in Nepal are consistent with observations made elsewhere in the EGB that shallow groundwater is more contaminated than deep groundwater, and shallow layers of soils have more As accumulation than deeper layers.

Impact on the food chain and human health

Enrichment of As in agricultural soils from contaminated irrigation water followed by As-uptake crops has been the cause of the increasing number of health issues in this region. Rice, being the crop of high water use, contains a high concentration of As compared to other crops (Das *et al.*, 2004). Arsenic in rice is mostly in an inorganic form. Rice cultivation is mostly carried out under flooded conditions (anaerobic), which enhances the availability and mobility of arsenite (As [III]), which is easily absorbed by the rice plant from the soil solution (Xu *et al.*, 2008). In contrast, in aerated soils, arsenic is present as As(V) and is unavailable for the plant due to adsorption followed by precipitation with Fe hydroxides. As-rich irrigation water reduces plant height, rice yield, the number of filled grains, grain weight and root biomass. Arsenic concentrations in root, straw and rice husk increased significantly when irrigated with As-rich water. Arsenic accumulates more in rice straw and roots, which may cause an indirect threat to human beings through cattle meat and milk (Abedin, 2002).

It is now evident that inorganic arsenic exposure deactivates the function of enzymes, some important anions, cations, transcriptional events in cells and causes other direct or indirect effects. Approximately 75 per cent of the population in the EGB are exposed to groundwater with As concentration greater than 100 μg/L. The WHO in 2001 estimated that between 35 and

77 million people in Bangladesh (30–60 per cent of the total population) are at risk of drinking contaminated water. Approximately 1 in 100 people who drink water containing 0.05 mg arsenic per litre or more for a long period may eventually die from arsenic-related cancers (WHO, 2001).

Most of the population affected by arsenic-related diseases are from poor socio-economic backgrounds and suffering from malnutrition. Infants and children may be uniquely susceptible to toxic exposures. The possibility of age-specific effects is compounded by, particularly, high intake of arsenic-contaminated drinking water; the daily intake, in volume per unit of body weight (mL per kg per day), of infants is three to four times more than that of the adult and the toxic risk should increase accordingly.

Arsenic ingestion through food in addition to water may cause further damage. What is of further concern is the fact that the harvested crops from these belts are sold off and uncontaminated regions get caught in the whirlpool as well. Animals, especially domestic animals, also become vulnerable as water and straw are laced with arsenic. People who consume the meat of animals infected with arsenic in turn ingest it.

Arsenicosis has an economic impact that feeds back through the health and human well-being of both the individual and the family (Hanchett, 2004; Roy, 2008). The people affected by arsenic poisoning suffer from health hazards and economic barriers, unhealthy and strained 'within' family relations and also confront severe mental stress at the social front.

There is a social cost of arsenic pollution. With the symptoms of arsenicosis becoming evident, more often various agonizing social issues arise, such as affected persons are sacked from their jobs, arsenic-affected children are sometimes debarred from school, the affected girls are denied marriage and often married women face problems in their marital life and even divorce ensues. Affected people are quite often debarred from social functions. The arsenic-affected persons are being boycotted socially, mistaking them as patients infected with contagious diseases, and their condition precludes the simplest social interactions and often leads to depression and suicide. There is even evidence that burial rites have been refused to some victims (Hassan *et al.*, 2005).

With arsenic poisoning setting in and spreading over gradually, the total agricultural hours would decrease. Thus, there would be an obvious drop in the income of the affected person or family. The affected persons gradually lose their work capability and mostly retire early due to physical illness. Also, illness forces the person out of work and many are caught in the debt trap of the local money lender who charges them exorbitant interest rates ranging between 5 and 10 per cent per month (Chowdhury *et al.*, 2001).

Bihar

Groundwater is the major source of water for irrigation in Bihar to grow rice, wheat, maize, pulses, vegetables, etc. Arsenic accumulation in these crops through contaminated groundwater will cause health issues. However, very few studies focused on this issue in Bihar. Only a few districts were surveyed for health risk and very few studies concentrated on As accumulation in the food chain. Singh and Ghosh (2011) collected soil, rice (Sonam Dhan), wheat, maize and lentil samples in the Patna district and reported the arsenic concentrations in wheat, maize, rice grain, husk and lentil. Chakraborti *et al.* (2003) carried out a medical examination of 550 people (390 adults and 160 children) in the Bhojpur district, and reported that 13 per cent of the adults and 6.3 per cent of the children had typical skin lesions. Singh and Ghosh (2012) surveyed 264 people and reported that 1.1 per cent, 5.6 per cent, 2.6 per cent and 0.4 per cent of the population had been affected by diarrhoea, gastric problems, body itching and pigmentation, respectively. In Haldichapra, 0.9 per cent, 5.8 per cent, 4 per cent and 2.3 per cent of the population were suffering from diarrhoea, gastric problems, body itching and pigmentation on their bodies, respectively. These studies demonstrate the range of diseases caused by arsenic, irrespective of age and gender.

West Bengal

Based on a review, Rahman and Hasegawa (2011) reported that the arsenic concentration in rice varied from 0.04 to 0.43 mg/kg in the arsenic-affected areas of some districts in West Bengal. Pal *et al.* (2009) studied the arsenic load in raw and cooked rice in the populations of arsenic-affected and unaffected areas and Kolkata City, and reported that there is a significant correlation between irrigation water and rice, and soil and rice for both *boro* (groundwater) and *aman* (rainwater) rice varieties. The average arsenic contents in *boro* and *aman* raw rice collected from the contaminated area are 0.25 mg/kg and 0.82 mg/kg, respectively, whereas in the uncontaminated area, they are 0.05 mg/kg and 0.04 mg/kg, respectively. Likewise, the mean arsenic content in the raw rice collected from Kolkata City is 0.14 mg/kg. These studies illustrate that plants absorb soil arsenic during growth and maturity.

Roychowdhury *et al.* (2002) investigated total arsenic food composites collected from the Murshidabad district in West Bengal, where groundwater is a major source for irrigated agriculture. In the study site the arsenic concentrations in the tube well water and soil were 85 μg/L and 11.35 mg/kg, respectively. They reported that the As content is generally high in the skin of most of the vegetables. Furthermore, high As was also observed in the cooked food stuff compared to the raw variety. Based on food stuff arsenic content, Roychowdhury *et al.* (2002) estimated the daily dietary intakes of As for adults (171.20 μg) and children (91.89 μg). Samal *et al.* (2003) analysed the arsenic content in crops and vegetables in Uttarpara and Doulatpur villages. They reported that the arsenic accumulation in arum is higher than in green gram, and it is present in a decreasing order in arum>rice>spinach>potatoes. In another study, Roychowdhury *et al.* (2008) evaluated the impact of sedimentary arsenic through irrigated groundwater on soil, plants, crops and human well-being in the Murshidabad and Nadia districts of West Bengal. Moreover, they reported that the daily dietary intake of inorganic arsenic by an adult from rice grain alone (2.32 μg/kg body weight/day) is higher than the WHO recommended levels.

Samal (2005) reported the arsenic concentration in rice (0.16–0.19 mg/kg) and crops (0.08–0.33 mg/kg) in the Haringhata Block of West Bengal. Roychowdhury *et al.* (2005) reported that the mean arsenic concentrations in surface, plants root, top soil (0–30 cm) and the root zone are 14.2 mg/kg, 13.7 mg/kg, 14.8 mg/kg and 14.2 mg/kg, respectively. The mean arsenic content in root, stem, leaf and all parts of plants are 1.0 mg/kg, 0.3 mg/kg, 0.25 mg/kg and 0.51 mg/kg, respectively. These results show only a fraction of soil arsenic is found in plants.

Bhattacharya *et al.* (2010a) reported that the total arsenic concentrations varied between < 0.0003 and 1.02 mg/kg in tubers and vegetables, and that potato had the highest arsenic concentration (0.65 mg/kg). They documented the mean arsenic content of other grains, tubers and vegetables as: *boro* rice grain (0.45 mg/kg), arum (0.41 mg/kg), amaranth (0.37 mg/kg), radish (0.34 mg/kg), *aman* rice grain (0.33 mg/kg), lady's fingers (0.30mg/kg), cauliflower (0.29 mg/kg), brinjals (0.28 mg/kg), wheat (0.134 mg/kg), garlic (0.13 mg/kg), lentil (0.1 mg/kg), beans (0.11 mg/kg), green chili (0.09 mg/kg), tomato (0.08 mg/kg), bitter-gourd (0.02 mg/kg), lemon (0.01mg/kg) and turmeric (0.003 mg/kg). Overall tubers and grains have higher concentration of arsenic than vegetables.

Bhattacharya *et al.* (2010a) reported that arsenic uptake in rice varied with the season (pre-monsoon > post-monsoon), and arsenic accumulation also varied with rice varieties, e.g. white mini kit (0.31±0.005 mg/kg), IR 50 (0.29±0.001 mg/kg) and Jaya rice variety (0.14±0.002 mg/kg). Besides, they found maximum arsenic accumulation in the straw (0.89±0.019 – 1.65±0.021 mg/kg) compared to the husk (0.31±0.011 – 0.85±0.016 mg/kg) and the grain (0.14±0.002 – 0.31±0.005 mg/kg).

In West Bengal, several studies have concentrated on human health due to arsenic contamination. Chowdhury *et al.* (2001) screened 86,000 people in 306 villages belonging to seven arsenic-affected districts in West Bengal. They reported that the mean arsenic content (μg/kg) in hair, nail, urine and skin scale was 1,480 (n=7,135), 4,560 (n=7,381), 180 (n=9,795) and 6,820 (n=165), respectively.

In the Jalangi Block, West Bengal, Rahman *et al.* (2005) reported that 88 per cent of biological samples (hairs, nails and urine) contained As levels above normal values, i.e. 0.08–0.25 mg/kg (hair); 0.43–1.08 mg/kg (nail); 0.005–0.04 mg/day/1.5l (urine). In addition, they reported that 21 per cent of the people were affected with skin lesions.

Uchino *et al.* (2006) evaluated the main intake source of arsenic by the villagers by eliciting information from arsenic-affected families in the Mushidabad district, West Bengal, India. They reported that the drinking water contributed 6.07 per cent, 26.7 per cent and 58.1 per cent of the total arsenic intake of mild, moderate and high arsenic-affected families, respectively. They also reported that As-contaminated food composites are a major source of As intake for mild and moderate As-affected families. Roychowdhury *et al.* (2002) concluded that arsenic intake of adults and children (10-year-olds) via foods was approximately 180 and 97 μg/d, respectively. The intake through drinking water is 400 and 200 μg/d for adults and children, respectively.

Mondal and Polya (2008) reported that the contribution of drinking water, rice and cooked rice for cancer risk is 48 per cent, 44 per cent and 8 per cent, respectively. Similarly, Samal *et al.* (2011) reported that the total intake of As from foodstuffs by adults and children is 560 μg/day and 393 μg/day, respectively, in the Nadia district. They estimated the As in urine to vary between 154 and 276 μg/L, and also highlighted the potential risk of As exposure to local inhabitants through the food chain.

Bangladesh

Arsenic concentration varied from <0.01 to 5mg/kg in plants and highest concentrations were recorded in old leaves and roots (Pendias and Pendias, 1992). In addition, leafy vegetables contain higher concentrations compared to fruits (Hossain, 2006). Das *et al.* (2004) reported that arsenic levels in vegetables like kachu sak (*colocasia antiquorum*), potato (*solanum tuberosum*) and kalmi sak (*ipomea reptans*) exceeded the food safety limits of 1.0 mg/kg. Based on a study of 37 vegetables, pulses and spices, Williams *et al.* (2006) reported maximum arsenic content in radish leaves (0.79 mg/kg), arum stolons, spinach and cucumber; and a lowest of 0.2 mg/kg in most fruits, vegetables and spices. Farid *et al.* (2003) carried out a detailed study on arsenic-contaminated irrigation water and its effect on vegetables in Bangladesh, both in arsenic-contaminated and non-contaminated areas, and reported the arsenic contents of vegetables is in the order of: potato (0.103 mg/kg) > brinjal (0.049 mg/kg) > cauliflower (0.011 mg/kg).

In Bangladesh, most arsenic found in rice is inorganic and cooked rice possesses 10 to 35 per cent higher arsenic content compared to raw rice. Meharg and Rahman (2003) reported that the arsenic content in rice grain collected from the arsenic-contaminated western part of Bangladesh varies from 0.03 to 1.84 mg/kg. Williams *et al.* (2006) documented that arsenic level in the rice grains ranged from 0.04 to 0.92 mg/kg and 0.04 to 0.91 mg/kg during *aman* and *boro* seasons, respectively. Islam *et al.* (2004) reported that the arsenic in the *boro* rice collected from the Gopalganj, Rajbari and Faridpur districts in Bangladesh is 0.05–2.05 mg/kg. In the Satkhira district, the arsenic content in raw rice collected from a highly arsenic-contaminated area is 0.57–0.69 mg/kg (Rahman *et al.*, 2006a). According to Rahman *et al.* (2007), the arsenic concentration in rice fractions is in the order of husk > bran > whole rice > polished rice.

In many areas in Bangladesh, high arsenic is encountered in food composites due to groundwater usage, and approximately 35 million people are exposed to As poisoning, while another 80 million are at risk (Safiuddin *et al.*, 2011). In 2000, the School of Environmental Studies (SOES) and the Dhaka Community Hospital collected more than 10,000 hair, nail, urine and skin-scale samples from 210 As-affected villages and found that 83.2 per cent of hair samples, 93.8 per cent of nail samples, 95.1 per cent of urine samples and 97.4 per cent of skin-scale samples exceeded toxic levels (Arnold *et al.*, 1990; Farmer and Johnson, 1990). Chowdhury *et al.* (2000) collected hair (3,332 samples), nail (3,321 samples), urine (1,043 samples) and skin-scale (373 samples) from As-affected villages and reported that 81, 94 and 95 per cent of hair, nail and urine samples exceeded acceptable levels. People are mostly suffering from arsenicosis[1] and skin cancer (Safiuddin *et al.*, 2011). In addition, a higher rate of fetal loss and infant deaths is also reported from areas where people consume high As through groundwater (Sohel *et al.*, 2010). Rahman *et al.* (2006b) reported that males were more susceptible to developing skin lesions than females when exposed to As contamination.

Nepal

Arsenic content in the food chain was estimated by Dahal *et al.* (2008a) in the Nawalparasi district, where the As content in rice and vegetables was estimated as: rice (0.06–22.20 mg/kg); potato (<0.01–0.79 mg/kg); cauliflower (0.09–3.45 mg/kg); onion (<0.01–1.20 mg/kg); and brinjal (<0.01–1.14 mg/kg). They documented that the As content in different parts of rice and vegetables decreases in the order of roots > shoots > leaves > grains (without husk)/fruits.

Between 2.5 and 3.5 million people are exposed to As levels between 10 and 50 μg/L (Pokharel *et al.*, 2009; Shrestha *et al.*, 2003a). Shrestha *et al.* (2003a) identified arsenicosis-related dermatosis (skin disease) in 1.3 to 5.1 per cent of the population. Most of them have symptoms of melanosis and keratosis on the palms, torso and soles of the feet. Suspected arsenicosis patients were identified in the villages of Tilakpur and Thulokunuwar in the Nawalparasi district, where the groundwater As concentration is very high (271 and 473 μg/L, respectively). Arsenic content in hair samples collected from As-affected areas indicated that 95 per cent of samples exceeded acceptable levels (0.080–0.250 mg/kg), whereas 62 per cent of samples exceeded the acute toxicity levels (1 mg/kg).

Ahmad *et al.* (2004) conducted clinical observations in Goini and Thulo Kunwar villages in the Nawalparasi district and identified melanosis (95.6 per cent of patients), keratosis (57.8 per cent of patients) and leuco-melanosis (3.3 per cent of patients) in these villages. Maharjan *et al.* (2006) reported that the prevalence of arsenicosis ranged from 0.7 per cent in the Kapilavastu district to 3.6 per cent in the Nawalparasi district with an average of 2.2 per cent. In addition, they reported that the highest prevalence (18.6 per cent) of arsenicosis was found in the Patkhouli village of Nawalparasi, where 95.8 per cent of tube wells were contaminated with As.

Overall, arsenic load in the food chain and its impact on human health in Nepal Terai is not understood properly due to lack of data and studies. These studies confirm observations made in Bihar, West Bengal and Bangladesh that As contamination has led to a range of skin and cancer diseases, and in some villages in Nepal almost everyone has been affected.

Discussion

Chakraborti *et al.* (2010) and Chakraborti *et al.* (2015) provide extensive reviews of studies on arsenic contamination of groundwater in Bangladesh. Only a limited amount of research related to As contamination in Bihar and Nepal has been reported.

Although numerous studies have been carried out to understand As contamination in groundwater, soil and crops in the EGB, long-term studies are rare. There are numerous groundwater wells that remain to be tested in order to determine the magnitude of the problem in the EGB. Long-term monitoring will help to understand temporal variation in As contamination.

Despite the plethora of studies, there is still a debate to identify the exact processes linked to As contamination. A comprehensive method should be developed to conduct a complete and thorough evaluation of As contamination in soil, water, foods (grains, vegetables, fruits, etc) and its impact on human health. Such studies should include standardized As testing methods. Some studies recommended alternative methods or potential water sources to tackle As issues in the EGB (Adeel, 2002; Safiuddin *et al.*, 2011). Some strategies to mitigate the health impacts of As are:

1 Noting that the depths at which As occurs vary across the EGB, the installation of pumps at As-free depths is highly recommended for domestic purposes.
2 Usage of rain water for drinking and domestic usage instead of groundwater. Implement rainwater harvesting programmes – roof top rainwater harvesting – at household level.
3 Crops that require less water should be grown in the affected area. Alternate rice water management strategies need to be promoted to minimize anaerobic conditions.
4 Distribution of treated groundwater and surface water through pipe lines.
5 Encourage community participation and awareness campaigns in the As-affected areas to manage the As crisis.

Conclusions

Arsenic contamination is a major issue in districts bordering the Ganges, where 75 per cent of the total population lives. It is widely accepted that As migrated with fluvial sediments from the Himalayas to the EGB. As levels widely exceed acceptable limits in soils, groundwater, food and human organs. Groundwater in 15 districts in Bihar, 14 districts in West Bengal, 61 districts in Bangladesh and 20 districts in Nepal Terai is contaminated (As > 10 μg/L). The link between arsenic in irrigation water, in soils and in crops is unmistakably evident. There is substantial evidence that food and forage grown in these districts have As in excess of acceptable limits. Arsenic uptake in rice varied with the season (pre-monsoon > post-monsoon), and arsenic accumulation varied with rice varieties. This is possibly due to rice being cultivated under flooded conditions, which create anaerobic conditions conducive for the absorption. The arsenic concentration in rice fractions is in the order of husk>bran>whole rice>polished rice. Alternate rice water management strategies need to be promoted to minimize anaerobic conditions.

In several studies, it has been noted that daily dietary intake of inorganic arsenic by an adult from rice grain itself (2.32 μg/kg body weight) is higher than the WHO recommended level. As content is generally high in the skin of most vegetables. High As was observed in cooked foodstuff compared to the raw variety. Leafy vegetables contain higher concentrations of arsenic compared to fruits. Inhabitants in As-contaminated districts are often affected by arsenicosis, a common term to describe a range of illnesses. Often arsenic content in hair, nail, skin and urine samples exceeded acceptable limits.

Although numerous studies have been carried out to document As contamination in groundwater, soil and crops in the EGB, long-term studies are rare. Long-term monitoring will help to understand temporal variations in As contamination.

Note

1 Drinking water rich in arsenic over a long period leads to arsenic poisoning or arsenicosis. This term is collectively used to refer to a range of skin diseases, such as dermatosis, melanosis, keratosis and leuco-melanosis.

References

Abedin, M.J. 2002. *Arsenic accumulation in paddy soils, grain and straw*. Ph.D. Thesis. Department of Plant and Soil Science, University of Aberdeen, Scotland, UK.

Acharyya, S.K., Shah, B.A., Ashyiya, I.D., and Pandey, Y. 2005. Arsenic contamination in groundwater from parts of Ambagarh-Chowki block, Chhattisgarh, India: Source and release mechanism. *Environmental Geology* 49(1): 148–58.

Adeel, Z. 2002. The disaster of arsenic poisoning of groundwaters in South Asia – A focus on research needs and UNU's role. *Global Environmental Change* 12: 69–72.

Ahmad, S.A., Maharjan, M., Watanabe, C. and Ohtsuka, R. 2004. Arsenicosis in two villages in Terai lowland Nepal. *Environ Sci* 11(3): 179–88.

Arnold, H.L., Odam, R.B., and James, W.D. 1990. *Diseases of the Skin: Clinical Dermatology*, 8th edn. Philadelphia, PA: WB Sanders.

BGS (British Geological Survey). 1999. *Groundwater Studies for Arsenic Contamination in Bangladesh*. Main Report and Supplemental Volumes 1–3, Government of the Peoples Republic of Bangladesh, Ministry of Local Government, Rural Development and Cooperatives, Department of Public Health Engineering, Dhaka, Bangladesh and Mott MacDonald International Ltd., UK. Available at www.bgs.ac.uk/research/groundwater/health/arsenic/Bangladesh/reports.html (accessed in February 2014).

Bhattacharya, P., Samal, A.C., Majumdar, J., and Santra, S.C. 2010a. Arsenic contamination in rice, wheat, pulses, and vegetables: A study in an arsenic-affected area of West Bengal, India. *Water Air Soil Pollution* 213: 3–13.

Bhattacharya, P., Samal, A.C., Majumdar, J. and Santra, S.C. 2010b. Accumulation of arsenic and its distribution in rice plant (*Oryza sativa L.*) in Gangetic West Bengal, India. *Paddy and Water Environment* 8: 63–70.

Bissen, M. and Frimmel, F.H. 2003. Arsenic – A review. Part I: Occurrence, toxicity, speciation, mobility. *Acta Hydrochim Hydrobiol* 31: 9–18.

Brammer, H. and Ravenscroft, P. 2009. Arsenic in groundwater: A threat to sustainable agriculture in South and South-east Asia. *Environ Int* 35: 647–54.

Chakraborti, D., Ghorai, S.K., Bhaskar, D., Pal. A., Nayak, B. and Shah, B.A. 2009. Arsenic exposure through groundwater to the rural and urban population in the Allahabad-Kanpur track in the upper Ganga plain. *J. Environ. Monit.* 11: 1455–1459.

Chakraborti, D., Mukherjee, S.C., Pati, S., Sengupta, M.K., Rahman, M.M., Chowdhury, U.K., Lodh, D., Chanda, C.R. and Chakraborty, A.K. 2003. Arsenic groundwater contamination in Middle Ganga Plain, Bihar, India: A future Danger? *Environmental Health Perspective* 111: 1194–1200.

Chakraborti, D., Rahman, M.M., Das, B., Murrill, M., Dey, S., Mukherjee, S.C., Dhar, R.K., Biswas, B.K., Chowdhury, U.K., Roy, S., Sorif, S., Selim, M., Rahman, M. and Quamruzzaman, Q. 2010. Status of groundwater arsenic contamination in Bangladesh: A 14-year study report. *Water Research* 44: 5789–5802.

Chakraborti, D., Rahman, M.M., Mukherjee, A., Alauddin, M., Hassan, M., Dutta, R.N., Pati, S. Mukherjee, S.C., Roy, S., Quamruzzman, Q., Rahman, M., Morshed, M., Islam, T. Sorif, S., Selim, M., Islam, M.R. and Hossain, M. 2015. Groundwater arsenic contamination in Bangladesh – 21 years of research. *Journal of Trace Elements in Medicine and Biology* 31: 237–48.

Chowdhury, U.K., Biswas, B.K.,Chowdhury, T.R.,Samanta, G. and Mandal, B.K. 2000. Groundwater arsenic contamination in Bangladesh and West Bengal, India. *Environmental Health Perspectives* 108: 393–97.

Chowdhury, U.K., Rahman, M.M., Mandal, B.K., Paul, K., Lodh, D. and Biswas, B.K. 2001. Groundwater arsenic contamination and human suffering in West Bengal, India and Bangladesh. *Environmental Sciences* 8: 393–415.

Dahal, B.M., Fuerhacker, M., Mentler, A., Shrestha, R.R. and Blum, W.E.H. 2008. Screening of arsenic in irrigation water used for vegetable production in Nepal. *Archives of Agronomy and Soil Science* 54(1): 41–51.

Dahal, B.M., Fuerhacker, M., Mentler, A., Karki, K.B., Shrestha, R.R. and Blum, W.E.H. 2008. Arsenic contamination of soils and agricultural plants through irrigation water in Nepal. *Environ Pollution* 155(1): 157–63.

Das, H.K., Mitra, A.K., Sengupta, P.K., Hossain, A., Islam, F. and Rabbani, G.H. 2004. Arsenic concentrations in rice, vegetables, and fish in Bangladesh: A preliminary study. *Environ Int* 30: 383–87.

Dhar, R., Biswas, B., Samanta, G., Mandal, B., Chakraborti, D., Roy, S., Jafar, A., Islam, A., Ara, G., Kabir, S., Khan, A.W., Ahmed, S.A. and Hadi, S.A. (1997). Groundwater arsenic calamity in Bangladesh. *Current Science* 73(1): 48–59.

Dittmar, J., Voegelin, A., Roberts, L.C., Hug, S.J., Saha, G.C., and Ali, M.A. 2007. Spatial distribution and temporal variability of arsenic in irrigated rice fields in Bangladesh. 2. Paddy soil. *Environ Sci Technol* 41: 5967–5972.

Duxbury, J.M. and Zavala, Y.J. 2005. *What are safe levels of arsenic in food and soils?* CIMMYT/USGS. Symposium on the Behaviour of Arsenic in Aquifers, Soils and Plants: Implications for Management, Dhaka, 16–18 January. Centro Internacional de Mejoramiento de Maíz y Trigo and the U.S. Geological Survey.

EU (European Union). 1998. Council directive 98/83/CE of 3 November 1998 on the quality of water intended for human consumption. *Off J Eur Union L* 330: 32–54.

FAO (Food and Agriculture Organization). 1985. Water quality guidelines for maximum crop production. Food and Agriculture Organization/UN. Available at www.fao.org/docrep/T0551E. 2006/9/13 (accessed in February 2014).

Farmer, J.G. and Johnson, L.R. 1990. Assessment of occupational exposure to inorganic arsenic based on urinary concentrations and speciation of arsenic. *Br. J. Ind. Med.* 47(5): 342–48.

Fazal, M.A., Kawachi, T. and Ichion, E. 2001. Extent and severity of groundwater arsenic contamination in Bangladesh. *Water Int.* 26(3): 370–79.

Fitz, W.J. and Wenzel, W.W. 2002. Arsenic transformations in the soil-rhizosphereplant system: Fundamentals and potential application to phytoremediation. *J Biotechnol* 99: 259–78.

Gurung, J.K., Hiroaki, I. and Khadka, M.S. 2005. Geological and geochemical examination of arsenic contamination in groundwater in the Holocene Terai Basin, Nepal. *Environ Geol* 49(1): 98–113.

Gurung, J.K., Ishiga, H., Khadka, M.S. and Shrestha, N.R. 2007. The geochemical study of fluvio-lacustrine aquifers in the Kathmandu Basin (Nepal) and the implications for the mobilization of arsenic. *Environmental Geology* 52(3): 503–17.

Hanchett, S. (2004). *Social aspects of the arsenic contamination of drinking water: A review of knowledge and practice in Bangladesh and West Bengal.* Dhaka: Arsenic Policy Support Unit, Government of Bangladesh.

Hossain, M.F. 2006. Arsenic contamination in Bangladesh – An overview. *Agriculture, Ecosystems and Environment* 113: 1–16.

Hassan, M.M., Atkins, P.J. and Dunn, C.E. (2005). Social implications of arsenic poisoning in Bangladesh. *Social Science & Medicine*, 61(10): 2201–2211.

Islam, M., Jahiruddin, M. and Islam, S. 2004. Assessment of arsenic in the water-soil-plant systems in Gangetic floodplains of Bangladesh. *Asian J Plant Sci* 3: 489–93.

Kunar, S., Jain, S.K., Shekhar, S. and Sharma, V. 2009. Arsenic in groundwater in India: An overview. *Bhu-Jal News*, 24(2&3): 1–9.

Maharjan, M., Roshan, R., Shrestha, R., Ahmad, S.K.A., Watanabe, C. and Ohtsuka, R. 2006. Prevalence of arsenicosis in Terai, Nepal. *J Health Popul Nutr* 24(2): 246–52.

Mahimairaja, S., Bolan, N.S., Adriano, D.C. and Robinson, B. 2005. Arsenic contamination and its risk management in complex environmental settings. *Adv Agron* 86: 1–82.

Mallick, S. and Rajagopal, N.R. 1996. Groundwater development in the arsenic-affected alluvial belt of West Bengal – Some questions. *Current Science* 70(11): 956–8.

Mandal, B.K. and Suzuki, K.T. 2002. Arsenic round the world: A review. *Talanta* 58: 201–35.

Meharg, A.A. and Rahman, M. 2003. Arsenic contamination of Bangladesh paddy field soils: Implications for rice contribution to arsenic consumption. *Environ Sci Technol* 37: 229–34.

Meharg, A.A., Williams, P.N., Adomako, E., Lawgali, Y.Y., Deacon, C. and Villada, A. 2009. Geographical variation in total and inorganic arsenic content of polished (white) rice. *Environ Sci Technol* 43: 1612–1617.

Mondal, D. and Polya, D.A. 2008. Rice is a major exposure route for arsenic in Chakdaha block, Nadiadistrict, West Bengal, India: a probabilistic risk Assessment. *Appl Geochem* 23: 2987–2998.

Norra, S., Berner, Z.A., Agarwala, P., Wagner, F., Chandrasekharam, D. and Stuben, D. 2005. Impact of irrigation with as rich groundwater on soil and crops: A geochemical case study in West Bengal Delta Plain, India. *Appl Geochem* 20: 1890–1906.

Pal, A., Nayak, B., Das, B., Hossain, M.A., Ahameda, S. and Chakraborti, D. 2007. Additional danger of arsenic exposure through inhalation from burning of cow dung cakes laced with arsenic As a fuel in arsenic-affected villages in Ganga–Meghna–Brahmaputra plain. *J Environ Monit* 9: 1067–1070.

Pal, A., Chowdhury, U.K., Mondal, D., Das, B., Nayak, B., Ghosh, A., Maity, S. and Chakraborti, D. 2009. Arsenic burden from cooked rice in the populations of arsenic-affected and non-affected areas and Kolkata city in West Bengal, India. *Environmental Science & Technology*, 43: 3349–3355.

Pendias, K.A. and Pendias, H. 1992. *Trace Elements in Soils and Plants.* 2nd edn. CRC Press, Boca Raton, Ann Arbor, London, pp. 203–9.

Pokharel, D., Bhandari, B.S. and Viraraghavan, T. 2009. Arsenic contamination of groundwater in the Terai region of Nepal: An overview of health concerns and treatment options. *Environment International* 35: 157–61.

Rahman, M.A. and Hasegawa, H. 2011. High levels of inorganic arsenic in rice in areas where arsenic-contaminated water is used for irrigation and cooking. *Sci. Total Environ.* 409(22): 4645–4655.

Rahman, M.A., Hasegawa, H., Rahman, M.A., Rahman, M.M. and Miah, M.A.M. 2006. Influence of cooking method on arsenic retention in cooked rice related to dietary exposure. *Sci Total Environ.* 370: 51–60.

Rahman, M.A., Hasegawa, H., Rahman, M.M., Rahman, M.A. and Miah, M.A.M. 2007. Accumulation of arsenic in tissues of rice plant (*Oryza sativa L.*) and its distribution in fractions ofrice grain. *Chemosphere* 69: 942–8.

Rahman, M.M., Sengupta, M.K., Ahamed, S., Chowdhury, U.K., Hossain, M.A. and Das, B. 2005. The magnitude of arsenic contamination in groundwater and its health effects to the inhabitants of the Jalangi – One of the 85 arsenic-affected blocks in West Bengal, India. *Sci Total Environ* 338: 189–200.

Rahman, M., Vahter, M., Sohel, N., Yunus, M., Wahed, M.A. and Streatfield, P.K. 2006. Arsenic exposure and age- and sex-specific risk for skin lesions: A population-based case-referent study in Bangladesh. *Environ Health Perspect* 114: 1847–1852.

Rajmohan, N. and Prathapar, S.A. 2014. Extent of arsenic contamination and its impact on the food chain and human health in the eastern Ganges Basin: A review. *IWMI Working Paper* (161).

Ravenscroft, P., Brammer, H. and Richards, K.S. 2009. *Arsenic pollution: A global synthesis.* UK: Wiley-Blackwell.

Roberts, L.C., Hug, S.I., Dittmar, J., Voegelin, A., Saha, G.C. and Ali, M.A. 2007. Spatial distribution and temporal variability of arsenic in irrigated rice fields in Bangladesh. 1. Irrigation water. *Environ Sci Technol* 41(17): 5960–5966.

Roy, J. (2008). Economic benefits of arsenic removal from groundwater – A case study from West Bengal. *India. Sci. Total Environment* 397: 1–3.

Roychowdhury, T. 2008. Impact of sedimentary arsenic through irrigated groundwater on soil, plants, crops and human continuum from Bengal delta: Special reference to raw and cooked rice. *Food and Chemical Toxicology* 46: 2856–2864.

Roychowdhury, T., Uchino, T., Tokunaga, H. and Ando, M. 2002. Survey of arsenic in food composites from an arsenic-affected area of West Bengal, India. *Food Chem Toxicol* 40: 1611–1621.

Roychowdhury, T., Tokunaga, H., Uchino, T. and Ando, M. 2005. Effect of arsenic-contaminated irrigation water on agricultural land, soil and plants in West Bengal, India. *Chemosphere*, 58: 799–810.

Safiuddin, M.D., Shirazi, S.M. and Yusoff, S. 2011. Arsenic contamination of groundwater in Bangladesh: A review. *International Journal of the Physical Sciences* 6(30): 6791–6800.

Saha, D., Dwivedi, S. N. and Sahu, S. 2009. Arsenic in groundwater in parts of middle Ganga plain in Bihar – An appraisal. *Bhu-Jal News*, 24(2–3): 82–94.

Saha, D., Sarangam, S.S., Dwivedi, S.N. and Bhartariya, K.G. 2010. Evaluation of hydrogeochemical processes in arsenic-contaminated alluvial aquifers in parts of Mid-Ganga Basin, Bihar, Eastern India. *Environmental Earth Sciences* 61(4): 799–811.

Samal, A.C. 2005. *An investigation on Accumulation of Arsenic in Ecosystem of Gangetic West Bengal and Assessment of Potential Health Risk*, Ph.D. Thesis, University of Kalyani, West Bengal, India.

Samal, A.C., Kar, S., Bhattacharya, P. and Santra, S.C. 2011. Human exposure to arsenic through foodstuffs cultivated using arsenic contaminated groundwater in areas of West Bengal, India. *Journal of Environmental Science and Health, Part A* 46: 1259–1265.

Samal, A. C., Chakraborty, S., Choudhuri, P., Kar, S. and Santra, S. C. 2003. Groundwater arsenic and its impact on agriculture and health: A case study in West Bengal. In Santra, S.C. (ed.). *Proceeding of National Conference On Recent Environmental Changes – Its Impact On Health, Agriculture and Ecosystem*, 6–7 August, University of Kalyani, West Bengal.

Shrestha, B. R., Whitney, J. W. and Shrestha, K. B. (eds) 2003. *The State of Arsenic in Nepal*. National Arsenic Steering Committee, Environment and Public Health Organization, Kathmandu, Nepal.

Shrestha, R.R., Shrestha, M.P., Upadhyay, N.P., Pradhan, R., Khadka, R. and Maskey, A. 2003. Groundwater arsenic contamination, its health impact and mitigation program in Nepal. *J Environ Sci Health* A38(1): 185–200.

Singh, S.K. and Ghosh, A.K. 2011. Entry of arsenic into food material – A Case Study. *World Applied Sciences Journal* 13(2): 385–90.

Smedley, P. 2005. Arsenic occurrence in groundwater in south and east Asia: Scale, causes and mitigation. Towards a more effective operational response volume I, environment and social unit South Asian Region, Water and Sanitation Program-WSP. World Bank, pp. 17–80.

Sohel, N., Vahter, M., Ali, M., Rahman, M., Rahman, A., Streatfield, P.K., Kanaroglou, P.S. and Persson, L.A. 2010. Spatial patterns of fetal loss and infant death in an arsenic-affected area in Bangladesh (Paper No.53). *Int. J. Health Geogr.* 9: 53.

Styblo, M., Razo, L.M.D., Vega, L., Germolec, D.R., LeClyse, E.L., Hamilton, G.A., Reed, W., Wang, C., Cullen, W.R., Thomas, D.J. 2000. Comparative toxicity of trivalent and pentavalent inorganic and methylated arsenicals in rat and human cells. *Arch. Toxicol.* 74(6): 289–99.

Sun, G., Li, X., Pi, J., Sun, Y., Li, B. and Jin, Y. 2006. Current research problems of chronic arsenicosisin China. *J Health Popul Nutr* 24: 176–81.

Thakur, J.K., Thakur, R.K., Ramanathan, A.L., Kumar, M. and Singh, S.K. 2011. Arsenic contamination of groundwater in Nepal – An overview. *Water* 3: 1–20.

Uchino, T., Roychowdhury, T., Ando, M. and Tokunaga, H. 2006. Intake of arsenic from water, food composites and excretion through urine, hair from a studied population in West Bengal, India. *Food Chem Toxicol* 44(4): 455–61.

US Environmental Protection Agency. 2002. Drinking water regulations for arsenic and clarifications to compliance and new source contaminant monitoring. EPA-816-K-02–018, August 2002. Available at www.epa.gov/safewater/arsenic/index.html (accessed in February 2014).

van Geen, A., Zheng, Y., Versteeg, R., Stute, M., Horneman, A., Dhar, R., Steckler, M., Gelman, A., Small, C., Ahsan, H., Graziano, J.H., Hussain, I. and Ahmed, K.M. 2003. Spatial variability of arsenic in 6000 tube wells in a 25 km^2 area of Bangladesh. *Water Resour. Res.* 39(5): 1140.

WHO (World Health Organization). 2004. *Guidelines for Drinking-water Quality*, vol 1, 3rd edn. Recommendations WHO, Geneva.

WHO (World Health Organization). 2011. *Arsenic in Drinking-water*. Background document for development of WHO Guidelines for Drinking-water Quality. WHO/SDE/WSH/03.04/75/Rev/1.

Williams, P.N., Islam, M.R., Adomako, E.E., Raab, A., Hossain, S.A. and Zhu, Y.G. 2006. Increase in rice grain arsenic for regions of Bangladesh irrigating paddies with elevated arsenic in groundwater. *Environ Sci Technol* 40: 4903–4908.

World Bank. 2005. *Towards a more effective operational response. Arsenic contamination of groundwater in South and East Asian countries*. Volume II – Technical Report, Water and Sanitation Program, South Asia Region.

Xu, X.Y., McGrath, S.P., Meharg, A.A. and Zhao, F.J. 2008. Growing rice aerobically markedly decreased arsenic accumulation. *Environ Sci Technol* 42: 5574–5579.

Yamazaki, C., Ishiga, H., Ahmed, F., Itoh, K., Suyama, K. and Yamamoto, H. 2003. Vertical distribution of arsenic in Ganges delta sediments in Deuli Village, Bangladesh. *Soil Sci Plant Nutr* 49: 567–74.

13

Ganga water quality

Dirty past, promising future?

Javier Mateo-Sagasta and Vinod Tare

Key messages

- Despite previous efforts to clean the river the main Ganga stream still directly receives at least 2.7 billion liters of sewage from medium and big cities every day, of which at least 74 per cent is untreated.
- Industrial effluents are in the range of 10–20 per cent of the total volume of wastewater directly reaching the Ganga. Although a relatively low proportion, they are a cause for major concern because they are often toxic and non-biodegradable.
- The river is also impacted by non-point source pollution resulting from uncontrolled faecal sludge disposal, open defecation and agricultural pollution including livestock. Nevertheless, the contribution from these sources to the river pollution has not been quantified.

- With an ever increasing amount of pollution loads the river remains heavily polluted. Faecal contamination is a case in point as coliform levels are high all along the river and make the water generally unsuitable for traditional bathing and drinking purposes, with the exception of a few upstream locations.
- In the Ganga River Basin there are not complete quantifications of the impacts from water quality degradation and the economic value of these impacts, but the few partial assessments that do exist suggest that the impacts of water pollution are in the billions of dollars every year.
- Addressing the formidable and multidimensional challenges that the Ganga is facing will require new knowledge, good governance, capacity development, massive financial resources, appropriate technologies, strong political will, partnerships and stakeholder engagement. Future solutions need to be comprehensive and innovative, ambitious but realistic, and learn from the strengths and weaknesses of previous efforts to clean the river, and from other national and international experiences of river restoration.

Introduction

Since time immemorial, the River Ganga has been revered by Indians and people from all over the world, who come with a belief that taking a dip in her holy waters will rid them of their sins. Ganga is not only a holy river but a lifeline for the 600 million people living in her basin, who rely on her waters for irrigating, drinking, generating energy and manufacturing. With time and because of unsustainable urbanization, industrialization and agriculture intensification, the holy Ganga has become heavily polluted. All forms of pollutants including discarded offerings used in various religious acts are released into the river, a river that in many stretches has very reduced flows with extremely low capacity to carry the pollution loads received. The resulting water quality degradation is having significant impacts on human health and ecosystems, and is limiting the development of the whole basin. It is therefore of utmost importance to take urgent action to prevent pollution loads into the river and tributaries, and ensure enough water flows so that the health of the River Ganga is maintained at a desired level. Future solutions need to be comprehensive and innovative, take into account the strengths and weaknesses of previous efforts to clean the river, and consider other national and international experiences of river restoration.

The water quality challenges are not homogeneous along the river: (i) in the upper segment the river flows on a steep and narrow rocky bed, carries cold water and is subjected to much less anthropogenic pollution but has a highly sensitive and fragile ecosystem and biodiversity; (ii) the river in the middle segment enters and flows in plains, meandering mostly on a bed of fine sand, has a wide river bed and flood plain, and is greatly modified through human interventions in terms of huge quantities of water diversion/abstraction and subjected to a high degree of pollutant loads from domestic, industrial and agricultural activities; (iii) the river in the lower segment has experienced considerable changes in sediment transport and deposition, causes widespread flooding, undergoes frequent changes in her channel path, and is subjected to international disputes on flows and interventions made and/or being carried out/planned (IITC, 2010).

This chapter provides an analysis of water quality problems in the Ganga and options for improvement. The chapter has been structured using the DPSIR[1] assessment framework (Figure 13.1) and reviews the main trends in urbanization, industrialization and agriculture intensification (including livestock and aquaculture) in the basin and analyses how all these drivers have translated into more water abstraction and pollution loads beyond sustainable limits. The chapter also reviews the resulting trends in terms of water quality degradation along the river and the associated risks

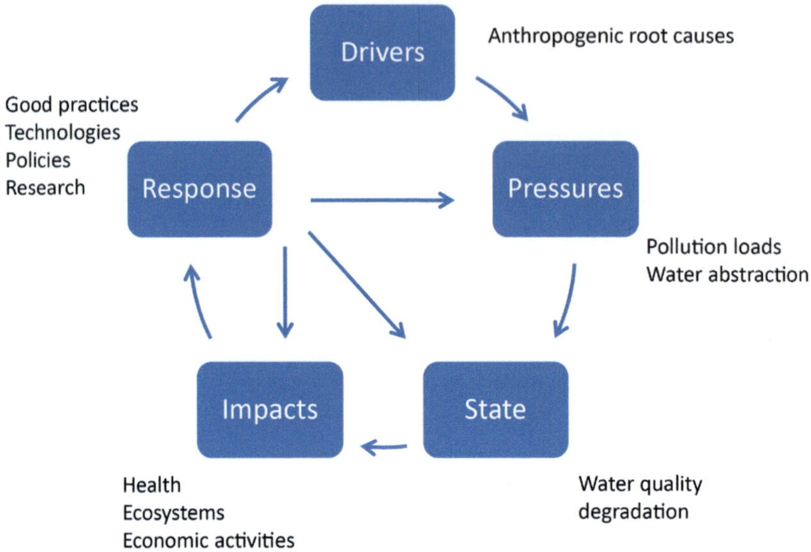

Figure 13.1 Drivers-Pressures-State-Impacts-Responses (DPSIR) framework for water quality management

for human health, ecosystems and economic activities. Finally, it analyses the responses in the last decades to address the problem and suggest new pathways towards a more promising future with a healthier Ganga.

Pollution: the root causes

Sewage, septage and municipal solid waste

During the last decades a rampant growth together with massive rural–urban migration has shaped the increasingly urbanized landscape of the Ganga river banks. Today the basin is considered the world's most populous river basin and is home to more than 600 million Indians (accounting for both urban and rural), or about half the country's population. The Ganga's main stream runs through 36 Class I cities (with population over 100,000), 14 Class II cities (with population between 50,000 and 100,000) and about 50 smaller towns with population above 20,000. According to the Central Pollution Control Board (CPCB) (2013a) these Class I and II cities generate more than 2.7 billion litres of sewage every day, although this figure may be underestimated as it is calculated as a portion of the municipal water supplied to towns and cities, which also tends to be underestimated (CSE, 2014), neglects urban run-off, and does not consider the wastewater generated in smaller towns. From the wastewater generated in these 50 Class I and II cities along the Ganga main stream only a fraction is treated before it reaches the river: while the installed treatment capacity is only 1.2 billion litres per day, actually, according to a CPCB inspection of 51 of the 64 sewage treatment plants (STP) identified along the Ganges, only about 60 per cent of the total installed capacity is actually used as many STPs are non-operational or mis-functional (CPCB, 2013b). This means that in the best case scenario 26 per cent of the wastewater generated along the Ganga is treated, but this percentage is likely to be significantly smaller since, as discussed before, the generated wastewater is underestimated. Additionally, the number of towns and cities that discharge their wastewater to the tributaries

Table 13.1 Sanitation arrangements in urban areas in Ganga Basin states

States	*Total number of households*	*Households with latrine*				*Households lacking latrines*		
		Total	*Connected to piped sewer*	*Connected to septic tank*	*Connected to others (e.g. pits)*	*Total*	*Using public latrine*	*Practising open defecation*
Bihar	2,013,671	69%	7%	53%	9%	31%	2%	29%
Jharkhand	1,495,642	67%	14%	49%	4%	33%	2%	31%
Uttar Pradesh	7,449,195	83%	28%	47%	8%	17%	2%	15%
Uttarakhand	592,223	94%	32%	53%	9%	1%	2%	5%
West Bengal	6,350,113	85%	14%	45%	26%	15%	4%	11%

Source: Census (2011)

of the Ganga, and not only to the main stream, is substantially bigger, which aggravates the pollution problem (for example, through heavily polluted tributaries such as the Rivers Ramganga, Kali and Gomati). The CPCB has identified 138 drains discharging 6 billion litres per day of polluted water into the Ganga (CPCB, 2013a).

Another issue is the lack of septage management systems – the 2011 census data (Table 13.1) indicates that 45–53 per cent of the urban households in states across the Ganga Basin are connected to septic tanks, rather than piped sewers. While there is growing interest in septage management, treatment of septage does not currently exist in the Ganga Basin. The septage from the septic tanks, if and when collected, is disposed of in an unregulated manner and without any treatment, most commonly in peri-urban farms, informal landfills or the nearest drain, contributing to diffuse pollution and further adding to the pollution challenge in the Ganga, particularly during the rains, when this pollution runs off. This situation is prevalent in the smaller cities and towns, which lack even the most basic sewerage or treatment infrastructure.

The Ganga River and banks also suffer from inadequate and improper solid waste management. Dumping of domestic solid wastes, including plastics, glass and organic waste, affect the river water quality, aesthetics and aquatic life. The collection capacity of municipal solid waste in the Ganga states is limited and the treatment capacity is almost nil. During festival seasons, over 70 million people bathe in the Ganga over a few weeks to clean themselves from their past sins. Some materials like floral offerings and other materials used for religious purposes, washing of clothes, wallowing of animals, throwing un-burnt/partially burnt dead bodies are simply left in the river.

Industrial wastewater

Economic growth in the Ganga Basin has come hand-in-hand with a powerful industrialization process. Now there are 764 industries in the main stem of the Ganga and the main tributaries the Kali and the Ramganga, most of them within Uttar Pradesh state (CPCB, 2013a). Table 13.2 shows the sectorwise distribution of industrial units and their water consumption and wastewater generation.

Industrial effluents are in the range of 10–20 per cent of the total volume of wastewater directly reaching the Ganga. Although a relatively low proportion, they are a cause for major concern because they are often toxic and non-biodegradable.

Table 13.2 Industrial water consumption and wastewater generation in the Ganga Basin

Type of Industry	*Total Units*	*Water Consumption (MLD)*	*Wastewater Generation (MLD)*
Chemical	27	210.9	97.8
Distillery	23	78.8	37.0
Food, Dairy & Beverages	22	11.2	6.5
Pulp & Paper	67	306.3	201.4
Sugar	67	304.8	96.0
Textile, Bleaching & Dyeing	63	14.1	11.4
Tannery	444	28.7	22.1
Others	41	168.3	28.6
Total	764	1123	501

Source: CPCB (2013a)

Agricultural waste

There are voices that claim that agro-chemicals are a key polluter in the river (e.g. Ghosh, ND), and others that believe that it is not a significant source of pollution (e.g. Trivedi, 2010). The first group blames agro-chemicals for damaging the riverine ecosystems to an extent that these lose their self-treatment capacity. The other group suggests that pollution from agriculture is negligible as compared to industrial or urban pollution. Nevertheless, these are more perceptions rather than evidence-based facts. Actually there are no comprehensive nor scientifically sound studies that have assessed the contribution of agro-chemicals to the pollution in the Ganga, and not only agro-chemicals: organic pollution from livestock and aquaculture remain largely unassessed in the basin. However, the observed trends in terms of expansion and intensification of the agricultural sector, including the sharp increase in fertilizer consumption (Figure 13.2) and pesticides use, and the booming development of livestock farms in the Ganga Basin call for comprehensive studies to understand the actual contribution of the sector to the degradation of the river.

Use of pesticides in India began in 1948 while manufacturing of DDT and benzene-hexachloride (BHC or Lindane) began in the year 1952 (Gupta, 2004). In 1958, India was producing over 5,000 metric tonnes of pesticides. Currently, there are approximately 145 pesticides registered for use, and production has increased to approximately 85,000 metric tonnes. The first report of poisoning due to pesticides in India came from Kerala in 1958 where over 100 people died after consuming wheat flour contaminated with parathion. Despite the fact that the consumption of pesticides in India is still very low, about 0.5 kg/ha of pesticides against 6.60 and 12.0 kg/ha in Korea and Japan, respectively, there has been widespread contamination of food commodities with pesticide residues. In India, 51 per cent of food commodities are contaminated with pesticide residues and out of these, 20 per cent have pesticide residues above the maximum residue levels on a worldwide basis (IITC, 2011b).

Pesticide contamination poses significant risks to the environment and non-target organisms ranging from beneficial soil microorganisms to insects, plants, fish, birds and humans. A limited

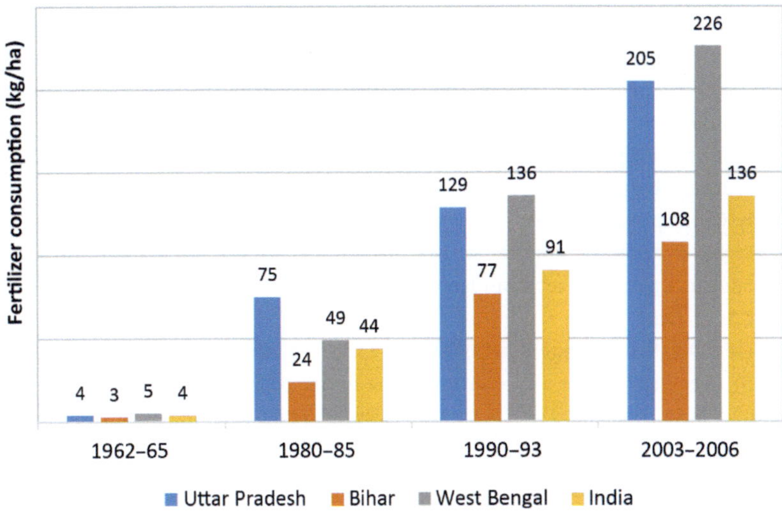

Figure 13.2 Evolution of fertilizer consumption (kg/ha) across states in the Ganga River Basin
Source: IITC (2011a)

number of studies have been conducted in the last three decades in the Ganga River Basin where pesticide concentrations have been monitored in various environmental media. In most of these studies, total DDT and total HCH concentrations have been reported in detectable and significant concentrations. Pesticide levels in water samples from the Ganga River Basin ranged from non-detectable to a maximum total pesticide concentration of 13.1 micro-g/l in the city of Varanasi. Only four of the fifteen studies reviewed showed total pesticide concentrations higher than the permissible levels of 1 micro-g/l. Pesticide levels in sediments ranged from 0.05 ng/g to 240 ng/g. Significant and high pesticide levels have been reported in fish tissue with maximum total DDT concentration of 7.5 micro-g/g in catfish and maximum total HCH concentration of 3.5 micro-g/g in catfish. These results indicate that pesticides like DDT and HCH bioaccumulate to significant levels in fish tissue. Health risks associated with these levels of pesticides in food need to be examined in greater detail (IITC, 2011b).

Reduced water flows

There is disruption to high velocity flow due to several hydroelectric projects and dams in the highly sensitive and fragile ecosystems in the upper reaches. The middle stretch of around 1,080 km from Haridwar to Varanasi is the most degraded due to significant irrigation canal and groundwater abstractions and a high degree of pollutant loads from different sources. Flow estimates after the canal diversions at Haridwar, Bijnor and Narora indicate that original Ganga River flow is almost completely lost, with little or no capacity to perform its ecosystem services and assimilate the large pollution loads (Mateo-Sagasta, 2015).

Surface water quality trends

With an ever increasing amount of pollution loads from the cities, industries and farms along the Ganga and her tributaries, the river remains heavily polluted, posing great health risks to humans and ecosystems. Faecal contamination is a case in point (Figure 13.3). Indeed, coliform levels are high all along the river and make the water generally unsuitable for traditional bathing (Class B waters), not to speak about drinking purposes (Class A waters), barring a few upstream locations. What is even more alarming is that the situation has not improved in the last decades; on the contrary, coliform levels are on the rise and substantially increased from 1996 through 2010, particularly at the middle stretch of the River Ganga (IITC, 2012).

According to the available data (IITC, 2012) the concentration of faecal coliform along the river is substantially bigger during the monsoon than during the dry period. This could seem a paradox as with increased flows one would expect reduced pollution concentration in the river, but actually reflects the importance of non-point sources of faecal pollution such as open defecation and uncontrolled faecal sludge disposal on land which are washed out to the river during periods of rain.

The average Dissolved Oxygen (DO) (Figure 13.3) and Biological Oxygen Demand (BOD_5) (Figure 13.3) over the years at all locations in the entire course of the Ganga River appear to have remained fairly constant. The DO values over the years, in general, are on the higher side and conform to the designated best-use classification (Class A). Organic pollution as measured by BOD_5, however, does not conform to the limit specified for Class A waters at some of the locations. The DO values during the wet season are lower as compared to the dry season although the level of organic pollution (BOD_5) remains fairly constant during both seasons at most of the locations. The level of organic pollution (BOD_5) is substantially high in the middle stretch of the river (particularly from Fatehgarh to Tarighat) over the years despite the fact that the DO level is on the higher side. The higher level of organic pollution in the Fatehgarh–Tarighat stretch is due to the fact that the stretch is subjected to severe anthropogenic pollution in the form of discharge of partially treated or untreated sewage and industrial effluents (IITC, 2012). The fact that the large organic pollution loads in this stretch do not translate into proportional DO drops suggests a reasonable self-purifying capability for the River Ganga at least in this stretch.

Impacts on human health and ecosystems

Waste and wastewater from cities, industries and agriculture contain pathogens and chemicals that concentrate in the Ganga waters and sediments and accumulate along the trophic chain, posing serious risks for human health, the environment and productive activities. These impacts may include (Hernandez Sancho *et al.*, 2015):

- Increased burden of disease due to reduced drinking or bathing water quality
- Increased burden of disease due to unsafe food (e.g. contaminated fish or vegetables irrigated with polluted waters)
- Increased burden of disease when working or playing in wastewater-irrigated area
- Decreased biodiversity
- Lost aesthetic value or bad odours
- Diminished recreational opportunities
- Increased GHG emissions
- Reduced fish and shellfish catches, or reduced market value of fish and shellfish
- Reduced industrial productivity, or increased costs of treating water for industries
- Reduced agricultural productivity
- Reduced market value of harvested crops, if unsafe wastewater irrigation
- Reduced number of tourists, or reduced willingness to pay for recreational services

In the Ganga River Basin, due to data limitations and methodological complexities, there are not complete quantifications of the impacts from water quality degradation, and the economic value of these impacts, but the few partial assessments that do exist suggest that the impact is in the billions of dollars every year.

An assessment conducted by the Water and Sanitation Program of the World Bank (WSP, 2011) quantified the economic value of some of the impacts of inadequate sanitation for India. The study assessed: (i) health-related impacts (premature deaths; costs of treating diseases; productive time lost due to people falling ill; time lost by care givers who look after them); (ii) domestic water-related impacts (household treatment of water; use of bottled water; a portion of costs of obtaining piped water; time costs of fetching cleaner water from a distance); (iii) access time impacts (cost of additional time spent accessing shared toilets or open defecation sites; absence of children [mainly girls] from school and women from their workplaces); and (iv) tourism impacts (potential loss of tourism revenues and economic impacts of gastrointestinal illnesses among foreign tourists).The total economic impacts amounted to a loss of US$53.8 billion in 2006, with the health-related economic impacts having the biggest share (US$38.5 billion) (WSP, 2011) which is not surprising as diarrhoea is the third leading cause of childhood mortality in India, and is responsible for 13 per cent of all deaths/year in children under five years of age (Lakshminarayanan and Jayalakshmy, 2015). These economic impacts are conservative estimations as they do not consider a number of impacts of inadequate sanitation of economic importance, such as impacts on ambient water quality and the resulting effect

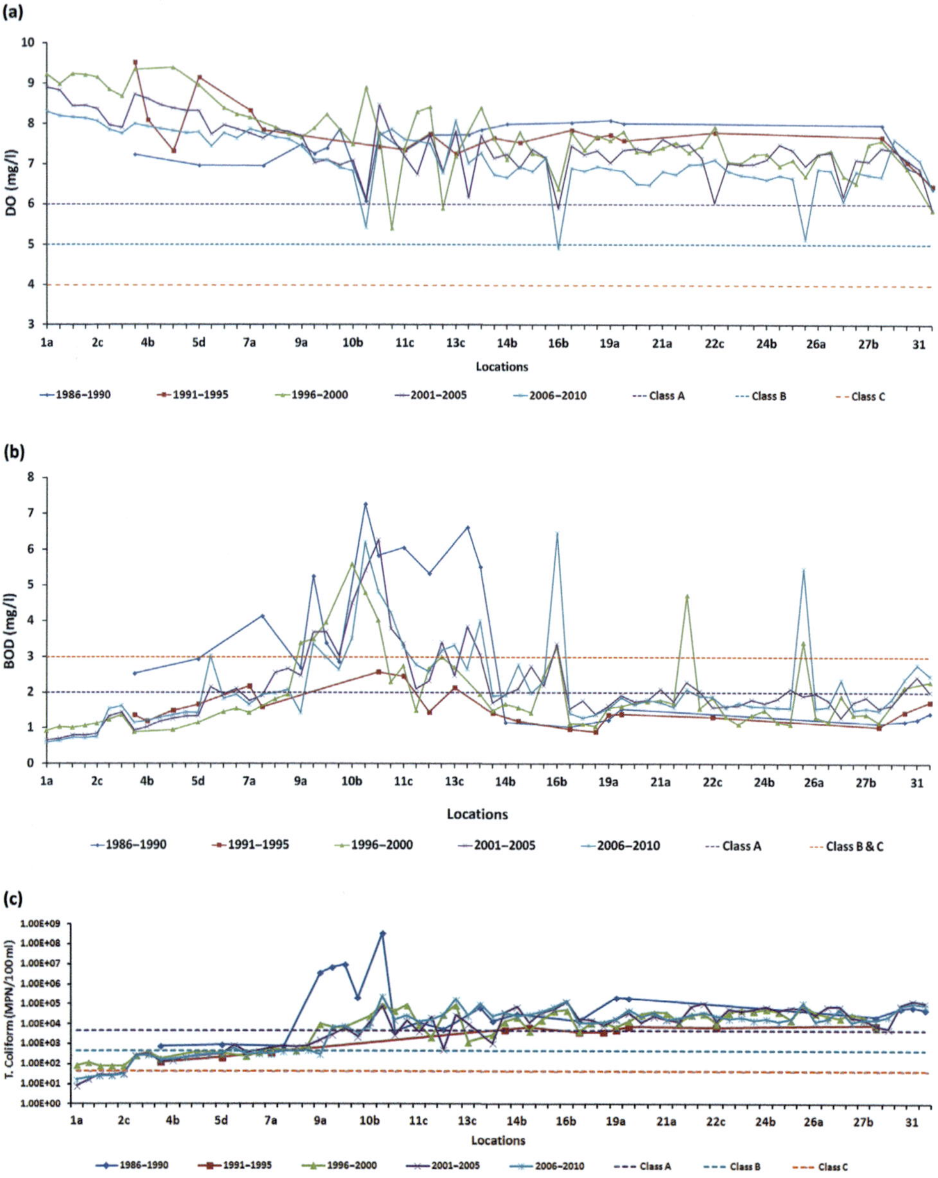

Figure 13.3 Locations along the Ganga River for sampling average Dissolved Oxygen (DO) and Biological Oxygen Demand (BOD)

Source: IITC (2012)

Locations

1a	Uttarkashi u/s (Bhagirathi)	**10a**	Kanpur u/s (Bithoor)	**21a**	Barh u/s
1b	Uttarkashi d/s (Bhagirathi)	**10b**	Kanpur d/s (Shuklaganj)	**21b**	Barh d/s
2a	Deoprayag u/s (Bhagirathi)	**10c**	Kanpur d/s (Jane Village)	**22a**	Mokama u/s
2b	Deoprayag u/s (Alaknada)	**11a**	Allahbad u/s(ujahni) Fatehpur	**22b**	Mokama d/s
2c	Deoprayag d/s (Ganga)	**11b**	Bathing Ghats at Sangam	**22c**	d/s Bata-McDowell
3a	Ranipur u/s (Ganga)	**11c**	d/s Deehaghat	**23a**	Munger u/s
3b	Ranipur d/s (Ganga)	12	Vindhyachal Pakka Ghat	**23b**	Munger d/s
4a	Rishikesh u/s	**13a**	Varanasi u/s	**24a**	Sultanganj u/s
4b	Rishikesh d/s	**13b**	Dashashwmedh Ghat	**24b**	Sultanganj d/s
5a	Haridwar u/s	**13c**	d/s Kaithy	**25a**	Bhagalpur u/s
5b	Har-Ki-Pauri	**13d**	Near Malviya Bridge	**25b**	Bhagalpur d/s
5c	Lalta Rao	**13e**	Tarighat	**25c**	d/s Champanala
5d	Dam Kothi	**14a**	Buxar u/s	**26a**	Kahalgaon u/s
5e	Mishrpur	**14b**	Buxar d/s	**26b**	Kahalgaon d/s
6a	Bijnor u/s (Ganga)	**15a**	Chapra u/s (Ghaghra)	**26c**	d/s NTPC Drain
6b	Bijnor d/s (Ganga)	**15b**	Chapra d/s (Ghaghra)	**27a**	Sahebganj u/s
7a	Garh Mukteshwar u/s	**16a**	Arrah u/s (River Gangi)	**27b**	Sahebganj d/s
7b	Garh Mukteshwar d/s	**16b**	Arrah (d/s River Gangi)	**28**	Rajmahal d/s
8a	Anoopshahar u/s (Ganga)	**17**	Koelwar (River Sone)	**29**	Berhmpore (middle)
8b	Anoopshahar d/s (Ganga)	**18a**	Hajipur u/s (River Gandak)	**30**	Palta (middle)
9a	Fatehgarh u/s	**18b**	Hajipur d/s (River Gandak)	**31**	Dakshineswar (middle)
9b	Kannauj u/s (a/c R'ganga & b/c Kali)	**19a**	Patna u/s	**32**	Uluberia (middle)
9c	Kannauj d/s (a/c Kali)	**19b**	Patna d/s		
		20a	Fatuha u/s		
		20b	Fatuha d/s		

Note: Class a: Water for use as drinking water source without conventional treatment but after disinfection; Class b: waters for use for organized outdoor bathing; Class c: waters for use as drinking water source with conventional treatment followed by disinfection

Figure 13.3 (continued)

on fisheries and the environment. Nevertheless they provide a good indication of the extent of the problem, also for the Ganga Basin, which concentrates half of the population of India and has a sanitation status below the national average.

The River Ganga supports one of the largest and most diverse fish populations in the world, comprising around 20 per cent of all of the freshwater species found in India (Sarkar *et al.*, 2012). It hosts an estimated 143 species belonging to 11 orders, 72 genera and 32 families. However, data on the population dynamics of four major fish groups suggests that this is changing significantly. Over the period 1958–1997, for example, the total catch of major carp species has significantly decreased in many fish centres and the proportion of carp in the fishery declined from 44 per cent in 1976 to 13 per cent in the current decade (Payne *et al.*, 2004). However, other species of low economic value have increased. There are a number of factors that have damaged the state of the fisheries in the last decades including flow interception due to large hydroelectric projects in upper Ganga, abstraction of large quantities of water for irrigation in the form of canals (Upper, Madhya and Lower Ganga canals), transport of large quantities of industrial pollution in rivers (Ram Ganga and Kali), increase of pollution load from urban areas along the river (domestic, industrial, treated and untreated), increase in the sediment load due to deforestation, indiscriminate fishing in middle and lower Ganga and fishing of important juvenile species, especially of Indian major carp (Mathur *et al.*, 2012). Pollution affects fisheries especially in the middle and lower stretches of the river.

In the middle segment most of the threatened and migratory species are under severe threat because of the impacts of industrial effluents disposed into the river (Sarkar *et al.*, 2012). Bioaccumulation of heavy metals has been observed in fish in this stretch of the river (e.g. in

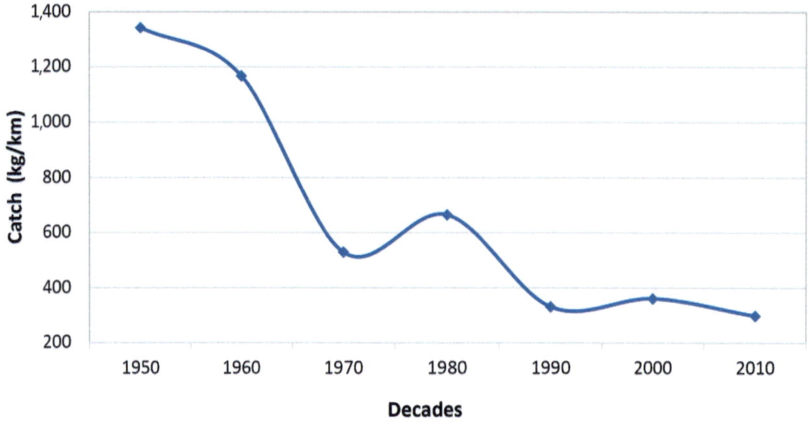

Figure 13.4 Decline of fish catch/km at Allahabad during 1950 to 2010
Source: Sarkar *et al.* (2012)

Varanasi [Sinha, 2004]). In Allahabad, a water pollution hotspot, a steady decline of all economic fish species has been observed during the past six decades (Figure 13.4). The catch/km declined from 1,344 to 300 kg between 1950 and 2010 (Sarkar *et al.*, 2012).

In the lower segment of the Ganga, a large amount of sediment is transported downstream and is distributed across the floodplains during the period of inundation which results in sedimentation of fish spawning sites and impacts on breeding. Wetlands are also heavily impacted by sediments and siltation with impacts on fisheries (Sarkar and Dubey, 2015). At Patna centre the total fish catch has reduced from 109 tons (1961–66) to 47 tons (1986–93) and at Bhagalpur centre from 113 tons (1961–70) to 86 tons (1981–88) (Sarkar *et al.*, 2012). Nevertheless, water pollution (including siltation) is only one of multiple factors affecting fisheries and the correlation between pollution increase and fisheries reduction is not always positive.

Mangroves in the delta are also impacted by pollution. Oil or gas exploration, petroleum production and accidents involving large oil tankers cause significant damage to mangrove ecosystems, causing defoliation of trees, mortality of all sessile and benthic organisms and contamination of many water fowls with a minimum recovery period of 10 years (Kathiresan, 2008). In the Ganga Sundarbans, several industrial effluents are released into the adjacent coastal water bodies. For example, Haldia port-cum-industrial complex releases wastes that contain several complex ions and different organic and inorganic compounds. When these compounds are mixed with the coastal water bodies they enter in to the sediment through percolation and the pneumatophores contact with the sediment and also take up the organic and inorganic compounds. These compounds hamper the circulatory system of the mangroves species such as *H. fomes* (Behera, 2012).

Past responses and new approaches

Ganga Action Plan (GAP) I and II

With a view to improving the water quality in the river, the first-ever multi-state, national-level substantial effort for abating the pollution of the river was launched in the form of the Ganga Action Plan (GAP) in 1985. Two assessments of pollution issues in the river from the Central Pollution Control Board (CPCB, 1982; CPCB, 1984) formed the basis for the GAP.

These CPCB reports found that most pollution from point-sources was coming from the sewage generated in 25 Class I cities in Uttar Pradesh, Bihar and West Bengal, all located on the banks of the Ganga. Therefore, the main focus of the plan was on interception, diversion (I&D) and treatment of sewage generated from these 25 cities. Implementation of phase I of the plan (GAP I) ended in 2008 (with 10 years of delay). GAP II was started in 1993 when implementation of GAP I was not even half way (IITC, 2011c) and is still under progress in five states, Uttarakhad, Uttar Pradesh, Bihar, Jharkhand and West Bengal. Together, GAP I and GAP II targeted interception, diversion and treatment of sewage of more than 37 cities located on the banks of the River Ganga. The plan also identified grossly polluting industries and mandated the installation of effluent treatment plants. Up to 2011 an amount of Rs. 1612.38 crores (US$245 million) has been spent on the GAP (MoEF, 2011).

While it is believed that interventions under GAP I and II resulted in checking the further deterioration of the river (which would have been exacerbated in the absence of these schemes) and had different other strengths (as described by IITC, 2011c), some of the weaknesses and shortcomings of GAP I and II have been identified as follows (IIT, 2009; IITC, 2011c):

- Limited scope of pollution issues addressed
- Lack of ownership among urban local bodies (ULB)
- Challenges of land acquisition, legal delays and other implementation issues
- Lack of adequate attention to generating and allocating funds for operation and maintenance (O&M) of assets
- Under-utilization of STPs, in some instances due to non-conveyance of the sewage to the STPs, particularly in the absence of upstream systems such as branch sewer and house connections

Namami Gange

The Government of India recently approved the flagship Namami Gange programme, which integrates the efforts to clean and protect the Ganga River in a comprehensive manner. The programme has a budget outlay of INR 200 billion (more than US$3 billion) for the next five years which incorporates the remaining budget for GAP II, the budget (US$1 billion) allocated to the World Bank National Ganga River Basin Project, and more. The Namami Gange programme has a greater scope than the GAP and is designed to focus on pollution abatement interventions, namely: interception, diversion and treatment of wastewater flowing through the open drains through bioremediation; appropriate in situ treatment; use of innovative technologies; sewage treatment plants (STP); industrial effluent treatment plants (ETP); rehabilitation and augmentation of existing STPs and immediate short-term measures for arresting pollution at exit points on the river front to prevent inflow of sewage, etc.

Citing inadequate focus on O&M under previous schemes, the Government of India plans to provide funds for O&M of the assets for a minimum 10-year period, and to adopt a public private partnership (PPP)/special purpose vehicle (SPV) approach for pollution hotspots.

The Namami Gange budget is to be disbursed by the National Mission for Clean Ganga (NMCG), under the recently renamed Ministry of Water Resources, River Development and Ganga Rejuvenation. The NMCG is the operational wing of the National Ganga River Basin Authority (NGRBA) which has been set up by the central government as the planning, financing, monitoring and coordinating authority for strengthening the collective efforts of the central and state government for effective abatement of pollution and conservation of the River Ganga. The NGRBA adopts a river basin approach, which is a substantial change as compared to the GAP. State program management groups (SPMG) have been established in all five main stem states. Projects are being implemented by multiple executing agencies, mainly parastatal organizations such as the Uttar Pradesh Jal Nigam, Bihar Urban Infrastructure Development Corporation, Kolkata Metropolitan Development Authority and some urban local bodies (ULB).

Ganga River Basin Management Plan (GRBMP)

One of the important functions of the NGRBA was to prepare and implement a Ganga River Basin Environment Management Plan (GRBEMP), which was prepared by a Consortium of Seven Indian Institutes of Technology and associate partners. The main (Tare *et al.*, 2015) and supporting documents were submitted to the Government of India in January 2016. The GRBMP is based on secondary data and a broad stakeholder involvement. Based on a comprehensive analysis of multiple issues affecting the river, which were documented in thematic reports, the plan makes suggestions and recommendations in the form of eight missions: Aviral Dhara (continuous flow), Nirmal Dhara (unpolluted flow), Ecological Restoration, Sustainable Agriculture, Geological Safeguarding, Basin Protection Against Disasters, River Hazards Management and Environmental Knowledge-Building and Sensitization. The plan is very ambitious and one of its main recommendations is to implement a 'Zero liquid discharge' policy for all Class I cities and grossly polluting industries; the effluents from these cities and industries will need to be recycled and reused, but not discharged (treated or untreated) to surface waters. The capital expenditure (capex) needed for the adoption of such a recommendation has been estimated at US$100 billion during the coming 25 years,[2] raising questions on what realistic financial strategy could make it possible.

Promising future?

Addressing the formidable and multidimensional challenges that the Ganga is facing will not be an easy task and will require new knowledge, capacity development, massive financial resources, appropriate technologies, strong political will, partnerships and stakeholder engagement. Future solutions need to be comprehensive and innovative, ambitious but realistic, and learn from the strengths and weaknesses of previous efforts to clean the river, and from other national and international experiences of river restoration (Mateo-Sagasta, 2015).

Addressing knowledge gaps

The actual contribution of agriculture, livestock and aquaculture to the water quality degradation in the Ganga River is not known. In countries such as China, the United States and others in Europe these sectors have been identified to be major contributors to water pollution. The trends in agro-chemical use as well as the density of livestock suggest that these pressures could also be important in the Ganga Basin. To understand the extent of the problem a sustained research and modelling effort would be needed to better understand the pathways and loads of nutrients and organic matter from these sources to water bodies. Similarly, the contribution of other non-point-sources of pollution, such as faecal sludge or open defecation, to the degradation of the Ganga is not well understood and will need further research.

The hydrological links between groundwater and surface water in the Ganga Basin have not been properly assessed and modelled, therefore it is not possible to estimate the contribution that groundwater pollution may have to the Ganga and tributaries, and vice versa. Understanding this is particularly important for pollutants such as nitrate, pesticides and salinity.

Overall, a comprehensive water quality model at the basin scale that allows simulation of the cost-effectiveness of solutions will be a must for planning and assessments. Rejuvenating the Ganga and bringing it back to its past glory will require massive investments. In a context of relevant but still limited financial resources it will be critical to select the most cost-effective combination of solutions to meet the required water quality standards and improve river health, as opposed to selecting solutions at whatever cost. These solutions need to include pollution abatement from different sources, the restoration of appropriate water flows and, better, a combination of both. Understanding how the implementation of these solutions at scale translates into reduced pollution loads, enhanced water flows and, consequently, improved water quality along the river in different seasons will require complex water quality modelling, an exercise that has not been done comprehensively in the Ganga Basin.

Any water quality assessment and modelling will require a sufficient amount of good quality data. The current water quality monitoring network along the Ganga and its tributaries is very poor and will need to be strengthened with a substantially larger number of stations, monitoring more parameters and with a greater frequency.

Finally, a better understanding of how pollution translates into health and environmental impacts, and the costs of such impacts, will help to raise awareness of the size of the problem and will help justify the massive investments needed to restore the river.

Developing institutional capacities and addressing policy gaps

The enormity and complexity of options and position papers, as well as the different and sometimes divergent interests of the stakeholders involved in the Ganga clean-up efforts, has caused what may be termed a 'policy paralysis', in which decisions come slowly, if at all. For example, as of March 2016 only US$65 million has been disbursed from the World Banks US$1 billion budget for Ganga restoration. For this reason, there is a need for a simplification of choices through an approachable 'menu-based' system that can be standardized for all decision makers in all Ganges River Basin regions. The GRBMP is a step towards this direction (Tare *et al.*, 2015).

Mobilizing financial resources

As stated elsewhere in this chapter, great investments are required to ensure improving river water quality. The Government of India has already allocated over INR 200 billion for this purpose through the Namami Gange programme. With this proposed investment the Indian government intends to create a market for treated sewage and industrial effluents through a reuse and recycle approach, thereby keeping rivers free from sewage and industrial effluents. This strategy would in turn lead to internalized impact of discharge of sewage and industrial effluents into urbanization and industrial development processes, respectively. This way the 'polluters pay' principle will also become operative. Much more investment can be attracted from local urban bodies and industries through public private partnership (PPP) as this would also mean water security for urban centres and industries. This would indirectly motivate polluters to pay for cleaning up their act and thereby keep water bodies free of pollution.

Partnering and stakeholder engagement

The Ganga Basin is transboundary in nature and heavily populated with people of diverse socio-cultural and religious beliefs. There are numerous formal and informal institutions, religious bodies, NGOs and environmental groups, and a vocal academia and media who all have a major stake in the basin and the river. On the positive side all these bodies have vast knowledge of the traditional systems and river behaviour, and a large public following. An efficient mechanism for knowledge sharing and effective communications will be vital for constructive and quick stakeholder engagement, increased awareness and informed decisions.

Adopting appropriate technologies and practices

Decision makers will need to choose validated and context-specific technical solutions that can work at scale (e.g. Libhaber and Orozco-Jaramillo, 2012) and create the enabling environment for these solutions to be sustainable. This implies institutional, financial and technical capacity development. Fortunately, there are various well-established centralized as well as decentralized solutions available which have a high potential to support progress if the lessons from institutional failure (see e.g. Murray and Drechsel, 2011) are taken on board.

Another positive sign is the emerging understanding that for scaling up resource recovery and safe reuse (RRR) of wastewater and sludge, economics is as much needed as technical solutions. Several agencies have come up with techno-commercial models for a variety of situations. For example, IWMI and its partners have developed scalable RRR business models based on the study of 200 empirical cases that bridge the sanitation and agricultural sectors and are being tested now for their feasibility in 10 cities across the globe (Otoo and Drechsel, 2015). They also support public and private entities through the creation of innovative low-cost technologies that can work at scale in low-income countries, support livelihoods, enhance food security, support green economies and contribute to cost recovery in the sanitation service sector. However, resource recovery from wastewater and sludge has to go hand in hand with the development of guidelines and policies on safe reuse, and be complemented with broader programmes for water pollution control and impact mitigation, as supported by WHO, UNEP and FAO (WHO, 2006).

Notes

1 Drivers, Pressures, State, Impacts and Responses.
2 www.spml.co.in/mediaroom/2015/Global-Water-Intelligence-June-2015.pdf.

References

Behera M.D. (2012). The Status of Sundari (*H. fomes*) an indicator species in the Sunderbans. The Lower Ganga River Basin. In *Ganga River Basin Management Plan, Interim Report*. Indian Institutes of Technology of Bombay, Delhi, Guwahati, Kanpur, Kharagpur, Madras, Roorkee.

CPCB (1982). 'Basin Sub-Basin Inventory of Water Pollution: Part I – The Yamuna Sub Basin' Assessment and Development Study of River Basin Series: ADSORBS/6/1981–82, Central Board for the Prevention and Control of Pollution, New Delhi.

CPCB (1984). 'Basin Sub-Basin Inventory of Water Pollution: Part II – The Ganga Basin' Assessment and Development Study of River Basin Series: ADSORBS/7/1982–83, Central Board for the Prevention and Control of Pollution, New Delhi.

CPCB (2013a). *Pollution Assessment: River Ganga*. Central Pollution Control Board. Delhi.

CPCB (2013b). *Performance Evaluation of Sewage Treatment Plants under NRCD*. Central Pollution Control Board, India.

CSE (2014). *Ganga: The river, its pollution and what can we do about it*. A Center for Science and Environment briefing paper. Delhi.

Ghosh, D. (2010). Pollution from non-point sources in Ganga Basin: A community-based potential for managing the unmarked crisis. Available at: http://dhrubajyoti.net/publ/Ganga%20Pollution_DGhosh.pdf. Last accessed 21 May 2016.

Gupta P.K. (2004). Pesticide exposure – Indian scene. *Toxicology* 198: 83–90.

Hernandez-Sancho, F., Lamizana, B., Mateo-Sagasta, J. and Qadir, M. (2015). *Economic Valuation of Wastewater: The Cost of Action and The Cost of No Action.* United Nations Environmental Program, Nairobi.

IIT (2009). *Status Paper* on River Ganga State of Environment and Water Quality. National River Conservation Directorate Ministry of Environment and Forests Government of India, Indian Institute of Technology, Roorkee.

IITC (2010). River Ganga at a Glance: Identification of Issues and Priority Actions for Restoration; 001_GBP_IIT_GEN_DAT_01_Ver 1_Dec 2010.

IITC (2011a). Agriculture in the Ganga river Basin: An Overview; 015_GBP_IIT_SEC_ANL_ 01_Ver 1_Dec 2011.

IITC (2011b). Emerging Contaminants in Ganga River Basin with Special Emphasis on Pesticides; 013_GBP_IIT_EQP_SOA_02_Ver 1_Dec 2011.

IITC (2011c). Strength Weakness, Opportunity and Threat (SWOT) Analysis of Ganga Action Plan (GAP). In *Ganga River Basin Management Plan,* Interim Report. Indian Institutes of Technology of Bombay, Delhi, Guwahati, Kanpur, Kharagpur, Madras, Roorkee.

IITC (2012). Water Quality in the Ganga River and Efficacy of Sewage Treatment Processes in A Case for Adopting Tertiary Treatment; 023_GBP_IIT_EQP_ANL_01_Ver 1_June 2012.

Kathiresan, K. (2008). *Threats to Mangroves. Degradation and destruction of mangroves.* Centre of Advanced Study in Marine Biology. Annamalai University, India, pp. 476–83.

Lakshminarayanan, S. and Jayalakshmy, R. (2015). Diarrheal diseases among children in India: Current scenario and future perspectives. *Journal of Natural Science, Biology and Medicine* 6(1): 24–8. Available at: www.ncbi.nlm.nih.gov/pmc/articles/PMC4367049/. Last accessed 21 May 2016.

Libhaber, M. and Orozco-Jaramillo, A. (2012). *Sustainable Treatment and Reuse of Municipal Wastewater for Decision Makers and Practicing Engineers.* IWA, London.

Mateo-Sagasta, J. (2015). *Towards a healthy Ganges.* Global Water Forum. Available at: www.globalwaterforum.org/tag/ganga/. Last accessed 21 May 2016.

Mathur, R.P., Kapoor, V., Shivam, A. and Verma, J. (2012). 'Status of fish and fisheries of Ganga River'. In *Ganga River Basin Management Plan,* Interim Report. Indian Institutes of Technology of Bombay, Delhi, Guwahati, Kanpur, Kharagpur, Madras, Roorkee.

MoEF (2011). Status of Programs Under GAP-II as on 30.12.2009. Ministry of Environments and Forests, Government of India. Available at: http://moef.nic.in/modules/recentinitiatives/NGRBA/progress.htm. Last accessed 21 May 2016.

Murray, A. and Drechsel, P. (2011). Why do some wastewater treatment facilities work when the majority fail? Case study from the sanitation sector in Ghana. *Waterlines* 30(2): 135–49.

Otoo, M. and Drechsel, P. (2015). *Resource Recovery from Waste: Business Models for Energy, Nutrient and Water Reuse.* London: Earthscan-IWMI.

Payne, A. I., Sinha, R., Singh, H. R. and Huq, S. (2004). 'A review of the Ganges Basin: Its fish and fisheries'. In R. Welcomme and T. Petr (eds). *Proceedings of the Second International Symposium on the Management of Large Rivers for Fisheries.* Volume 1, pp. 229–251. FAO, Office for Asia and the Pacific, Bangkok, Thailand.

Sarkar, U.K. and Dubey, V.K. (2015). Threats to fish and fisheries in the Ganga Basin: Consequences of changing habitats and altering diversity patterns. Global Water Forum. Available at: www.globalwaterforum.org/tag/ganga/. Last accessed 21 May 2016.

Sarkar, U.K., Pathak, A.K., Sinha, R.K., Sivakumar, K., Pandian, A.K., Pandey, A., Dubey, V.K. and Lakra, W.S. (2012). Freshwater fish biodiversity in the River Ganga (India): Changing pattern, threats and conservation perspectives. *Rev. Fish. Biol. Fish.* 22: 251–72.

Sinha, R.K. (2004). *Monitoring of heavy metal load in the River Ganga at Varanasi.* Final Tech. Rep. Submitted to National River Conservation Directorate, MOEF, Govt of India, New Delhi.

Tare, V., Roy, G. and Bose, P. (2015). *Ganga River Basin Management Plan (2015).* Main document. Indian Institutes of Technology of Bombay, Delhi, Guwahati, Kanpur, Kharagpur, Madras, Roorkee.

Trivedi R.C. (2010). *Water quality of the Ganga River – An overview.* Aquatic Ecosystem Health & Management.

WHO (2006). *Guidelines for The Safe Use of Wastewater in Agriculture.* WHO: Geneva.

WSP (2011). Economic Impacts of Inadequate Sanitation in India. WSP. Available at: www.wsp.org/sites/wsp.org/files/publications/WSP-esi-india.pdf. Last accessed 21 May 2016.

PART III

Governance and livelihoods

14

Institutions and policies governing water resources management in the Ganges River Basin

M. Dinesh Kumar, Dwarika N. Dhungel, M. Monirul Qader Mirza and Diana Suhardiman

Key messages

- The diverse institutional approaches to water resources management in each of the riparian countries pose significant institutional and policy challenges towards holistic water resource management in the basin.
- The way in which water policies and institutions are currently set up, following a fragmented and sectoral approach, has resulted in poor coordination of planning and management at all levels, from local to transboundary.

- While a single agency with multiple functions might raise transparency and accountability at a national level, there is currently little cooperation both at regional and national levels. This in turn hinders attempts to develop water resources cooperatively across the basin.
- Decisions concerning water allocation from major sources are often based on political interests, which do not take into account different types of water use. This can lead to ineffective water development strategy and inefficient resource use.
- The absence of well-defined property rights regimes creates uncertainty about actual water-use arrangements.

Introduction

The Ganges River rises in the Western Himalayas, before entering Uttarakhand. It flows south and east through the Gangetic Plain of north India into Bangladesh, where it empties into the Bay of Bengal. Shared by Nepal, India and Bangladesh, the Ganges is the most densely populated river basin in the world. It supports the lives of over 500 million people and presents great opportunities for social and economic development, as well as challenges in the conservation of ecosystems (World Bank, 2014). Opportunities include drinking water provision, water for agriculture, hydropower, navigation services, recreation, fishery and a wide variety of ecosystem services. Riparian countries benefit from the river, and also suffer from the hydrological extremes of floods and droughts (Mirza, 2011; Sharma *et al.*, 2010).

Perhaps because the basin has the largest concentration of the world's rural poor, coupled with food insecurity, past research on water management has focused almost entirely on irrigation development and enhanced agricultural productivity. Additionally, there has been research into how institutions and policy regimes could be strengthened in order to promote the intensive use of water resources. This is undertaken with the aim of increasing agricultural production, while promoting equity in access to irrigation, in order to reduce poverty (Molden *et al.*, 2001; Sadoff *et al.*, 2012).

While such research was useful in directing agricultural development in the basin, it tended to overlook the structural constraints and challenges the basin faces, from improving agricultural productivity to reducing pollution. There has been little focus on cooperation among the basin countries with regard to the optimum use of the Ganges regarding economic development in relation to potential hydropower development and flood management. Consequently, benefits from agricultural modernization and hydropower development remain largely untapped, while problems related to flood and drought management systems continue to be inadequately addressed, resulting in repetitive failures to protect people's lives and livelihoods (Mirza, 2013).

Better management of water in the Ganges is required to protect the basin's hydrological integrity, sustain and enhance the socio-economic and ecological benefits it yields, and mitigate the mounting costs of natural disasters. This requires a basin-wide understanding of the physical and socio-economic systems and existing institutions and policies, and how these could influence the development and use of water in the basin.

Existing institutions related to water resources in the basin

Nepal

In Nepal, the relevant water-related institutions can be divided into four categories: planning and coordinating bodies; sectoral policy planning institutions; regulatory bodies; and autonomous service providers including local governance bodies.

Water resources-related bodies in charge of policy, planning and coordination include: the National Development Council (NDC); the National Water Resources Development Council (NWRDC); Investment Board, Nepal (IBN); the National Planning Commission (NPC); the Environment Protection Council (EPC); and the Water and Energy Commission (WEC). The NWRDC and WEC are water-related policy planning and coordinating bodies, whereas the NDC and NPC are responsible for formulating national development policy. The EPC is the policy, planning and coordinating body for environmental issues only.

Water resources management is governed through a sectoral approach, where each ministry is responsible for the formulation, implementation and monitoring of (sub) sectoral policies, plans and programmes. At present, there are eight sector ministries working on water-related issues:

- The Ministry of Energy (MoE), which is responsible for hydroelectricity generation and overall power-sector development
- The Ministry of Irrigation (MoIR), which is in charge of irrigation development and responsible for water-induced disaster
- The Ministry of Urban Affairs (MoUA), which is responsible for drinking water supply and water sanitation provision
- The Ministry of Agriculture and Cooperatives (MoAC), which is responsible for agricultural development
- The Ministry of Forest and Soil Conservation (MoFSC)
- The Ministry of Science, Technology and Environment (MoSTE), which covers hydrology and climatology aspects of water resource management
- The Ministry of Federal Affairs and Local Development (MoFALD), which is in charge of local infrastructure development in rural areas

For the implementation of (sub) sectoral policies, plans and programmes, these ministries are supported by their respective departments. For example, while the Department of Irrigation is responsible for the implementation of large irrigation projects (e.g. infrastructure development) countrywide, the Department of Local Infrastructure Development & Agricultural Roads (DoLIDAR) can also develop small-scale irrigation schemes as part of local infrastructure development.

Furthermore, Nepal has various regulatory bodies such as the Electricity Tariff Fixation Commission (ETFC), and the Water Supply Tariff Fixation Commission (WSTFC). At district level, the District Water Resources Committee (DWRC) is in charge of the registration of Water User Associations (WUA), which can use water resources for drinking water, irrigation and electricity production. There are also autonomous bodies focusing on service provision, like the Nepal Electricity Authority (NEA), responsible for electricity generation, transmission and distribution, the Groundwater Development Board (GDB), in charge of planning and management of groundwater provision, and the Nepal Water Supply Corporation (NWSC) which supplies water to 20 cities across the country. While Kathmandu Valley Drinking Water Limited is responsible for the supply of drinking water in the valley, a few drinking water boards established by the government are in charge of drinking water supply to urban areas outside Kathmandu.

India

In India, among the most important water-related institutions are the Central Water Commission (CWC) and Central Ground Water Boards (CGWB). Both are technical agencies in charge of water resource evaluation and planning at the national level. The CWC assesses the surface water potential of the 20 major basins.

The CGWB's assessment of groundwater recharge, net groundwater balance and overall groundwater development status is crucial for framing groundwater development and management polices by both the State and Central Ground Water Departments. The State Groundwater Departments come under the State Water Resources Department and are independent of the CGWB or the Ministry of Water Resources (MoWR). In India, water is a state subject except for the inter-state river basins. Therefore, water-related agencies at the national level only play advisory roles for matters concerning the states. While the CWC could carry out techno-economic appraisals of development projects proposed by state governments, it lacks a mandate to regulate the appropriation of water in the country's basins. This falls under the responsibility of state-level agencies, and includes the inter-state basins. Similarly, the CGWB works closely with State Groundwater Departments to carry out resource assessments at district and block level, but lacks any mandate to regulate the actual allocation and use of groundwater resources within the states and union territories of India.

Apart from water resources planning, there are various government agencies in charge of development, management and use of water. For example, operating at both state and inter-state levels, the Ganga Flood Control Commission (GFCC) is an inter-state government body that deals with flood management for eleven states. It comes under MoWR and is headquartered in Patna, Bihar. It acts as the secretariat and executive wing of the Ganga Flood Control Board (GFCB), which provides overall governance and policy, and is headed by the Union Minister for Water Resources. Like the GFCC, the GFCB also comes under MoWR. Since 1972, the GFCC has prepared plans for flood management on the 23 river systems of the Ganga Basin.

The MoWR also has its own Flood Management Programme (FMP). As part of this programme, respective states formulate and submit detailed flood management schemes to the GFCC. In practice, however, the utilization of funds from the GFCC has been very poor. This is due to delays in the completion of schemes and a lack of detailed project reports. Moreover, as the flood management plans developed by GFCC are not statutory documents, it is not legally binding on the state water resources departments to follow them.

Each basin state has various line agencies for water provisioning, such as the State Water Supply Departments and the Industrial Development Corporations for water supply in rural areas and industrial estates. Under the Water Resources Department (WRD) in each state, there is the Minor Irrigation Department (MID), which plans and implements schemes of less than 1,000 ha. It is primarily concerned with maximizing irrigation and, to a lesser extent, flood control. Recently, MIDs in some states have stopped implementing minor irrigation schemes, because private well irrigation has increased greatly over the past three to four decades. At present, these departments only carry out water level and quality monitoring, and collaborate with CGWB for five-yearly groundwater assessments.

The State Pollution Control Boards (SPCB) and Central Pollution Control Board (CPCB) are in charge of water quality monitoring and enforcement of pollution control at state and national levels. While the CPCB periodically monitors water quality in the Ganges and its tributaries, lakes, wells and tanks (CPCB, 2008), the SPCB is responsible for the enforcement of pollution control measures including those concerning the discharge of effluent from industries and municipalities. Watershed management agencies in various states have also set up watershed development programmes focusing on soil and water conservation. Finally, the National Hydroelectric Power Corporation (NHPC) is engaged in the planning, design, execution and operation of hydroelectric projects across the basin, particularly in the sub-Himalayan region of Uttaranchal.

Bangladesh

Like India and Nepal, water institutions in Bangladesh have evolved around five fundamental problems. These are floods, droughts, food insecurity, pollution and transboundary challenges. Water institutions are focused on efforts to alleviate these problems. After the partition of India in 1947, the then East Pakistan inherited part of the Bengal Irrigation Department whose primary functions included canal construction, irrigation management, drainage, flood embankment construction, maintenance and the collection of fees from the irrigation beneficiaries. The East Pakistan Water and Power Development Authority (EPWAPDA) was created in 1959 for coordinated water and power development (Ali, 2002). During that time, the government also attempted to increase food production through expanding irrigation and flood control facilities. To this end, the East Pakistan Agriculture Development Corporation (EPADC) was formed in 1961.

Over the decades, Bangladesh created many public sector organizations to oversee water resources development and management. The principal public organizations, under the Ministry of Water Resources (MoWR), include the Water Resource Planning Organization (WARPO) and the Bangladesh Water Development Board (BWDB). The first is in charge of the preparation of a Water Master Plan as well as the overall strategies and policies for water development and conservation, while the latter is responsible for flood control, drainage and irrigation.

The MoWR has four research bodies. The River Research Institute (RRI) is responsible for conducting multidisciplinary and problem-oriented research in the fields of irrigation, coastal hydraulics and soil mechanics. The Institute of Water Modelling (IWM) focuses on water modelling to enhance the quality of planning and maintenance of various water resources projects. The Centre for Environment and Geographic Information Services (CEGIS) is responsible for natural resource management planning, system development and capacity building. Finally, the Joint Rivers Commission (JRC) maintains liaison between India and Bangladesh through coordinated research on flood control, flood forecasting and cyclone warnings. Organizations in other sectors that interlink with the water sector are the Public Health Engineering Department (PHE), the Local Government Engineering Department (LGED), the Water and Sewerage Authority (WASA), the Department of Fisheries (DOF), the Inland Water Transport Authority (BIWTA), the Roads and Highways Department (R&H) and the Department of Environment (DOE). The National Water Council presides over all these institutions in matters of water policy and legislation, especially with regard to the formulation of water policy and inter-sectoral coordination.

Review of policies in water, land and energy sectors

Nepal

In Nepal, the Constitution (CAS, 2015) is the main policy document concerning the use of natural resources, including water. It states that:

> the State shall pursue a policy of conserving the natural resources available in the country by imbibing the norms of inter-generation judicious use of it and for the national interest. The policy shall ensure the fair distribution of the benefits generated by it by giving local people the priority and preferential rights.
>
> *(Article 51.g.1)*

It also states that 'the State shall pursue a policy of prioritizing national investment in water resources based on people's participation and making a multi-utility development of water resources' (Article 51.g.2). In addition, Article 279 has also made it obligatory for the government to submit any accession to, acceptance of or approval of treaties or agreements related to, among others, 'natural resources and the distribution of their use to the parliament' for ratification. Such agreements are to be approved by two-thirds majority of the total members in both houses of federal legislature.

Since 1956, when the first five-year plan was launched, the government has been following a sectoral approach to water resources management. This is borne out by the preparation of separate chapters for different water-related subjects (e.g. energy, irrigation, drinking water and sanitation) in the national periodic and long-term plans. It was only in 1992 that the government framed an overarching Water Resources Act (1992), which later translated into two major policies. The Water Resources Strategies (WRS, 2002) recognized water use for different purposes, such as domestic, irrigation, industrial and generating hydropower. The National Water Plan (WRS, 2005) aimed to implement the WRS in a timely manner.

According to the WRS (2002), water resources development should 'contribute to the national objectives of social and economic development'. The WRS incorporated 10 strategic outputs[1] with specific activities in three different time frames. These outputs covered short- (5 years), medium- (10 years) and long- (20 years) term time frames. Prepared by the Water and Energy Commission Secretariat (WECS) and approved by the government in 2005, the NWP recognized the broad objectives of the WRS and aimed to contribute 'in a balanced manner to the overall national goals of economic development, poverty alleviation, food security, public health and safety, decent standards of living for the people and protection of the natural environment'.

The NWP also highlights the need to further develop Nepal's water resources, due to growing demand. It states that while 'Nepal has utilized mainly medium and small rivers for different uses such as drinking water, irrigation and hydropower, the larger and perennial Himalayan Rivers, except for a few run-of the-river schemes, have been virtually left untapped'. According to the plan, only a small percentage (15 BCM of 225 BCM) of annually available water has been used for economic and social purposes. The growing water demand is also noted in the WRS, which estimates that total demand for water will increase to 98,980 million m^3 (MCM) in 2027 compared to 14,880 million m^3 in 2000. Even after meeting the increased demand, there would still be substantial untapped water in Nepal (WRS, 2002).

In line with the WRS and NWP, the government was to prepare an integrated National Water Resources Policy (NWRP). In 2012, the WECS prepared a draft NWRP, which is yet to be endorsed by the government. In the absence of an integrated national water policy, the sectoral policy approach continues to be the basis for programme planning. While the NWP highlights the need for an integrated approach, the possibility of government coming up with an integrated national water policy seems to be remote. Once all three levels of government become fully operational, only the federal government may formulate the NWRP. It will, however, have to seek the recommendations of the National Natural Resources and Fiscal Commission (Article 250 of the Constitution of Nepal 2015). The Commission is mandated to:

> determine the base for computing the share of investment and return for the federal, provincial and local government while exploiting natural resources, and make necessary recommendations. Also to make recommendation about coordination and mitigation of disputes likely to arise regarding distribution of natural resources between the federation and province, between the provinces, between a province and local level entity, or between local level entities.
>
> *(Article 251 of the Constitution of Nepal 2015)*

India

In India, there are at least three major policies with regard to the development, allocation and management of water resources in the basin. These policies include state policies, pricing policies for irrigation, urban and rural water supply, and policies promoting the use of equipment to improve irrigation efficiency.

In alignment with the National Water Policy (NWP) (2002; 2012) the State Water Policies give priority to drinking water in water allocation. The policy for pricing of drinking water supply in rural areas aims to improve access for poor communities via public systems. The general policy for pricing of irrigation water is to make it affordable to all, rather than generating revenue. Cost recovery and economic efficiency are not considerations in pricing of water in these two sectors.

In large urban centres, though, pricing policy reflects the actual volumetric consumption in order to encourage the efficient use of water. The policy is not intended to recover the actual cost of production and supply of water from domestic consumers, and instead it applies cross-subsidization with higher charges for industrial and commercial users. Metering is partial in most cities for domestic water connection. In smaller towns, the extent of metering and pro rata pricing is very poor (Kumar *et al.*, 2014).

Regarding energy pricing policy, agriculture is the sub-sector which consumes the largest amount of energy. Energy pricing policy in agriculture influences the way water is used in irrigating crops. The Indian part of the basin is characterized by the presence of millions of wells, mostly shallow tube wells, either electric or diesel. These are used for irrigating millions of small farms. The pricing policy for both electricity and diesel, which are used for pumping groundwater, aims to make energy available at very low prices so as to bring down irrigation costs (Kumar *et al.*, 2010; Kumar *et al.*, 2011). Cross-subsidization, with very high unit charges for industrial and commercial users, reduces the revenue loss to the utility.

In the electricity sector, farmers get electricity at highly subsidized rates. Only in West Bengal is electricity charged on a pro rata basis, through the metering of all agricultural connections. Diesel is also subsidized, though farmers who use diesel engines incur much higher costs for production and supply of irrigation water compared to electric well owners. A study in eastern Uttar Pradesh and South Bihar showed that farmers using diesel engines for irrigation used water more efficiently than their counterparts using electric pumps in the same area (Kumar *et al.*, 2010).

Bangladesh

India's participation in the freedom struggle of Bangladesh in 1972 opened up avenues for transboundary cooperation. The Indo-Bangladesh Joint Rivers Commission (JRC) was created in 1972 to foster cooperation between the two countries in the areas of flood control and irrigation. The JRC in Bangladesh comes under the MoWR and officers are usually deputed from the BWDB. Although the JRC was created with good intentions, it lacks executive power (Mirza, 2013). Despite decision-making limitations, the JRC has assisted in successful negotiations of agreement (1977) and Memoranda of Understanding (1982 and 1985), and a Treaty (1996) on the Ganges water-sharing problem.

In 1983, following massive construction efforts to build 1,963 km of embankments, 8,000 hydraulic structures and 1,000 river closures (Hughes *et al.*, 1994), the government put a priority on water sector development and created the National Water Resources Council, an inter-ministerial body headed by the President (now headed by the Prime Minister). The Master Plan Organization (MPO) was launched in the same year, with the World Bank appointed as its executive agency. The MPO completed the National Water Plan (NWP) in 1986. The plan assessed the country's available surface and groundwater resources and was highly criticized for its excessive emphasis on groundwater resources.

After the disastrous floods of 1987 and 1988, efforts were launched to initiate cooperative programmes with the upstream countries – India, Nepal, Bhutan and China. In the following three years, bilateral reports were produced that emphasized advance flood forecasting and warnings as well as multi-purpose storage construction. A substantial volume of hydrological, hydraulic, climate and socio-economic information was generated at national level, as the country was divided up (initially) into 173 catchments. However, apart from some improvements in flood forecasting and warning from India, no significant progress was made in terms of actual development. While the FAP study (1989–1996) recommended a wide range of action plans under its 26 components, it was highly criticized by NGOs and civil society organizations for not having open dialogue and giving environmental, social and fishery interests little consideration. The FAP did not proceed to project implementation level. The government formulated the Bangladesh Flood and Water Management Strategy (BWFMS) in 1996. Providing policy guidelines for water resources development and management, it ensured that public participation, environmental impact assessments and multi-criteria analysis during the planning process were made mandatory in all future projects (WMO, 2003).

In 1998, the government produced a draft of the National Water Policy (NWP). The NWP adopted the principles of integrated water resources management (IWRM) which was a significant policy shift. While the policy states that 'water will be considered an economic resource and priced to convey its scarcity value to all users and provide motivation for its conservation' (Alam and Quevauviller, 2014), it also emphasized participatory water management and highlighted the importance of stakeholder participation. The NWP further recognized the importance of protection, restoration and preservation of the environment, its biodiversity and water quality.

In line with the NWP, the National Water Management Plan (NWMP) was prepared in 2001 and approved in 2004. It had three major objectives: (1) rational management and wise use of Bangladesh's water resources; (2) people's quality of life improved by equitable, safe and reliable access to water for production, health and hygiene; and (3) clean water in sufficient and timely quantities for multi-purpose use and preservation of the aquatic and water-dependent ecosystem (WARPO, 2001). The plan period was divided into three phases: short term (5 years), medium term (10 years) and long term (25 years). The NWMP has 84 programmes that cut across 11 different sectors of the economy. Access to safe drinking water and sanitation has been given priority. In practice, the plan has failed to address these problems (WMO, 2003).

In 2012, the government prepared its '2010–2021 Perspective Plan' where the IWRM concept has become the focus of the plan. In general, the plan envisages redesigning and equipping the government's water sector agencies and institutions towards more effective implementation of policies and strategies. For example, in focusing on 'water application efficiency', it attempts to reduce water demand through an increase in water-use efficiency to 50 per cent from the current 30 per cent. Another significant policy shift is to reduce dependence on groundwater. The plan also emphasizes the importance of dialogue for more effective flood and drought management across the basin (GED, 2012).

Institutional and policy challenges in achieving basin-wide water management

Water institutions and the manner in which they produce rules determine the region's ability to manage its water resources. As noted by Biswas (2008), institutions responsible for implementing water policies suffer from serious deficiencies in the basin, lacking efficiency with respect to legal and regulatory aspects, implementation of rules, accountability and responsiveness to the needs of users.

Water resources management in the Ganges Basin has always been conducted in an unintegrated manner, due to challenges in existing institutional structure and policy rules. The way different agencies arrange their different types of water use, almost independently of each other, shows how development is seldom based on integrated planning, both at national and basin level. At basin level, the current institutional and policy regimes have resulted in water security problems (Babel and Wahid, 2008). In this section, we discuss and analyse the institutional and policy set-up that contributes to these problems.

Fragmented, sectoral and supply-side approach

Government agencies in each country of the basin carry out different roles and functions in water management, often without any coordination between each other. While task division is needed to ensure clear roles and responsibilities, the sectoral approach often leads to undesirable outcomes. For example, in constructing embankments for flood control, the WRD in Bihar did not take into account how low floods can benefit the fishing and agricultural communities in terms of nutrient inputs and fish catch (MoWR, 2014). Similarly, the Minor Irrigation Department of West Bengal and Bihar promotes well irrigation almost entirely without recognizing the potential mineralization of groundwater and its likely impact on drinking water. Moreover, these agencies use supply-side approaches to meet the growing demand for water, often relying entirely on engineering/technical interventions. For example, in the area of flood control, there has been too little focus on non-structural interventions such as flood preparedness, land use regulations for flood plains and issuing of flood warnings.

Supply-side interventions in water management are not only ineffective in meeting the demand for water, but often have serious negative social, economic and environmental consequences (Frederick, 1992). Uncontrolled withdrawal of water from the river and its tributaries for industrial and municipal use, and disposal of effluent into the same river without proper treatment in large cities, are not only keeping demands unchecked, but also reducing freshwater availability.

In Nepal, a fragmented, sectoral approach to water resource management prevails because the coordinating bodies are inactive. In theory, the formation of the NWRDC in 1993 was an extremely important milestone. The NWRDC was expected to function as a mini-parliament where water-related issues would be debated in order to decide how this resource should be used by different sectors in a coordinated manner at home and in neighbouring countries. Input from the NWRDC would have been valuable both to Parliament in legislating policy and to the NDC and the NPC in formulating national plans and policies. However, during its 18 years of existence, the NWRDC met only four times, with the last meeting occurring in 2002 (Pun, 2012).

Similarly, chaired by the Minister for Energy and comprised of the representatives from various government and non-government agencies,[2] the WEC is another coordinating body which hardly functions. It lacks a role in programme planning or budgetary allocations to the various water-related sub-sectoral programmes, and the NPC and the Ministry of Finance continue to be the sole decision makers on these matters. Because of this, the WEC has not monitored the implementation of the NWP either. As the sectoral agencies only look to the NPC for programme planning and budgetary allocations, the government has not seen the WEC as the central water planning body. The NPC will have to be restructured in terms of institutional design and function once the federal structure comes into force, according to the 2015 Constitution.

Single agency and multiplicity of functions

Organizing agencies along line functions is one of the most important principles of a sound water resource management programme (Frederiksen, 1998). In the Ganges Basin countries, the same line agency is often entitled to carry out a multiplicity of functions. For example, in Bihar, the WRD builds irrigation and flood control schemes, undertakes flood forecasting, assesses flood losses, coordinates with the CWC and other central agencies and issues flood warnings (MoWR, 2014). This reduces the effectiveness of the agency because they have to prioritize some activities, and also raises transparency and accountability issues.

These issues are most evident in the way the SPCB restricted its data access on water quality to the wider public. As discussed earlier, the SPCB is mandated to enforce pollution control norms relating to water bodies, as well as periodically monitoring water quality in rivers, groundwater, streams and lakes. The performance of the agency should be evaluated in terms of its ability to protect water quality by enforcing pollution control. This should lead to legal action against offending industries and municipalities. Obviously, as the periodic monitoring of water quality and publishing of water quality data can open the agency's performance to public scrutiny, the agency lacks any incentive to make this data available.

While a single agency with many functions raises transparency and accountability issues at the national level, such agencies may be preferred for inter-governmental negotiations at the basin level. For example, while Bangladesh relies on the NWRC to discuss transboundary water issues, Nepal faces an institutional dilemma as to which sectoral ministry should represent the country. Following the abolition of the MoWR in 2009, the Ministry of Energy (MoE) and the Ministry of Irrigation (MoIRR) were formed. Following the creation of these two ministries, many question how the Government of Nepal can represent itself in various inter-governmental forums in the basin.

Centralized nature of institutions and lack of well-defined property rights regimes

The state-level institutions dealing directly with water are centralized. They use top-down approaches for data collection, planning and implementation. This can reduce the effectiveness of water management actions for several reasons. First, as planning is based on macro-level data, it often does not involve local communities in identifying important issues for water planning. Consequently, the designed plans are often inadequate as they fail to capture the local resource availability, conditions and problems (MoWR, 2014).

Second, since decisions regarding the allocation of water are taken at the ministerial level, often influenced by political interests, they are rarely based on integrated water-use assessment, incorporating the different types of water use.

The absence of well-defined property rights regimes creates uncertainty about actual water-use arrangements (Pearce and Warford, 1993). In India, *de facto* rights to use groundwater are attached to land ownership rights (Saleth, 1993; Singh, 1995). The dominant heritage principle implied in the Transfer of Property Act IV of 1882 and Land Acquisition Act of 1894 views groundwater as an easement of land (Saleth, 1993). But, there is no limit on the quantity of water that can be abstracted by a well owner (Kumar, 2007). Thus, a large farmer in West Bengal, who owns an electrified deep tube well, can irrigate 8 ha of land, whereas farmers owning less than one acre have to purchase water from his well at a prohibitive price (Mukherji *et al.*, 2012). The unrestricted access that rich well owners enjoy over groundwater disadvantages the poor, as it increases the monopoly power of the former.

In Nepal, people willing to make use of water resources for collective benefits on an institutional basis may form a WUA and have it registered. The District Water Resources Committee approves and registers WUAs so that they can provide drinking water, irrigation and electricity to their members. Coordination at local levels is still an issue mainly due to the absence of elected officials in local governance bodies. No elections to these bodies have taken place since 2002.

Inadequate institutional regime to deal with transboundary issues

Little regional cooperation on water resources development and management exists between the neighbouring countries (Biswas, 2008; Brichieri-Colombi and Bradnock, 2003; Verghese, 2007; Dhungel, 2013). The WRS of Nepal recognizes shared navigational use of Ganges river waters and the benefits of cooperation for hydropower generation. Similarly, the NWP looks towards developing a better framework for cooperation with India and Bangladesh. To this end, the government is encouraging the formation of a regional committee in water resource management under SAARC. While medium- and long-term targets have been set regarding water-sharing treaties, with power-sharing agreements and the resolution of riparian issues in international water courses, the current scope of regional cooperation is limited to power trading between Nepal and India. In this context, the signing of the Framework Agreement for Energy Cooperation in 2014 may be a step towards increased cooperation, providing the framework is implemented in good faith.

The Farakka Treaty of 1996, signed between India and Bangladesh, and concerned with Ganges water sharing, has not been successful in resolving all issues between the two countries (Rahman, 2005).

Conclusions

The fragmented sectoral decision-making set-up, as well as the supply-side approach, poses great challenges for basin-wide water resources management. Centralized decision making leads to ineffective programme planning and policy implementation, often resulting in conflict and inefficient water management.

At basin level, fragmented and sectoral decision-making structures and processes result in weak transboundary institutions and policy networks. Consequently, the existing institutions and policies perform sub-optimally. The current institutional set-up for water resources management needs to be re-oriented towards a basin-wide arrangement for the entire region's benefit. Also, international regimes for cooperation in water resources management, flood control and energy security need to be strengthened. They should incorporate compliance mechanisms for water- and power-sharing treaties through a wide range of activities, such as joint planning, and knowledge and information sharing.

Additionally, the pricing policies for electricity and water in the farm sector have to change in order to reflect the costs of production and supply. Current pricing policies only encourage inefficient use of these two resources. There is currently unequal access to groundwater in the farm sector which is a serious concern, as hundreds of millions of farmers face economic scarcity of water. The pricing policies in urban water sectors, combined with poor metering, encourage wasteful use of water. Overall there is a lack of alignment in institutions and policies in different sub-sectors. This makes it difficult to achieve good water management outcomes.

Acknowledgements

In the preparation of the Nepali portions of the chapter, the second author has received help and inputs from a number of professionals and colleagues associated with the different agencies of the Government of Nepal. All of them deserve sincere thanks. But we would especially like to remember and thank Eng. Sriranjan Lacoul, former Secretary Government of Nepal; Mr Jagat Bhusal, Chairperson, Electricity Tariff Fixation Commission and Engineers; Basudev Dahal, Deputy Director General, Department of Irrigation; Sunil Kumar Dhungel, Deputy Managing Director, Nepal Electricity Authority; Hari Prasad Sharma, former Chairperson, Water Supply Tariff Fixation Commission; and Mahendra Raj Sapkota for their support in polishing and improving the contents of the Nepali portions.

Notes

1 The 10 strategic outputs are broadly divided into three categories: (i) security-related;(ii) use-related; and (iii) mechanisms-related. For elaborated descriptions of these strategic outputs, see WRS Executive Summary (2002) and NWP (2005).

2 The members of the WEC are: secretaries of various related ministries (eleven secretaries), Dean, Institute of Engineering TU, President Nepal Engineering Association, representative of FNCCI and two experts on water resources and energy from non-governmental organizations nominated by the government for a period of two years. The secretary at the WECS is its Member Secretary.

References

Alam, M.M. and P. Quevauviller. (2014). An Evaluation of Integrated Water Resources Management (IWRM) Activities in Bangladesh. *Asia Pacific Journal of Energy and Environment* 1(1): 22–38.

Ali, M.L. (2002). *An Integrated Approach for the Improvement of Flood Control and Drainage in the Coastal Belt of Bangladesh.* CRC Press, Netherlands.

Aryal, R.S. (2011). *The Law on Ownership and Right to Water in Nepal.* Submitted to Jalsrot Vikas Sanstha/Global Water Partnership Nepal.

Babel, M.S. and S.M. Wahid (2008). Freshwater Under threat: Vulnerability Assessment of Freshwater Resources to Environmental Change – Ganges-Brahmaputra-Meghna River Basin Helmand River Basin Indus River Basin, United Nations Environment Programme (UNEP, Nairobi), Kenya. Available at www.unep.org/pdf/southasia_report.pdf (accessed 28 August 2013).

Biswas, A.K. (2008). 'Management of Ganga Brahmaputra Meghna System Ways Forward'. In O. Varis, A.K. Biswas and C. Tortajada (eds). *Management of Trans boundary Rivers and Lakes Water Resources Development and Management.* Berlin: Springer.

Brichieri-Colombi, S. and R.W. Bradnock (2003). Geopolitics, Water and Development in South Asia: Cooperative Development in the Ganges-Brahmaputra Delta. *The Geographical Journal* 169(1): 43–64.

Central Pollution Control Board (2008). *Status of Water Quality in India 2007.* Central Pollution Control Board, Ministry of Environment and Forests, Government of India, New Delhi.

Central Water Commission (2012). Integrated Hydrological Data Book, Hydrological Data Directorate, Information Systems Organization, Central Water Commission, Ministry of Water Resources, Government of India, March 2012.

CAS (Constitutional Assembly Secretariat) (2015). Constitution of Nepal. Unofficial translation by Nepal Law Society, IDEA and UNDP, Singha Durbar, Kathmandu.

Dhungel, D.N. (2013). *Regional cooperation on the Ganga basin: Yet a mirage?* Observer Research Foundation, New Delhi and the Asia Foundation, Issue Brief, September.

Dhungel, D.N. and K. Ghising (2014). *Nepal and India energy cooperation: Prospects.* A paper in a Dialogue on a South Asian Cooperative Architecture: The Indo-Nepal Dialogue (I) jointly organized by the Delhi Policy Group and Nepal Institute for International and Strategic Studies, held at India Habitat Center, New Delhi, 21–23 December 2014.

Frederick, K.D. (1992). Balancing Water Demands with Supplies: The Role of Management in a World of Increasing Scarcity. *World Bank Technical Paper* 189. Washington, DC: The World Bank.

Frederiksen, H.D. (1998). Institutional Principles for Sound Management of Water and Related Environmental Resources. In A.K. Biswas (ed). *Water Resources: Environmental Planning, Management, and Development.* New Delhi: Tata McGraw-Hill.

General Economic Division (GED) (2012). *Perspective Plan of Bangladesh: 2010–2021.* General Economic Division, Planning Commission, Government of the People's Republic of Bangladesh, Dhaka.

Government of Nepal (2015). The Constitution of Nepal, 2015. Ministry of Law, Justice, Constituent Assembly – Legislative Parliament, Law Book Management Committee.

Hindustan Times (2012). Ganga receives 2900 million ltrs of sewage daily. *Hindustan Times*, 17 April.

His Majesty's Government of Nepal (2002). Water Resources Strategy Nepal. Water and Energy Commission Secretariat, Singha Durbar, Kathmandu Nepal.

His Majesty's Government of Nepal: Executive Summary (2002). Water Resources Strategy Nepal. Water and Energy Commission Secretariat, Singha Durbar, Kathmandu Nepal.

His Majesty's Government of Nepal (2005). National Water Plan. Water and Energy Commission Secretariat, Singha Durbar, Kathmandu Nepal.

Hughes, R., Adnan, S. and Clayton, B.D. (1994). *Floodplains or Flood Plans? A review of approaches to water management in Bangladesh.* Research Advisory Service, Dhaka.

Kumar, M.D. (2007). *Groundwater Management in India: Physical, Institutional and Policy*. New Delhi: Sage Publications.

Kumar, M. Dinesh, Singh, O.P. and Sivamohan, M.V.K. (2010). Have diesel price hikes actually led to farmer distress in India?, *Water International* 35(3): 270–84.

Kumar, M.D., Scott, C.A. and Singh, O.P. (2011). Inducing the Shift from Flat Rate or Free Agricultural Power to Metered Supply: Implications for Groundwater Depletion and Power Sector Viability in India. *Journal of Hydrology* 409 (1–2): 382–94.

Kumar, M.D., Viswanathan, P.K. and Bassi, N. (2014). 'Water Scarcity and Pollution in South and South East Asia: Problems and Challenges'. In P. Harris and G. Lang (eds). *Routledge Handbook of Environment and Society in Asia.* London: Routledge.

Molden, D., Sakthivadivel, R. and Habib, Z. (2001). *Basin Level Use and Productivity of Water. Examples from South Asia.* International Water Management Institute, New Delhi.

Ministry of Water Resources (2014). Operational Research to Support Mainstreaming of Integrated Flood Management under Climate Change. Final Report Phase I, Central Water Commission, Ministry of Water Resources, financed by Asian Development Bank and co-financed by DFID, New Delhi.

Mirza, M.M.Q. (2013). Restructuring the Bangladesh Joint Rivers Commission for Conflict Resolution and Water Cooperation. *South Asia Journal* 1(9): 28–49.

Mirza, M.M.Q. (2011). Climate Change, Flooding in South Asia and Implications. *Regional Environmental Change* 11(1): 95–107.

Mukherji, A., Shah, T. and Parthasarathy, B. (2012). Kick-starting a Second Green Revolution in Bengal, Commentary. *Economic and Political Weekly* 47(18): 27–30.

Pearce, D.W. and J. Warford (1993). *World Without End: Economics, Environment, and Sustainable Development.* New York: Oxford University Press.

Pun, S.B. (2012). 'Nepal's Water Resources in 2030'. In. S.R. Sharma, B.R. Upreti and K. Pyakuryal (eds). *Nepal 2030: A Vision for Peaceful and Prosperous Nation.* South Asia Regional Office of the Swiss National Centre of Competence in Research (NCCR North – South) and Department of Development Studies, Kathmandu University, Kathmandu.

Rahman, M.M. (2005). *Integrated Water Resources Management in the Ganges Basin: Constraints and Opportunities.* Department of Civil and Environment Engineering, Laboratory of Water Resources, Helsinki University of Technology, Finland.

Sadoff, C., Rao, H., Blackmore, D., Wu, X., McDonnell, A., Jeuland, M., Lee, S. and Whittington, D. (2012). Ten Fundamental Questions for Water Resources Development in The Ganges: Myths and Realities. *Water Policy* 15: 147–64.

Saleth, R.M. (1993). Water Rights System: Is it Feasible for India? *Journal of Indian School of Political Economy* 5(3): 578–94.

Sharma, B., Amarasinghe, U., Xueliang, C., Condappa, D., Shah, T., Mukherji, A., Bharati, L. Ambili, G., Qureshi, A. Pant, D. Xenarios, S. Singh, R. and Smakhtin, V. (2010). The Indus and the Ganges: River Basins Under Extreme Pressure. *Water International* 35(5): 493–521.

Singh, K. (1995). 'Co-operative Property Rights as an Instrument of Managing Groundwater'. In M. Moench (ed.). *Groundwater Law: The Growing Debate.* Monograph, VIKSAT– Pacific Institute Collaborative Groundwater Project, Ahmedabad.

The Economic Times (2010). China dams Brahmaputra river in Tibet, 16 November.

The Hindu (2013). Nepal, India and Bangladesh to make most of Ganga water, hydropower, 15 April.

Varis, O. and Tortajada, C. (2008). *Management of transboundary rivers and lakes*. Berlin: Springer.

Verghese, B.G. (2007). *Waters of Hope: Facing New Challenges in Himalayan Ganga Cooperation*. 4th edn. India Research Press.

Water and Energy Commission Secretariat (2002). *Water Resources Strategy Nepal 2002*, WECS, Government of Nepal, Singhdurbar, Kathmandu.

Water and Energy Commission Secretariat (2005). *National Water Plan 2005*, WECS, Government of Nepal, Singhdurbar, Kahmandu.

Water Resources Planning Organization (WARPO) (2001). National Water Management Plan (NWMP) 1. WARPO, Dhaka.

World Bank (2014). *Ganges Strategic Basin Assessment: A Discussion of Regional Opportunities and Risks*. Report No. 67668-SSA, South Asia Water Initiatives.

World Meteorological Organization (WMO) (2003). *Integrated Flood Management – Case Study Bangladesh: Flood Management*. WMO, Geneva.

15

Poverty, inequalities and vulnerability of the rural poor

Upali A. Amarasinghe, Fraser Sugden and Floriane Clément

Key messages

- Income poverty in the Ganges River Basin (GRB) has reduced considerably in the last two decades. However, the basin is still one of the largest contiguous hotspots of poverty; 26 per cent of the 600 million basin population are income poor, but 75 per cent are multidimensional poor, deprived of the basic human requirements of education, health and assets required for decent living. Spatially targeted interventions are required to address the income and multidimensional poverty.
- Vulnerability and chronic poverty are also prevalent in the basin. Both ex-post and ex-ante poverty measures are required for forward-looking interventions for poverty reductions.

Accurate measures of vulnerability to poverty are not only useful for the addressing of interventions for reducing both vulnerability and chronic poverty, but also help with enhancing the resilience of poor people against the shocks and risks.
- Rural poverty reduction efforts have often focused on increasing agricultural productivity. However, efforts to raise agricultural productivity alone might not benefit the chronically poor. An entitlement-based approach can reduce the incidence of rural poverty faster. Increasing access to irrigation, especially groundwater, could play an important role in reducing poverty, but that alone is not sufficient and needs to be coupled with other measures that fall outside of the water sector.
- Access to groundwater is tightly linked to land tenure and ownership patterns. The latter are critical structural constraints preventing agricultural growth. Policy supporting a radical overhaul of land tenure and ownership patterns, which change at a painfully slow pace now, could reduce poverty based on unequal agrarian relations. This can be the catalyst for rapid investments for agricultural intensification and productivity growth in the eastern Ganges.

Introduction

The Ganges River Basin (GRB) is one of the world's largest contiguous hotspots of poverty. A popular measure of poverty – the head count ratio (HCR), the percentage of people living below the national or international income poverty lines (Foster *et al.*, 1984) – is high in the riparian countries of India, Bangladesh and Nepal of the GRB (World Bank, 2015). High HCR and high depth and severity of poverty leave large pockets of chronic poverty in the riparian countries (Metha and Shah, 2003).

However, poverty in the basin is multidimensional. Deprivations of the basic human needs of education, health and living standard – the other dimensions of poverty – have featured prominently in recent poverty assessments (Cohen, 2009; Alkire and Santos, 2011). The multidimensional poverty index (MPI), a composite measure of various aspects of human well-being (education, health, and standard of living), of the riparian states/countries within the basin is worse than that of many of the poorest countries in the world (Alkire and Santos, 2011).

The most commonly quoted causes of poverty for the basin stem from two conditions. The prevailing extreme climatic conditions, which have become increasingly variable (Gosain *et al.*, 2006) with recurrent floods and droughts, and the high population density, where the rural population is largely dependent on agriculture and characterized by low levels of literacy, skills and knowledge, and affected by inadequate access to land and water resources (Sharma, 2010).

These factors lay the basis for a popular policy and scientific narrative that calls for a second green revolution to eradicate poverty in the basin. This raises a few questions though, including 'why has the first green revolution not been as successful as expected in this region?' Is raising agricultural productivity sufficient to reduce multidimensional poverty and entrenched inequalities? This chapter proposes to re-examine the proximate causes and underlying drivers of poverty in the basin as of today within a rapidly changing social, economic and biophysical context. We first explore the temporal and spatial trends of poverty, inequality and vulnerability in the basin for the last two decades, using secondary data. Then, we delve into the proximate and structural factors that have contributed to produce and reproduce poverty and vulnerability in the basin, based on a review of the state-of-the-art literature.

We distinguish the proximate causes of chronic poverty in rural areas and its underlying causes, which we call 'structural factors'. Our focus, when exploring the proximate causes, is the distribution of assets and access to agricultural inputs and natural resources. Dwelling into structural factors allows exploring the economic and political structures that constrain and support individuals and social groups to access and use assets and resources 'to be and to act' (Bebbington, 1999), looking at issues of recognition and governance. Lastly, based on our analysis, we propose policy options to support reduction of poverty, inequality and vulnerability in the basin.

Poverty trends

Incidence of poverty

The HCR, which shows the percentage of the population below the income poverty line, has halved in India and Bangladesh and by 40 per cent in Nepal since the mid-1990s (Table 15.1). In spite of these promising trends, the level and variation of poverty in the GRB is still high:

- In 2005, the GRB had about 214 million poor people, nearly 40 per cent of its total population; by 2011, the total poor population had reduced to 158 million, about 26 per cent of the total population.
- Rural poverty has decreased from 175 to 127 million people between 2004–05 and 2010–11, or from 44 per cent to 29 per cent of the total rural population.

The extent of poverty varies considerably within the basin. The district-level rural poverty map (Figure 15.1) shows that the eastern part of the GRB, including Nepal, is the poverty hotspot of the GRB.

- The Indian states of Uttar Pradesh, Bihar, West Bengal and Uttarakhand, which lie entirely inside the GRB, hold 70 per cent of the poor population of the GRB (about 111 million) in 2010, and 82 per cent of them live in rural areas.
- Nepal and Bangladesh have another 9 per cent of the total poor in the GRB, and 74 per cent of them live in rural areas.

Depth and severity of poverty

The poverty gap index estimates the depth of poverty as the average distance income of the poor from the poverty line, and the squared poverty gap index estimates the severity of poverty among the poor as average squared distance of income from the poverty line (Foster *et al.*, 1984). Reductions in depth and severity of poverty show the extent to which the poorest of the poor were able to improve their livelihood over time, although this does not necessarily translate into reduction in HCR. Policies that combine and address all three indices can reduce poverty faster in all segments of the poor population.

The depth and severity of poverty in the riparian countries are slowly reducing. However, they are still substantially high in the basin. For example, the HCR of Jharkhand, Chhattisgarh and Bihar shows a three to four times higher incidence of poverty than that of the western region of the GRB, but the depth and severity of poverty are seven to eight times higher, respectively. Slow rates of reduction of depth and severity coupled with high population growth show that there is a sizable chronically poor population in the basin.

Table 15.1 Trends of HCR, poverty gap index and squared poverty gap index of the riparian countries in the GRB

Indicator	*Sector*	*India*			*Bangladesh*			*Nepal*			*GRB*	
		1993–1994	*2004–2005*	*2011–2012*	*1993–1994*	*2004–2005*	*2011–2012*	*1993–1994*	*2004–2005*	*2011–2012*	*2004–2005*	*2011–2012*
HCR	Rural	50.1	41.8	25.7	39.5	28.6	21.1	43.3	34.6	27.4	43.9	28.9
	Urban	31.8	25.7	13.7	13.7	14.6	7.7	21.6	9.6	15.5	26.5	18.1
	Total	45.3	37.2	21.9	35.2	25.1	17.6	41.8	30.8	25.2	39.7	26.2
Poverty gap index	Rural	5.6	3.3	–	14.6	5.3	3.7	12.1	8.5	5.9	–	–
	Urban	–	–	–	8.6	2.6	1.3	6.5	2.2	3.2	–	–
	Total	4.8	2.7	–	13.6	4.6	3.1	11.8	7.6	5.8	–	–
Squared poverty gap index	Rural	2.8	1.6	–	–	1.5	1.0	–	3.1	2.0	–	–
	Urban			–	–	0.7	0.4	–	0.7	1.0	–	–
	Total	2.8	2.5	–	–	1.3	0.8	–	2.7	1.8	–	–

Sources: Planning Commission of India (2007, 2014a), BBS (2012), CBS (2012)

Notes: Data for Bangladesh refers to National Living Standards Surveys of 1995–96, 2005 and 2010, and data for Nepal refers to National Living Standards Surveys of 1995–96, 2003–04 and 2010–11, respectively

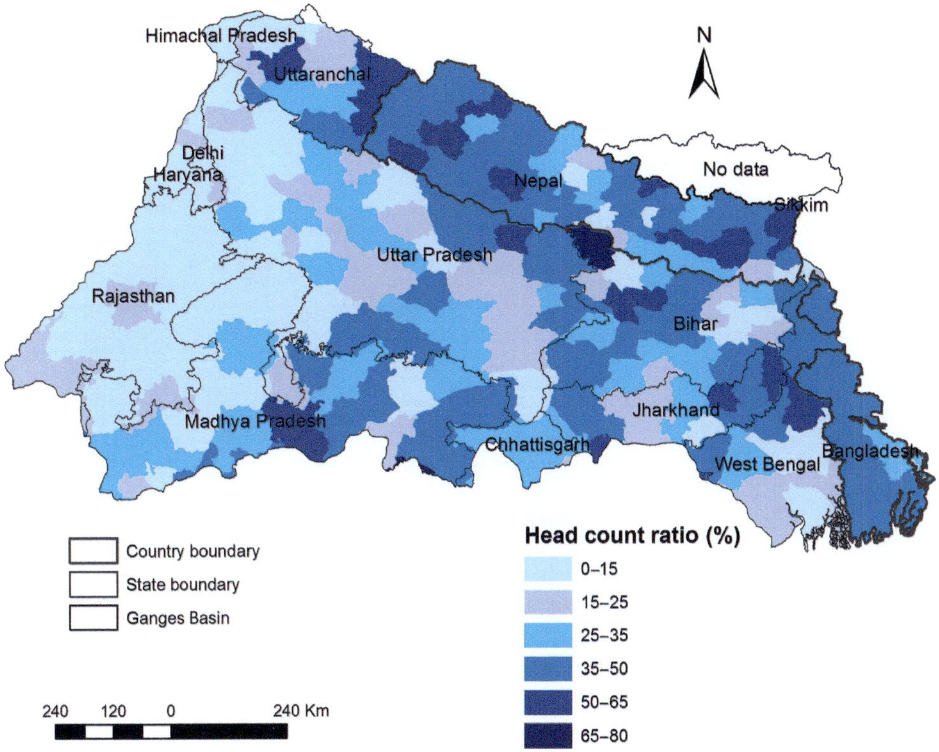

Figure 15.1 Spatial variation of HCR in 2004–05
Source: Sharma *et al.* (2010)

Chronic poverty

The chronic poor are those who experience poverty in an extended period of time (Hulme, 2003). Understanding chronic poverty requires data observed from the same individuals or households over several years. This data is not available for India. However, according to Mehta and Shah (2003), there is a close correlation between high depth and severity of poverty and chronic poverty. Severity of poverty was persistently high in the regions of southern, eastern and central Uttar Pradesh, south-western Madhya Pradesh, Chhattisgarh, Bihar and Jharkhand, where substantially high tribal and socially disadvantaged communities live. These people are physically and socially isolated and hence are unable to access resources and entitlements. Moreover, they have a low level of education and social networking, which makes it difficult for them to find and stay in areas with better job opportunities (Mehta and Shah, 2003).

Unlike in India, Bangladesh and Nepal have panel data coverage for the whole country. In Bangladesh, 31 per cent of the 2000 poor population, and 93 per cent of the small land-holders (less than 1 ha) were chronically poor, of whom the poverty status has not changed since 1987 (Sen, 2003). In Nepal, 20 per cent of the total population, and 43 per cent of the poor population, are chronically poor, i.e. poor in both 1995–96 and in 2003–04 (UNDP, 2015). High chronic poverty in Nepal could be due to low agricultural wages and low financial and human capital (educational achievement) (Bhatta and Sharma, 2006), which Erenstein *et al.* (2010) finds a marked decline from the western to eastern gradient of the GRB.

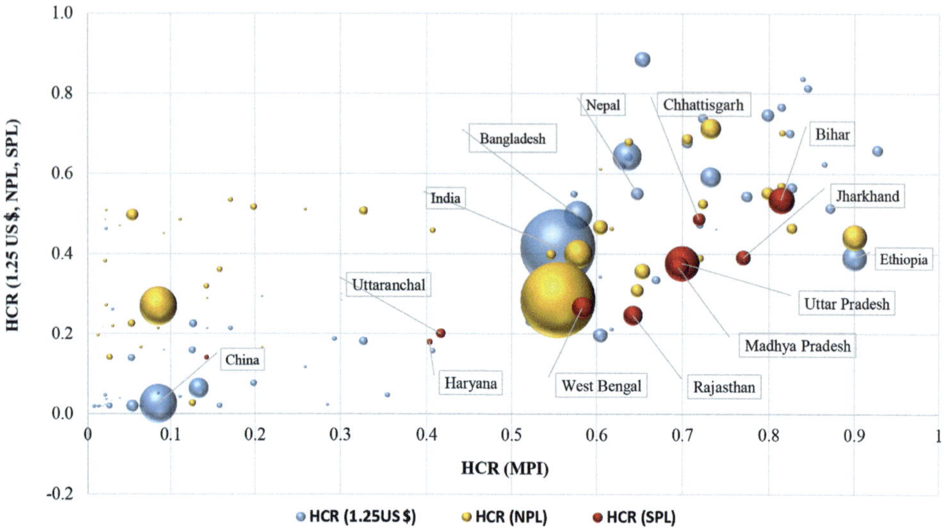

Figure 15.2 MPI poor and HCR below 1.25 US$/day

Sources: HCR (MPI), HCR (1.25 US$) and HCR (IPL) data are from Alkire and Santos (2011), HCR (SPL) are from the Planning Commission of India (2014b)

Note: MPI is the composite of three primary dimensions: inadequate health, education, standard of living, and 10 sub-indicators (Table 15.2). Each primary dimension receives equal weight of 1/3, and each sub-indicator within them receives equal weights. The aggregate value of the 10 indicators is the average intensity of deprivation. The MPI is the product of HCR (MPI) and the average intensity of deprivation, where HCR (MPI) is the percentage of population with average intensity of deprivation indicator below 30 per cent

Multidimensional poverty

Figure 15.2 reveals the depth of multidimensional poverty in the GRB. It shows the scatterplots of HCR of Alkire and Santos's (2011) multidimensional poverty indicator HCR (MPI) versus various other HCRs: per cent of population below the national poverty line (HCR [NPL]), below the international poverty line of 1.25US$ (HCR [IPL]), and below the state poverty line (HCR [SPL]). The three riparian countries of India, Bangladesh and Nepal have 760 million multidimensionally poor people, which is significantly higher than the number of multi-dimensionally poor people in all countries (mainly in Asia and Africa) that are poorer than India and Bangladesh. Moreover, many of the riparian Indian states are some of the poorest according to MPI. The Indian states of Uttar Pradesh, Uttaranchal, Bihar and West Bengal, and Nepal, are among the 25 poorest according to MPI – they have about 286 million multidimensionally poor people.

Although widely used in poverty assessment, HCR (NPL) is not a good indicator for identifying multidimensional poverty. The HCR (MPI) for many poor countries is substantially higher than HCR (NPL), and overall they are only moderately correlated. This is due to nutritional adequacy in the first dimension of HCR (MPI) and food poverty in the HCR (NPL). The HCR (MPI) is also generally lower than the HCR (US$ 1.25), but they are significantly correlated. This is because the international poverty line (US$ 1.25) captures both food and non-food consumption of most poor countries.

The poor standard of living is the largest contributor to multidimensional poverty in the GRB (Table 15.2). In particular, the inadequate access to sanitation, housing, cooking fuel and

Table 15.2 MPI and components of the riparian countries and regions in the GRB

Riparian states/ countries of GRB	*Education*		*Health*		*Standard of living*						*MPI*		
	Schooling	*Child school attend-ance*	*Mortality*	*Nutri-tion*	*Elec-tricity*	*Sanita-tion*	*Drinking water*	*Floor*	*Cooking fuel*	*Assets*	*MPI value*	*HCR (MPI)*	*Average intensity of deprivation*
Weights	*1/6*	*1/6*	*1/6*	*1/6*	*1/18*	*1/18*	*1/18*	*1/18*	*1/18*	*1/18*			
India	0.18	0.25	0.23	0.39	0.29	0.49	0.12	0.40	0.52	0.38	0.30	0.55	0.54
Uttar Pradesh	0.18	0.36	0.37	0.46	0.48	0.62	0.07	0.58	0.66	0.41	0.39	0.70	0.55
Bihar	0.35	0.52	0.35	0.61	0.65	0.74	0.04	0.70	0.79	0.57	0.50	0.81	0.61
West Bengal	0.25	0.23	0.19	0.42	0.41	0.47	0.07	0.48	0.57	0.43	0.32	0.58	0.54
Uttarakhand	0.08	0.10	0.15	0.30	0.15	0.33	0.10	0.30	0.37	0.27	0.40	0.47	9.60
Madhya Pradesh	0.22	0.32	0.31	0.50	0.25	0.65	0.31	0.57	0.67	0.52	0.39	0.70	0.56
Jharkhand	0.26	0.45	0.30	0.56	0.55	0.73	0.42	0.63	0.76	0.55	0.46	0.77	0.60
Chhattisgarh	0.21	0.29	0.31	0.52	0.24	0.69	0.22	0.64	0.70	0.48	0.39	0.72	0.54
Rajasthan	0.21	0.32	0.28	0.44	0.31	0.60	0.24	0.36	0.61	0.47	0.35	0.64	0.55
Haryana	0.08	0.20	0.15	0.30	0.08	0.34	0.08	0.24	0.39	0.25	0.20	0.42	0.48
Delhi	0.04	0.09	0.07	0.09	0.00	0.10	0.03	0.02	0.05	0.06	0.06	0.14	0.44
Bangladesh	0.24	0.09	0.24	0.37	0.39	0.48	0.03	0.54	0.57	0.45	0.29	0.58	0.50
Nepal	0.29	0.16	0.30	0.40	0.43	0.56	0.14	0.60	0.63	0.47	0.35	0.65	0.54

Source: Alkire and Santos, 2011

assets contribute the most. Additionally, much of the eastern part of the basin also fared very poorly in terms of nutrition, child school attendance and access to electricity. Demographically, the rural areas and socially disadvantaged groups have higher multidimensional poverty. In India, the marginalized groups – the scheduled tribes, scheduled caste and other backward class – are multidimensionally poorer than the other social groups (Alkire and Santos, 2011; Mehta and Shah, 2003). Inadequate access to nutritional diet, sanitation, housing, cooking fuel and assets are the major contributing factors to high MPI in this group.

Because poverty is usually measured at the household level, they often leave out intra-household inequalities. However, many studies have shown that women, the elderly and the disabled are more affected by chronic poverty than others. They face multiple forms of deprivation which tend to be mutually reinforcing, making it more difficult to escape from the poverty trap (Mehta and Shah, 2003).

Inequality and poverty

Poverty and inequality have no clear relationship (Beteille, 2003). Income growth reduces poverty, but can increase inequalities between the powerful and the poor and marginalized. In the GRB, there are conflicting views on the equitable nature of economic growth in the riparian countries. While some report increased regional or sectoral income inequalities in India since the start of the economic reforms in the 1990s (Sen and Himanshu, 2004; Pal and Ghosh, 2007), others show no significant change (Bhalla, 2003; Singh *et al.*, 2003).

Estimates of the Gini coefficient show that income inequality among the rural population has decreased in many of the major riparian states of India (Planning Commission of India, 2014b), while it has increased in Bangladesh except in the Dhaka region (Uz Zaman and Akita, 2012). While income growth and other government programmes have contributed to decreasing inequality and poverty in the Indian GRB (Tiwari, 2014), higher income growth has contributed to increased inequality in rural areas in Bangladesh (Uz Zaman and Akita, 2012), which could be due to high chronic poverty in Bangladesh.

Vulnerability to poverty

Vulnerability to poverty can be seen as the risk that a household will suddenly reach a position in which it is unable to cope, leading to a catastrophe. Unlike poverty, the measures of vulnerability to poverty are ex-ante risk (Chaudhuri, 2003), and show the likelihood that people remain poor or escape out of poverty, given the present status and future exogenous or endogenous shocks. Therefore, accurate measures of vulnerability and their causes are at least as useful as the measures of poverty for effective targeting of interventions for reducing both vulnerability to poverty and chronic poverty.

The assessments for measuring vulnerability have surged recently (Chaudhuri, 2003; Calvo and Dercon, 2005; Landau *et al.*, 2012). They are important for two reasons:

1 Vulnerability is different from poverty, where many non-poor now can be poor or those who are poor now remain poor in the future for many reasons. While the poverty measures are ex-post assessments of the state of human well-being, the vulnerability to poverty measures are ex-ante assessments of future risks to human well-being.
2 Vulnerability to poverty increases with increasing exogenous risks from floods and droughts, sharp economic fluctuations, or impacts of natural hazards such as earthquakes, flash floods and cyclones, or from idiosyncratic factors exacerbated by poverty. Therefore, understanding

the present level of poverty and the influence of risks and shocks in the future is useful for enhancing resilience and adaptation.

Although the GRB has large poverty hotspots, there have not been any detailed econometric assessments of the dynamics of vulnerability to poverty. This is due to inadequate data, especially that of large panel surveys. However, the few available analyses using limited panel and cross-sectional data sets across the riparian countries show interesting poverty–vulnerability dynamics.

Analysis of the village-level panel surveys covering rural semi-arid areas of Andhra Pradesh and Maharashtra in India shows that current poverty level is only one component for vulnerability to poverty in the future (Gaiha and Imai, 2008). The aggregate risks (due to crops and weather factors) and idiosyncratic risks (due to health and unemployment), conditions prevalent in the GRB, contribute to vulnerability more than does poverty.

In Bangladesh, the substantial non-poor population at present is vulnerable to poverty in the future. An assessment using recent data (Azam and Imai, 2009) shows that more than 50 per cent of the rural population in 2005 is vulnerable to poverty in the future, although income poverty in 2005 was only 29 per cent. Vulnerability is high in the riparian regions of the GRB and particularly among agriculturally dependent households and those without education.

In Nepal, analysis of the panel living standards survey data in 1993–94 and 2000 showed that two-thirds of the population in the mountain region and close to half the population in the Tearai and rural hills are vulnerable and chronically poor. Major factors associated with vulnerable groups are inadequate human capital (i.e. illiteracy), high dependency on agricultural wage as a primary source of income, low value of the agricultural holding, violence and conflict-related deaths, and regional factors such as locations with access difficulty.

Detailed assessments of vulnerability are useful for enhancing the resilience of poor people to risks and shocks. The concepts of resilience and vulnerability offer two complementary analytical frameworks to inform policy responses that support poor households to mitigate risks and shocks (Miller *et al.*, 2010). Yet their dynamics and complementarities are two aspects least understood in the GRB.

Causes of chronic rural poverty

> *Poverty comes to be seen as a lack of resources rather than an absence of entitlements, as an 'economic' rather than a political problem*
>
> *(Green and Hulme, 2005)*

There are many different theories and schools of thought on the causes of chronic poverty and vulnerability, ranging from the effects of global capitalist development to individual characteristics and psychology (Hulme and Shepherd, 2003). In this chapter we adopt an entitlement-based approach to explain why people have become or remained chronically poor in the basin, which we combine with an analysis of broader political and economic factors and social relations (Bebbington, 1999; Leach *et al.*, 1999).

Proximate causes: the role of agricultural productivity

The causes of poverty in the basin are spatially heterogeneous. Mehta and Shah (2003) distinguished the proximate causes of poverty in less-favoured areas in India. The drylands are affected by frequent failure of crops and by an acute lack of employment opportunities, which also affects vulnerability; the forest-based regions are characterized by the erosion of entitlements of poor people to access, use and benefit from natural resources, lack of infrastructure and markets, as well as social exclusion. Yet, many analysts and a large number of rural development programmes have focused on one single cause: low agricultural productivity.

The rationale for increasing agricultural productivity to alleviate poverty is commonly justified by three factors: low gross domestic product (GDP), low contribution by agriculture to total GDP, and a large agriculturally dependent population in the riparian states and countries. A few decades ago, agricultural GDP growth was a major contributor to overall economic growth of the riparian countries, which in turn had a major impact on reducing rural poverty (Fan and Haque, 2000; Mellor, 2001). However, the contributions from the agriculture sector to total GDP have gradually decreased in all riparian countries. Moreover, the agriculturally dependent populations have decreased slowly between 1990 and 2013 (29 per cent to 21 per cent in Bangladesh, and 24 per cent to 22 per cent in India; but in fact increased in Nepal from 34 per cent to 41 per cent [FAO, 2015]).

Figure 15.3 shows that many riparian regions with a large agriculturally employed population (Uttar Pradesh, Bihar, Madhya Pradesh and Chhattisgarh) still have low agricultural and total GDP per person and a high incidence of poverty. The width of the bubbles in Figure 15.3 is proportionate to the total GDP per person, with Bihar having the lowest (US$295/person) and Haryana having the largest (US$1,493/person). Such figures suggest that improving agricultural output to increase GDP can still be a quick way of reducing rural poverty in the region.

There is potential for the riparian states to increase agriculture GDP through cropping intensification. In the Indian part of the GRB, land use intensity is less than 75 per cent in both the *Kharif* (June to October) and *Rabi* (November to March) seasons. Cropping in the summer season hardly exists. Intensity of irrigation is even lower – less than 50 per cent in each season.

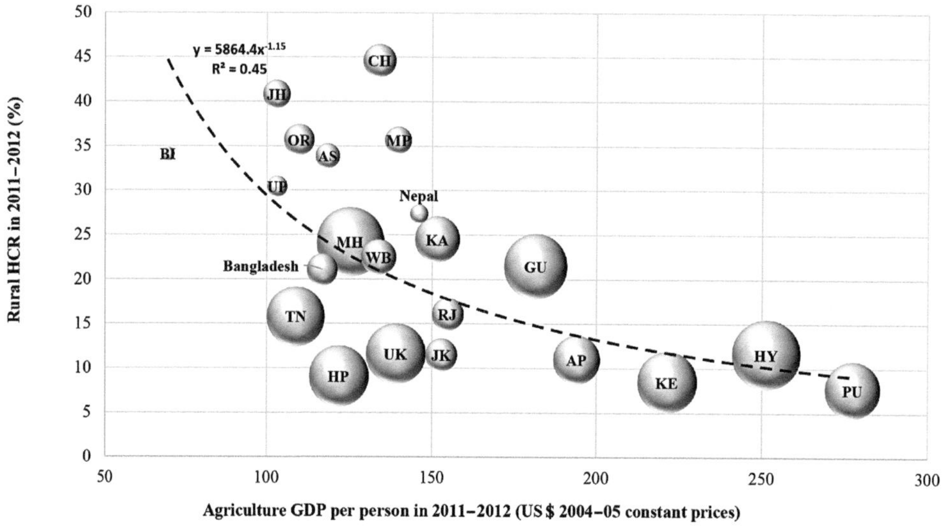

Figure 15.3 Rural HCR and agricultural GDP/person of Indian states, Nepal and Bangladesh

Notes: RGB riparian states (BI – Bihar, CH – Chhattisgarh, HY – Haryana, JH –Jharkhand, MP – Madhya Pradesh, RJ – Rajasthan, UP – Uttar Pradesh, UK – Uttarkhand, WB – West Bengal), other states (AS – Assam, HP – Himachal Pradesh, JK – Jammu and Kahmir, MH – Maharashtra, KA – Karnataka, KE – Kerala, PU – Punjab, TN – Tamil Nadu)

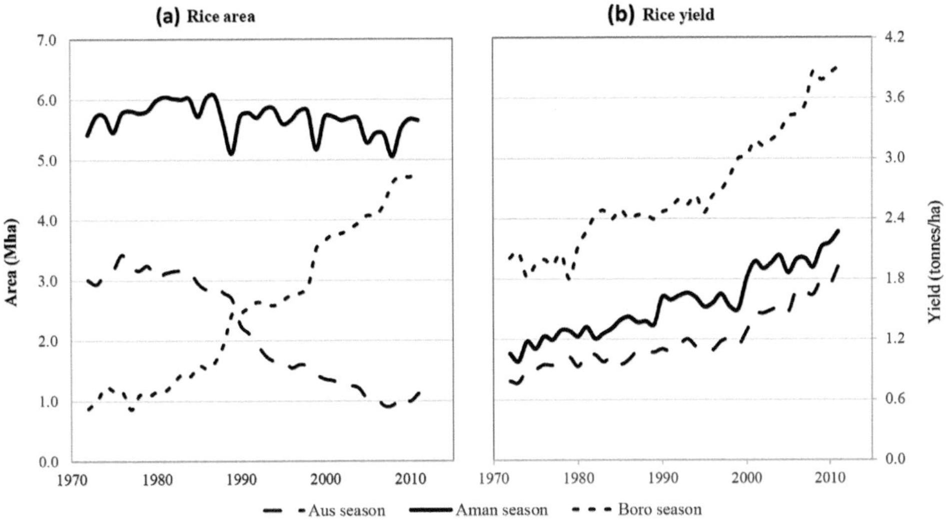

Figure 15.4 Changing rice crop patterns and yield in Bangladesh
Source: Amarasinghe *et al.* (2014)

With proper cropping pattern and irrigation use, potential exists to increase annual cropping intensity from less than 150 per cent now, to 200–300 per cent. Such changes are already happening in some parts of the basin.

Changes in cropping and irrigation patterns of rice are major contributors to increasing irrigated area, cropped area and production of rice in Bangladesh (Figure 15.4). Changes in boro rice area, mainly irrigated in the summer season (March–June) have helped the country to double its rice production since 1990 (Amarasinghe *et al.*, 2014). This has also helped crop diversification and growth in farmers' income, which subsequently improved the rural economy and reduced rural poverty.

Structural causes of poverty

Many studies investigating farm productivity and household incomes found negative correlation between enhanced agricultural productivity and reduced poverty (e.g. Huang *et al.*, 2006; Minten and Barrett, 2008). However, many increasingly challenge the nexus of productivity and poverty too, notably so in India. In the latest draft of the national policy for farmers, the National Commission on Farmers in India emphasized the need to 'focus more on the economic well-being of the women and men feeding the nation than just on production' (National Commission on Farmers – India, 2006, p. 2).

In this respect, it is worth examining the effects of the green revolution on poverty (Lipton, 1989; Harris, 1991; Beck, 1995; Das, 2002). Many agree that the green revolution resulted in an increase in inequality among farmers. On the one hand, the green revolution has been said to enrich the better-off farmers at the expense of the poorest (Beck, 1995); it also increased the reliance of poor farmers on external inputs and pushed them into a vicious circle of debt and made them more vulnerable to droughts (Shiva *et al.*, 1999). On the other hand, the gains in productivity of the technology adopters have resulted in increased agricultural employment opportunities and reduction in food prices due to increased production. However, the latter arguments have also been criticized based on economic arguments (e.g. lowered food prices have resulted in reduced benefits for farmers located in other areas left untouched by the green revolution and thereby reduced employment opportunities for labourers).

Agrarian class structure and land ownership

To better understand chronic poverty and rising inequality, despite improvements in inputs and agricultural technology, it is important to look at the structural causes of poverty, rooted in the multiple axes of class, caste and gender. One of the most critical axes of structural inequality is land ownership, which is intricately connected to historical axes of exclusion rooted in caste and gender ideology. In the Gangetic Plains in particular, one's position in a deeply inequitable agricultural land ownership system is central to the inter-generational reproduction of poverty and social exclusion.

The three systems of land administration are the foundation for contemporary inequalities in the region (Banerjee and Iyer, 2002). These included the *raiyatwari* system where farmers were directly accountable to the colonial rulers to pay tax, the *mahalwari* system where communities had responsibility, and the *zamindari* system where the traditional proprietors from the Mughal era were given rights to collect revenue.

Most severe inequalities are in the domains where the zamindari system predominated, where there was continuity in Mughal era social relations. It is predominant across Bihar, Bengal and the Lucknow region (Oudh). The zamindars became the administrators of all property rights to land under their jurisdiction and could collect fixed revenue from the tenants according to customary rates (Raihan *et al.*, 2009). This would be transferred to the colonial authorities, after retaining a portion for themselves (Hartmann and Boyce, 1983). In lowland Nepal, a similar tax appropriation system was in place, known as the *jimidari* system, although rather than revenue being channelled to the colonial authorities, it was appropriated by the feudal Rana regime (Regmi, 1978). Across the region, these landlords were able to expand their estates by appropriating the land of indebted peasants.

After independence a wave of land reforms took place across the region and the zamindar's political power was terminated in India, Nepal and East Pakistan (now Bangladesh) (Adhikari, 2006). Despite the stated objectives of the reforms, there was limited political commitment to change, the ceilings on land-holdings introduced were weakly enforced, and the reforms failed to create real transformations in agrarian relations (Regmi, 1976; Adhikari, 2006). Landlords were integrated into the state agencies implementing reforms, and acted as an impediment to change (Sugden and Gurung, 2012).

However, the land reforms that took place in West Bengal in the 1970s achieved relative success in breaking up many of the zamindari estates. The West Bengal Land Reforms Act of 1977 imposed strict regulations on ceilings, and restricted landlords' capacity to evict tenants. It gave tenants permanent and inheritable tenure, and enforced a 25 per cent limit to the crop share. By 1993, more than 65 per cent of an estimated 2.3 million sharecroppers had been registered, facilitating significant productivity increases (Banerjee *et al.*, 2002). As of 2005, 445,503 ha of agricultural land had also been redistributed in the state (Bakshi, 2008).

Recent research shows declining authority of landlords in Bihar and southern Nepal (Sugden *et al.*, 2014a). The main reason for this is that the new economic opportunities available to the younger generation from the educated landed elite mean that the pressure to hold onto the land has declined. Nevertheless, inequality remains acute. In some regions, landlords retain large estates and relations of production remain 'semi-feudal' in character. The only change includes reduced political and ideological bondage to landlords. Furthermore, in many cases, the

Table 15.3 Distribution of land-holdings

State	*Year*	*Percentage of households with:*				
		No land	*< 1 ha*	*1–2 ha*	*2–4 ha*	*>4ha*
Bihar	1971–72	4.3	71.7	15.1	9.2	4.0
	1982	4.1	76.6	12.4	7.8	3.1
	1992	8.6	80.6	11.1	6.0	2.3
	2003	7.6	89.4	7.1	2.7	0.8
West Bengal	1971–72	9.8	77.6	12.6	7.3	2.4
	1982	16.2	81.6	11.5	5.5	1.4
	1992	11.0	85.9	9.5	3.9	0.7
	2003	6.2	92.1	5.7	1.4	0.2

Source: NSSO (2006)

break-up of larger estates entails the redistribution of land not to the tenant and labouring majority, but to large and medium farmers with purchasing power (Sugden *et al.*, 2014b). The decline in large estates is also linked in some instances to fragmentation within the family in the context of population growth.

Statistics from Bihar show that inequality in land distribution remains high (Table 15.3). Over the last three decades, the percentages of landless households and those with less than 1 hectare have increased steadily, a process likely to be linked to both population growth and distress sales of land.

With regard to tenancy data, the NSS suggests that 12.8 per cent of the owned area is leased by other farmers, who represent 12.3 per cent of the sample. Out of these tenants, 93 per cent have less than 0.5 ha of their own land, while 1.94 per cent lease out land. In the east and central Terai of Nepal the proportion of farmers owning land has actually declined significantly, while landlessness and the proportion of pure tenants has increased (Nepal National Sample Census of Agriculture, 1992–2011). In Bangladesh, land inequality has intensified over the last few decades. Between 1984 and 2005, absolute landlessness has increased from 8.8 per cent to 14.0 per cent, and households that own less than 0.5 acres have increased from 56.5 per cent to 68.0 per cent. The percentage of pure tenants has also increased from 1.1 per cent in the 1960s to 6.51 per cent in 2005.

In sum, across Bihar, the Nepal Terai and Bangladesh, at the base of the agrarian structure is a large class of marginal and tenant farmers, while at its apex is a small class of large farmers and landlords. These historically entrenched land inequalities are central to the reproduction of poverty in the region (Sugden *et al.*, 2014a). A study from across Bangladesh, West Bengal, Bihar and Nepal (Sugden *et al.*, 2015) noted that the most marginalized population are tenants and part tenants, who have to surrender up to 50 per cent of their harvest as rent. The rent burden, combined with rising living costs and monetization of the economy, had an unprecedented impact on livelihoods. For many households, the rent burden not only contributes to food insecurity, it increases the cost of production, impeding their capacity to invest in agricultural inputs, education and healthcare. And many of these households are trapped in cycles of indebtedness to both landlords and private money lenders.

Access to water resources has been identified as a key determinant of reduced rural poverty (Hussain *et al.*, 2004; Narayanamoorthy, 2007). However, this relationship hides issues of inequity in entitlements to water across class, caste and gender, which are, in the region, closely related to entitlements to land. Often the land title is in the name of the head of the household, often men. In such a case, when women take over irrigation after men migrate, they cannot claim access to public subsidies for groundwater irrigation.

The relationship between groundwater and poverty might not hold true under high diesel prices and unreliable and inadequate public electricity supply. The latter have particularly affected the poor who depend on pump irrigation services to irrigate their crops (Shah *et al.*, 2009). Hussain *et al.* (2004) also found high incidence of poverty within the command areas of large-scale irrigation systems. The major reasons for this are the inequitable distribution of water between head, middle and tail reaches, exacerbated by unequal land distribution. Lastly, uncontrolled groundwater development across the basin has led to serious environmental and health issues (Natarajan and Prathapar, 2014) which in the long-term may accentuate vulnerability.

In this context, male out-migration for menial wage labour in urban centres is increasingly a necessity for the most marginal farming households to subsist (Sugden *et al.*, 2014a). Paradoxically, in the case of the Nepal Terai, migration itself worsens the debt crisis, as farmers seeking overseas opportunities must take significant loans to pay to manpower agents (THRD Alliance, 2012).

Policies and governance

Addressing the structural causes of poverty entails changes in governance. In India, agricultural water interventions supposedly largely rely on decentralized planning. Generally, decentralization and participation have often been presented as institutional panaceas for poverty reduction. One of the underlying hypotheses is that democratic participation will reinforce accountability mechanisms, which in turn will improve equity in the distribution of government services and public goods (Johnson, 2003). Yet there is no consistent evidence of the causality between political and administrative decentralizations and poverty reduction (Crook and Sverrisson, 2001).

First, top-down decentralization policies in the field of natural resource management have often been more recentralizing than decentralizing (Ribot *et al.*, 2006). After two decades of the panchayat raj system, Singh describes India as a 'centralized representative democracy' (2007, p. 7) which has not been able to address rural poverty. When comparing the outcomes of decentralization in South Asia and West Africa, Agrawal and Ribot (2000) showed that a high level of downward accountability, and creating institutions that allow durable devolution of decision-making powers to local authorities, are key to pro-poor outcomes. Civil society can play a crucial role in enhancing accountability by representing poor people in democratic institutions and influencing policy processes and institutional reforms.

However, addressing rights and entitlements (e.g. to land) that underlie political structures in rural areas is critical to ensure that devolved authority does not simply empower local elites and aggravate existing poverty and inequality (Johnson, 2003). Opening of markets, reduced state support following liberalization of the economy, and the fluctuating price of cash crops are the main factors behind farmers' indebtedness and suicides in India (Kennedy and King, 2014). The latter observe that a large share of rural population has not benefited from the economic growth of the past 20 years. There should be support to ensure that intermediation between small farmers and new market institutions such as supermarkets, grades standards and export horticulture are fair for smallholders (Lipton, 2004).

Increasing intensity and improving productivity alone would not be sufficient to reduce rural poverty. There is clear evidence now that a vibrant non-farm economy can reduce rural poverty much faster than just the growth of farm income alone (De Janvry *et al.*, 2005; Lanjouw and Shariff, 2004; Himanshu *et al.*, 2011). A vibrant non-farm economy already absorbs part of the rural workforce, contributing to overall household income and well-being (Deshingkar, 2010).

Policies should support minimizing the risks and costs of existing migration, provide information for increasing access to non-farm opportunities, and develop the necessary infrastructure, markets and technology for creating more non-farm income opportunities.

Conclusions

Although the rate of income poverty has declined in the GRB in recent decades, the absolute number of poor, especially in rural areas, is still large. In fact, the GRB has more rural poor than most of the poorest countries in Africa. Multidimensional poverty is even worse, where more than 450 million people – three-quarters of its population – in 2011 are deprived of basic needs of education, health and assets.

Given the depth and severity of poverty, the GRB perhaps has the most vulnerable and chronically poor in the basin. Many of the poor as well as non-poor could be vulnerable to poverty in the future. However, there are no detailed quantitative assessments in the basin on understanding structural trends, economic, social-demographic and right-based, and their relationships that contribute to vulnerability to poverty. This is an urgent requirement for designing forward-looking interventions for reducing poverty given the increasing frequency of shocks and risks due to climate variation, economic fluctuations and high population density in the basin.

Poverty maps provide useful tools for showing various aspects and causes of poverty at sub-national scale. Such maps have been used by many countries for designing and implementing spatially targeted pro-poor interventions (Henninger and Snell, 2002). Sub-national multi-dimensional poverty maps over time could be even more useful for designing context-specific pro-poor interventions for the GRB.

Although the contribution of agriculture to GDP of the riparian countries is decreasing, the riparian regions of the GRB still have low total GDP per person and a large agriculturally dependent population. It is therefore widely held that improving agricultural output through crop intensification and productivity growth will help reduce poverty in the short term, given the right mix of institutions, knowledge, technology and financial assistance. However, increasing productivity should not be set as a goal, but as one of the multiple means to improve household well-being and reduce poverty and vulnerability, together with other forms of support to local livelihoods. Recent studies show that efforts that promote productivity increase can also increase inequalities in access to natural resources.

Fragmented lands, land tenure patterns and unequal agrarian relations are major constraints to change poor farmers' entitlements and increase agricultural productivity in the region. Although changes in land tenure patterns are taking place, the pace is slow. Policy interventions to implement fair reforms in land ownership and tenure patterns, although politically sensitive and difficult, can support reducing entrenched inequalities and unfair production relations in the region. It is equally crucial that local government institutions are increasingly more downwardly accountable to local men and women farmers, notably through the development of vibrant farmers' organizations representative of class, caste, gender and ethnicity in the region and the devolution of resources to local government bodies.

References

Adhikari, J. 2006. *Land Reform in Nepal: Problems and Prospects*. Kathmandu: Nepal Institute of Development Studies.

Agrawal, A. and Ribot, J.C. 2000. *Analyzing Decentralization: A Framework with South Asian and West African Environmental Cases*. Washington, DC: World Resources Institute.

Alkire, S. and Santos, M.E. 2011. *Acute Multidimensional Poverty: A New Index for Developing Countries.* Proceedings of the German Development Economics Conference, Berlin 2, No. 3.

Amarasinghe, U. Sharma, A., Bharat, Muthuwatta, R., Laland Khan, Z.H. 2014. *Water for Food in Bangladesh: Outlook to 2030.* IWMI Research Report 158.

Azam, M.S. and Imai, K.S. 2009. Vulnerability and poverty in Bangladesh. *Chronic Poverty Research Centre. Working Paper* 141.

Bakshi, A. 2008. Social Inequality in Land Ownership in India: A study with particular reference to West Bengal. *Social Scientist*, pp. 95–116.

Banerjee, A.V. and Iyer, L. 2005. History, institutions and economic performance: The legacy of colonial land tenure systems in India.

Banerjee, A.V., Gertler, P.J. and Ghatak, M. 2002. Empowerment and efficiency: Tenancy reform in West Bengal. *Journal of Political Economy*, 110: 239–80.

BBS (Bangladesh Bureau of statistics) 2012. Household Income and Expenditure survey (HIES) 2010. Accessed via http://203.112.218.66/WebTestApplication/userfiles/Image/HIES-10/Chapter-06.pdf.

Bebbington, A. 1999. Capitals and Capabilities: A framework for analyzing peasant viability, rural livelihoods and poverty. *World Development*, 27(12): 2021–2044.

Beck, T. 1995 The Green Revolution and poverty in India: A case study of West Bengal. *Applied Geography*, 15(2): 161–81.

Beteille, A. 2003. Poverty and inequality. *Economic and Political Weekly,* 4455–4463.

Bhalla S.S. 2003. Recounting the poor: Poverty in India, 1983–99. *Economic and Political Weekly*, 338–49.

Bhatta, S.D. and Sharma, S.K. 2006. *The Determinants and Consequences of Chronic and Transient Poverty in Nepal.* CPRC Working Paper 66. Chronic Poverty Research Centre, UK.

Calvo, C. and Dercon, S. 2005. Measuring individual vulnerability. Discussion Paper 229, Department of Economics, University of Oxford. Accessed via http://economics.ouls.ox.ac.uk/ 13354/1/Item.pdf.

CBS (Central Bureau of Statistics) 2012. *Poverty in Nepal.* Accessed via http://cbs.gov.np/wp-content/uploads/2012/Surveys/CBS%20view%20on%20poverty%20in%20Nepal.pdf.

Chaudhuri, S. 2003. *Assessing Vulnerability to Poverty: Concepts, Empirical Methods and Illustrative Examples.* Department of Economics, Columbia University, New York.

Cohen, A. 2009. *The Multidimensional Poverty Assessment Tool. Design, development and application of a new framework for measuring rural poverty.* International Fund for Agriculture Development: Rome.

Crook, R. and Sverrisson, A. (2001). Decentralisation and poverty-alleviation in developing countries: A comparative analysis or, is West Bengal unique? *IDS Working Paper 130.* Brighton, UK: Institute of Development Studies.

Das, R.J. 2002. The Green Revolution and Poverty: A theoretical and empirical examination of the relation between technology and society. *Geoforum,* 33(1): 55–72.

De Janvry, A., Sadoulet, E. and Zhu, N. 2005. *The Role of Non-Farm Incomes in Reducing Rural Poverty and Inequality in China.* Department of Agricultural & Resource Economics, UCB.

Deshingkar, P. 2010 *Migration, Remote Rural Areas and Chronic Poverty in India.* London: ODI

Erenstein, O., Hellin, J. and Chandna, P. 2010. Poverty mapping based on livelihood assets: A meso-level application in the Indo-Gangetic Plains, India.*Applied Geography*, 30(1): 112–25.

Fan, S. Hazell, P. and Haque, T. 2000. Targeting public investments by agro-ecological zone to achieve growth and poverty alleviation goals in rural India. *Food Policy* 25: 411–28.

FAO (The Food and Agriculture organization). 2015. FAOSATAT database. Accessed via faostat.fao.org.

Foster, A.D. and Rosenzweig, M.R. (2003) Agricultural productivity growth, rural economic diversity, and economic reforms: India, 1970–2000. D. Gale Johnson Memorial Conference. October 25.

Foster, J., Greer, J. and Thorbecke, E. 1984. A class of decomposable poverty measures. *Econometrica: Journal of the Econometric Society*, 761–6.

Gaiha, R. and Imai, K. 2008. Measuring vulnerability and poverty estimates for rural India, Research paper/UNU-WIDER, No. 2008.40, ISBN 978–92–9230–088–3. Accessed via http://hdl.handle.net/ 10419/45160.

Gosain, A.K., Rao, S. and Basuray, D. 2006. Climate change impact assessment on hydrology of Indian River basins. *Current Science, 90*(3): 346–53.

Green, M. and Hulme, D. 2005. From correlates and characteristics to causes: thinking about poverty from a chronic poverty perspective, *World Development,* 33(6), pp. 867–79.

Harris, J. 1991. 'The Green Revolution in North Arcot: Economic trends, household mobility, and the politics of an 'awkward class'. In Baltimore, P.B. (ed.). *The Green Revolution Reconsidered: The Impact of High-Yielding Rice Varieties in South India.* Johns Hopkins University Press, pp. 57–84.

Hartmann, B. and Boyce, J. K. 1983. *A Quiet Violence: View from a Bangladesh Village*. London: Zed Books.

Henninger, N. and Snell, M. 2002. *Where are the Poor? Experiences with the Development and Use of Poverty Maps*. Washington, DC: World Resources Institute.

Himanshu, H., Lanjouw, P., Mukhopadhyay, A. and Murgai, R. 2011. Non-farm diversification and rural poverty decline: A perspective from Indian sample survey and village study data. *Asia Research Centre Working Paper 44*. London: London School of Economics and Political Science.

Huang, Q., Rozelle, S., Lohmar, B., Huang, J. and Wang, J. 2006. Irrigation, agricultural performance and poverty reduction in China. *Food Policy*, 31(1): 30–52.

Hulme, D. 2003. Chronic poverty and development policy. An introduction. *World Development*, 31(3): 399–402.

Hulme, D. and Shepherd, A. 2003 Conceptualizing chronic poverty, *World Development*, 31(3): 403–23.

Hussain, I., Giordano, M. and Hanjra, M.A. 2004. 'Agricultural water and poverty linkages: Case studies on large and small systems'. In Bank, A.D. (ed.). *Water and Poverty: The Realities. Experiences from the field*. Manilla: Asian Development Bank, pp. 52–73.

Johnson, C. 2003. *Decentralisation in India: Poverty, Politics and Panchayati Raj*. London, UK: Overseas Development Institute (ODI).

Kennedy, J. and King, L. 2014. The political economy of farmers' suicides in India: Indebted cash-crop farmers with marginal landholdings explain state-level variation in suicide rates, *Globalization and Health*, 10(16).

Landau, K., Klasen, S. and Zucchini, W. 2012. Measuring vulnerability to poverty using long-term panel data, SOEP papers on Multidisciplinary Panel Data Research, No. 481. Accessed via www.econstor.eu/bitstream/10419/68443/1/725702885.pdf.

Lanjouw, P. and Shariff, A. 2004. Rural non-farm employment in India: Access, incomes and poverty impact. *Economic and Political Weekly*, 4429–4446.

Leach, M., Mearns, R. and Scoones, I. 1999. Environmental Entitlements: Dynamics and institutions in community-based natural resources management, *World Development*, 27(2): 225–47.

Lipton, M. 1989. *New Seeds and Poor People*. London: Unwin Hyman.

Lipton, M. 2004. New directions for agriculture in reducing poverty: the DFID Initiative. Department for International Development. Accessed via http://dfidriculture consultation.nri.org/launchpapers/ michael lipton.html.

Mehta, A.K. and Shah, A. 2003. Chronic poverty in India: incidence, causes and policies. *World Development*, 31(3): 491–511.

Miller, F., Osbahr, H., Boyd, E., Thomalla, F., Bharwani, S., Ziervogel, G., Walker, B., Birkmann, J., Van der Leeuw, S., Rockström, J., Hinkel, J., Downing, T., Folke, C. and Nelson, D. 2010. Resilience and vulnerability: Complementary or conflicting concepts? *Ecology and Society*, 15(3): 11.

Minten, B. and Barrett, C. B. 2008. Agricultural technology, productivity, and poverty in Madagascar, *World Development*, 36(5): 797–822.

Mellor, J. 2001. Employment Multipliers from Agricultural Growth and Poverty Reduction. *The Pakistan Development Review*. 40(4, Part I, Winter): 371–400.

Narayanamoorthy, A. 2007. Does groundwater irrigation reduce rural poverty? Evidence from Indian states, *Irrigation and Drainage*, 56(2–3): 349–62.

Natarajan, R. and Prathapar, S.A. 2014. *Extent of Arsenic Contamination and Its Impact On the Food Chain and Human Health in the Eastern Ganges Basin: A Review*. Colombo, Sri Lanka: International Water Management Institute (IWMI).

National Commission on Farmers – India, M.o.A., Government of India 2006. *Serving Farmers and Savig Farming, Jai Kisan: Revised Draft National Policy for Farmers*. New Delhi.

Nayak, P.K., Chattopadhyay, S.K., Kumar, A.V. and Dhanya, V. 2010. Inclusive Growth and its Regional Dimension. *Reserve Bank of India Occasional Papers* 31(3).

NSSO (National Sample Survey Organization) 2006. *Household Ownership Holdings in India, 2003*. National Sample Survey Organisation (NSSO), Ministry of Statistics and Programme Implementation, Government of India.

Pal, P. and Ghosh, J. 2007. Inequality in India: A survey of recent trends. *Economic and Social Affairs Working Paper*, 45.

Planning Commission of India 2007. *Poverty Estimates for 2004–05*. New Delhi: Government of India, Press Information Bureau.

Planning Commission of India 2014a. *Report of the Expert Group to Review the Methodology for Measurements of Poverty*. New Delhi, India: Government of India, Press Information Bureau.

Planning Commission of India 2014b. *The Data Book Compiled for Use by the Planning Commission*. New Delhi, India: Government of India, Press Information Bureau.

Raihan, S., Fatehin, S. and Haque, I. 2009. Access to land and other natural resources by the rural poor: The case of Bangladesh. Accessed via http://mpra.ub.uni-muenchen.de/38621.

Regmi, M. 1976. *Landownership in Nepal*. Delhi: Adroit Publishers.

Regmi, M. 1978. *Land Tenure and Taxation in Nepal*. Delhi: Bibliotheca Himalaya.

Ribot, J.C., Agrawal, A. and Larson, A.M. 2006. Recentralizing while decentralizing: How national governments reappropriate forest resources, *World Development*, 34(11): 1864–1886.

Sen, B. 2003. Drivers of escape and descent: changing household fortunes in rural Bangladesh. *World Development*, 31(3): 513–34.

Sen, A. and Himanshu 2004. Poverty and inequality in India: II: widening disparities during the 1990s. *Economic and Political Weekly*, 39(39): 4361–4375.

Shah, A. 2013. The agrarian question in a Maoist guerrilla zone: Land, labour and capital in the forests and hills of Jharkhand, India. *Journal of Agrarian Change*, 13: 424–50.

Shah, T., Hassan, M.U., Khattak, M.Z., Banerjee, P.S., Singh, O.P. and Rehman, S.U. 2009. Is irrigation water free? A reality check in the Indo-Gangetic Basin, *World Development*, 37(2): 422–34.

Sharma, B., Amarasinghe, U., Xueliang, C., de Condappa, D., Shah, T., Mukherji, A. and Smakhtin, V. 2010. The Indus and the Ganges: River basins under extreme pressure. *Water International*, 35(5). 493–521.

Shiva, V., Emani, A. and Jafri, A.H. 1999. Globalisation and threat to seed security: case of transgenic cotton trials in India, *Economic and Political Weekly*, 34(10/11): 601–13.

Singh, N., Bhandari, L., Chen, A. and Khare, A. 2003. Regional inequality in India: A fresh look. *Economic and Political Weekly*, 15: 1069–1073.

Singh, S. 2007. 'Introduction'. In Singh, S. and Sharma, P.K. (eds). *Decentralization. Institutions and Politics in Rural India*. New Delhi: Oxford University Press, pp. 1–36.

Sugden, F. 2013. Pre-capitalist reproduction on the Nepal Terai: Semi-feudal agriculture in an era of globalisation. *Journal of Contemporary Asia*, 43: 519–45.

Sugden, F. and Gurung, G. 2012. *Absentee Landlordism and Agrarian Stagnation in Nepal: A Case from the Eastern Terai*. Kathmandu, Nepal Institute of Development Studies.

Sugden, F., Silva, S. d., Saikia, P., Maskey, N. and Kumar, A. 2015. *Irrigation and Water Management Constraints for Marginal and Tenant Farmers in the Eastern Gangetic Plains*. Report compiled for project: Sustainable and Resiliant Farming Systems Intensification.

Sugden, F., Maskey, N., Clement, F., Ramesh, V., Philip, A. and Rai, A. 2014. Agrarian stress and climate change in the Eastern Gangetic Plains: Gendered vulnerability in a stratified social formation. *Global Environmental Change*, 29: 258–69.

Sugden, F., Shrestha, L., Bharati, L., Gurung, P., Maharjan, L., Janmaat, J., Price, J. I., Sherpa, T. Y. C., Bhattarai, U., Koirala, S. and Timilsina, B. 2014. *Climate Change, Out-migration and Agrarian Stress: The Potential for Upscaling Small-scale Water Storage in Nepal*. IWMI Research Report 159, Colombo: Sri Lanka.

THRD Alliance. 2012. *Impact of Remittances in Madhesh: Study Report*. Kathmandu: Tarai Human Rights Defenders Alliance.

Tiwari, S. 2014. Assessment of growth and poverty and inequality in Uttar Pradesh: A long term Analysis. *Journal of Economic & Social Development*, 10(1).

UIS (UNESCO Institute for Statistics) 2015. Education data. Accessed via http://data.uis.unesco.org/Index.aspx?DataSetCode=EDULIT_DS&popupcustomise=true&lang=en.

UNDP (United Nations Development Programme) 2015. *Human Development Report 2014*. Accessed via http://hdr.undp.org/sites/default/files/hdr14-report-en-1.pdf.

Uz Zaman, K.A. and Akita, T. 2012. Spatial dimensions of income inequality and poverty in Bangladesh: An analysis of the 2005 and 2010 household income and expenditure survey data. *Bangladesh Development Studies*, 35(3): 19.

World Bank 2015. Poverty data. Accessed via http://data.worldbank.org/indicator/SI.POV.DDAY.

16

Gender, agricultural investment and productivity in an era of out-migration

Fraser Sugden, Panchali Saikia, Niki Maskey-Amatya and Paras Pokharel

Key messages

- Male out-migration represents a critical transformation in the Ganges Basin over the last two decades, driven by climatic, economic and agrarian stress.
- For the rural majority, rather than transforming livelihoods, out-migration is just one among many livelihood strategies which is necessary for households to meet their subsistence needs, particularly when households are already subject to the burden of climate stress, debt and exploitative agrarian relations.

- Remittances are mostly diverted to non-productive and subsistence uses, and not into agriculture.
- The 'positive' impact that migration has on livelihoods is often restricted to better-off households who have access to land and capital, and particularly those that are male-headed. Such households often earn more from their migrant family members, can divert remittances into investments, and can offset the labour burden by hiring in labour.
- Migration is placing a further constraint on agricultural productivity by placing stress on community institutions who manage irrigation canals and other infrastructure.
- While there have been some positive impacts upon women's control over resources and decision-making power for women-headed households, this group also faces by far the most considerable work burden and economic stress after migration.

Introduction

There has been an unprecedented rise in male out-migration over the last decade in South Asia. The South Asian portion of the Ganges Basin, which encompasses Nepal and parts of India and Bangladesh, has been at the forefront of this transformation. The root of the migration boom lies in the fact that this is a region with some of the highest concentrated poverty in the world and deeply entrenched semi-feudal agrarian relations (Kishore, 2004; Rodgers and Rodgers, 2001; Sugden, 2013). With the marginal farmer majority unable to subsist under existing relations of production, there is a growing articulation between the economy of the countryside and the urban pockets of western India (Sugden, 2013). In the case of Nepal, overseas migration is particularly prominent, with peasant labour supporting the petro-economies of GCC countries and newly industrialized countries such as Malaysia (Graner and Gurung, 2003).

This report seeks to identify the impact of out-migration on agricultural decision making and investment patterns amongst the poorest households. Of particular concern is the so-called 'feminization of agriculture', defined by Deere (2005) as an increase in women's levels of participation in the agricultural sector, either as self-employed or as agricultural wage workers, and an increase in the proportion of women in the agricultural labour force relative to men. This could be either because more women are working or because fewer men are working in agriculture (due to migration), as is the case in the Ganges Basin.

The feminization of agriculture has brought some opportunities in terms of gender empowerment, with enhanced decision-making power and access to resources in some cases (Masika, 2002b). However, it has also been a source of considerable vulnerability for women, with implications for agricultural productivity as well as the well-being of those who stay behind. Of particular interest are the transformations in the internal production relations within the household (Adhikari and Hobley, 2012; Hadi, 2001; Lahiri-Dutt, 2012; Paris *et al.*, 2005; Sugden *et al.*, 2014a, 2014b). Women are increasingly obliged to play a more prominent role in core on-farm activities to compensate for the outflow of male labour, leading to increased workloads (Adhikari and Hobley, 2012; Paris *et al.*, 2005). While women have long been the backbone of the agrarian labour force, not to mention their prominent role in the reproductive sphere, they are also increasingly in charge of managing the land and household affairs.

Demographic change in India, Nepal and Bangladesh

The Ganges Basin extends over an area of 1,086,000 km and is among the most populated areas in the world. The population density in this basin, however, varies from less than two persons per square kilometre in Nepal to more than 500 persons per square kilometre in downstream

India in the states of Bihar and West Bengal, and nearly 964 persons per square kilometre in Bangladesh (BBS, 2011; Hosterman *et al.*, 2009).

Migration has been a significant trend across the basin over the last two decades. According to the census of India, the rural population declined from 72.19 per cent in 2001 to 68.84 per cent in 2011, and the level of urbanization has increased from 27.81 per cent to 31.16 per cent (Census of India, 2001, 2011). Thirty per cent (307 million) of the total population (1.02 billion) were reported as migrants by place of birth according to the 2001 census, while in 1991 it was just 27.4 per cent (Census of India, 2001). Migration is both intra-state and inter-state, mostly from low-wage to high-wage areas. The migration of people across state boundaries increased by 54.5 per cent from about 27.2 million in 1991 to about 42.1 million in 2001. According to the 2001 Census of India, labour migration has significantly increased in the two most populous states in the Ganges Basin – Uttar Pradesh with 2.6 million migrating out of the state, and Bihar which has 1.7 million leaving. In India, migration has been overwhelmingly male-led, and as of 2007–08 data female migrants were estimated to constitute just 14.53 per cent of total labour migrants, with a large proportion restricted to the primary agricultural/forest sector (Mazumdar *et al.*, 2013).

In Bangladesh the census data from 2010 states that the rate of internal migration is approximately 35.3 per 1,000 population and overseas migration 36.1. There has been considerable increase in rural to urban migration in the country, from 11.7 per 1,000 population in 2000 to 24.5 in 2010 (BBS, 2011). The international migration rate in the country has seen both upward and downward trends since 1976 when overseas migration became officially sanctioned in the country. Nearly 90 per cent of migration is to the Middle East and Malaysia, and it is mostly men who migrate for employment. According to the Bureau of Manpower, Employment and Training (BMET) in Bangladesh the number of individuals migrating abroad in 1999 was 268,182 (Islam, n.d.). This saw an increase in the fiscal year 2010–11 to 439,000 migrants and 691,000 in 2011–12, but a decline in 2012–13 of 36.18 per cent with only 441,000 migrants (Miah *et al.*, 2014).[1] Female international migration in Bangladesh has become more prominent since 1996, but with several bans by the government on women migrating for lower level jobs, the percentage compared to male migrants remains low. It was only in 2008 with the withdrawal of major bans, and following on from adoption of the Overseas Employment Policy in 2006 to organize the overseas employment sector and to ensure the welfare of Bangladeshi workers abroad, that more than 80,000 women migrated to overseas employment (Islam, n.d.). Compared to male migration, this still remains low at 2–4 per cent of total international migration (Islam, n.d.).

In Nepal, the rural population has decreased from 86 per cent in 2001 to 83 per cent in 2011. As of the 1981 census at a national level, there were 402,977 household members in Nepal classed as 'absentee'. This had increased more than fourfold to 1,921,494 by the 2011 census. In 1980, 93.1 per cent of migration was to India (Khatiwada, 2014), although from the 1990s onwards there was a shift in the pattern of migration towards overseas destinations such as the GCC countries and Malaysia. By 2011, 90.1 per cent of migration was to overseas destinations and just below 10 per cent was to India (Khatiwada, 2014), although the latter is likely to be much higher if one considers seasonal migration which is often not captured in censuses. According to the 2011 census, 10 per cent of the population are absentee. As of the 2010–11 Living Standards Measurement Survey, 26.9 per cent of the absentee (migrant) population were female.

Figure 16.1 displays the percentage of women-headed households across the Ganges Basin. Given the tendency for males to be the primary labour migrants, this offers some insights into the levels and character of migration across the region. As one would expect Nepal and the

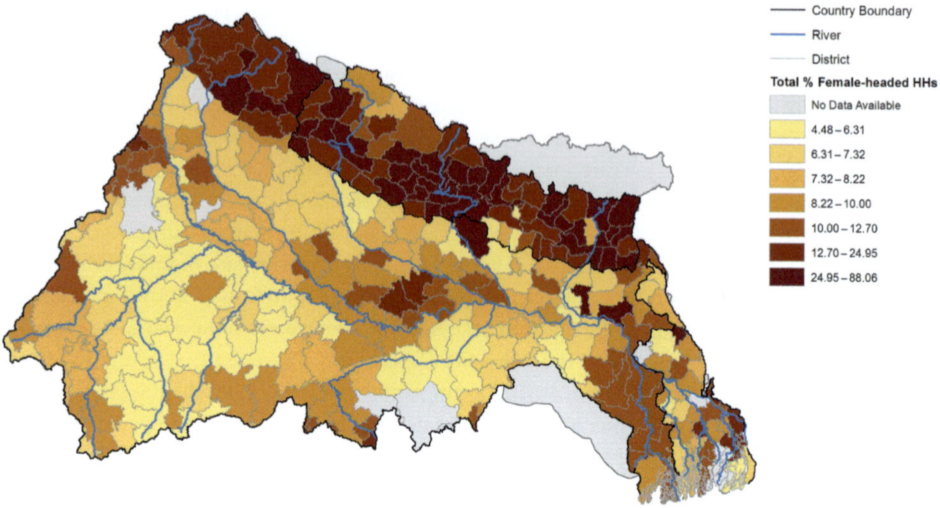

Figure 16.1 Percentage of female-headed households in the Ganges Basin
Source: Water Land and Ecosystems Gender and Development Basin Maps, available at http://maps.vista-info.net/gis/pmapper-4.2.0/map_default.phtml?resetsession=ALL&config=IWMI_GANGES_composite&type=iwmi

Himalayan regions display the highest proportion of female-headed households. This is not only due to the high levels of migration from this region; there is also more longer term migration in the case of Nepal, mostly to the GCC, which causes a rise in female-headed households. The percentage of female-headed households is also relatively higher in Bangladesh and eastern India, the traditional labour-sending region.

Although migration is high in this region, it creates fewer women-headed households due to the greater incidence of shorter term or seasonal migration (Sugden *et al.*, 2014a).

The cause of the increase in migration from the Ganges Basin is complex, yet is rooted in 'agrarian stress'. A number of studies have highlighted the role of climate change and ecological stress in accentuated migration. While people migrate from their homes in response to rapid onset of disasters like flooding, a phenomena widespread in Bangladesh, slow onset pressures such as drought and land degradation are equally significant (Cell, 2009; WEN, 2010). Climate, though, is rarely the sole driver of migration from the Ganges Basin, with economic pressures, and the strain it puts on the most marginalized socio-economic groups, being equally important (Ahmad, 2012; Sugden *et al.*, 2014a). In a study from Nepal and Bihar in Sugden *et al.* (2014), however, it was emphasized that climate is not a sole driver of change, and out-migration was also intricately tied to economic pressures of globalization which is increasing the demand for cash. Rising cash needs are rooted in the rising costs of food and agricultural inputs, as is shown in studies by Kishore (2004) from Bihar, and Pant (2011) from Nepal. Meanwhile, a culture of consumerism is increasing the demand for commodities and is driving up the costs of cultural events such as marriage (Rankin, 2004; Sugden *et al.*, 2014a).

These economic pressures in the context of neo-liberal globalization perhaps in part explain the exponential increase in migration over the last two decades since the region underwent economic liberalization. Added to this are the new labour 'opportunities' outside. Post-liberalization urbanization has increased the capacity of India's booming cities to absorb the vast surplus labour pool from eastern India. In the case of Nepal and Bangladesh, migration opportunities to the GCC emerged in the neo-liberal context of the early 1990s and the aftermath of the Gulf War. Gulf governments sought a cheap workforce on short-term contracts who were culturally separated from citizens and non-aligned to the political fissures of the Arab world which threatened the ruling elites (Chalcraft, 2010; Hanieh, 2010), while in sending countries such as Nepal, migration overseas was actively encouraged (Graner and Gurung, 2003) as it sought an outlet to avert social unrest in the context of a stagnant industrial sector.

While the impact of migration on agriculture has been well covered in the literature, there are a number of pending questions. This includes the impact of migration on agricultural decision making and investment decisions. It is clear that migration provides a new source of cash income to the household. However, given the increase in workload for women, questions arise as to whether productivity gains can be realized. In a similar vein, there are unanswered questions relating to remittances themselves. It is questionable whether they are sufficient to invest significantly in agricultural inputs, not to mention whether this is considered a priority avenue of expenditure for families, particularly given the non-productive cash demands for cultural expenditures and debt servicing (Sugden *et al.*, 2014a). The pattern of remittance payments also merits further attention when assessing their role in strengthening livelihoods for families who stay behind. For example, overseas migration for two to three years at a time may yield higher incomes, yet the payments may be irregular. At the same time, seasonal migration, which often occurs for several months in a year, may yield lower income, yet the inflow of cash into the household may be more regular. Based upon a case study in the heart of the EGP (Bihar and the Nepal Terai), and an extensive collection of interviews with women farmers, it is hoped that this chapter will shed some light on some of these issues.

Methodology

Fieldwork was carried out between January and July 2014 in three districts in 'Mithilanchal', a broader cultural region which spans the Nepal–India border, encompassing the north-east of Bihar and parts of the central and eastern Nepal Terai, as part of a study funded by the CGIAR Programme on Climate Change and Food Security (CCAFS). In total 127 women farmers were interviewed, all of whom had experienced the migration of a family member. In Saptari in Nepal, two VDCs[2] were included in the sample, Odraha and Lalapatti, both in a fertile region on the edge of the Koshi river floodplains. Forty-one interviews were completed. In Dhanusha, Nepal, 44 interviews were completed in Thadi Jijha VDC, a large community in a remote and drought-prone region on the Nepal–India border. In Madhubani, Bihar, 42 interviews were carried out in one large Panchayat, Rakuwari, east of the Kamala River.

The interviews included both a quantitative section collecting basic information on agricultural production, remittances and income, and a qualitative section to assess perceptions and experiences of male out-migration. Respondents were introduced to the team by gatekeepers, who were representatives from local NGOs and farmer groups, who we had worked with previously. They were then invited to participate, although further participants were engaged through a 'snowballing' technique, whereby initial households we interviewed would in turn introduce us to other potential respondents such as neighbours or relatives; it is thus not a random sample. Across all three sites, 105 out of 127 respondents were small farmers with less than 1 ha of land, tenants or landless labourers. In an earlier study which covered both the Dhanusha and Madhubani field sites (Sugden *et al.*, 2014a), it was found that this group represents between two-thirds and three-quarters of the rural population, making the results relevant to the larger farming population across the region. With regards to large farmers and landlords, they represent a very small proportion of the rural population (at between 2 per cent and 10 per cent according to the earlier study), and although they are important in shaping agrarian relations, they were not a focus of this study, which is on how the small farmer majority cope with migration. Across the three sites, only five households with more than 2 ha of land were included in the sample, which is approximately the proportion of the population they represent.

Situation in the village and role of agriculture

The three sites have a lot in common in that they both fall within the greater Mithilanchal region, a cultural-linguistic area spanning the Nepal–India border. Diverse feudal state formations in the past, and a rigid caste system led to the development of a deeply hierarchical social formation in all three sites. Its origins lie primarily in the medieval tax collection system of the region[3] which propped up a class of local upper caste intermediaries who acquired large tracts of land for their services (Chaudhury, 1964; Regmi, 1976; Sugden, 2013). These inequalities were solidified throughout the nineteenth and early twentieth centuries in Madhubani under British colonialism, which propped up the pre-colonial *zamindar* class to collect revenue (Ram, 1997). Similarly, in Dhanusha and Saptari, Nepal's feudal Rana regime ensured the reproduction of caste and class disparities through supporting a complex network of landlord intermediaries who channelled a large share of agricultural surplus to Kathmandu (Regmi, 1976).

In the post-colonial era, long after agrarian tax had declined in significance as a source of state revenue (see Habib, 2002; Regmi, 1976), the descendants of these intermediaries remained important landlords. Despite some cosmetic attempts at land reforms, these lacked political will in their implementation and failed to produce meaningful changes in the distribution of land (Alden-Wily *et al.*, 2008; Kishore, 2004; Sugden and Gurung, 2012). In spite of some recent changes in the composition of the land-owning class with the buying and selling of estates within the upper and middle castes (Sugden, 2014), land inequality remains deeply entrenched, and the predominantly low caste base of the agrarian structure has not seen any significant change to its resource base.

National Sample Survey data from Bihar noted an increase in the proportion of farms smaller than 1 ha from 80 per cent to 89 per cent between the early 1990s and early 2000s (NSSO, 2006), while another report from the region suggested tenancy and landlessness had increased (Karan, 2009). In a similar vein, the National Living Standards Measurement survey data from the east and central Nepal Terai has indicated that over the last two decades, there has been an increase in the proportion of landless households from 10 per cent to 27 per cent between 1995–96 and 2010–11 (Central Bureau of Statistics, 1996, 2011).

In the case study communities today, most households are small farmers, tenants and labourers, who constitute a significant majority in the region (Sugden, 2013; Sugden *et al.*, 2014a). In the Madhubani sample in particular, there is a very high proportion of tenants or part-tenants (60 per cent), most of whom work for a small number of landlords. In Dhanusha and Saptari, the proportion of tenant or part-tenant farmers in the sample is less, but there is a greater proportion of landless labourers (16 per cent and 10 per cent respectively), many of whom are from the Dalit community and work for large farmers or landlords. There is also a greater proportion of marginal farmers with less than 0.5 ha (23 per cent and 39 per cent).

In terms of the caste composition of the sample, 44 per cent of the sample in Dhanusha are from the marginalized Dalit community, while most respondents in the other two communities are from the middle castes, known in India as 'other backward castes'. Upper caste Brahmins represent just a small proportion of the sample, yet they represent most of the larger farmers (see Figure16.2).

In all three regions, migration was present in low levels from the 1970s to the 1990s, mostly to India. Agrarian stress, due to climate change and rising costs of inputs, combined with rising

Table 16.1 Composition of sample by farmer land-ownership category

Farmer category	*Dhanusha*		*Saptari*		*Madhubani*		*Total No.*
	No.	*%*	*No.*	*%*	*No.*	*%*	
Tenant or part-tenant	11	25	14	34	25	60	50
Owner cultivator with <0.5 hectares	10	23	16	39	12	29	38
Owner cultivator with >0.5 hectares	16	36	7	17	5	12	28
Landless labourer	7	16	4	10	0	0	11
Total	44		41		42		127

living costs to cause an unprecedented rise in male out-migration from the 1990s onwards (Karan, 2003; Sugden *et al.*, 2014a). To give an idea as to how widespread this is, the survey conducted in the same part of Madhubani and Dhanusha for an earlier study (Sugden *et al.*, 2014a) indicated that migration was widespread across all categories, although it was particularly high amongst land-poor households. In Thadi Jijha in Dhanusha, 68 per cent of tenants, part-tenants and landless labourer households and 56 per cent of owner cultivators had migrants. In Madhubani, these figures were 70 per cent and 60 per cent respectively. Out-migration has long been the domain of males, particularly in the 30s to 40s age group. In the earlier study in Madhubani, for example, 87 per cent of long-term migrants (away more than a year) and 99 per cent of seasonal migrants (away part of the year only) in the surveyed households were male. In Dhanusha males constituted 98 per cent of long-term migrants respectively, and all seasonal migrants. Figure 16.3 shows that just under half of the sampled households are women-headed, the vast majority of which comprises households whereby, due to migration, the women are in charge of all household affairs.

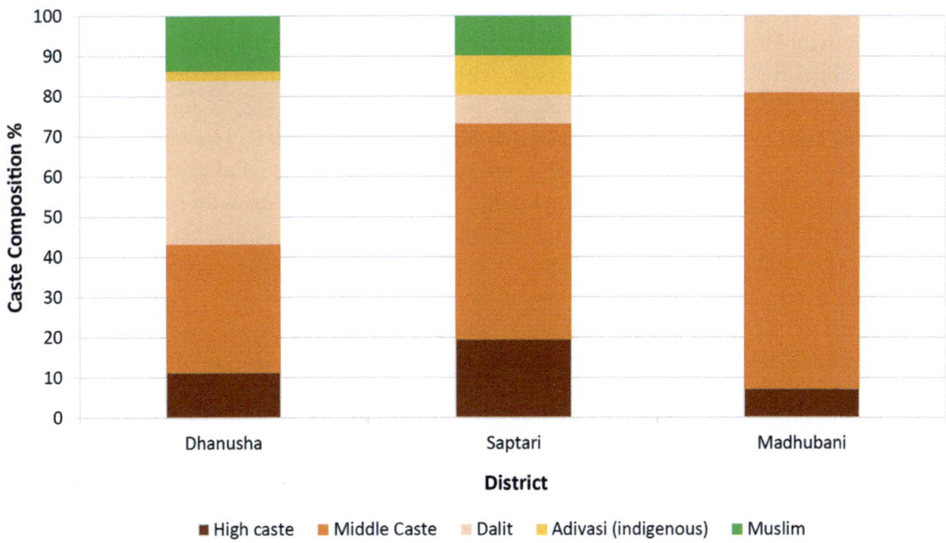

Figure 16.2 Caste composition of sample by district

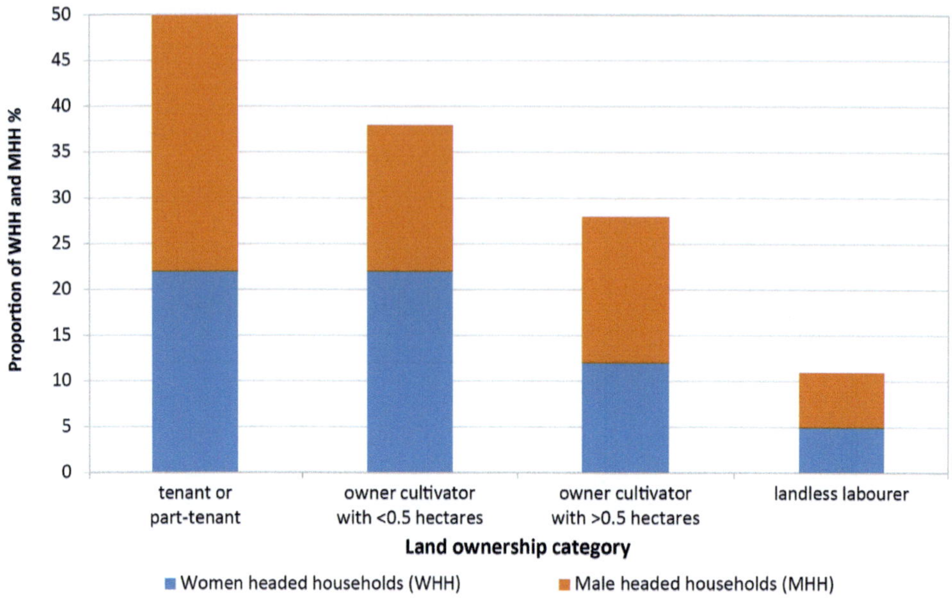

Figure 16.3 Proportion of WHH and MHH in sample by land-ownership category

Migration, remittances and changed agricultural practices

A critical question is how remittances are invested, and whether migration is actually contributing to socio-economic progress among the marginal and tenant farmer majority in north Bihar and the Nepal Terai, and contributing to agricultural development.

To begin, it is important to emphasise that levels of remittances are highly variable, and appear to be related to the households' *pre-existing* position in the agrarian structure. There is considerable variation, with some families in all farmer categories and both women and male headed households (see Table 16.2), reporting a significant increase in incomes. When looking at the averages though, differences emerge, with poorer households who are tenants or who own less than 0.5 ha of land reporting an average increase of $1,343 for women-headed households (WHH) and $1,948 for male-headed households (MHH). Households with more than 0.5 ha of land, in contrast, have reported an increase of $1,835 and $2,069 respectively.

This disparity is in part likely to be due to the high levels of indebtedness for poorer households, an issue which will be covered below, while it may also be due to the nature of the remittances themselves. Migrants from better-off households usually have a higher level of education, and thus could potentially receive higher wages in their host country or city. Figure 16.4 shows how 50 per cent of migrants from landless labourer households have no education, and the remainder have only studied below or have not passed the important class 10 exams. It is also low for tenants/part-tenants and poorer owner cultivators, when compared to their middle and large farmer counterparts. There was a perception that even if the work is unskilled for both better educated and less educated migrants, the former were often able to secure jobs in companies which offered better conditions of employment and wages. All cases of households being cheated or deceived by employers were from the poorer (owning < 0.5 ha and tenant) farmer categories.

Even when one considers that there has been an increase in incomes, migration was rarely considered a source of 'accumulation'. When asked an open-ended question about the primary

Table 16.2 Average increase in income between male- and women-headed households from different land-ownership categories

Land ownership and gender of household head	*Average increase in income (US$)*
WHH (tenant or owning less than 0.5 ha land)	1,343
WHH (owning more than 0.5 ha land)	1,835
MHH (tenant or owning less than 0.5 ha land)	1,948
MHH (owning more than 0.5 ha land)	2,069

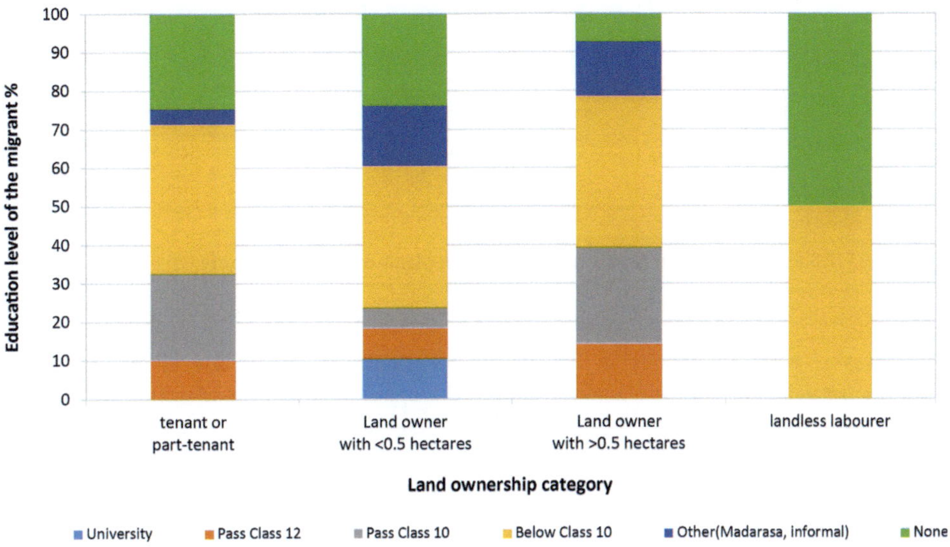

Figure 16.4 Land ownership status and education of migrants

reasons for migration of their family members, 74 per cent of women noted economic insecurity and the difficulty to meet immediate food needs as the main reasons, factors which echoed men's views, as expressed in focus groups. This was followed by indebtedness (9 per cent), land scarcity (4 per cent), the need to pay for children's education (5 per cent), payments for marriage and dowry (4 per cent) and medical costs (2 per cent). Interestingly, fewer than 2 per cent of the women respondents noted that family members had migrated to generate funds to invest in land or to invest in agricultural inputs, with just 12 per cent noting it as a secondary reason.

The importance of migration for meeting basic subsistence needs echoes findings from other literature. Adhikari and Hobley (2013), in a study from the eastern hills of Nepal, noted how most remittances are for basic household expenditure, with women from poor and medium households often saving just 25 per cent of their remittance income for investment in education, house improvements or purchase of land. Another study in Bihar noted how 96 per cent of respondents cite food items as the most important remittance expenditure, followed by health care (78 per cent), while productive reinvestment in agriculture is listed as third best for remittance investment (30 per cent) (Tsujita and Oda, 2012). A study from the Terai-Madhesh identified food and clothing as the most important avenue for investment, followed by education and healthcare (THRD Alliance, 2012).

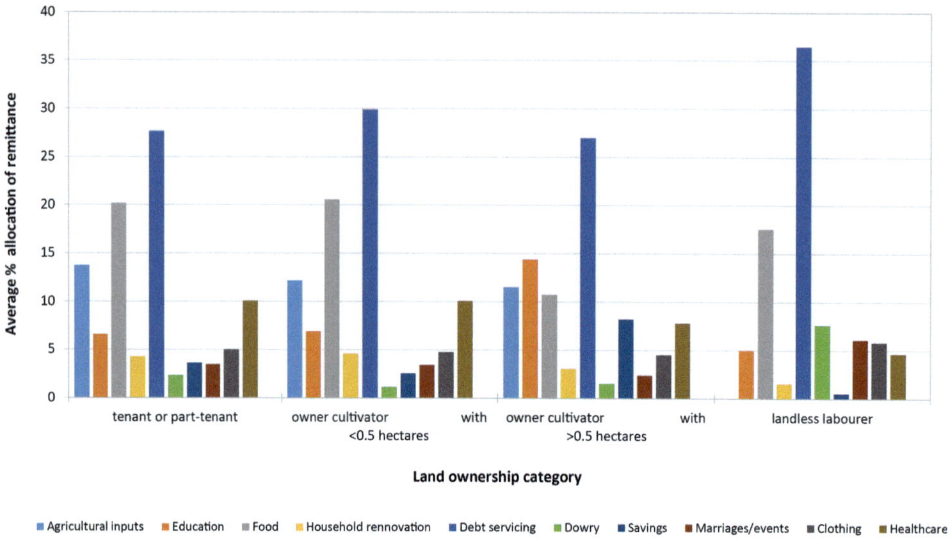

Figure 16.5 Allocation of remittances according to farmer land-ownership category

However, when one looks at the estimated expenditure of remittances by households over the last year, what is particularly striking is the significance of debt servicing, which is far higher than expenditure on food (see Figure 16.5). For landless labourers, it is close to double expenditure on food, and it remains significantly higher for all other farmer categories. Debt and migration are intricately connected, albeit in diverse ways. First, many household members migrate primarily to repay past debts. This can include consumption loans to purchase food, as well as debt incurred while paying for dowry and weddings, both of which were listed as important secondary reasons for migration themselves. In Madhubani, in particular, 19 out of 29 poorer (tenant or owning < 0.5 ha) households cited loan repayment as one of the reasons for migration, and one focus group, during a ranking exercise, even cited it as the primary reason for migration. Similarly, in Dhanusha, eight out of 27 of these poorer households cited loan repayment as a reason for migration. The cycle of debt for these households does not necessarily end following out-migration.

A second phenomena is households taking loans to facilitate the migration process, particularly in the context of overseas migration which is common in Saptari and Dhanusha, and involves large upfront costs. To go abroad, households are expected to pay extortionate fees to manpower agents and middlemen of around Rs150,000 ($1,500), often far beyond the actual costs of tickets and visa processing. Out of the interviewees in Dhanusha, all but two households had taken loans to meet these costs, and the two which had not had sold land and a shop to pay the intermediaries. THRD Alliance's report on remittances in the Terai-Madesh noted how families are often trapped in repeated cycles of debt, whereby migrants return from overseas having just repaid one loan, only to take another to migrate again, with the hope of better income next time around (THRD Alliance, 2012).

Nearly all loans in the three sites are taken from private informal lenders, and interest is generally high, varying from 35 per cent to 48 per cent, and occasionally as high as 60 per cent per year. Households who are unable to repay loans using remittances are often trapped in perpetual indebtedness. This often occurs if the migrant falls ill and is unable to work, or when they are cheated by their employers, underpaid or not paid at all. Similarly, the women who stay behind reported sometimes having to take new consumption loans while their husbands were away, if, for example, the remittance money is insufficient for day-to-day needs or there is a gap for a few months. In two separate focus groups in Madhubani, it was highlighted how life is very insecure in these contexts, and women often face harassment and threats from money lenders if they cannot repay these loans on time.

The majority of remittance money is being used for basic food needs at a time of rising cash needs, and most significantly, to service debts. It is also evident that while remittances have provided households with a much needed source of cash, they are by no means being used as a source of wealth 'accumulation'. This phenomenon is epitomized by one farmer's testimony in an earlier IWMI-led participatory video project in a neighbouring VDC of Dhanusha:

> There are 10 members in my family. Some need medical treatment, some need clothing. Every member has his/her own needs. If only one person is abroad, how can he fulfil all members' requirements? Remittance is not a sustainable solution because a person may have maximum salary 20,000NPR per month but the household expenses may be 40,000NPR. In this case, can the remittance solve the problems?

While remittances have met some of the subsistence needs of the marginal farmer majority – in particular, by providing them a source of cash – the farmer in question notes that there is still a shortfall, and showed scepticism that they can ever replace agriculture as a source of livelihood security.

While it is clear that the majority of respondents do not consider migration a source of wealth, is there a minority of stay-behind women (or their families) who have been able to increase their asset base, and translate this into improved agricultural productivity? Figure 16.6 suggests

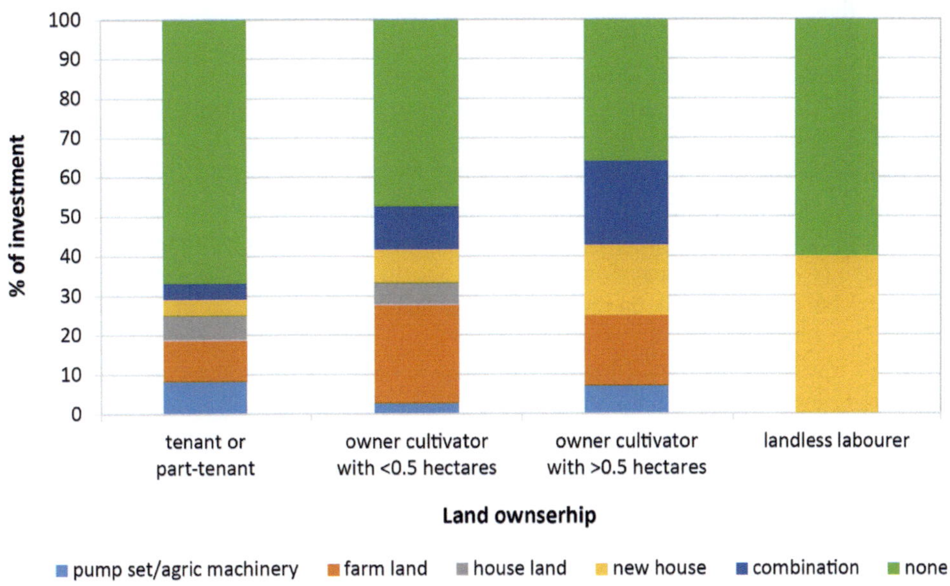

Figure 16.6 Investments following migration of family member

Note: 'Combination' refers to investments in more than one of the above categories that the majority of households had made no large investments following the migration of a family member. As one would expect, however, this varies according to the land-ownership status of the household, for whom (as Table 16.2 suggested) the increase in income is often much higher. The lowest overall levels of investment were among tenants/part-tenants and landless labourers, with only 33 per cent having made large expenditures in assets. Fifty-two per cent of smaller and 63 per cent of larger owner cultivators had made investments.

While a handful of households are clearly increasing their asset base, particularly those with more land in the first place, are these assets actually being used to boost agricultural production? Figure 16.5 suggests that investment in agricultural inputs is similar across farmer categories, and is well below what is invested in food and debt servicing. Nevertheless, this data refers mostly to recurring costs such as fertilizer and irrigation, and does not capture large one-off expenditures using remittance money saved over a long period of time. This could include investments in agricultural equipment such as pump sets. According to Figure 16.6, the number of respondents who had invested in agricultural machinery or pump sets was less than 10 per cent.

To explore this issue further, data was collected on irrigation use before and after migration. Even if households had not invested in high-value irrigation equipment such as pumps, they may have increased the irrigated area as they have more cash to rent irrigation equipment such as pump sets and tube wells. Again, Figure 16.7 suggests that for the majority of farmers, there has been no change. For 18 per cent of tenants and 14 per cent of smaller owner cultivators, there had actually been a *decrease* in irrigation use. This may be due to the tendency for farmers to reduce the cultivable area in the dry season due to labour shortages, particularly for poorer households such as rent-paying tenants, who cannot afford to buy in labour (Sugden *et al.*, 2014a). There are, however, a handful of households who are now irrigating land which was previously rainfed, although this included only 10 per cent of tenants as opposed to 24 per cent of small owner cultivators and 28 per cent of larger owner cultivators, again highlighting the constraints faced by those without land.

No relationship was found between male- and women-headed households with regards to changes in irrigation practices, although some respondents from women-headed households in Saptari and Madhubani complained in the focus groups that gender ideologies created some barriers in accessing irrigation. Managing the rental of a pump set and tube well for irrigation has traditionally been the male domain, and with the males absent, some women complained of discomfort in having to visit the pump owners' house to negotiate or receive the equipment. They often faced difficulties locating someone to rent from, and finding time to oversee the application of water to the land. Some women in a focus group in Saptari also complained that they were often overcharged, on the pretence that they 'could afford it' as they had a husband abroad.[4]

While it is clear only a small number of households are making productive investments in agriculture, another question is whether remittances can be used to increase the cultivable area by buying land, reducing dependence on tenancy and increasing food security. When asked to note the advantages of migration, a significant 18 respondents in Saptari and eight in Dhanusha noted the capacity to invest in land. Nevertheless, this aspiration may not be reflected in reality, as Figure 16.8 suggests that for the vast majority of households, there has been no change in land ownership since migration, and a handful of households have actually sold land since their family member migrated, usually to pay off migration loans. A study by Tsujita and Oda (2012) in Bihar suggested that remittances are generally not large enough to improve farmers' land-holding status or to change their position in the village hierarchy.

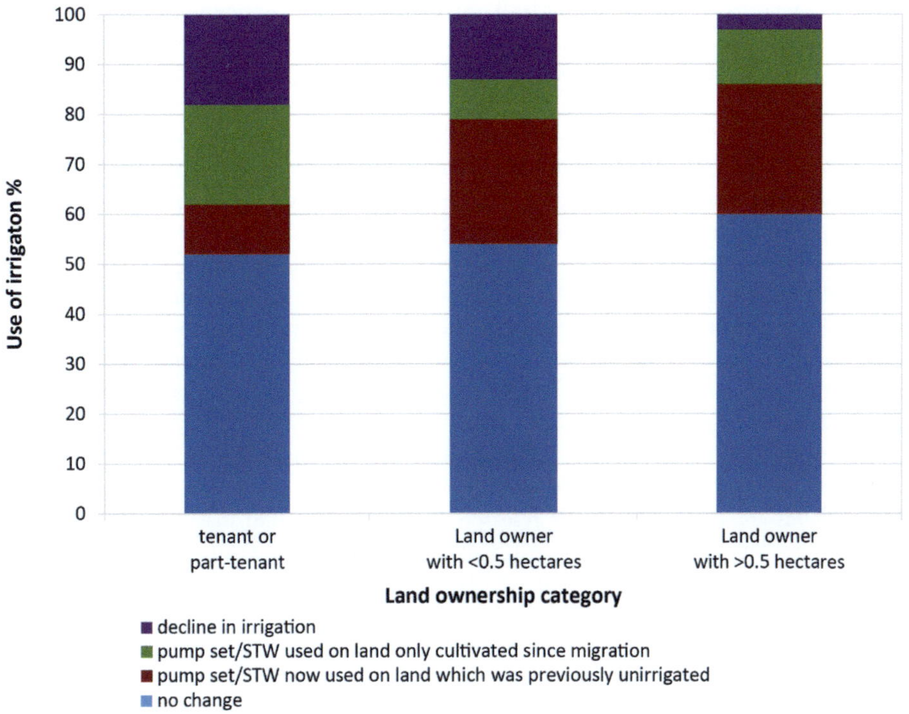

Figure 16.7 Irrigation use before and after migration

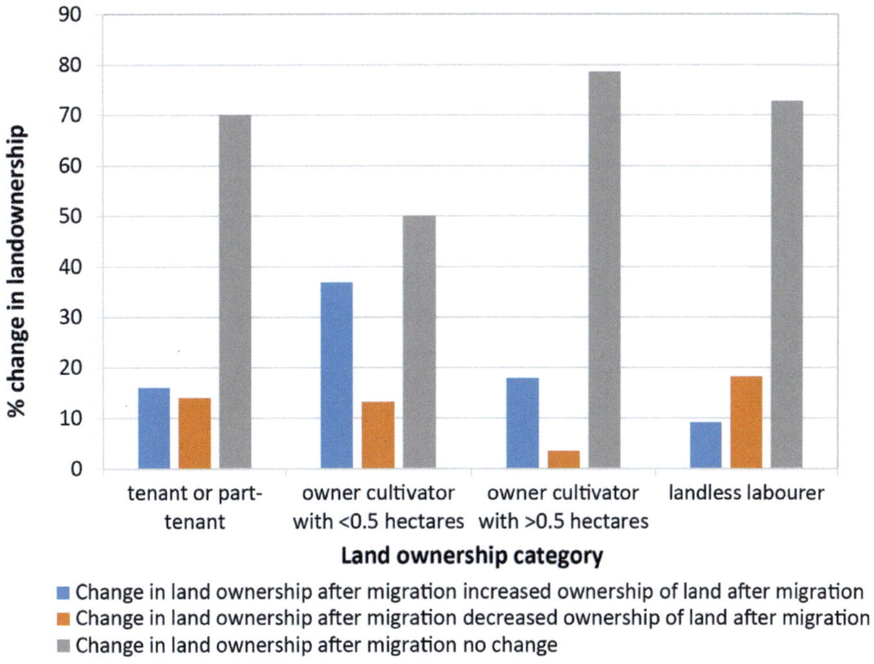

Figure 16.8 Percentage of households who purchased or bought land after migration of family member

Of those who have actually bought land, this appears greatest among small owner cultivators with less than 0.5 ha. The areas bought were generally small, and for most respondents were less than 10 kata (1 kata = 0.034 ha). Purchases by tenants or part-tenants, however, are considerably lower, which is unsurprising given the burden of debt and high rents which mean the land market is inaccessible for this group. Purchases are also low for larger owner cultivators, perhaps because they already have some livelihood security from their holdings, and it is less of a necessity to expand their holdings. For these households who had bought land, it would be a mistake to assume they had all increased their overall wealth due to migration. As noted, land is important for the livelihoods of those who stay behind, and some households had not paid for the land with remittances, but with loans, suggesting these investments are a coping rather than accumulation strategy. One of the poorer farmers had actually sold land to send the husband abroad, but later taken a Rs300,000 loan to buy land again. Another respondent had bought land using the same large loan of Rs 500,000 which was also used to facilitate migration in the first place.

Interestingly, purchases of land were approximately a third higher for women-headed households, and this appears to be part of a strategy to ensure some livelihood security while their husbands are away. The earlier research in the region found that women are likely to become more dependent upon the land than before due to the change in the frequency of income flows, as shown in Sugden *et al.* (2014a). For example, overseas migrants may only send money every few months, particularly in Madhubani, where migration is mostly to Indian cities and is much more informal. Four of the women-headed households in Dhanusha had actually been landless before, and had purchased land to meet their subsistence needs after the migration of their husbands.

Work burden and labour allocation in the context of out-migration

It has been shown how stay-behind populations face persisting challenges in investing in land and agricultural inputs following male out-migration, despite the increase in remittances. Nevertheless, there is another set of constraints these populations now face, even affecting overall agricultural productivity. The most significant issue is the high work burden of women, an issue raised in the literature on migration and agriculture (Gardner, 1995; Hadi, 2001; Sugden *et al.*, 2014a). With the migration of males, a large number of women have to take up new work responsibilities aside from the usual agricultural tasks (such as transplantation, weeding and harvesting) and reproductive activities in the home. This includes tasks formerly in the male domain such as managing water for irrigation, buying fertilizers and seeds, overseeing labourers, negotiating with landlords, sowing and taking grains to the mills for cleaning and polishing.

While some households reported that they had left land fallow in the dry season due to labour shortages within the household, the larger data set showed no solid evidence of reduced cropping intensity due to migration. Agriculture remains critical for those who stay behind to ensure household food security, and thus 80 per cent noted no reduction in the area cultivated of different crops since the migration of their family member. The inevitable outcome, however, was a significant increase in the workload of women on the land.

As one would expect, rising workload is most significant for women-headed households. When asked about the challenges of household and agricultural work associated with migration (see Table 16.3), 60 per cent of women-headed households who were tenants or owned less than 0.5 ha noted high workload as the primary problem faced during household reproductive work, with 70 per cent noting it to be the main problem for agricultural work.

Table 16.3 Whether high workload is listed as a challenge of household/agricultural work in context of migration

Household head and wealth categorization	*High workload key challenge of household/ reproductive work*	*High workload key challenge of agricultural work*	*Total no. of respondents*
Women-headed household (tenant or owning < 0.5ha)	60	70	43
Women-headed household (owning > 0.5ha)	54	54	11
Male-headed household (tenant or owning < 0.5ha)	54	61	28
Male-headed household (owning > 0.5ha)	36	32	22

Table 16.4 Productivity per ha by gender of household head and land ownership

Gender of the household head and land ownership	*Average productivity of paddy (kg/ha)*	*Average productivity of wheat (kg/ha)*
WHH (tenant or owns <0.5ha)	1,571.27	714.09
WHH (owns>0.5ha)	1,293.48	414.55
MHH (tenant or owns <0.5ha)	1,304.23	737.07
MHH (owns >0.5ha)	2,650.81	1,176.74

The increase in work burden for reproductive activities is of interest, as despite the fact that this was traditionally considered the female domain, men did support women with some of these activities prior to migration. For example, seven out of 20 women-headed household respondents in Saptari and 12 out of 21 in Dhanusha noted that their husbands previously used to support them with reproductive tasks such as collection of fuel, livestock rearing, buying consumables, child care, or looking after less economically active elders.[5] Therefore, with the absence of these males, women felt a greater pressure, particularly at times when household work was particularly high, for example, if a family member was sick. This latter point is interesting, as while 10 households noted that ability to meet health expenses was a benefit of migration, the ability to cope during the illness of a family member was raised as a constraint for 18 respondents. Illness would often mean the women would need to care for the family member in question while also managing their labour responsibilities on the farm.

With regards to male-headed households, workload is an important constraint also, although it is less significant than for women-headed households, as one would expect.

While a majority of poorer male-headed households had also cited workload as a primary challenge of agricultural and reproductive work, it was a much lesser concern for better-off male headed households (36 per cent and 32 per cent respectively listed it). This group not only benefits from greater labour availability, but a stronger socio-economic status and enhanced capacity to buy in labour from outside the household. When workload was raised as a concern by respondents from this group, the discussion usually focused on the challenges of finding outside labourers willing to work at the normal wage rate, or the extra burden of monitoring excess land which is sharecropped out. In fact, Table 16.4 suggests that productivity per bigha (the local unit of land measurement)[6] is significantly higher for male-headed households with more than 0.5 ha. The capacity to invest on the land is highest, and labour shortages are less likely to negatively affect agricultural yields.

Community breakdown, gender relations and upkeep of irrigation infrastructure

Another more complex reason for a potential negative impact of migration on agriculture relates to changes in cooperative relations within the community. Adhikari and Hobley (2011) note that migration has led to increased individualization and reduced engagement in public good activities, particularly as a rising work burden drives households to prioritize which public activities will actually provide them benefits.[7] In the study sites, the levels of community mobilization for public goods was limited, even before migration. For example, only nine respondents in Dhanusha, six from Saptari and just one from Madhubani noted that their husbands were engaged in community activities prior to migration, and most of these were from male-headed households. Past community engagements included contributing to road construction, temple maintenance, the renovation of a dam in Dhanusha, youth clubs, or membership of canal user committees.

Migration-associated problems were most notable with regards to irrigation canal management committees. There are a number of canal systems in the study area, particularly in Saptari. The management committees who coordinate local level repair, maintenance and water allocation have traditionally been the male domain. However, the migration of male members has in some instances placed stress on these committees. The replacement of a male member needs to be done through a formal meeting, which usually only takes place if there are major repairs to take place. As meetings are not regular or scheduled, often the committees continue functioning without such members. Similar challenges were evident with respect to pond management committees, whereby the local fisheries development office complained of high turnover of executive committee members, with chairperson positions often lying vacant, and thus impeding decision making and management. For example, fisheries cooperatives are entitled to government financial support, and other members of the committee often cannot apply for funds, given the absence of executive members.

What is likely to be a growing problem, however, is that women are not taking over from men in the leadership positions of any of these canal or pond committees, resulting in positions laying vacant, and undermining the effectiveness of committees. The water user association in Odraha VDC did not have a single woman on their executive committee, and the government-set quota for 33 per cent womens' representation has not been fulfilled. This also applied to four out of 10 associations looked at as part of a comparative study in Morang district, further east. Even for the groups which did have women on the committee, it was reported that they rarely participate in meetings. In some cases, they are there in name only to fulfil the government quota, and the husbands take part in the meetings instead. There was a perception during interviews with committee members that this was due to cultural norms which restricted women's engagement in public activities, including travel to cities for meetings in divisional irrigation or fisheries offices. It was claimed by male members that women are often not familiar with the procedures necessary for running the committees, such as the accounting system or how to deal with transactions at the divisional irrigation office.

Another claim was that women cannot meet the time commitment that being a member of a committee entails. This assertion becomes problematic, however, if one considers the widespread assertion that women's involvement in committees has not changed following migration, yet their involvement in physical labour on the canals has increased. According to the association members, it can take up to two weeks to clean the branch canals each year, depending on the command area, and each farming household must provide a labourer. In Odraha in Saptari this has increased due to frequent monsoon flood inundation of canals which requires the building of new branch channels. Males cannot give the time due to migration, or, in the case of Morang district, due to the fact that many work in full-time factory jobs. In this context, women have taken on these roles, adding to their work burden. Tasks for women mostly involve removing the extracts or mud and cleaning the *kulos* (channels). Women from female-headed households who cannot give the time are often expected to hire labourers to compensate.

Aside from the fact that women are not playing an active role in managing irrigation systems, a more significant reason for gender inequalities undermining agricultural productivity relates to women's access to government programmes and services. These may include, for example, subsidized agricultural equipment or tube wells, or participation in agricultural training programmes. In Madhubani, interviewees claimed that it was very difficult to get information on government programmes as women would often only hear about opportunities through their migrant husbands over the phone, or via inlaws. Due to cultural norms, many women also would not participate in Panchayat (village government) meetings or those organized by NGOs, and thus they would remain unaware of ongoing development projects. In Dhanusha and Saptari women also complained that the high work burden makes it difficult for them to attend public meetings. The capacity to participate in the public sphere was worsened by the harassment women would often face just completing simple tasks such as travelling to the town to collect remittances, an issue also raised by Adhikari and Hobley (2011) in their study from upland Nepal. One quotation by a woman in Dhanusha, interviewed as part of an IWMI-led participatory video project in Thadi Jijha, illustrated these challenges:

> People widen their eyes with ill intention while we are going to collect money. Especially women gossip about other women. They comment, 'there must be some story of going to the town. Why did she go there?' People natter when a woman talks to any other man. They don't try to understand the condition. If I need any help from others then how can I do without talking to them? But people do not think so.

In all sites, this was raised as a problem, with women often being subjected to gossip regarding their participation in day-to-day public activities which break traditional gender norms.

Gender empowerment

Distribution of remittances within the household

The data thus far suggests that out-migration has had limited (and sometimes negative) impact on agricultural incomes and productivity, and on the distribution of assets and food security for stay-behind populations, particularly women-headed households. It is also important, however, to look beyond the household economic status and investigate how migration has affected intra-household relations.

As was already noted, an increased workload is a primary outcome of migration for the women who stayed behind, and this is a source of considerable social, economic and emotional stress (Sugden *et al.*, 2014a). Masika (2002a), however, questions this widely held view that male out-migration makes women more vulnerable because in some instances male migration can give women greater decision-making powers, and open up new livelihood opportunities. This is true to some extent with regards to women-headed households. In the study sites, respondents from all but one woman-headed household noted how there was a transfer of decision-making

Table 16.5 Person responsible for collecting and deciding on use of remittances for male- and women-headed households

Gender of household head	*Person responsible*	*Who collects remittances (%)*	*Who decides how remittances are used (%)*
Women-headed households	No remittances received yet	1.6	1.6
	Herself	68.9	63.9
	Joint	NA	32.8
	In-law	19.7	1.6
	Husband	1.6	.0
	Other	8.2	.0
Male-headed households	No remittances received yet	0	.0
	Herself	31.8	.0
	Joint	NA	45.5
	In-law	31.8	28.8
	Husband	12.1	21.2
	Other	19.7	.0

power on household expenditure, education of children and agricultural cropping practices to themselves following the migration of their husbands. Most women considered this new independence a positive change, despite the increase in work burden. This increase in decision-making power parallels other findings from the region (Adhikari and Hobley, 2012; Tamang, 2011), although these studies also note that this authority often does not extend to decisions over high-value investments such as purchases of land or house construction.

For male-headed households only five women noted some transfer of decision making after migration, despite the increase in workload, all of whom were in Madhubani (with none in Saptari and Dhanusha). In most cases decision making was instead transferred to a father or brother-in-law. In interviews, some women from male-headed households even felt their decision making had been constrained even further. While they had in the past consulted their husbands regarding decisions, now they were dependent upon consulting with fathers-in-law or other senior males, with whom they felt less comfortable.

Another question relates to the control over remittance cash. According to Table 16.5, it is clear that the majority (68 per cent) of those from women-headed households collect the remittance money themselves, while 64 per cent also decide how is is used. However, it is surprising that even though such women have a considerable labour burden and manage most affairs on the farm, a substantial 20 per cent still rely on an in-law to collect the cash, while 32 per cent still consult their husbands before deciding on how it is spent. This is of course much higher for male-headed households, whereby a significant 28 per cent of women respondents have no say on how remittances are used, and 45 per cent must consult their husbands. Only 32 per cent of such women actually collect the remittances, with it more often being collected by in-laws or other family members.

Conclusions

Migration and feminization of agriculture is a critical transformation in the Ganges Basin – particularly in the eastern region which is one of the primary sources of low migrant wage labour in South Asia. This case study from the agricultural heartland of Bihar and the Nepal

Terai has highlighted that the impact of out-migration on the women who stay behind is complex. On the whole, the positive impact migration has on livelihoods is often restricted to better-off households who have access to land and capital, and particularly those that are male-headed. For the rural majority, though, out-migration has been shown to be just one among many livelihood strategies which is necessary for households to meet their subsistence needs, particularly when households are already subject to the burdens of climate stress, debt and – for tenants – rent. While there have been a selected number of respondents who have been able to accumulate some limited wealth to purchase a house or land, the majority have not expanded their asset base, and many have actually sold productive assets, often to fund the migration process. In fact, one outcome of migration which has a clearly negative impact is debt. Indebtedness drives migration, while also being worsened by migration itself as farmers take loans both to send a male family member away, or to meet short-term food needs for stay-behind populations.

Migration is placing a further constraint on agricultural productivity for the stay-behind population by placing stress on community institutions which manage irrigation canals and other infrastructure. Women have to contribute an increased work burden in maintaining these systems in the absence of men, but are not taking over from men in day-to-day management of the systems through committees. Furthermore, gender ideologies which restrict women's role in the public sphere, and limited efforts for projects to reach out to women, undermine their access to some of the government initiatives to support farmers.

While there have been some positive impacts upon women's control over resources and decision-making power for women-headed households, this group also faces by far the most considerable work burden and economic stress after migration.

In sum this report has cast doubt upon common assumptions that migration has led to a transformation of rural economies. While in some instances it may offer some short-term benefits or opportunities for accumulation, it is also a source of considerable vulnerability for stay-behind populations through the work burden, debt and a shrinking asset base.

It is clear from this case study that government and non-government actors in the Ganges Basin need a new approach to deal with gender issues in the context of demographic change. Of particular importance is the need to make labour-saving and productivity-improving technologies and resources *accessible* through direct targeting and engagement with women farmers, and through more favourable rules. New technologies are still disseminated through a traditional focus on male 'leader farmers', and some subsidies and resources still depend on documents rarely held by women, such as land ownership certificates (Sugden, 2014). Second, the spiralling cycle of debt must be broken through a more rigorous regulation of the informal money-lending sector, and of the manpower business in places like Nepal which charges extortionate fees from overseas migrants. Third, the drive to migrate is itself intricately connected with internal inequalities in the distribution of land and resources in an often caste-based and hierarchical social formation. It is thus critical not to lose sight of more radical transformations through land reform and other legislation to undermine the feudal social structures which drive migration in the first place.

Although migration is high in this region, it creates fewer women-headed households due to the greater incidence of shorter term or seasonal migration (Sugden *et al.*, 2014a).

Notes

1 This was possibly due to bilateral issues between the Government of Bangladesh and traditional migrant receiving countries which affected the number of new people taking up foreign employment (Hussein, 2014).

2 VDC is the local administrative unit in Nepal, equivalent to the Indian Panchayat.

3 This included the Mughal tributary Kingdom of Darbhanga in present-day Madhubani, and the Sen and then Gorkhali states of Nepal.
4 The study by Paris *et al.* (2005) in Uttar Pradesh also noted how it is difficult for women to contact owners of equipment such as tractors and threshers, often from other villages, in the absence of their husbands.
5 Curiously, though, only two women-headed households reported male labour input in such activities in Madhubani.
6 1 bigha in Nepal = 0.67ha, while in Bihar it represents 0.37ha.
7 This study notes that community forestry user groups, for example, are facing difficulties maintaining active members.

References

Adhikari, J. & Hobley, M. 2012. *Everyone is Leaving – Who Will Sow our Fields? The Effects of Migration from Khotang District to the Gulf and Malaysia*, Swiss Agency for Development and Cooperation (SDC) Nepal.

Adhikari, J. & Hobley, M. 2013. *Everyone is Leaving – Who Will Sow Our Fields?: The Effects of Migration from the Khotang District to the Gulf and Malaysia*, Nepal Institute of Development Studies.

Ahmad, N. 2012. 'Gender and climate change in Bangladesh: The role of institutions in reducing gender gaps in adaptation program'. *Social Development Paper*, 126.

Alden-Wily, L., Chapgain, D. & Sharma, S. 2008. *Land Reform in Nepal: Where is it coming from and where is it going?*, Kathmandu, DFID.

Bangladesh Bureau of Statistics (BBS). 2011. *Population and Housing Census 2011*. Government of the People's Republic of Bangladesh: Statistics and Informatics Division.

Census of India. 2001. *Population Totals – India Data Sheet.* Government of India: Office of the Registrar General Census Commissioner, India, Indian Census Bureau.

Census of India. 2011. *Population Totals – India Data Sheet.* Government of India: Office of the Registrar General Census Commissioner, India, Indian Census Bureau.

Central Bureau of Statistics 1996. *Nepal Living Standards Survey 1995/6*, Kathmandu.

Central Bureau of Statistics 2011. *Nepal Living Standards Survey 2010/11*, Kathmandu.

Chalcraft, J. 2010. *Monarchy, migration and hegemony in the Arabian Peninsula.*

Chaudhury, P.C.R. 1964. *Bihar District Gazetteers: Darbhanga*, Patna, Superintendent Secretariat Press.

Climate Change Cell. 2009. 'Climate change, gender and vulnerable groups in Bangladesh'. *DoE, MoEF, Component 4b, CDMP, MoFDM, Dhaka*, pp. 1–82.

Deere, C.D. 2005. *The Feminization of Agriculture?: Economic Restructuring in Rural Latin America*, United Nations Research Institute for Social Development.

Gardner, K. 1995. *Global Migrants, Local Lives: Migration and Transformation in Rural Bangladesh.* Oxford: Oxford University Press.

Graner, E. & Gurung, G. 2003. Arab ko Lahure: Looking at Nepali Labour Migrants to Arabian Countries. *Contributions to Nepalese Studies*, 30, pp. 295–325.

Habib, I. 2002. *Essays in Indian History: Towards a Marxist Perception; With, The Economic History of Medieval India: A Survey*, Anthem Press.

Hadi, A. 2001. International migration and the change of women's position among the left behind in rural Bangladesh. *International Journal of Population Geography*, 7, pp. 53–61.

Hanieh, A. 2010. Temporary migrant labour and the spatial structuring of class in the Gulf Cooperation Council. *Spectrum: Journal of Global Studies*, 1, pp. 67–88.

Hosterman, H.R., McCornick, P.G., Kistin, E.J., Pant, A., Sharma, Bharat, R. and Bharati, L. 2009. *Water, Climate Change, and Adaptation: Focus on the Ganges River Basin.* Durham, NC: Duke University, Nicholas Institute for Environmental Policy Solutions.

Hussein, Z. 2014. Explaining the Recent Decline in Remittances in Bangladesh. Available from: http://blogs.worldbank.org/peoplemove/governance/explaining-recent-decline-remittances-bangladesh.

Islam, M.N. n.d. *Gender analysis of migration from Bangladesh.* Dhaka, Bangladesh: Bureau of Manpower, Employment and Training (BMET). Available from: www.bmet.gov.bd/BMET/resources/Static%20PDF%20and%20DOC/piblication/Gender%20Analysis%20of%20Migration.pdf.

Karan, A. 2003. Changing pattern of migration from rural Bihar. *Labour and Development*, 9, pp. 155–195.

Karan, A. 2009. Changing Land–Caste–Class Relations and Rural Poverty in Bihar. *Agrarian Reforms, Land Markets, and Rural Poor*, p. 271.

Khatiwada, P.P. 2014. 'International Migration and Citizenship'. In C.B.O. (ed.). *Statistics*. Population Monograph of Nepal 2014. pp. 211–37.

Kishore, A. 2004. Understanding agrarian impasse in Bihar. *Economic and Political Weekly*, 39, pp. 3484–3491.

Lahiri-Dutt, K. 2012. Women-headed households in farming in India and Nepal: A baseline survey report, Australian Centre for International Agricultural Research.

Masika, R. 2002a. Editorial. *Gender & Development*, 10, pp. 2–9.

Masika, R. 2002b. Editorial, Special Issue: Climate change. *Gender & Development*, 10, pp. 2–9.

Mazumdar, I., Neetha, N. & Agnihotri, I. 2013. Migration and gender in India. *Economic & Political Weekly*, 48, pp. 54–64.

Miah, M., Khan, M., Mannan, A. and Rahman. M. 2014. Recent trends of international migration and remittance flows: An empirical evidence of Bangladesh. *IOSR Journal of Economics and Finance (IOSR-JEF)* 2(5), pp. 16–23.

NSSO 2006. *Household Ownership Holdings in India, 2003*, National Sample Survey Organisation (NSSO), Ministry of Statistics and Programme Implementation, Government of India.

Pant, B.D. 2011. *Soaring Food Prices in Nepal: Causes and Consequences*, Kathmandu, IIDS.

Paris, T., Singh, A., Luis, J. & Hossain, M. 2005. Labour outmigration, livelihood of rice farming households and women left behind: a case study in Eastern Uttar Pradesh. *Economic and Political Weekly*, 40, pp. 2522–2529.

Ram, B. 1997. *Land and Society in India: Agrarian Relations in Colonial North Bihar*, Orient Blackswan.

Rankin, K.N. 2004. *The Cultural Politics of Markets*, London, Pluto Press.

Regmi, M. 1976. *Landownership in Nepal*, Delhi, Adroit Publishers.

Rodgers, G. & Rodgers, J. 2001. A leap across time: when semi-feudalism met the market in rural Purnia. *Economic and Political Weekly*, 36, pp. 2–8.

Sugden, F. 2013. Pre-capitalist reproduction on the Nepal Terai: semi-feudal agriculture in an era of globalisation. *Journal of Contemporary Asia*, 43, pp. 519–45.

Sugden, F. 2014. *Landlordism, Tenants and the Groundwater Sector: Lessons from the Terai-Madhesh of Nepal*, IWMI Research Report 162.

Sugden, F. & Gurung, G. 2012. *Absentee Landlordism and Agrarian Stagnation in Nepal: A Case from the Eastern Terai*, Kathmandu, Nepal Institute of Development Studies.

Sugden, F., Maskey, N., Clement, F., Ramesh, V., Philip, A. & Rai, A. 2014. Agrarian stress and climate change in the Eastern Gangetic Plains: Gendered vulnerability in a stratified social formation. *Global Environmental Change*, 29, pp. 258–69.

Sugden, F., Shrestha, L., Bharati, L., Gurung, P., Maharjan, L., Janmaat, J., Price, J.I., Sherpa, T.Y.C., Bhattarai, U., Koirala, S. & Timilsina, B. 2014. *Climate Change, Out-migration and Agrarian Stress: The Potential for Upscaling Small-scale Water Storage in Nepal*, Colombo.

Tamang, S. 2011. *Impact of remittance economy on agricultural dependent rural women: A study on impact of migration on Tamang women in Manedanda Lalitpur, Nepal*, report prepared for The Social Inclusion Research Fund.

THRD Alliance 2012. *Impact of Remittances in Madhesh: Study Report*, Kathmandu, Tarai Human Rights Defenders Alliance.

Tsujita, Y. and Oda, H. 2012. Caste, land, and migration: a preliminary analysis of a village survey in an underdeveloped state in India.

WEN 2010. *Gender and the climate change agenda: The impacts of climate change on women and public policy*, Women's Environment Network.

17

Can there be progress on transboundary water cooperation in the Ganga?

Mark Giordano, Dipak Gyawali, Ainun Nishat and Uttam Kumar Sinha

Key messages

- While cooperation on the Ganges has been disappointing, a change in direction is possible if the right steps are taken.
- These include (a) broader thinking on participation in water discussions, (b) a change in outdated mentality on data sharing, and (c) a more strategic approach to linking and de-linking water issues from broader political questions.

- As the largest power in the basin, India must also lead by example to facilitate trust, not simply out of altruism but also because of the gains it can accrue.

Introduction

History and culture tie together the states of modern South Asia. So too could and should their shared water resources. In the west, India and Pakistan do honour a long-standing agreement to divide (Gyawali, 2001), if not cooperate over, the Indus. But in the east, cooperation over the much larger and much more highly populated Ganga has languished along with overall relations between its four riparian states.

This failure to cooperate over the Ganga is unfortunate. Differences between each of the riparian states in terms of physical geography and economic need provide opportunities for the creation of shared benefits including those related to transportation, hydropower, irrigation, flood control and salinity management. Much of the area occupied by the Ganga Basin falls in what has been dubbed South Asia's 'poverty square', with substantially more people below the dollar-per-day poverty line than in all the countries of sub-Saharan Africa combined. Cooperation to improve the productive uses of Ganga waters, reduce their destructive impacts, and maintain water quality and ecosystem services should be one of the greatest engines anywhere for poverty reduction. Finally, failure to make even limited progress on the Ganga has made solution of the broader set of diplomatic challenges facing the region more difficult than necessary. At the same time, of course, broader political differences have slowed water cooperation.

While the history of water cooperation in the Ganga has been poor and impediments to collaboration continue, domestic political changes have opened up at least the possibility for new diplomatic and other efforts. Some progress has recently been made on festering border issues between India and Nepal and between India and Bangladesh, potentially opening the way for new discussions on water. Domestic and international interest in the Ganga is high, for example, in India's Clean Ganga campaign, but also through recognition of the potential threats imposed by climate change in the broader Himalaya region. China has emerged as a perceived water competitor on the interconnected Brahmaputra, but also as a potentially more engaged economic and political partner in South Asia as a whole. Thus while the challenges are great, the potential opportunity for positive transformation in water relations in the medium and long term is possibly as high as it has been since 1947.

This chapter provides an overview of the problems as well as the opportunities and possible next steps to achieving them. After first reviewing the limited and contentious history of transboundary cooperation on the Ganga, it describes the possible gains from improved relations, and recent changes in the political environment that may make those gains possible. Finally it outlines steps that might be taken to overcome remaining impediments.

A strained past

The history of international water relations on the Indus perhaps begins with the departure of the British in 1947 and the creation of independent Pakistan and India. The new border separating then West Pakistan from India cut through the Punjab, a key grain-producing region, as well as the Indus Basin and the irrigation systems that allowed Punjabi agriculture to thrive. Both Pakistan and India recognized the existential imperative in keeping the Indus irrigation systems working, but the broader animosity between the two countries and the notoriously difficult nature of the upstream–downstream dynamic created by partition made solution difficult. Nonetheless, the resolve of both sides, along with diplomatic, financial and technical support from the international community, kept discussion going and eventually resulted in the 1961 Indus Basin Agreement. While the core of the agreement simply partitions the Indus waters rather than providing a framework for broader water-sharing cooperation, its very creation is still remarkable, as is its endurance through ongoing hostilities and even outright war (Alam, 2002).

In the east, water was not a major issue in partition and the confusion that followed.[1] While the new borders cut across the Ganga and its tributaries, large-scale irrigation systems and hydropower had not been developed and therefore had not been disrupted. There was thus no immediate imperative to engage with water, in particular as there were other diplomatic problems to resolve. For example, India's unresolved border with the newly (1949) established People's Republic of China, as well as the status of Tibet, led to conflict in 1962. Border demarcation questions and sovereignty issues have likewise persisted between India and Nepal. Perhaps most importantly, East Pakistan was never fully integrated with West Pakistan. India supported East Pakistan's move for independence, leading to the Indo-Pakistan War of 1971. Despite Indian support, relations between it and independent Bangladesh declined sharply after the first Bangladeshi Prime Minister, Sheikh Mujibur Rahman, was assassinated in 1975, leaving a range of unresolved issues including additional border questions.

Nonetheless, while water has not been the primary focus of international relations among Ganga riparian states, a history of bilateral interaction does exist. We now provide a brief review of the water relations between India and Bangladesh and India and Nepal over the Ganga, as well as an overview of China's growing involvement in South Asian water.

India–Bangladesh

At Bangladesh's independence, a Friendship Treaty was signed with India that included the establishment of a Joint River Commission to address shared water issues. Despite the commission's broad mandate, the primary issue it has been asked to address relates to the Farakka barrage. India began construction of the barrage in 1965 to divert dry season water from the Ganga into the Hooghly tributary to flush sediment out of the port of Calcutta (now Kolkata) and to provide irrigation and drinking water. Bangladesh voiced initial concerns about the impact of possible flow reductions on its water availability and later over the impact on salinity, navigation and fisheries. When the barrage was completed in 1975, the two countries agreed on a trial operation. However, when relations worsened after the assassination of Sheikh Rahman, India unilaterally continued its trial operations. Bangladesh took the issue to the United Nations to force action. After creating a number of interim agreements, a 'permanent', 30-year agreement on barrage operations was reached in 1996. While the accord establishes a mechanism for maintaining minimum flows in the low season and resolving disputes, lack of detail and India's track record of implementation have continued to cause Bangladesh concern.

A second set of issues in the India–Bangladesh relationship have revolved around India's unilateral development of tributaries, in particular the Teetsa, in its Himalayan region, and the impact on flows to Bangladesh.[2] India is in the process of building a series of run-of-the-river hydropower dams on the Teetsa but in addition has a major barrage to divert irrigation water at Gazaldoba. An ad-hoc water-sharing agreement was reached between India and Bangladesh in 1983, but Bangladesh still accuses India of taking out too much water and endangering the dry season flows it needs for its own farmers in its drought-prone north.

As with Farakka, a series of interim agreements have been put in place to try to resolve the issue. A final agreement was to be concluded during a 2011 visit to Bangladesh by Indian Prime Minister Manmohan Singh. However, protests within India by the Chief Minister of West Bengal called into question the Indian central government's authority to sign. The agreement was withdrawn, though the central government continues to assure Bangladesh that it wants 'a fair, a reasonable' settlement. Still, the withdrawal was a source of embarrassment to the Indian central government and a further source of irritation for Bangladesh (Prasai and Surie, 2013). India's desire to build the Tipaimukh dam on the Barak River (in Manipur state) and projects on other rivers (the Feni, the Manu, the Muhuri, the Dharla and the Dudhkumar) have brought similar issues and tensions. Renewed discussion of India's proposed river-linking project and the potential impact on flows into Bangladesh is also a source of concern.

India-Nepal

The relationship between India and Nepal on the Ganga has been, if anything, even more contentious than that between India and Bangladesh. The Koshi Agreement, signed in 1954 to provide flood control as well as hydropower and irrigation, provides a case in point. Under the agreement, India provided funding for and undertook construction of a flood control sluice on the Koshi River just inside the Nepal border. India was also to compensate Nepal for inundated land and other damages and provide Nepal with half the hydropower produced as well as water for 29,000 ha of irrigation. From the start, many in Nepal felt that India was not living up to the agreement (e.g. not paying the full compensation), that it was inherently unfair (e.g. India received 1.5 million ha of irrigation) and that it violated Nepali sovereignty (the barrage is operated in Nepali territory by Indian engineers). While the agreement went through later modifications to address some of the grievances, the Nepal side has never been satisfied.

The severity of the problem was made clear when the embankment failed in August 2008. Neglect of the barrage by both sides and failure to agree on a needed second dam contributed to embankment collapse and a major flood that killed at least 250 people and destroyed 300,000 homes, displaced 3 million and damaged 800,000 ha of crop land. Each side blamed the other for the failure, as they have been doing for the last half century.

Similar issues of distrust have surrounded the original 1952 Agreement on the Gandak and its later modifications, and even the more recent 1996 Mahakali Treaty and its predecessors (Gyawali and Dixit, 1999; Gyawali, 2003). The core of that agreement is related to hydropower production as well as flood control and irrigation. The treaty specifically mentions that Nepal is entitled to 4 per cent of total flow but is silent on the Indian share. Some in Nepal fear that this could give India the potential to unfairly claim the remaining 96 per cent. New hydropower projects with India (often with private companies) are currently under development or discussion in Nepal. However, many in Nepal are similarly suspicious that claimed benefits will not materialize and that Nepal will face additional costs.

China

China's territorial control of the Tibetan Plateau also gives it control of the world's largest repository of freshwater. China has already completed multiple hydroelectric projects on many of the rivers emanating from the plateau. The opportunities for China on Ganga tributaries are relatively limited, but it has completed and planned substantial hydropower production on the Brahmaputra where it is also considering river-diversion schemes. Beijing reveals little about its projects, but according to observers it has already constructed 10 dams on tributaries of the upper Brahmaputra, with three more under construction. It is also constructing a 510-megawatt dam at Zangmu as part of a larger plan to build 28 dams on the Brahmaputra before the river enters India. Most worrisome for India and potentially Bangladesh is a gigantic 38-gigawatt hydropower project going forward at the 'great bend' at Motuo, where the Brahmaputra drops 2,500 metres into India.

Despite Indian officials having repeatedly raised their concerns about Chinese dam building on the Brahmaputra and its possible impacts on India, Beijing has refused to participate in any intergovernmental dialogue. As the upper riparian state, China has in the past even refused to provide flood alerts or warn India about water releases from dams. In June 2000 a breach in an upstream dam in Tibet raised the Brahmaputra's water levels in Arunachal Pradesh by 30 metres, leaving 26 dead and 35,000 homeless. This damage could have been contained if China had alerted India. Instead, Beijing refused to even acknowledge the dam burst for months after. Similar unannounced releases of excess water by China also caused flash floods in Himachal Pradesh in 2000, 2001 and 2005.

The existing and planned hydropower dams are run-of-the-river and therefore should have relatively small impacts on downstream flows. However, there are also plans, at least on paper, to divert Brahmaputra water as part of China's south to north water-transfer scheme. While the project seems unlikely given the potential costs of both development and operations (due to the substantial power required in lifting water over a number of mountain passes), Chinese action cannot be ruled out and therefore is a source of risk and distrust to both India and Bangladesh.

The potential

Differences between each of the Ganga riparian states in terms of physical geography and economic development suggest at least some promising opportunities for the creation of shared benefits including those related to hydropower, irrigation, flood control and salinity management. However, the Ganga is one of the largest river basins (1.7 million km^2) in the world and, if a country, would be the third largest with over 700 million people. Even under the best of circumstances, trying to follow current international norms and create a basin-scale governance structure such as a River Basin Organization that involved all four riparian states in meaningful overall decision making would be difficult. Even within countries, creating such a body is difficult. In India, for example, water is a state subject, and there is no institutional mechanism to bring its seven Ganga-sharing states (or any other states) together over water. As is discussed more below, this also means that legitimate differences within countries over the best national approaches to Ganga management have no mechanism for discussion. Given all these factors, the creation of meaningful basin-scale institutions is probably impossible. However, a fully basin-scale approach might not be necessary or even desirable.

In fact, while the potential direct benefits from large-scale cooperation have not been well studied, they may be smaller than commonly perceived. One of the few assessments, though with conclusions contested officially and unofficially in both Nepal and Bangladesh, was a 2014 World Bank report. The report challenges some common assumptions about the nature of the river system and potential gains from cooperation. It showed the opportunities to use storage in upstream Nepal to mitigate flooding in downstream India and Bangladesh to be much smaller than commonly assumed, though it also showed that targeted storage in Nepal could potentially increase critical low flows. An overall conclusion was that, given the volume of Ganga and the size of potential storage sites, vast numbers of multipurpose dams run to maximize water benefits across countries was not viable.

This finding does mean though that better water outcomes are not possible through more limited collaboration. The study, for example, also confirmed the tremendous untapped hydropower potential of Nepal, a potential beyond its foreseeable energy needs. The study estimated that if the 23 largest dams under consideration in Nepal were built, they would generate US$5 billion of power annually, a value equal to more than a quarter of Nepal's current GDP. Most of these dams would be run-of-the-river and so would not provide storage for flood control but also would not cause many of the negative externalities associated with large-scale storage. However, a key question is how a hydropower deal could be credibly structured in a way that brings satisfactory energy, financial and other benefits to Nepal.

Cooperation over hydropower (primarily on the Brahmaputra tributaries) of a somewhat different nature could bring benefits to India and Bangladesh. While India can unilaterally build dams in its eastern states, the political geography of the borders means it needs Bangladesh for distribution and a market. Cooperation between India and Bangladesh to ensure that Farakka operations provide benefits to both countries, or at least do not harm Bangladesh, is another obvious opportunity.

The creation of any large-scale water infrastructure of course brings with it negative consequences and needs to be done with extreme care. In transboundary contexts, a concern is that upstream countries will unilaterally take actions whose negative consequences are externalized downstream. But poorly planned projects can also have undue negative consequences within their country of origin, particularly in terms of displaced populations and the destruction of natural ecosystems and the livelihoods they support. More open cooperation and communication on shared rivers can also contribute to better national water decision making. While the focus of this chapter is on cooperation between countries over the Ganga, it is equally true that interests on the best approaches to the Ganga also vary widely within its riparian states.

While the direct value of greater water cooperation in the Ganga would be high, the indirect value beyond the water sector could be higher still. There is tremendous opportunity for gains through sub-regional connectivity in transportation, trade and power that can also be linked to environmental benefits (NWCF, 2013). Cross-border investment is far below its potential. Coordinated efforts to stop militants operating across borders is lacking. For more than 50 years the broader politics of the region have held back international cooperation on Ganga waters. But failure to resolve water issues is also holding back overall cooperation.

The promise

The history of water relations between Ganga riparian states has been less than ideal. The limited agreements in place have all been problematic, and even good-faith international efforts to address their shortcomings or move forward in other ways have often been undermined by domestic politics. The previously mentioned role of West Bengal in scuttling an agreement between India and Bangladesh is one high-profile example. However, a review of regional papers in any of the South Asian riparian states will show how nationalist rhetoric is commonly used to undermine central government authority to move forward on international water cooperation.

Nonetheless, a number of recent events suggest that the situation is changing and new opportunities are emerging. Indian Prime Minister Modi, elected in 2014, has prioritized a 'Neighborhood First' policy and early in his term visited all Ganga riparian states including China. Modi was the first Indian Prime Minister to visit Nepal in 17 years, and new agreements were signed in connection with his trip to improve transport links, provide new lines of credit for hydropower and irrigation, and expand power trade. Terms in the agreements were generally more favourable to Nepal than those in previous agreements. The two countries also began work to resolve remaining border issues, an especially important topic for Nepal. Unfortunately, though, much of the initial hope has already turned to disappointment, as the periodic cut of trade by India to Nepal in response to the new Nepali constitution shows.

Prime Minister Modi's visit to Bangladesh also resulted in multiple new agreements. While the trip did not produce a hoped-for agreement on the Teetsa or establish the next meeting of the Joint Water Commission, it did result in the critically important final resolution of their remaining border disputes. West Bengal Chief Minister Mamata Banerjee, who had earlier stopped the Teetsa Agreement, supported the agreement and even travelled to Bangladesh for the signing.

While China's actions have not so far been directly related to the Ganga, it is riparian to the river and plays a significant and growing role in regional politics, perhaps especially in terms of the India–Nepal relationship. India has concerns over China's role in Nepal (as does Nepal over India's role in Nepal's constitutional process). However, China has now also become India's largest trading partner. Signalling the importance of Sino-Indian relations, Li Keqiang made India his first foreign visit as Chinese Prime Minister, with the stated purpose of building mutual trust and cooperation.

The path

The potential for gain through increased international cooperation on the Ganga, both direct and indirect, is substantial, as is the current level of goodwill. However, given past history, progress will be exceedingly difficult. We outline here some steps that could be taken to help move towards sustained long-run cooperation on the Ganga that will provide benefits for all of the riparian states and their people.

Expand stakeholder definitions and increase internal transparency

One impediment to increased international cooperation on the Ganga has been internal, domestic politics. The Nepali opposition once used the terms of a seemingly minor water agreement with India to bring down its government. They then changed the constitution to make future agreements with India more difficult. Still today, international collaboration with India is used to rally opposition support. The role of the state government of West Bengal in stopping the national government from entering an agreement with Bangladesh has already been mentioned. Within Bangladesh, Indian actions on water are often presented as a demonstration of the national government's ineffectiveness.

One reason international water cooperation can so easily be used for internal political gain is that national governments alone have typically been defined as the primary stakeholders in decision making and negotiation. This narrow approach hinders the development of public understanding of the potential gains to cooperation and can cause the legitimate concerns of those who may be negatively impacted to be bypassed or dismissed. The result is limited buy-in and the opportunity for later critique.

Consultation by national governments with a broader stakeholder coalition on the issues and options would help overcome this problem. The coalition might include affected state and local governments, civil society representation from directly affected populations and environmental organizations. In some cases, stakeholder groups might themselves be cross-border in nature. For example, 'upstream' populations in both India and Nepal may have interests more in common with each other than with their national governments, at least concerning water resources development. Understanding the perspectives of such groups can help to forge better agreements and, as importantly, better ensure their support after the negotiation process has ended.

A broader, stakeholder-based approach would, though, also require a change in the transparency of decision-making processes. While transparency has not been an endemic part of South Asian water management, domestic or international, there are steps that can be taken to help change the tradition.

Increase data sharing, externally and internally

A major element of increased transparency is the sharing of data. Noted practioner and scholar (the late) Ramaswamy Iyer has said that one of the key issues in water disputes between India and Bangladesh was the lack of reliable data in terms of flow rates and water utilization, as well as measures of the impact and damage of Farakka withdrawals. In fact, flow rates on the Ganga are considered secrets, a problem that makes management decisions difficult and adds to overall levels of distrust.

Failure to collect and share data also limits the ability to analyse what the benefits and costs of cooperation might be, how those benefits and costs can be equitably shared, and the best directions for cooperation strategies. One of the main points of the World Bank's Ganga study was that any assessment of opportunities on the Ganga was problematic because of limited data and the failure to share what data was available. Unfortunately, the problem of data sharing is not just between countries but also within. In India, for example, there is no common database for the Ganga that governmental agencies can use in domestic decision making.

Data is a form of power and so reluctance to share is understandable. However, the nature of hydrologic data collection is rapidly changing. Space-based technologies coupled with more powerful software analytics are increasingly making it possible to monitor rainfall, flow, evapotranspiration and even groundwater withdrawal remotely. Furthermore, this data can be collected by essentially anyone, especially as collection and analysis costs continue to decline. While ground-truthing will remain important, control of physical measurements will be increasingly devalued, as will the value of maintaining data secrecy. Rather than holding onto past approaches, a move by all basin states to share data and cooperate on better data collection and analysis will provide a platform for goodwill and create, rather than destroy, political capital.

Building collaboration at other levels

In addition to sharing data, sharing the use of that data can be a step forward in building a shared understanding of the issues and opportunities, networks of experts across basin states, and overall cooperation. The role of scientific collaboration in further diplomatic cooperation has been widely recognized. At the same time, it is legitimate for countries to maintain their own propriety analyses and analytical tools. A way to bridge this dichotomy could be to increase scientific collaboration on issues related to the Ganga but not directly to negotiations. For example, in a report on the Indus Water Treaty, Suba Chandran suggested a broad collaboration between India, Pakistan, Nepal, Bangladesh and China to jointly study Himalayan glaciers. Track II dialogues in which national NGOs or think tanks work within and across borders to establish dialogue on issues and options can be another approach.

Strategically link and de-link water issues from broader diplomacy

A barrier to water cooperation has been the difficulty in disentangling water disputes from other bilateral problems and disassociating them from national political struggles. Discussions on water between India and Bangladesh are influenced as much by issues of Bangladeshi migrant influxes to India's north-east, Chakma refugees, porous borders and insurgent groups as they are about water itself.

Bundling issues can be a way to expand negotiating spaces and allow positive-sum solutions to what would have been a series of smaller zero-sum problems. Though it failed, the Teesta accord would have been favourable to Bangladesh and was offered to earn Dhaka's support in clamping down on two of the most lethal insurgent groups in India's north-eastern region operating out of Bangladesh. In other instances, progress is best made by de-linking water from other issues. The border agreement between India and Bangladesh was de-linked from water. Its success may now provide the opportunity for progress on water and other issues. Strategically linking and de-linking water and other issues depending on circumstances will be a key to overall progress.

Indian reassurance

By any measure, India is a large country in terms of its physical size, its population, its economy and its military. This makes it a legitimate cause for concern among its much smaller neighbours. Its past actions on Farakka and dam construction on other waters shared with Bangladesh, as well as implementation of water treaties with Nepal, have not helped to alleviate these concerns. However, recent efforts to resolve long-standing issues and push forward on new cooperation gave an indication that a new environment was possible. In terms of water, this could be furthered by ensuring that the terms of any agreements with Nepal are perceived as equitable, especially in a period when Nepal has been weakened through an earthquake and political upheaval. Unfortunately, though, the current impasse over trade has not moved this opportunity in the right direction. For Bangladesh, India can provide reassurances that it will follow international norms and not move forward on any aspects of its river-linking project that could have impacts outside its border without first having full consultations. India could also consider discussion forums that included all Ganga riparian states as at least a sign that it is cognizant of the power imbalance in bilateral meetings. The point is that India would not need to forfeit its power but rather signal its goodwill so that it too can benefit from greater cooperation.

Conclusions

Despite the potential direct and indirect benefits, cooperation between the riparian states of the Ganga has been limited, and in fact conflict and rancour has more been the norm. Part of the problem lies within broader political problems, problems that will have to be remedied for water cooperation to flourish. At the same time, though, improved water relations can contribute to broader political cooperation, and there are steps that can be taken now to move in that direction. These include a broader inclusion of stakeholders within national discussions of water priorities, a change in the increasingly outdated mentality behind hydrologic data secrecy, improved informal collaboration across borders both to build trust and develop alternative options for problem solving, and more strategic approaches to linking and de-linking water issues from broader political questions. As the largest power in the basin, India also needs to lead by example to facilitate trust, not simply out of altruism but also because of the gains it can accrue.

Notes

1 The first formal international relations over the Ganges were likely surveys starting in 1905 for the 1918 Sarada Treaty between independent Nepal and British India. The Mahakali Treaty discussed later in this chapter builds in part from the Sarada agreement. While not international in nature, early hydraulic work by the British on the Ganga also played a role in the later development of international

relations. The Ganga barrage, built between 1848 and 1854, was the first major modern engineering project on the Ganga and was designed as a single purpose irrigation project to use Himalayan waters to provide 3,000 km^2 of supplemental irrigation to the Ganges Plains. The narrow purpose and ethos of the project, its technology, and the first Indian engineering school at Rorkee that was connected to it can be argued to have set the stage for some of the fundamental conflicts that still plague national and international management of the Ganga. The approach is seen by some as plains and agriculture-biased and caters neither to upstream hill nor downstream coastal concerns; and it also does not answer the needs of other river uses such as fisheries or inland navigation.

2 The tributaries enter the Ganga via the Brahmaputra and are thus technically beyond the scope of this book. However, since the Teesta has in recent geological past been connected to the Mahananda and the Ganga before joining the Brahmaputra (Bandyopadhyay and Gyawali, 1994), and it contributes to the overall question of water diplomacy on the Ganga, it is also discussed here.

References

Alam, U.Z. (2002). Questioning the water wars rationale: A case study of the Indus Waters Treaty. *The Geographical Journal*, 168(4), 341–53.

Bandyopadhyay, J. and Gyawali, D. (1994). Himalayan water resources: Ecological and political aspects of management. *Mountain Research and Development*, 14(1), 1–24.

Gyawali, D. 2001. 'Pluralist Politics under Monistic Design: Water Accords in South Asia'. In Samaddar, R. and Reifeld, H. (eds). *Peace as Process: Reconciliation and Conflict Resolution in South Asia.* New Delhi: Manohar for Konrad Adenauer Stiftung.

Gyawali, D. (2003). *Rivers, Technology and Society*. London: Zed Books.

Gyawali, D. and Dixit, A. (1999). Mahakali impasse and Indo-Nepal water conflict. *Economic and Political Weekly*, 34(9): 553–64.

NWCF (Nepal Water Conservation Foundation) (2013). *Re-imagining Regional Cooperation on Transboundary Ganga: Nepali Academic and Civil Society Concerns*. Kathmandu, Nepal.

Prasai, S. and Surie, M.D. (2013). *Political Economy Analysis of the Teesta River Basin.* The Asia Foundation, New Delhi.

18

Water challenges of the Ganges Basin

An agenda for accelerated reform

Tushaar Shah, Prachanda Pradhan and Golam Rasul

Key messages

- High population, acute poverty and growing pressure on land and water resources will shortly breach the limits of the overall carrying capacity of the Ganges Basin. Accelerated reforms in governance and management can bring quick improvements.
- Reforms in irrigation bureaucracy and institutions, investments in operation, maintenance and management, rationalizing of irrigation service fees and completion of the last mile projects can rapidly expand irrigation benefits of the large public irrigation systems.

- To accelerate agricultural growth in the Ganges Basin, governments need to provide reliable and affordable energy to all the smallholder groundwater pump owners and service providers. Solar power is best suited to energize the Ganges Water Machine and will stimulate a vibrant, deep and broad irrigation service market in the entire basin.
- Municipal authorities and public utilities need to make massive investments in the treatment and orderly distribution of domestic and industrial wastewater to peri-urban farmers and related sectors.
- The basin countries can make rapid strides in water security for the basin population by emphasizing management improvements and institutional reforms rather than just public investment in water infrastructure.
- A Ganga university needs to be tasked to build management capacities in irrigation and water supply bureaucracy in all the basin countries.
- Developing the basin as a whole in a cooperative framework would be the most desirable course for all the states of the Ganges Basin but this cannot materialize unless they are able to find a common ground to resolve their conflicting priorities.

Introduction

With a human population of 510 million in 2011, the Ganges Basin is one of the most densely populated basins in the world. Another distinct aspect is the poverty of people: the basin is home to more dollar-a-day poor than all of sub-Saharan Africa. The predominance of an agrarian livelihood system implies growing pressure on land and water resources and the overall carrying capacity of the basin. With basin population projected to grow to 868 million by 2050 (Amarasinghe *et al.*, 2007), the Ganges Basin is destined to witness inexorable pressure on this carrying capacity and likely some of the most strident water conflicts.

The basin is shared primarily by Nepal, India and Bangladesh, each having its own problems, perceptions and priorities. Nepal, the upper riparian, would like to develop its massive hydropower potential to alleviate the poverty of its people and kick-start economic growth. India, the middle riparian, would like to minimize recurrent damage to lives and livelihoods by annual floods while also increasing its food, water and livelihoods security. For Bangladesh, the lowest riparian ensconced in the floodplains of the Ganges, Brahmaputra and Meghna river systems, the Ganges is critically important for agriculture, food security, livelihoods, transportation, environmental security and maintaining healthy ecosystems. *Boro* rice, the dry season irrigated crop with high yield, contributes a large share of rice production and is the bedrock of food-security for its land-short farming households. All the Ganges countries are urbanizing rapidly, and need to prepare to meet growing urban and industrial demand for water. Nepal wants to develop its hydropower and irrigation. India and Bangladesh want Nepal to build large storages to reduce flood damage; but Nepal wants to build decentralized small storages. For eastern India and Bangladesh, managing floods, too, is a persistent challenge.

Earlier chapters of this book have provided detailed accounts of the socio-economic, environmental and hydrological dimensions of water resources management in this vast basin. In this last chapter, our focus is to tease out a clutch of actionable line items that water policy leaders in these three basin states can consider for implementation. While there are specific proposals for each basin state to respond to its unique problems and issues, there is much that is common and needs a common response. The most critical need is for joint development of the entire basin's potential which can be possible only in a cooperative framework. Developing such a framework is going to be slow and arduous; but cooperating on harvesting some low-hanging fruits may be the way to start a collaborative enterprise.

Institutional evolution in the Ganges Basin

Global discourse on water governance has generated a stylized format for water institutional reforms in a river basin. However, for these to take root, these reforms need to be set in the context of each river basin. Understanding the physical, social and political context is critical. For example, the level of economic modernization of a society is a critical consideration (see Table 18.1). The overarching aspect of the Ganges Basin is the poverty of its people which determines the nature and type of institutional arrangements that dominate the water economy. Poor countries have highly informal water economies that are hard to regulate and govern while rich countries have highly formal water industries that function within a robust and relatively good legal-regulatory framework. Most water users in a highly formal water economy are secondary users, connected with the water governance regime through organized service-providing primary users amenable to regulation. Those in a predominantly informal water economy are mostly primary users, drawing water directly from nature to meet their personal as well as productive water requirements. Implementing formal approaches to water management through laws and higher level institutions will not work without a basic level of infrastructure and intermediation being in place.

Institutional reforms arising from global discourse often intend to transform, all at once, a predominantly informal water economy into a predominantly formal one – something that would normally be the result of a long process of economic growth and the transformation that comes with that. Formalizing informal water economies is often seen as a route to improving water governance. But evidence across the world suggests that there is no shortcut for a poor society to morph its informal water economy into a formal one; the process by which this happens is organically tied to wider processes of economic growth. When countries try to force the pace of formalization, as they will no doubt do, interventions come unstuck. Interventions are more likely to work if they aim to improve the working of an informal water economy.

In the Ganges Basin, thanks to poor management, the relative share of public systems in riparian countries has been declining. The proportion of rural households that are served irrigation service or domestic water supply by public or community institutions is less than 15 per cent. This implies that an overwhelming majority of the rural population either self-provides its water needs or is served by informal arrangements such as local water markets. High-level officials glibly assert that the water challenge is not a challenge of resource scarcity but one of resource governance and management. Yet, given the vast challenges the basin faces, India, Nepal and Bangladesh have invested little in creating water management capacities, structures and systems.

Closing the management deficit

By far the most promising among the low-hanging fruit is closing the serious management deficit in the basin water economy. The Ganges Basin water economy is managed so far in a 'resource development' mode; the biggest challenge facing the basin states now is to make a transition to 'resource management' mode. This requires a focus on managing three components of the water economy: water resources, water infrastructure and water services. In managing its water infrastructure, Ganges Basin states are in a perpetual 'build-neglect-rebuild' cycle, a sure indicator of poor management. As the region urbanizes in the coming decades, it will invest massively in water infrastructure, especially urban and non-agricultural. However, without management capacities, these investments will yield poor returns, private or social. A major folly is to assume that the skills required to build an infrastructure are the same as those required to manage it. Bangladesh, India and Nepal do little to reskill engineers for successful management

Table 18.1 Ganges Basin in global trajectory of institutional evolution

Evolution from primitive to modern basin water economies				
	Stage I *Fully informal*	*Stage II* *Largely informal*	*Stage III* *Rapidly formalizing*	*Stage IV* *Fully formalized water industry*
Example	*Congo Basin*	*Ganges Basin*	*Olifants, South Africa*	*Rhine, Europe; Murray Darling, Australia*
Dominant mode of water service provision	Self-provision	Public and self-provision	Public provision with self-supply declining	Modern water industry with zero self-supply
Rural population as percentage of total	80–90	50–80	20–50	5–20
Agricultural water use as percentage of total managed water	>90	80–90	70–90	60–70
Percentage of total water use self-supplied	>90	70–90	20–70	0–20
Water management capacities	+	++	+++	+++++
Utilities' cost to serve water	+	++	++++	++++
Institutional arrangement in water sector	Informal self-help and mutual help; community institutions dominate	Informal exchange institutions dominate; water markets co-exist with community institutions	Organized service providers crowd out mutual help and community institutions	With rise of modern water industry, self-help and community institutions decline
Priorities of water governance	Infrastructure creation and operation in a welfare mode	Improve service management without cost recovery	Improve infrastructure and service management with cost recovery	Integrated management water service, infrastructure and resource with full cost recovery

of water systems. India has invested more than Bangladesh and Nepal in creating training infrastructure, but of little import. For example, India has over a dozen state-level water and land management institutes. There are similar institutes for training in water supply and sanitation. India's National Water Academy at Pune is supposed to provide higher level skills. Several of India's Institutes of Technology (IITs) have strong hydrology departments. However, none of these offer training and education in *managing* water resources, infrastructure and services.

Given the size and complexity of its water economy and its growing water challenges, the Ganges Basin needs a much larger – and better – knowledge infrastructure than it presently has for research, training and education. A first step would be a world class Ganges university, established jointly by the basin states, with a focus on *managing* the resources and challenges of the basin's water economy, emphasizing all water uses – irrigation, urban and rural water supply and sanitation, and hydropower. Equally, the basin states should invest in performance monitoring and benchmarking of public systems in irrigation, water supply and sanitation.

Improving the performance of public irrigation systems

Recent decades have been witnessing tectonic shifts in the structure of the Ganges basin water economy and its organization. While governments and donors have continued to invest in gigantic irrigation and hydropower projects, the overall management and performance of public and community irrigation systems suffered rapid deterioration while the decentralized informal groundwater economy has boomed. The British in India built and managed irrigation systems as commercial enterprises which earned for the irrigation department 12 per cent return on capital investment until 1945. Irrigation managers levied and ruthlessly collected irrigation service fees at non-trivial rates, amounting to 25–40 per cent of the value of output. Farmers complained about extortionate irrigation service fees (ISF), but were assured irrigation service of a certain level by managers regularly investing in the maintenance and upkeep of the system.

After 1950, the situation in India began changing. The new welfare state relaxed norms of cost recovery. Some states found the cost of collection too high and abolished ISF altogether. Others refused to revise and rationalize ISF for decades. Gradually, collection discipline also deteriorated with most states collecting just 10–15 per cent of ISF demand based on highly subsidized ISF rates. Budgets given to irrigation departments were barely enough to cover salaries, with little or nothing for operation and maintenance (O&M). By the late 1980s, many governments banned new recruitment in irrigation departments. Regional governments with thousands of hectares under public irrigation command areas were left with irrigation departments that were ageing and shrinking. Gradually, as deferred maintenance mounted, irrigation commands began to shrink. Bihar, West Bengal and Uttar Pradesh witnessed the sad prospect that investment in new projects and rehabilitating old projects kept apace, but the area actually served by public canals declined. When ISF was vigorously collected, the paper trail created an excellent management information system. Now that ISF remains uncollected, there is no reliable information on how much area a system irrigates and the quality of service offered. Bihar abolished ISF in the 1960s; Uttar Pradesh, the heartland of canal irrigation, also recently abolished ISF altogether. Increasingly, mega irrigation projects are driven by rent-seeking politicians whose focus is on the construction of irrigation systems, which offers opportunities for giving away lucrative contracts, and not their management.

In Nepal, large-scale irrigation systems were developed only in the 1970s and 1980s. However, it is expected to have improved performance and productivity of these systems. Out of 1.3 million ha of irrigated area, Nepal has a unique situation where the farmer-managed irrigation systems (FMIS) share is almost 70 per cent and the contribution of the public irrigation system is only 30 per cent. Hence, irrigation bureaucracy has to have a two-pronged approach of improving public sector irrigation systems with farmers' participation and providing appropriate assistance programmes to FMIS where their autonomy is maintained in management of the systems (Pradhan, 2012).

This situation calls for a major effort to revitalize the public and community irrigation sectors in the Ganges Basin. The Indian portion of the Ganga Basin has some 20 million hectares (mha) of irrigation potential created from public irrigation systems. Against this, only 6–7 mha is actually used. The infrastructure already created can serve 2.5 times the area currently served simply by concentrating resources on clearing the backlog of deferred maintenance, tightening the management of irrigation systems, restoring the accountability relationship between farmers and irrigation staff, rationalizing the irrigation service fee and vigorously collecting it, and improving management capacities all-round in irrigation departments.

Proactively managing aquifer storage

While public irrigation has declined during the past five decades, private irrigation has experienced a veritable boom throughout the Ganges Basin, based mostly on shallow tube wells. Between Bangladesh, India and the Terai areas of Nepal, the Ganges Basin today has 12–15 million shallow tube wells (STW) irrigating some 30 mha. Table 18.2 provides a broad picture of the colossal groundwater economy that has emerged as the mainstay of this vast basin's agrarian economy. This groundwater boom financed by private capital has arguably given the Ganges Basin three times more irrigated area in 50 years than public irrigation investments did in 200 years. Groundwater proved a 'democratic resource', taking the benefit of irrigation to every corner of this vast basin, far outside the reach of public canals. STW also became the mainstay of small and marginal farmers eking out a living from their shrinking holdings by working their land more intensively. Farmers prefer well irrigation over canal irrigation because it is more reliable, available on demand and offers greater water control, allowing them to adopt high-value farming systems. Bangladesh has 1.53 million pumps in use for irrigation, of which 0.18 million are electric and 1.33 million (> 90 per cent) are diesel. Despite subsidies for electricity, farmers prefer diesel pumps due to their low capital cost and ease of moving around within small and fragmented farmlands.

In sum, then, over the past 60 years, the Ganges Basin's agrarian economy has shifted its reliance for irrigation from surface storages – large dams, small tanks, streams and canals – to aquifer storage. This is a positive development. In the 1970s, scientists talked about the Ganges Water Machine. They argued that with its vast and porous aquifer systems, intensification of dry season groundwater irrigation can not only accelerate agricultural growth but also reduce the intensity of monsoon floods that regularly devastate thousands of villages in Nepal Terai, Bihar and Bangladesh every year. Rapid development of groundwater irrigation during recent decades has set the Ganges Water Machine in operation. However, it can deliver much more than it has so far.

The key constraint is that although groundwater availability is relatively abundant, villages often lack electricity supply. Farmers here are forced to depend on diesel and this creates an economic paradox. In groundwater-scarce states, farmers face no economic water scarcity and pump like there is no tomorrow thanks to free or near-free power supply. In groundwater-abundant states, farmers are forced to economize on irrigation, resulting in low productivity. To add to their woes, farmers in the groundwater-surplus Ganges Basin on the Indian side are

Table 18.2 Agricultural groundwater economy of the Ganges Basin (2014–15)

Parameter	*India*	*Bangladesh*	*Nepal Terai*
Number of wells (million)	21	1.53	0.6
Total groundwater used (km^3)	230–250	45–55	9–15
Approximate value of groundwater used per year in billion US$	15–18	2.5–3	0.5–0.7

Source: Authors' estimates

often unable to market their produce to the government at minimum support price (MSP) due to limited procurement and storage infrastructure. The net result is that water-scarce north-western and peninsular India exports 'virtual water' (water embedded in the production process) to water-abundant eastern and tribal India. According to one estimate, Punjab, Haryana and Andhra Pradesh export nearly 56 billion cubic metres (BCM) of virtual water in the form of agricultural products. At the same time, Bihar, Jharkhand and Orissa import 29 BCM of virtual water to meet their food requirements (Verma *et al.*, 2009).

India has ambitious plans to link Himalayan rivers with peninsular ones at a massive investment of US$120 billion to physically transport water to water-scarce regions under its mammoth National River Linking Project. However, India's food policies must be aligned with water (and land) endowments to correct the perverse direction of domestic virtual water trade. Doing this would require encouraging farmers in eastern India to use their abundant groundwater resources to enhance their agricultural productivity. The first step to do this is to quickly expand the food procurement machinery. The second is to reduce the dependence of groundwater irrigation on hydro-carbons. For Bangladesh, for example, energy use in groundwater has a large role in the country's overall energy balance. Especially during summer, energy demand for *boro* rice irrigation can widen the electricity demand supply gap to 2,000 Mw as estimated by the Bangladesh Power Development Board (BPDB). Irrigation demand for diesel is ever more of an issue in the country's energy management. One possible response to the energy challenge of groundwater irrigation can be to solarize the Ganga Basin's groundwater economy. This needs to be accompanied by improved availability of fertilizers, seeds and other inputs. Even more important is to liberate the Ganges Water Machine from the high cost of diesel energy it needs to operate by introducing a more efficient and affordable energy source. Solar pumps, for long considered unviable, suddenly seem to offer a big opportunity here.

Breaking the energy–irrigation nexus

Many solutions to the problems afflicting the Ganges Basin water economy have to do with the nexus between energy costs and groundwater development. There is a sharp contrast between the energy–groundwater nexus of the Indus Basin versus the Ganges. Over several decades, the Indus Basin states have made rapid progress in rural electrification as well as groundwater irrigation with electric pump sets. As has been analysed by several researchers, the hydro-politics of irrigation here has witnessed tube well owners forming into formidable vote banks to demand and secure farm power subsidies for groundwater irrigation. In Pakistan Punjab and Sind, as well as in the Indian states of Punjab, Haryana, Uttar Pradesh and Rajasthan – which comprise the bulk of the lower Indus Basin – persistence of farm power subsidies has led to explosive growth in groundwater irrigation in a basin that had limited fresh groundwater in the first place. The nexus between energy subsidies and groundwater irrigation in the Indus Basin has thus created frighteningly perverse outcomes.

The situation in the Ganges Basin states is very nearly the opposite. The Ganges Basin has excellent alluvial aquifers abundantly recharged on an annual basis by snow-melt as well as monsoon rain. If there is any basin in the world appropriate for large-scale groundwater development for agriculture without any serious threat of depletion, it is the Ganges Basin, with abundant recharge and limited irrigable farm lands. Yet, the region has not been able to make full use of this potential thanks to poor rural electricity supply and high diesel prices. The nexus that hurts the Ganges Basin is the one between high energy prices and the inability to use abundant groundwater resources to alleviate agrarian poverty. Bangladesh will have to depend mostly on thermal sources and diesel for supporting groundwater irrigation. Nepal, of course, is the best placed among the Ganges Basin states to overcome the energy scarcity, given its massive unexploited hydro-potential. India, too, has unexplored hydro-potential in its north-eastern region. However, both Nepal and India will need years, even decades, to fully develop their hydro-power potential.

A big opportunity for quickly cracking the Ganges Basin's energy–groundwater nexus is offered by the precipitous and continuing fall during recent years in global prices of solar photovoltaic (PV) cells. Once considered prohibitively costly for irrigation pumping, solar power is destined to increase its presence in South Asia's pump irrigation economy in the years to come. Capital costs are still beyond the reach of the farmers. As a result, governments of basin states – especially in India and Bangladesh – are aggressively promoting solar pumps by offering 70–90 per cent capital cost subsidies.

However, the thinking underlying the current solar pump schemes is patchy and piecemeal. The need is to promote solar pumps as an integrated energy-water-sustainable farming solution. Agricultural growth in the Ganges Basin is seriously constrained by energy shortages and costs; the region has neither much grid electricity to offer to farmers nor proper transmission and distribution infrastructure. The region is floating on one of the world's best aquifer systems; yet, its agriculture is unable to benefit from intensive groundwater use for lack of energy. The Ganga Basin's small farmers, ever dependent on expensive diesel to run pumps, are unable to benefit from irrigation even though they have no shortage of groundwater. Solar pumps offer the potential to quickly energize eastern India's groundwater economy. Solar pumps have a high fixed cost, but near zero operating cost; thus solar pump owners will have strong incentives to sell irrigation services to neighbouring farmers. The solar pump promotion strategy in the Ganges Basin states should thus aim to stimulate vibrant, deep and broad irrigation service markets at the village level. To do this, there is need for programmes that: (a) offer a pro-rata subsidy of say US$1,000/kWp (instead of per cent of capital cost which leads to gold-plating and corruption); (b) allocate substantial funds (US$10–15 billion over a 5-year period for the basin states) so that a large number of suppliers compete to get a share in the market; (c) include support to establish a buried pipe network to neighbouring farmers to stimulate water selling; (d) avoid putting undue restrictions on the size of the pump and PV array that farmers can acquire; (e) oblige suppliers to provide a 5-year maintenance guarantee; and (f) establish rigorous third-party monitoring to ensure that suppliers keep their side of the deal. Such a programme can install 15–20,000Mw of groundwater-pumping capacity in the Ganges Basin, equivalent to four to five million 5-horsepower electric pumps, and create gross irrigation potential of 28–32 mha in a relatively short period of time.

Transforming water markets into equitable irrigation institutions

The institution of decentralized, village-level irrigation service markets (ISM) has served as a powerful instrument of smallholder irrigation development throughout South Asia, but particularly in the Ganges Basin where it is extensive and exploitive. When tube well irrigation technology first came to the region during the 1960s, it was accessible only to large farmers because of its lumpiness and high capital intensity. Small farmers found the technology beyond their reach. With their small land-holdings fragmented into several parcels, they found that even if they invested in a tube well in their largest and best parcels, their tube wells would remain grossly under-utilized while other parcels of their holdings would remain unirrigated. Many state governments therefore invested in heavy duty public tube wells to provide broad-based access to groundwater irrigation for the poor.

Many of these public tube well programmes failed, however, primarily because small farmers had created an ingenious solution for the lumpiness of tube well investment. They began selling irrigation services to their neighbouring farmers and buying it from other tube well owners in their other distant parcels. Thus emerged village-level ISMs that have transformed the irrigation business model in India, Pakistan, Bangladesh and Nepal Terai. ISMs got a strong fillip during the 1970s when many state governments began charging for electricity used in pumping groundwater on a flat tariff instead of metered consumption. Flat tariffs, linked to horsepower of the pump rather than metered electricity consumption, reduced the irrigation service provider's (ISP) marginal cost of pumping to near-zero levels and incentivized them to maximize water sales by lowering their prices and improving service quality. Where flat tariff was kept high, as in West Bengal until a few years ago, ISPs were under even greater pressure to maximize sales in order to recover their high fixed costs. Although it strained the finances of electricity utilities, ISM under flat tariff proved a great boon for millions of resource-poor water buyers who got a top-class irrigation service at affordable prices.

The poor in much of the Ganges Basin, which is flush with groundwater, remained deprived of the benefit from such a competitive ISM boom because ISPs here commanded a strong monopoly power. Because there is little electricity available, ISPs here mostly use diesel pumps which entail a low capital cost but high and rapidly rising pumping cost. The dieselized ISM of eastern India are thus low-volume-high-price markets in which resource-poor small farmers are the worst off. The only way to catalyse high-volume-low-price ISM in eastern India is to electrify the countryside and charge electricity to tube wells at a reasonably high flat tariff. However, this may not happen any time soon; it takes a long time to establish a rural grid network. Also, governments in eastern states will learn from western experience and insist on metered tariffs from the start, as West Bengal has done.

Solar irrigation pumps offer a great opportunity to catalyse highly competitive and equitable ISM in groundwater-abundant eastern India. An ISP with a solar pump faces high fixed cost (of depreciation and interest on solar panels) and very low operating cost; as a result, he would have a strong incentive to maximize the 'utilization factor' of his irrigation asset to recover his fixed costs by selling irrigation services to neighbours. A solar ISP is very likely to mimic the behaviour of ISPs in West Bengal when they were subject to high flat electricity tariffs: they maximized water sales even when competition among ISPs drove down the prices of irrigation services to the benefit of resource-poor water buyers. This is already evident from the limited experience of Bihar government's programme to solarize some 35 heavy duty public tube well programmes in partnership with Clero Energy. Despite several design flaws in the arrangement, the tube well operators are emerging as aggressive ISPs at competitive prices.

Water supply and sanitation: integrating informal and formal water economies

Three-quarters of residents of the Ganges Basin self-supply their domestic water needs directly from nature with minimal treatment; this figure is smaller and shrinking for urban areas, but vast numbers of urban Indians still use self-supply to supplement public supply (Shah and van Koppen, 2006). Self-provision is the result of the inability of the public systems to meet demand; the informal economy is a response to the opportunity to profitably serve a demand. Tanker water markets in many urban centres are pervasive and booming. However, private service providers are beginning to play an active role, even in the sanitation space.

As the Ganges Basin urbanizes, the municipal water infrastructure is unable to keep pace with the growing population. As towns grow into cities, they need massive investments in upgrading and expanding their water supply and sewerage network. An even more pressing problem, however, is of the revenue and business model. The critical mistake often made is of trying to replicate levels of service obtaining in rich, industrialized countries. This is throwing good money after bad. Copenhagen (Denmark), Glasgow (UK) and Essen (Germany) have among the best WSS service regimes in the world; but the tariff they collect is US\$6.3–7.6/m^3. At the other end are cities like Ashkabat (Turkmenistan), Damascus (Syria), Khulna (Bangladesh) and Chandigarh (India) where WSS services are poor by global standards. There may be several reasons for this; but an important one is that utilities raise less than US\$0.05/m^3 as water tariff. In many cities of the Ganges Basin, even as the value of real estate soars, public water supply is heavily subsidized while tanker water supply used for drinking by slum dwellers often costs Rs. 50–80/m^3. It will take decades before urban utilities are able to raise tariffs to sustain world-class water supply and sanitation services. Until then, governments and utilities need to find ways of improving WSS services even while keeping average tariffs at affordable levels.

Treating and recycling growing volumes of urban wastewater in a cost-effective manner is emerging as a critical challenge in towns and cities. According to India's Central Pollution Control Board, urban India generated more than 42,000 million litres of wastewater every day in 2011–12 (GoI, 2011). Our current sewage treatment capacity can handle less than a third of this; and to make things worse, only 55 per cent of the treatment capacity is operational due to poor upkeep and maintenance (CPCB, 2007). In most urban centres, treatment plants act merely as accumulators and large quantities of untreated wastewater are released. No wonder the Ganga, Yamuna and most Indian rivers look and smell like drains as they pass cities.

Several municipalities have planned investments in primary, secondary and even tertiary treatment plants for recycling wastewater. However, the cost of tertiary treated water is high and only industries can afford to pay for it. A largely ignored potential customer of wastewater is the peri-urban farmer. Each city can be re-imagined as a high-class irrigation system that releases near-constant volumes of water on a daily basis, year-round. Much 'waste' in wastewater is valuable nutrients that farmers pay a great deal to buy in the market. The 15 BCM of wastewater that Indian cities release annually[1] can potentially irrigate 3 mha, nearly twice the command area of the Sardar Sarovar Project.

Thousands of farmers in peri-urban areas already use municipal wastewater to irrigate crops. Palrecha *et al.* (2012) estimated that nearly 18,000 ha were being irrigated by wastewater in and around seven cities in Gujarat. The limited availability of freshwater, rapid pace of urbanization, and farmers' preference for nutrient-rich wastewater are the key drivers contributing to rapid expansion of wastewater irrigation. Several municipalities have devised formal or informal arrangements with peri-urban farmers. Rajkot Municipal Corporation, for instance, has been supplying water to two registered wastewater cooperatives since 1962.

The South Asian discourse on WSS involves leapfrogging from the current highly rationed public urban water supply to 24/7 supply, from near-zero tariff to full metered tariff, from minimal wastewater treatment to 100 per cent treated wastewater being released into rivers. However, a more realistic scenario for the Ganges Basin states is that they will go through a long intermediate phase before reaching this ideal state. It seems very likely that self-provision, informal markets and wastewater irrigation will grow in this phase. The challenge is to create WSS governance regimes that meaningfully integrate self-provision, informal economy and public provisioning into a viable and sustainable system.

Much of wastewater irrigation is happening by default, rather than by design, because wastewater is seen mostly as a problem and not as a resource. In the near future, a greater proportion of our freshwater resources will need to be diverted for municipal and industrial uses. Wastewater recycling and irrigation present an opportunity in urban and peri-urban planning. If done well, municipal and industrial wastewater systems can act as reliable and efficient irrigation systems for millions of hectares of cultivable land.

The shared concern: laying the foundation for cooperative basin development

A critical concern for all Ganges Basin states is to evolve a sustainable protocol for cooperative basin development work. Given the hydrological and ecological links between the upstream and downstream areas in the Ganges river system, cooperative development of the Ganges water resources is indispensable for addressing the growing challenges of water energy and food security in the three riparian countries. To overcome the existing problems, the focus could be shifted from sharing water to sharing the benefits derived from jointly developed projects in the Ganges. Abundant rain-fed and snow-fed water resources and a topography with favourable relief provide an excellent opportunity to generate hydropower in the Ganges river systems to meet the energy requirements of Nepal, India and Bangladesh. It is estimated that the GBM river systems have about 200,000 Mw hydropower potential, of which half or more is considered to be feasible for harnessing (Chalise *et al.*, 2003, cited in Rasul, 2015). Joint development of the huge untapped hydropower resources in Nepal and north-east India could help in addressing future energy demand and demand for irrigation for increased food production, as well as promote industrialization and economic growth.

Promoting a basin-wide approach for optimum use of Ganges water resources to store monsoon water, augment dry season water availability, generate hydropower and increase irrigation coverage is thus imperative. In the past, unilateral and bilateral approaches have hindered maximizing the benefits of water resources management. Moving to a basin-wide approach, and from single-objective to multi-objective development, will help maximize benefits throughout the basin. For example, upstream hydropower plants can also support downstream irrigation and flood control, benefiting both upstream and downstream countries. This is easier said than done. India and Bangladesh would like massive storages in Nepal which will bear all the costs of displacement and loss of land without sharing the benefits commensurately. Only a proposition that is considered by all three riparian countries will make any headway. And such a proposition if it ever materializes, will likely be founded on a much wider axis of cooperative action than just water sharing and will include a range of economic, trade and even political cooperation factors.

This will not happen by default. Institutional arrangements are needed to coordinate, facilitate and strengthen cooperation in water, hydropower and flood management in the Ganges

Basin, and to bring together scientists, development practitioners and policy makers to share, discuss, negotiate and develop a common understanding. The starting point could be cooperation in flood control, as flooding is a common issue for all three countries of the basin. A basin-wide data bank should be established with a system for timely sharing of meteorological, hydrological, economic and environmental data and information among the countries of the basin.

Water discussions among the three countries in the Ganges Basin are marked by a sense of mistrust and conflicts which hinder further utilization of this vast resource for the benefit of the entirety of the basin population (Upadhyay, 2012). The future of the basin depends on how the head-end and tail-end countries behave. A concerted effort is required to build trust and confidence among the three riparian countries. Mistrust is partly due to poor understanding of the benefits and costs of collaborative development. There is a need to promote joint research on transboundary water management issues; on integrated hydrological-economic modelling to explore alternative development options; and on quantification of the benefits and costs of regional cooperation, and the social, economic and environmental costs of non-cooperation. Efforts should be made to facilitate cross-border exchange among civil society organizations, NGOs, academic and scientific communities, and government officials to deepen relationships and foster informal intergovernmental cooperation processes. The three riparian countries should undertake joint research to produce credible information and knowledge, explore development potential and options, and assess risks, costs and benefits of cooperative management to support sound decision making. This can happen more easily if the countries in the basin expand the playing field of exploring opportunities for fruitful cooperation, including trade and economic ties.

There is a need for realistic understanding of the political economy of internal and transboundary water use among the three basin countries which are in different stages of water resources and infrastructure development. Climate change has added new challenges in water resource management in Nepal and elsewhere. Water endowment among these three countries also differs. Similarly water-induced disasters like floods or water shortages are to be collectively resolved. Further development and management improvements are to be based on cooperation and collaboration among the basin countries on an equal basis. Further basin-related infrastructures and management are to be promoted. In future, such cooperation and collaboration of head-end and tail-end countries are to be reviewed in the context of paradigm shift. If the need arises, the existing unequal agreements of river water utilization among these countries are to be reviewed in the context of paradigm shift. The future of the basin is predicted to be water scarcity, hence appropriate water use and management are to be ensured by a number of reform measures, and the collaboration of these countries is very important.

A start has to be made with a more nuanced and informed understanding of the hydrology of the basin. There needs to be sharing of water-related data and joint assessment of real-time-based weather and rain changes to predict and show possible water surge and flooding, in order to provide advance alerts and warnings and relocation/evacuation, which would benefit in preventing massive loss of life and property among the basin countries. Not being able to share information on water surge and river rise, and rapid rise in precipitation, may result in unnecessary casualties, both human and environmental. All the three above points of engagement have policy implications pertaining to natural resource governance, fiscal and transport policies, subsidy and loan policies, water management, and policies promoting sharing of water-related data and disaster prevention.

Conclusion and ways forward: country perspectives on water reforms in the basin

Given its unique set of water issues, Bangladesh needs to consider specific priorities in its water sector planning, with emphasis on these specificities. There is both a seasonal concentration of rainfall and high spatial variation in rainfall distribution in Bangladesh as elsewhere in the basin. Increasing the storage of water when it is abundant, and distributing the stored water when and where it is required, will be critical for addressing future water challenges in the Ganges Basin. One solution that has been widely discussed is building surface storages. Nepal has excellent sites for storage of monsoon water. Construction of large storage reservoirs on the Ganges tributaries originating in Nepal would benefit Nepal, India and Bangladesh. It is estimated that seven large dams in Nepal could increase the dry season (January–May) water availability of the Ganges by 70 BCM (Chaturvedi, 2012, cited in Rasul, 2015). The proposed Sapta Koshi high dam in Nepal would have enough storage capacity to provide both north Bihar (India) and Bangladesh with both a flood cushion and augmented dry season flows after meeting Nepal's full irrigation requirements. Water storage is not only a mechanism for adaptation to climate change, but also critical for ensuring water, food and energy security. Efforts should be made to explore joint projects for augmenting dry season water availability, with hydropower development of transboundary water resources for short-, medium- and long-term measures based on mutual understanding and priorities. Joint action plans could focus on both structural and non-structural measures. A strategic environmental assessment, including detailed studies of technical and economic feasibility, is required to identify the potential. For Bangladesh, it is also critical to invest in developing surface water irrigation wherever possible, and to improve its management, in order to reduce the pressure on groundwater abstraction for irrigation. The optimum utilization of surface water and groundwater for irrigation should be investigated. Urgent action is required to enhance the performance of irrigation systems by constructing and developing efficient irrigation channels, investing in an improved distribution system, and improving irrigation management. It is important to upgrade irrigation technologies and improve management systems to increase the efficiency of irrigation and energy use and minimize the wastage of water and energy resources. Policies need to be reformed in order to increase fees for water and electricity, and the system for collecting irrigation charges should be improved. At present irrigation fees are low and collection rates poor, thus the revenue cannot cover even a small part of the cost of operation and maintenance. The result is poor operation and maintenance and a poor irrigation service. For these, institutional arrangements are needed for sustainable groundwater management involving water users. These need to include investments in improved water and agricultural technologies, charging for water on a volumetric basis or crop-specific pricing, fixing quotas for groundwater extraction, facilitating markets for non-rice crops, promotion of alternative cropping patterns, and extra support for farmers making a transition to less water-demanding crops. Finally, Bangladesh needs to take a serious view of the energy implications of its groundwater economy. Expensive and unreliable energy constrains the supply of irrigation water and thus food production. Diesel irrigation pumps need to be replaced with solar PV. At present 1.33 million diesel pumps are being used for irrigation and about 1 billion litres of diesel are imported annually to operate them. The average solar radiation in Bangladesh is 3.85 kWh/m^2/year, which could be used to supply the energy required for irrigation during the monsoon. Although a number of efforts have been made to promote solar power irrigation, the contribution remains low. A proper policy and institutional arrangements are needed to integrate solar pumps in the national grid system and support selling of surplus power to the national grid, in order to overcome barriers such as high installation costs and encourage large-scale adoption of solar pumps. Introduction of a solar power irrigation system could save 760 Mw of power and 800 million litres of diesel every year.

The opportunities in the water sector that Nepal would emphasize are hydropower, tourism and environmental amenity. When fully developed, hydropower and eco-tourism have the potential to be major growth centres for its economy. To support this, the country also needs to invest in urban water supply and maintenance of clean river systems around urban areas. Irrigation management and services improvement of both public and farmer-managed irrigation systems (FMIS) is an area of priority too, as are issues related to ground water improvement and development through rural electrification; electric pumps will be promoted for groundwater irrigation. There are big opportunities for safe urban wastewater use in peri-urban agriculture after treatment. Nepal also needs to learn to better manage seasonal floods. All these are suffering for lack of resources, but more critically for lack of techno-managerial capacities. Appropriate investment in management capacity development along with investment for physical infrastructure is needed. Nepal has to improve its capacity in water resources management, and to that end, it needs to establish an Integrated Water Resources Institute for water resources planning and management. With proactive resource planning, Nepal needs to focus on water infrastructure development for different sectors including hydropower, irrigation and inter-basin water transfer. India's experience has shown that emphasizing infrastructure creation by itself is of little help. There is a need to invest in improving water services in a sustainable format. The new institute also needs to help improve Nepal's negotiation capacity for improving its position on regional versus bilateral issues of water management. The provision of knowledge centres like Ganga University (keeping in mind what is happening to SAARC University) – to be jointly established by riparian countries – can be a step towards new knowledge and skills to address the challenges of the massive and complex water economy of the region. However, we have to take into consideration existing national and regional research institutes and integrate with them (Pradhan, 2015).

Indian discussion on water priorities in the Ganges Basin states has a great deal in common with the discussions in Nepal and Bangladesh. However, India is a diverse country and has experienced different levels of economic dynamism in different parts. For southern and western states of India, the Ganges Basin is likely to emerge over the years as an exporter of food, water and labour; its economy is likely to remain agrarian longer than most southern states. This is perhaps why Indian politicians cannot take the river-linking project off their mental map, with the Ganges being the only large, surplus basin with no scope for storage. Economic growth, however, may rapidly change this calculus.

Overall, much as the global community may demand deep institutional reforms in the Ganges Basin water economy, governments here will be under pressure to respond to the here-and-now priorities of the people of the basin who are among the poorest in the world and lead predominantly agrarian lives. In Table 18.1, we explored how water institutions of a basin morph as the basin economy undergoes transformation. Policy goals and reform priorities also differ in basins with poor water economies from those with rich ones, as shown in Table 18.3. Global policy advocacy exhorts policy makers in Ganges Basin states to emulate water-governance regimes in the Murray Darling Basin in Australia or the Rhine in Europe. But in emulating these models, Ganges Basin policy makers will be regularly confronted by their own socio-economic realities which are so vastly different from those facing the Rhine and Murray-Darling Basins.

In conclusion, the Ganges Basin faces myriad challenges in improving the governance and management of its water economy; however, resolute action by basin states can bring quick improvements by picking several low-hanging fruits. In the public irrigation sector, rationalizing ISF, investing in improved maintenance, operation and management (MOM) of irrigation systems,

Table 18.3 Water sector priorities in river basins at different stages of socio-economic evolution

Goals of water policy and priorities for reform	*Stage I: Completely informal water economies (>90% of water users in the informal water economy)*	*Stage II: Largely informal water economies (>50% of water users in informal water economy)*	*Stage III: Rapidly formalizing water economies (>50% of water users served by formal service providers)*	*Stage IV: Highly formalized water economics (>90% of water users served by formal service providers)*
Exemplar basin	*Congo*	*Ganges*	*Olifants*	*Rhine/Murray Darling*
Investment priorities	Invest in local infrastructure to improve water access	Invest in meso-level water infrastructure for sustainable development of water resources	Invest in improving water productivity, wastewater recycling	Invest in 100% coverage in high-quality water service provision
Institutional reforms	Make informal water institutions equitable	Integrate informal water institutions with formal ones in private or public sector	Create meso-level participatory water institutions	Create a full-fledged water industry with proactive regulator
Water policy and legal regime	Establish basic water information system	Establish water policy and legal regime	Establish basin-level water allocation mechanism	Full-fledged basin management authorities
Financial sustainability of water infrastructure and services	Establish the principle of water as a social and economic good	Subsidy on O&M cost limited to 50%	75% service fee recovery for O&M of water infrastructure	100% water service as well as resource fee recovered for MOM

prioritizing the last mile projects and managing systems in a conjunctive use mode can rapidly expand irrigation benefits. Irrigation bureaucracies are in need of serious reform; and management audit of irrigation systems needs to be made regular. A Ganges Water University needs to be tasked to build management capacities in irrigation and water supply bureaucracy in all the basin states.

Conjunctive management of surface and groundwater can have a massive pay off in the Ganges Basin. A lot of this is happening, but by default rather than by design. To maximize the area under conjunctive management of surface and groundwater, canal system managers should be rewarded for spreading surface water across as large an area as possible.

Ganges Basin states also need to pilot and showcase as highly reliable, peri-urban irrigation systems with treated wastewater. Municipal authorities and/or private utilities need to be supported to invest in the treatment and orderly distribution of domestic and industrial wastewater to peri-urban farmers on full or partial cost recovery. When it comes to improving domestic water supply, urban and rural, we should be circumspect in leapfrogging to levels of service provision – such as 24/7 supply – that are way beyond the present management capacities of the Ganga Basin's local governance bodies. Instead, new initiatives need to incrementally build on what exists. Massive investments in the upgradating and modernization of water infrastructure and services are planned; however, these will materialize only when water utilities – public and private – figure out feasible approaches to pricing and cost recovery at levels that would make investments worthwhile.

The water challenge of the basin, as elsewhere in the low-income world, can be blamed on serious management deficit – management of the resource, of infrastructure, as well as of services. In the normal course, the management will improve but only to the extent that water becomes an economic good and when consumers are able to bear the bulk of the cost of supplying quality WSS services. It is no surprise that Singapore, one of the most water-scarce places in the world, is able to offer one of the best WSS services to its citizens who can afford to pay US$10–12 per m^3 of water. India's challenge is to provide better services at an affordable tariff. This is hard, but not impossible.

To accelerate agricultural growth in the Ganges Basin, governments need to allocate substantial resources to stimulate vibrant, deep and broad solar pump-based irrigation service markets in the entire basin. Solar power is best suited to energize the Ganges Water Machine.

States in the Ganges Basin have so far concentrated only on one option: public investment in infrastructure. They have placed very little emphasis on management improvements, governance reforms and institutional innovations. This is why returns on public investments in water infrastructure in the region have been poor and water projects have suffered from the build-neglect-rebuild syndrome. The basin countries can make rapid strides in water security by emphasizing management improvements and institutional reforms, rather than just public investment in water infrastructure. This shift of emphasis is the key challenge to be met by the governments in the Ganges Basin.

Note

1 42,000 million litres/day = 0.042 BCM ⋆ 365 days = 15.33 BCM/year.

References

Amarasinghe, U., Shah, T., Turral, H., and Anand, B.K. (2007). *India's water future 2025–2050: Business as usual scenario and issues.* Paper presented at the National Workshop on National River Linking Project of India: Analysis of hydrological, social and ecological issues, Delhi, India, 9–10 October.

CPCB. (2007). *Evaluation of Operations and Maintenance of Sewage Treatment Plants in India – 2007*. Central Pollution Control Board (CPCB), Ministry of Environment and Forests, Government of India.

Government of India. (2011). *Report of the Working Group on Urban and Industrial Water Supply and Sanitation for the Twelfth Five-Year-Plan (2012–2017)*. Submitted to the Steering Group on Water Sector, Planning Commission, Government of India, November.

Palrecha, A., Kapoor, D. and Malladi, T. (2012). *Waste water irrigation in Gujarat: An exploratory study*. IWMI-Tata Water Policy Program Highlight 30. Anand, India. Available online: www.iwmi.cgiar.org/iwmi-tata/PDFs/2012_Highlight-30.pdf (last accessed 30 December 2013).

Pradhan, P. (2012). Revitalizing Irrigation Systems for Food security: Vision and Approaches in Irrigation Systems. *Hydro Nepal*. Special Issue. April.

Pradhan, P. (2015). Role of Water Research Institutions in Nepal and its impact on Policy Changes. *Hydro Nepal*, 17.

Rasul, G. (2015). Water for growth and development in the Ganges, Brahmaputra, and Meghna basins: An economic perspective. *International Journal of River Basin Management*, 13(3): 387–400.

Shah, T. and van Koppen, B. (2006). Is India ripe for Integrated Water Resources Management (IWRM)? Fitting water policy to national development context. *Economic and Political Weekly*, 41(31): 3413–3421.

Upadhyay, S.N. (2012). *International Water Course Law and A Perspective on Nepal – India Cooperation*. Kathmandu: Ekta Books.

Verma, S., Kampman, D.A., Van der Zaag, P. and Hoekstra, A.Y. (2009). Going against the flow: A critical analysis of inter-state virtual water trade in the context of India's National River Linking Program. *Physics and Chemistry of the Earth*, 34(4–5): 261–69.

INDEX

Please note: *fig* indicates figure and *t* indicates table.